COMPLEXATION OF TRACE METALS IN NATURAL WATERS

DEVELOPMENTS IN BIOGEOCHEMISTRY

Complexation of trace metals in natural waters

Proceedings of the International Symposium, May 2–6 1983,
Texel, The Netherlands

edited by

C.J.M. KRAMER

Netherlands Institute for Sea Research
Den Burg, Texel
The Netherlands

and

J.C. DUINKER

Institute for Marine Research at
Kiel University
Kiel, FRG

1984 **MARTINUS NIJHOFF/DR W. JUNK PUBLISHERS**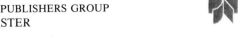
a member of the KLUWER ACADEMIC PUBLISHERS GROUP
THE HAGUE / BOSTON / LANCASTER

Distributors

for the United States and Canada: Kluwer Boston, Inc., 190 Old Derby Street, Hingham, MA 02043, USA
for all other countries: Kluwer Academic Publishers Group, Distribution Center, P.O.Box 322, 3300 AH Dordrecht, The Netherlands

Library of Congress Cataloging in Publication Data

Main entry under title:

Complexation of trace metals in natural waters.

(Developments in biogeochemistry ; 1)
Papers from the First International Symposium on the Complexation of Trace Metals in Natural Waters held at the Netherlands Institute for Sea Research, May 2-6, 1983.
1. Water chemistry--Congresses. 2. Complex compounds --Congresses. 3. Trace elements--Congresses. I. Kramer, C. J. M. II. Duinker, J. C. III. International Symposium on the Complexation of Trace Metals in Natural Waters (1st : 1983) : Nederlands Instituut voor Onderzoek der Zee) IV. Series.
GB855.C65 1984   551.48   84-8034
ISBN 90-247-2973-4

ISBN 90-247-2973-4 (this volume)

PRINTED IN THE NETHERLANDS

84 005075

CONTENTS

Invited lectures are marked with *

PREFACE

It is presently well recognized that total concentrations of trace
elements in any environmental compartment supply insufficient information
to understand important phenomena. The distinction and separate analysis
of specific chemical species are essential for understanding cycles in the
aquatic environment, involving identification and quantification of
sources, transport pathways, distributions and sinks, or, in the area of
interactions between trace elements and organisms to understand uptake,
distribution, excretion mechanisms and effects.
In the past, various ways have been developed to determine the nature
and extent of complexation of trace elements in natural systems.
Approaches have been followed along very different lines. These have not
always been fully appreciated by specialists working in even related
fields of complexation research.

The first International Symposium on the Complexation of Trace metals
in Natural Waters was held at the Netherlands Institute for Sea Research
(NIOZ, Texel, the Netherlands from 2-6 May 1983. The scientific programme
was planned by the chief organizers Drs. C.J.M. Kramer and J.C. Duinker (NIOZ)
together with Prof. Dr. H.W. Nürnberg (Kernforschungsanlage, Jülich,
Federal Republic of Germany) and Dr. M. Branica (Rudjer Bosković Institute,
Zagreb, Yugoslavia).

In this symposium it was tried to bring together a number of specialists
representing the main lines of approach, with the goal to evaluate the
analytical techniques and theoretical approaches with respect to complexation
of trace elements and complexing capacity in natural waters. The programme
was structured as follows: 13 specialists were invited to prepare and

present papers on subjects suggested by the organizers. Other contributions were presented as full papers or as posters, the latter in combination with short oral presentations. The distinction between oral presentations and posters was made as to obtain a balanced coverage of subjects within the time available.

It has been attempted to discuss and to compare the main techniques and approaches, such as electrochemical- (Differential pulse anodic stripping voltammetry and ion selective electrodes), chromatographic-, biological response-, solubilization-, ion exchange-, MnO_2 adsorption-, dialysis-, kinetic spectroscopic- and fluorescence quenching techniques. Several papers deal with the application to natural waters, theoretical approaches include modelling, kinetics of complexation and adsorption phenomena. Some papers cover the interaction with particles, others deal with the biological response of different organisms or the effects of the exudates, the interaction of trace metals with naturally occurring organics concern the complexation with humic- and fulvic acids and amino acids. This volume contains six parts in which we tried to combine papers of common interest:

- Techniques
- Theoretical approach
- Application to natural waters
- Interaction with particles
- Biological response

The organizers and editors want to express their gratitude to all participants for their involvement in the lively discussions, the results of which are reflected in the final form of various papers; to the invited speakers for their willingness to respond to the requests of the organizers to include specific subjects and approaches, and in several cases to accept a title that they would not have selected themselves; to all authors for keeping to the extremely tight time schedule for finalizing the papers, allowing completion of this work within a year after the symposium; to the Netherlands Institute for Sea Research for the hospitality and logistic and financial support; to the Royal Netherlands Academy for Arts and Sciences and the Commission of the European Communities, Directorate

General for Science, Research and Development for financial support; to the editorial board for their valuable comments on the submitted manuscripts; to the local organization committee, in particular Mrs. J. Hart for her patient and endless secretarial work before, during and after the symposium.

February 1984

C.J.M. Kramer   Texel

J.C. Duinker   Kiel

The Symposium was made possible by support of:

-the Commission of the European Communities

 Directorate General for Science, Research and Development

-the Royal Netherlands Academy of Arts and Sciences

-the Netherlands Institute for Sea Research (NIOZ)

LIST OF PARTICIPANTS AND (CO)AUTHORS

ALBERTS*, J.J., School of Fisheries, Univ. of Washington, Seatle Wash.
    98195 U.S.A.
ALLEN, H.E., Illinois Inst. of Technology, Pritzker Dep. of Env. Engineer
    Chicago Ill. 60616 U.S.A. Present address: Drexel University
    Environ. Studies Inst. Philadelphia PA 19104 U.S.A.
ALSTAD*, J., Dept. of Chem. University of Oslo, POBOX 1033, Blindern-Oslo
    3, Norway.
BALL*, J.W., Geological Survey Water Res. Div. 345 Middlefield Road, Menlo
    Park CA 94025 U.S.A.
BARTH, H., Comm. European Comm. DG XII: Sci. Res. & Developm. Rue de la
    Loi 200, Brussels, Belgium.
BERG, C.M.G. van den, Univ. of Liverpool, Dept. of Oceanography POB 147
    Liverpool L69 3BX UK.
BERNHARD, J.P., Dept. Chem. Université de Geneve, 30 Quai Ansermet 1211 -
    Geneve, Switzerland.
BOEF, G. den, Lab. Anal. Chem. Nieuwe Achtergracht 166, Amsterdam,
    the Netherlands.
BOS-VAN DER ZALM, C.H., Royal Melbourne Inst. of Tech. POBOX 2476 V GPO
    Melbourne Vic. Australia.
BOURG, A.C.M., French. Geol. Survey B.R.G.M., POB 6009, F-45060 Orleans,
    France
BRANICA, M., Rudjer Boskovic Institute Center for Mar. Res. POB 1016,
    41001 Zagreb Yugoslavia.
BREZONIK*, P.L., State Water Survey Div., POBOX 5050, Champaign Ill,
    61820 U.S.A.
BRIHAYE*, C., Inst. Chimie, Lab. de Chimie Anal. Univ. de Liège, Sar.
    Tilman 134000 Liège, Belgium.
BRUGGEMAN, W.H., Governm. Inst. Wastewater Treatm., Maerlant 134,
    Lelystad, the Netherlands.
BRÜGMANN, L., Inst. für Meereskunde, Seestrasse 15, 253 Rostock-Warnemunde
    DDR.
BRYAN*, G.W., Mar. Biol. Ass. UK the Lab. Citadel Hill, Plymouth
    PL1 2PB UK.
BUCKLEY*, P.M., Univ. of Liverpool, Dept. of Oceanography POBOX 147,
    Liverpool L69 3BX UK.
BUFFLE, J., Dept. of Inorg. & Analyt. Chem., 30 Quai Ernest Ansermet,
    CH-1211 Geneve 4 Switzerland.
BYRNE*, R.H., University of South Florida, USA.
CABANISS*, S.E., Dep. of Environm. Sci and Eng. University of North
    Carolina Chapel Hill NC 27514 U.S.A.
Campbell, P.G.C., Univ. du Quebec, INRS-ERL, C.P. 7500. Ste-Foy, Quebec
    Canada GIV 4C7
CASTILHO, P. del, Inst. for Soil Fertility, POBOX 30003, 9750 RA
    Haren, the Netherlands.
CAUWET, G., Lab. de Sédiment. et Géochim. Mar., Avenue de Villeneuve
    66025, Perpignan, France.
CLAYTON*, J.R. jr., School of Fisheries University of Washington
    Seattle Washington 98195, U.S.A.

---

* Not participating (co)authors are marked with an asterisk.

CLEVEN, R.F.M.J., Lab. Phys. & Colloidchem. L.H. de Dreyen 6, 6703 BC
    Wageningen, the Netherlands.
COALE, K.H., Center for Coastal Mar. Stud. University of California
    Santa Cruz CA 95064 U.S.A.
COFINO, W.P., Governm. Inst. Wastewater Treatm., Maerlant 134, Lelystad,
    the Netherlands.
COLLINS*, B.J., Dep. of Environm. Sci. and Eng. University of North
    Carolina, Chapel Hill NC 27514, U.S.A.
CORREIA DOS SANTOS, M.M., Centro de Quimica Estrutural Inst. Superior
    Technico 1096 Lisboa-Cedex Portugal.
CRESELIUS, E.A., Batelle Pacific NW Div., Marine Res. Lab. Sequim
    Washington, 98382, U.S.A.
DHARMVANIJ*, S., University of Liverpool, Dept. of Oceanography POBOX 147,
    Liverpool L69 3BX UK.
DRISCOLL, C.T., Dept. Civil Eng., Syracuse University, Syracuse NY
    13210, U.S.A.
DUINKER, J.C., Neth. Inst. for Sea Research, POB 59, 1790 AB Den Burg,
    the Netherlands. Present address: Inst. für Meereskunde,
    Dusternbrookerweg 20, D 2300 Kiel, BRD.
DYRSSEN, D., University of Goteborg, Dept. of Anal. & Mar. Chem.
    S 41296 Goteborg, Sweden.
ELBAZ-POULICHET, F., Ecole Normale Sup. Lab de Geol., 46 Rue d'Ulm,
    75230 Paris Cedex, France
FARDY*, F.F., CSIRO, Div. of Energy Chem., Private Mail Bag 7, Sutherland
    NSW 2232 Australia.
FLORENCE, T.M., CSIRO Div. of Energy Chem., Private Mail Bag 7, Sutherland
    NSW 2232 Australia.
FRIMMEL, F.H., Tech. University München Inst. Wasserchem., D-8000 München,
    BRD.
GAMBLE, D.S., Chem. & Biol. Research Inst., Agriculture Canada,
    Ottawa Ont. K1A OC6, Canada.
GERRITSE*, R.G., Inst. for Soil Fertillity, POB 30003, 9750 RA Haren,
    the Netherlands.
GILLAIN, G., Inst. Chimie, Lab. Oceanologie, Université de Liège,
    Sar. Tilman, B4000, Liège, Belgium.
GNASSIA-BARRELLI, M., Lab. de Physique et Chimie Mar. Station Marine
    la Darse, Villefranche-sur-Mer, France.
GONÇALVES, M.L.S., Centro de Quimica Estrutural Inst. Superior
    Technico 1096 Lisboa-Codex Portugal.
HAAN, H. de, Limnol. Lab. Tjeukemeer Dept. De Akkers 47, Oosterzee,
    the Netherlands.
HAEKEL, W., Chem. Dept. GKSS Research Center, D-2054 Geesthacht, BRD.
HAERDI*, W., Dept. of Inorg. & Analyt. Chemist., 30 Quai Ernest Ansermet,
    CH-1211 Geneve 4, Switserland.
HALL*. P., University of Goteborg, Dept. of Anal. & Mar. Chem. S41296
    Goteborg, Sweden
HANSON, A.K., jr., Graduate School of Oceanogr. University of Rhode
    Island, Narragansett, RI 02882, U.S.A.
HARALDSSON*, C., University of Goteborg, Dept. of Anal. & Mar. Chem.
    S 41296 Goteborg, Sweden.
HARDSTEDT-ROMEO, M., Lab. de Physique et Chimie Mar. Station Marine la
    Darse, Villefranche-sur-Mer, France.
HARMSEN, J., I.C.W., POB 35, Wageningen, the Netherlands.

HART, B.T., Water Studies Centre Caulfield Inst. of Techn., Caulfield, Vic. 3145 Australia.

HOLTKAMP, B.T.W., NIOZ, POBOX 59, Den Burg, Texel, the Netherlands.

IMBER, B., Royal Roads Mil. Coll., Dept. Chem. FMO Victoria BC Canada JOS 7BO.

IMMERZ*, A., Tech. University München, Inst. Wasserchem., D-8000 München, BRD.

IVERFELDT*, A., University of Goteborg, Dept. of Anal. & Mar. Chem. S 41296 Goteborg, Sweden.

JARDIM, W.F., Dept. of Botany, University of Liverpool, POBOX 147, Liverpool L 69 3BX, U.K.

JOHNSON, C.A., AGRG - Geol. Imp. College, Prince Consort Rd, London SW7, UK.

JONES, M.J., Water studies Center, C.I.T., Caulfield East, Vic. 3145, Australia.

KLAVENESS*, D., Dept. of Chem. University of Oslo, POBOX 1033, Blindern Oslo 3, Norway.

KOCK, W.C. de, MT-TNO Lab. Appl. Marine Res., POBOX 57, 1780 AB Den Helder, the Netherlands.

KRAMER, C.J.M., Neth. Inst. for Sea Research, POB 59, 1790 AB Den Burg, Texel, the Netherlands.

KREMLING, K., Institut für Meereskunde, Dusternbrookerweg 20, D2300 Kiel, BRD.

LAANE, R.W.P.M., BOEDE, POB 59, 1790 AB Den Burg, Texel, the Netherlands.

LAEGREID, M., Dept. of Chem. Univ. of Oslo, POBOX 1033, Blindern-Oslo 3, Norway.

LANGFORD, C.H., Concordia Univ. Dept. Chem. 1455 de Maisonneuve Blvd West, Montreal Quebec Canada H3G 1M8.

LANGSTON, W.J., Mar. Biol. Ass. UK the Lab. Citadel Hill, Plymouth PL1 2PB UK.

LEEUWEN, H.P. van, Lab, Phys. & Colloidchemie, Landbouwhogeschool, de Dreyen 6, 6703 BC Wageningen.

LUMSDEN*, B.B., CSIRO Div. of Energy Chem. Private Mail Bag 7, Sutherland NSM 2232 Australia. Present address: Oktedi Mining Corp. Port Moresby Papua-New Guinea.

LUND, W., Dept. of Chem. University of Oslo, Box. 1033 Blindern, Oslo 3 Norway.

MACKEY, D.J. POBOX 1538, Hobart 7001, Australia.

MANTOURA, R.F.C., Inst. for Mar. Envir. Res. Prospect Place, The Hoe Plymouth PL 1 3 DH, U.K.

MARTIN, J.M., Ecole Normale Sup. Lab de Geol., 46 Rue d'Ulm, 75230 Paris-Cedex, France.

MARQUENIE*, J.M. MT-TNO, Lab. Appl. Marine Res., POBOX 57, 1780 AB Den Helder, the Netherlands.

MILLERO, F.J., Rosenstiel School of Mar. Sci. University of Miami Rickenbacker Causeway FL 33149 U.S.A.

MOUVET*, C., French Geol. Survey B.R.G.M., POB 6009, F-45060 Orleans, France.

MURRAY, J.W., School of Oceanogr. University of Washington, Seattle, WA 98195 U.S.A.

NELSON, L.A., Inst. Mar. Environm. Res. Prospect Place the Hoe Plymouth PL1 3DH UK.

NICOLAS, E., Lab. de Physique et chimie mar. station marine La Darse, Villefranche-sur-Mer, France.

NIEDERMANN*, H., Tech. Univ. München, Inst. Wasserchem. D-8000 München, BRD.

NIET, G., de, Neth. Institute for Sea Research, POBox 59, 1790 AB Den Burg, the Netherlands.

NORDSTROM, D.K., Geological sur. water res. div. 345 Middlefield Road, Menlo Park CA 94025 U.S.A.

NÜRNBERG, H.W., Institut für Chemie Inst. 4 Angewandte Phys. Chemie, KFA Jülich BRD.

OSTERROHT, C., Inst. für Meereskunde, Dusternbrookerweg 20, D-2300 Kiel, BRD.

PALUMBO*, A.V., Atlantic Est. Fish Center, Beaufort NC 28516 U.S.A.

PLECHANOV, N., Dept. of Anal. Mar. Chem. Chalmers Univ. of Goteborg, S-41296 Goteborg, Sweden.

POLLEHNE*, F., Dept. of Biology, Univ. of Kiel, BRD.

RASPOR, B., Rudjer Boskovic Institute, Center for Mar. Res. POB 1016, 41001 Zagreb Yugoslavia.

ROBINSON*, M.G., Royal Roads Mil. Coll., Dept. Chem. FMO Victoria BC Canada Jos 7BO.

RUZIC, I., Rudjer Boskovic Institute Center for Mar. Res. POB 1016 41001 Zagreb Yugoslavia.

SALOMONS, W., Delft Hydr. Lab.-Haren Branch POBOX 30003, 4750 RA Haren, the Netherlands.

SANCHEZ*, A.L., School of Fisheries, Univ. of Washington, Seattle, Washington 98195 U.S.A.

SCHAFRAN*, G.C., Dept. Civil Eng. Syracuse Univ., Syracuse NY 13210, U.S.A.

SEIP*, H.M., Dept. of Chem. Univ. of Oslo, POBOX 1033, Blindern Oslo 3, Norway.

SHUMAN, M.S., Dep. of Environm. Sci.and Eng. University of North Carolina Chapel Hill NC 27514 U.S.A.

SIBLEY, T.H., School of Fisheries, Univ. of Washington, Seattle, Washington 98195, U.S.A.

SIGG, L., EAWAG, CH 8600 Dubendorf-Zurich, Switzerland.

SPERLING, K.R., Biologische Anstalt Helgoland, Lab. Sulldorf Wustland 2, 2000 Hamburg 55 BRD.

STOLZBERG, R.J., Dept. of Chem. Univ. of Alaska, Fairbanks Alaska 99701, U.S.A.

STRUYS, J., Nat. Inst. for Public Health, POBOX 1, Bilthoven, the Netherlands.

STUMM, W., Swiss Fed. Inst. for Water Res. and Water Pollution Control CH 8600 Dubendorf Switzerland.

SUGAWARA*, M., Institut für chemie, Inst. 4 KFA Jülich, BRD. Permanent address: Dept. of Chem. Faculty of Science, Hokkaido Univ. Sapporo Japan.

SUNDA, W., Atlantic Est. Fish Center, Beaufort NC 28516 U.S.A.

TESSIER*,A., Dept. of Inorg. & Analyt. Chemist, 20 Quai Ernest Ansermet CH-1211 Geneve 4 Switzerland.

TURNER*, D.R., Mar. Biol. Ass. UK The Lab., Citadel Hill, Plymouth PL1 2PB UK.

TUSCHALL, J.R., State Water Survey Div., POBOX 5050, Champaign Ill 61820, U.S.A.

UNDERDOWN*, A.W., Chem. & Biol. Res. Inst. Agriculture Canada Ottawa Ont., KIA OC6 Canada

VALENTA*, P., Institut für Chemie. 4 Angewandte Phys. Chemie KFA, Jülich, BRD.

VARNEY*, M.S., University of Liverpool, Dept. of Oceanography POB 147, Liverpool L69 3BX UK.

WEDBORG, M., Dept. of Anal. Mar. Chem., Chalmers Univ. of Goteborg, S-41296 Goteborg, Sweden.

WERF, M. van der, Biologisch Lab. Vrije Universiteit, 1007 MC Amsterdam, the Netherlands.

WESTERLUND*, S., University of Goteborg, Dept. of Anal. & Mar. Chem., S 41296 Goteborg, Sweden.

WEYDEN, C.H., van der, Inst. for Earth Sci., Budapestlaan 4, Utrecht, the Netherlands.

WHITFIELD*, M., Mar. Biol. Ass. UK the Lab., Citadel Hill, Plymouth PL1 2PB UK.

WILKEN, R. -D., GKSS-Forschungsz. Inst. Chemie, Max Planckstr. D-2054 Geesthacht, BRD.

WURTZ*, E.A., School of Fisheries, Univ. of Washington, Seattle, Washington 98195, U.S.A.

XU QING-HUI, 3rd Inst. of Oceanogr. National Inst. of Oceanogr. Xiamen, Peoples Republic of China.

YU GUO-HUI, 2nd Inst. of Oceanography, Hangzou, Peoples Republic of China.

ZINDER, B., EAWAG, CH-8600 Dubendorf-Zurich, Switzerland.

ZUEHLKE, R.W., Graduate School of Oceanogr. Univ. of Rhode Island, Kingston RI 02881 U.S.A.

to our late fathers

PART I TECHNIQUES

MEASUREMENT OF COPPER COMPLEXATION BY NATURALLY OCCURRING LIGANDS

W.F. JARDIM and H.E. ALLEN

1. INTRODUCTION

The characterization of metal forms in natural aquatic environments has become an important and attractive subject for research in the past decade. Metal speciation is required when assessing biological impact because bio-availability and toxicity seem to be directly related to the concentration of the metal ion species only (Sunda and Guillard, 1976; Anderson and Morel, 1978; van den Berg, 1979; Allen et al., 1980).

In natural water bodies, amelioration of copper toxic effects can occur either due to complexation or precipitation of the deleterious metal form by inorganic or organic matter. Humic substances, amino acids and extra-cellular products are the most common organic compounds that occur naturally and are capable of complexing trace metals (Singer, 1973).

Depending on their physico-chemical composition, natural waters will show different extents of complexation for the metal under study. Because natural ligands are present in very low concentrations, most of the traditional experimental approaches discussed by Rossotti and Rossotti (1961) and Beck (1970) for the determination of the stability constant for complexation cannot be applied with satisfactory accuracy.

Presently, several analytical procedures are used in determining the conditional stability constant, K', of these complexes and the ligand concentration, [L]. The total ligand concentration, which is frequently called the complexing capacity, is a measure of the amount of the metal under study which may be bound by sample components. Thus, it may include bonding by mechanisms other than complexation and will not include that portion of the binding sites already occupied by the metal used as the titrant or by other metals forming stronger complexes. Bioassay, voltammetry, ion exchange, selective ion electrodes and molecular separations are some of the techniques that have become powerful tools for speciation studies. Detailed reviews of these methods have been published by Hart (1981) and

Kramer, C.J.M. and Duinker, J.C. (eds.), Complexation of Trace Metals in Natural Waters. ISBN 90-247-2973-4
©1984 Martinus Nijhoff/Dr W. Junk Publishers, The Hague/Boston/Lancaster.
Printed in the Netherlands.

Neubecker and Allen (1983).

Chemical speciation of copper in natural waters has been carefully investigated by Sunda and Hanson (1979). Using ion selective electrodes (ISE), the authors detected the presence of at least three types of organic binding sites with remarkably different concentrations and conditional stability constants. McCrady and Chapman (1979) applied the same technique to several river water samples and compared the results with theoretical predictions. ISE have also been used in determining the binding capacities and stability constants of algal exudates (McKnight and Morel, 1979, 1980).

Anodic stripping voltammetry (ASV) has been widely used in the analysis of natural water samples to study their speciation. Chau et al. (1974) applied this technique to Hamilton Harbor and Lake Erie samples. Bender et al.(1970) studied metal complexation in secondary sewage effluent. Shuman and Woodward (1977) developed a titration method using ASV to evaluate conditional stability constants and ligand concentrations in natural waters. Tuschall and Brezonik (1981) discussed some limitations of this method when working with reducible metal-organic complexes.

In a 30-day experiment, Truitt and Weber (1981) used dialysis combined with metal titrations in order to obtain the complexing capacity of copper and cadmium with soil humic acid. "In situ" measurements have been carried out by Benes and Steinnes (1974) and a combined dialysis-ion exchange technique was applied to natural river water samples by Hart and Davies (1977).

Ion exchange using MnO_2 as a weak ion exchanger has been used for determining complexing capacity of fresh (van den Berg and Kramer, 1979) and seawater (van den Berg, 1982) samples as well as of algal exudates (van den Berg et al., 1979). The principle of the method is that the free metal ions present in the sample will distribute themselves between a small amount of added MnO_2 and the original amount of ligand present.

The quenching of the fluorescence of humic and fulvic acids by Cu^{2+} has been used by Weber and his co-workers (Saar and Weber, 1980; Ryan and Weber, 1982) to follow the course of complexometric titration. Saar and Weber (1980) obtained similar complexation capacities for samples monitored by this method and for samples monitored by a copper ion selective electrode.

Although many of these methods have been used for more than five years, the literature is very scant in terms of comparing values obtained by using these different analytical approaches. Tuschall and Brezonic (1981) have recently compared the results of ASV, ISE, fluorescence, ultrafiltration and competing ligand differential spectroscopy for the comparison of cadmium, copper, and zinc binding by humics from two swamp surface waters.

The aim of this paper is to further evaluate and discuss analytical methods. The organic ligands studied were humic acid, secondary sewage effluent and algal exudates because they are examples of naturally occuring ligands. The analytical methods were ASV, ISE, MnO_2 ion exchange, dialysis and fluorescence.

2. MATERIALS AND METHODS

2.1. Sample Pretreatment

2.1.1. Buffer. Because the conditional stability constant, K', is highly pH dependent, a good buffer system is required. The major concern in the choice of a buffer is to know the extent of metal complexation to permit correction of K' values. Piperazine-N-N'-bis (2-ethanesulfonic acid), PIPES, obtained from the Sigma Chemical Company, St. Louis, Missouri, was used because it is very effective at pH 6.60 and has a low stability constant with copper. Trace metals were removed from it by equilibrating a PIPES solution with MnO_2. PIPES was added to the samples to obtain a concentration of 7.5 mM and then the pH of the sample was adjusted to 6.60 ± 0.02.

Figure 1 shows the copper "background complexation" in the buffer system used. This complexation is caused not only by the buffer, but also by hydrolysis and possible metal loss onto the vessel walls. Calculations were made assuming that the metal available to complex ligands in the samples was the ionic fraction (Cu^{2+}) and not the total metal spiked (Cu_t).

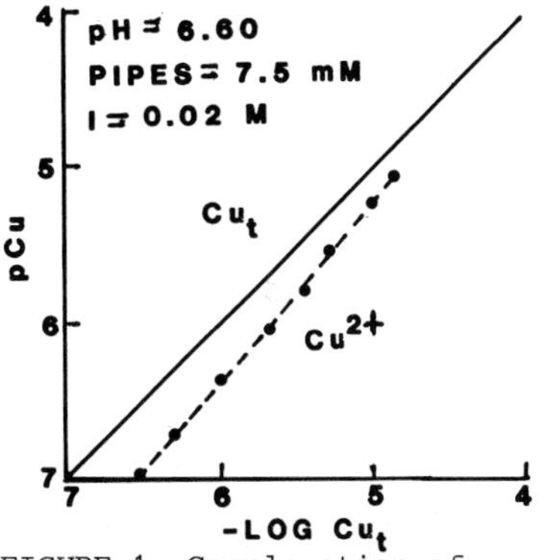

FIGURE 1. Complexation of copper by PIPES buffer. Data for pCu in the presence of PIPES buffer determined by ISE.

2.1.2. Humic Acid. Humic acid obtained from the Aldrich Chemical Co., Milwaukee, Wisconsin, was used to prepare a 20 mg/l solution. The ionic strength was adjusted to 0.02 M using $NaNO_3$ which had been previously equilibrated with MnO_2 to remove trace metal contamination. The sample was buffered and stored at 4°C.

2.1.3. Secondary Sewage Effluent. Secondary sewage prior to chlorination was collected at the John E. Egan Water Reclamation Plant (Schaumburg, Illinois). The sample was filtered by a Whatman GF/C glassfiber filter. The sample, whose initial pH was 6.95, was buffered with PIPES and stored at -4°C.

2.1.4. Algal Exudate. A cyanobacteria Oscillatoria sp. was isolated in
North Wales. The culture was unialgal but not axenic. Ml culture medium
(Jardim, 1982) was used with the macronutrient composition as follows: $NaNO_3$
(1500 mg/l), KH_2PO_4 (30.5 mg/l), $MgSO_4 \cdot 7H_2O$ (75 mg/l), $Ca(NO_3)_2 \cdot 4H_2O$
(23.6 mg/l, 10^{-4} M) and $FeCl_3 \cdot 6H_2O$ (0.27 mg/l, 10^{-6}M). After 15 days of
growth, the log phase culture was filtered through 0.45 µm membrane filters
and the filtrate was dried at 65°C. Prior to analysis the residue was re-
suspended in a PIPES buffer solution.

2.2. Experimental Procedures

All the glassware as well as the polycarbonate and polythylene flasks
were soaked in 10% HNO_3 (v/v) for 24 hours. Megapure double distilled
deionized water was used throughout.

2.2.1. MnO_2 ion exchange. The procedure of van den Berg and Kramer (1979)
was followed with some modifications. A 300 ml portion of the buffered sample
was spiked with 500 µl of 0.039 M MnO_2 suspension to obtain a final MnO_2
concentration equal to 65 µM. With constant stirring of the sample, 25 ml
aliquots were transferred to 50 ml polycarbonate centrifuge tubes. Copper
was then added to raise the concentration of the ten samples by 0.5 to 5 µM.
The samples were shaken for 24 h at 25°C and then centrifuged at 17,000 rpm
for 100 minutes. This centrifugation step, which replaced the filtration in
the original method, tended to minimize undesirable contamination. Copper
was analyzed in the supernatant using a Perkin Elmer model 305 flameless
atomic absorption spectrophotometer.

2.2.2. Amperometric titration (DPASV). A 25 ml portion of the previously
buffered sample was purged with high purity nitrogen for 10 minutes. A
water bath was used to control the temperature to 25± 1°C. Samples were
spiked with 0.5 to 5 µM copper and allowed to equilibrate for 30 minutes
after each addition prior to the voltammetric measurement of copper. The
measurements were made using a Princeton Applied Research model 174 polarographic analyzer equipped with a Metrohm E 410 hanging drop electrode. A
platium wire served as the counter electrode and a Ag/AgCl reference electrode
was used. Copper was plated at -1.0 V for 3 minutes with stirring followed
by 30 seconds quiescent and then stripped using a differential pulse wave
form. The instrumental conditions were modulation amplitude, 25 mV; scan
rate, 5 mV/sec and pulse rate 0.5 sec.

The equilibration time of 30 minutes was based on a preliminary study in which copper was added to each of the three samples to increase its concentration by 1 mM in the sewage effluent and algal exudate and 0.5 mN in the humic acid. As shown in Figure 2, the extent of reaction was at least 95% after the 30 minute euilibration time.

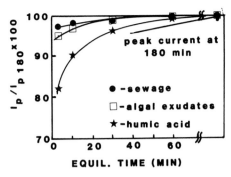

FIGURE 2. DPASV peak current as a function of the equilibration time of copper with the three types of ligands.

2.2.3. Ion Selective Electrode (ISE). An Orion Microprocessor Ion Analyzer model 901 was used in conjunction with a cupric ion selective electrode (Orion model 94-29A) and a junction Ag/AgCl reference electrode (Orion model 90-01). The electrode linear response was tested down to 10^{-12} M ionic copper by means of metal buffer solutions (Blum and Fogg, 1972). The ionic strength of the metal buffer solution was adjusted to 0.02 with KNO_3. The observed slope was 29.9 ± 0.5 mV per ten-fold change in ionic copper. The sample was continuously stirred with a reflon covered stirring bar and after each addition of copper the potential was determined after a twenty-minute period in which it did not change.

2.2.4. Dialysis. The method of Truitt and Weber (1981) was modified to permit the titration to be completed within a few days. Spectra/Por 6 dialysis bags (Spectrum Industries, Los Angeles, CA) with a 1000 Dalton molecular weight cutoff were soaked in warm HNO_3 (40°C and 1% v/v) for 24 h. Bags measuring 15 cm in length by 10 mm diameter were filled with a 4 ml of a 7.5 µM PIPES solution whose ionic strength had been adjusted to 0.02 M and whose pH had been adjusted to 6.60. Ten aliquots, each of 200 ml of previously buffered sample, were placed in 250 ml polythylene bottles and were spiked with copper. The dialysis bags were then submerged in the copper-containing sample solutions and allowed to equilibrate for 48 h at 25°C. Both the internal and external solutions had the same pH and ionic strengths. Copper was analyzed in both the retentate (ionic plus bound fraction) and the diffusate (ionic fraction) using flameless atomic absorption.

2.2.5. Fluorescence. The fluorescence spectrum was obtained using a Perkin-Elmer model 650-S fluorometer. NaOH was added to 200 ml of unbuffered sample to obtain a final pH of 6.60. One hour after spiking 20 ml aliquots with copper the decrease in fluorescence was determined. The excitation wavelength used was 360 nm and the emission wavelength was 460 nm. These wavelengths were selected based on scanning spectra to determine the maxima. It was assumed that for a spike of 5×10^{-4} M copper there would be 100% and for no added copper there was no quenching. This procedure is described in more detail elsewhere (Saar and Weber, 1980).

2.3. Data Treatment

There are two commonly applied means of data treatment used for the determination of complexation properties of environmental ligands. For both the molecular weight must be independently determined or the value of the equivalent concentration of ligand determined in the titration must be used in the computation of the stability constant.

In a system where the complexes formed are of a single stoichiometry, the results obtained from the titration with copper can be handled in a straightforward manner. Plotting $[Cu^{2+}]/[CuL]$ as a fraction of $[Cu^{2+}]$, where $[Cu^{2+}]$ is the concentration of ionic copper and $[CuL]$ is the concentration of copper bound to the ligand, will produce a straight line where the total ligand concentration is 1/slope and K', the conditional stability constant, is equal to the slope/ y-axis intercept (van den Berg et al., 1979).

Scatchard plots (Scatchard et al., 1957) are another frequently used means of data treatment. Here $[CuL]/[Cu^{2+}]$ is plotted versus $[CuL]$. The conditional stability constant is equal to the negative of the slope of the regression and the intercept on the ordinant is equal to the product of the total ligand concentration and the conditional stability constant.

For environmental water samples the simple titration curves described above are unlikely to be completely applicable. First, the diversity of organic ligands present can result in the formation of numerous different complexes. Second, an increase in the copper concentration can be followed by a change in the nature of the ligand sites available for binding copper. This more realistic situation requires a careful and detailed data treatment. For example, formation of a series of complexes (CuL_1, CuL_2, ..., CuL_n where L_1, L_2, ..., L_n are different ligands or are different types of sites on a macromolecule) can result in the line being curved. Also, various interactions such a adsorption, ion pair formation and polymerization will also cause a deviation in the linearity. This subject has recently been reviewed by van den Berg (1982) and Ruzic (1982).

3. RESULTS AND DISCUSSION

A Scatchard plot of the ISE data for the algal exudates is shown in Figure 3a. The curve has two distinct parts. The value of $10^{7.38}$ for K_1, is equal to the negative slope of the upper part of the curve ($[CuL]=$ 0-1.4 μM), whereas the value of $10^{6.50}$ for K_2, is the negative of the slope of the lower part using the remaining three points. The ligand concentration corresponding to K_1, is $[L_1]$ = 1.92 μM and the total ligand concentration, $[L_t]$ = $[L_1]$ + $[L_2]$ = 4.56 μM. Unless otherwise noted,

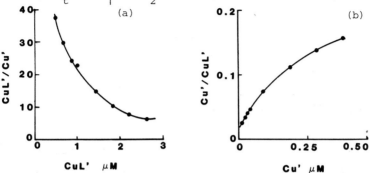

FIGURE 3. Algal exudates data for the ISE method. (a) $[CuL]/[Cu^{2+}]$ vs. $[CuL]$, Scatchard plot and (b) $[Cu^{2+}]/[CuL]$ vs. $[Cu^{2+}]$.

all results are from the treatment of data by the Scatchard method.

A plot of $[Cu^{2+}]/[CuL]$ vs. $[Cu^{2+}]$ for the same data is shown in Figure 3b. At low $[Cu^{2+}]$, up to about 0.12 μM, the graph is a straight line. The value of the stability constant (K_1') is $10^{7.38}$ for the first five points and $[L_1]$ = 1.52 μM. As the titration proceeds, either other complexes are formed or the number of binding sites per copper ion increases as indicated by a change in slope. Using only the last three data points for computation the value of K_2' is $10^{6.50}$. The total ligand concentration is 4.65 μM.

This example shows how concerned one must be when evaluating titration data for samples in which more than one type of ligand can be present. Unless the range of total copper concentration remains the same, it is meaningless to compare results obtained from different titrations. This restriction is true not only when comparing the results from different techniques but also for those obtained for different samples using the same technique.

Table 1 shows the variation of K' and [L] values as a function of the total copper concentration $[Cu_t]$ during the titration course using ISE. The value determined for the stability constant decreases and the ligand concentration increases as the concentration of metal added during the titration increases. When all the points are considered as a single curve, the Scatchard analysis gives an average stability constant, K', of $10^{7.04}$ and a ligand concentration, [L]; of 2.98 μM.

TABLE I. Variation of the conditional stability constant, K', and ligand concentration, [L], as a function of the total concentration of metal, Cu_t, during the titration.

Cu_t		[L]
µM	log K'	µM
0.5 - 1.0	7.48	1.68
0.5 - 1.5	7.34	2.04
0.5 - 2.0	7.25	2.27
0.5 - 2.5	7.14	2.61
0.5 - 3.0	7.04	2.98

3.1. Secondary Sewage Effluent

The study of organic compounds in sewage effluent is very important because of the eutrophying and pollutional effects of waste discharges to surface waters. Sewage effluents contain amino acids and many other low molecular weight compounds that have the ability to complex trace metals. This behaviour can help ameliorate the toxic effects of heavy metals but may also increase their soluble concentrations.

The characterization of organic compounds in secondary effluent has been done by Sachdev et al., (1976). As pointed out in their work, about 60% of the total soluble material present in the sewage had an apparent molecular weight less than 700 Daltons. Nevertheless, very little work has been done in terms of calculating the trace metal binding capacity of secondary sewage effluents. The study of Bender et al, (1970) of copper complexation by components of secondary sewage effluents showed that the total ligand concentration ranged from 10^{-6} to 8×10^{-5} moles/l and that small molecular weight fractions (500-1000 Daltons) were those predominantly responsible for the complexation.

Table II contains the results obtained for the secondary effluent sample. Using dialysis bags we confirmed that low molecular weight compounds (MW less than 1000 Daltons) were mainly responsible for the copper complexation. The copper concentrations of the retentate and diffusate solutions were found to be the same following the dialysis. A Scatchard plot of the ISE data for this sample is shown in Figure 4.

3.2. Algal Exudates

The evidence for metal complexation by algal exudates was first investigated by Fogg and Westlake (1955). They pointed out the ability of some polypeptides excreted by Anabaena cylindrica to complex cupric, zinc and ferric ions. More recently, Murphy et al., (1976) detected the presence of hydroxamate chelators in Lake Ontario during a blue-green algal bloom. These compounds are able to complex ferric and cupric ions and their production can be enhanced by iron starvation. Further work on algal siderophores characterization was done by Simpson and Neilands (1976).

TABLE II. Conditional stability constants, K', and ligand concentrations, [L], for the secondary sewage effluent.

METHOD	K'	[L] μM
ISE	$K_1' = 1.8 \pm 0.1 \times 10^7$	$L_1 = 4.24 \pm 0.14$
	$K_2' = 7.2 \pm 1.7 \times 10^5$	$L_2 = 12.96 \pm 1.86$
		$L_t = 17.20 \pm 2.00$
MnO$_2$	$K_1' = 3.6 \pm 0.4 \times 10^7$	$L_1 = 3.60 \pm 0.50$
DPASV	$K' > 8.0 \times 10^6$	$L_t = 2.05 \pm 0.15$

Because of their high affinity for copper, such extra-cellular compounds can affect metal speciation in both natural waters and synthetic media. This change in metal forms can be better studied if the conditional stability constant of the complexes formed has been evaluated.

Swallow et al., (1978) studied eight species of algae and, using ion selective electrodes, concluded that only one was able to produce copper complexing material.

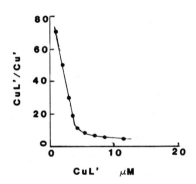

FIGURE 4. Secondary sewage effluent data for the ISE method.

Subsequently, McKnight and Morel (1979, 1980) reported the production of siderophores by a blue-green algal culture. Van den Berg and Kramer (1979), using the MnO$_2$ ion exchange technique, looked at the complexing material excreted by marine and freshwater algae. Sueur et al., (1982) applied the same approach to the study of copper complexing exudates in marine marcoalgae culture media.

Figure 3a shows the results obtained using ISE. Two distinctive types of binding sites (or ligands) are apparent. The first has a concentration $[L_1] = 1.52\ \mu$M and a conditional stability constant $K_1' = 10^{7.38}$. The

second ligand has a concentration $[L_2] = 3.04$ µM and a conditional
stability constant $K_2' = 10^{6.50}$. Table III provides a comparison of all
the results obtained for the algal exudate samples. Data from the dialysis
experiments indicated the ligands were able to diffuse across the membrane.
It is thus concluded that low molecular weight compounds (MW less than
1,000 Daltons) are responsible for the bulk of the copper ion complexation.
This evidence is in good agreement with the data obtained for diatoms by
Sondergaard and Schierup (1982). They showed that during a diatom bloom,
low molecular weight compounds (less than 700 Daltons) are the predominant
excreted material.

TABLE III. Conditional stability constants, K', and ligand concentrations,
[L], for the algal exudates.

METHOD	K'	[L] µM
	$K_1' = 2.4 \pm 0.3 \times 10^7$	$L_1 = 1.52 \pm 0.09$
	$K_2' = 3.2 \pm 0.6 \times 10^6$	$L_2 = 3.04 \pm 0.31$
		$L_t = 4.56 \pm 0.40$
MnO_2	$K_2' = 1.9 \pm 0.4 \times 10^6$	$L_t = 2.71 \times 0.33$
DPASV	$K' > 1.6 \times 10^6$	$L_t = 4.30 \pm 0.50$

3.3. Humic Acid

The chemistry of fulvic and humic acids including their interaction
with metal ions has received much recent attention (Schnitzer and Khan,
1972; Povoledo and Golterman, 1975; Gjessing, 1976; Reuter and Perdue, 1977;
Saar and Weber, 1982; Stevenson, 1982; Christman and Gjessing, 1983).
Originating from the decay of plants and animals, these geopolymers show
such a wide range of structures that, not surprisingly, they have not yet
been fully characterized. The relationship between humic material structure
and the biological components from which they are formed has been pointed
out; material originating from soil is generally more aromatic than is that
of marine origin. The predominant functional groups are OH and COOH.

In the aquatic environment, these compounds can complex cations such
as Ca^{2+}, Mg^{2+} and trace metals. The binding strength will be strongly
pH dependent because of the protonation of the functional groups (Buffle,
1980). Because copper complexation will increase the solubility and thus
the mobility of the metal, this subject has been investigated by several
workers (Mantoura and Riley, 1975; Shuman and Cromer, 1979; Buffle, 1980;

Buffle et al., 1980; Gamble et al., 1980).

TABLE IV. Conditional stability constants, K', and ligand concentrations, [L], for the humic acid sample.

METHOD	K'	[L] μM
ISE	$K'_1 = 2.6 \pm 0.3 \times 10^7$ $K'_2 = 9.2 \pm 3.7 \times 10^5$	$L_1 = 7.62 \pm 0.57$ $L_2 = 17.7 \pm 7.4$ $L_t = 26.0 \pm 8.0$
MnO$_2$	$K'_2 = 1.6 \pm 0.9 \times 10^6$	$L_t = 4.2 \pm 0.4$
DPASV	$K' = 5.3 \pm 0.2 \times 10^6$	$L_t = 0.73 \pm 0.11$
Fluorescence	$K' = 1.2 \pm 0.2 \times 10^6$	$L_t = 55 \pm 2$

Table IV presents the results of K' and [L] values obtained in this study. There is reasonable agreement between the conditional stability constants using different techniques, but the range covered by the ligand concentration values was large. Figure 5 shows the Scatchard plot for the humic acid data obtained using ISE. It confirms the previous investigations of Mantoura and Riley (1975) and Gamble et al., (1980) about the presence of multi-ligand sites in these compounds. As with the sewage effluent and algal exudate samples, the dialysis technique failed to show a significant difference in the copper concentrations of the retentate and diffusate.

4. SUMMARY AND CONCLUSION

The comparison of the conditional stability constants and binding capacities of copper by naturally occuring ligands must be carefully conducted. One of the reasons is that the characteristics of both secondary sewage effluents and humic and fulvic acids are dependent on their sources. Algal exudate composition varies according to the species used and the

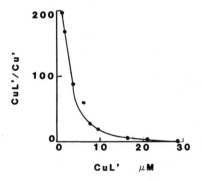

FIGURE 5. Humic acid data for the ISE method.

growth conditions. The complexation parameter values are dependent on ionic strength, pH, temperature and the presence of other compounds. Also, in a system with multiple ligands or ligands with multiple types of binding sites, the concentration range of metal used in the titration will drastically affect the results for the complexation parameters. Extrapolation of K' values assuming the linearity of K' as a function of pH should be done with care as this assumption may not hold for some samples (Shuman and Woodward, 1977).

We have obtained a reasonable agreement among K' and [L] values for both the algal exudate and sewage effluent samples. The humic acid sample, however, showed a greater variation among ligand concentration values. Figure 6 shows the variation of the ionic copper concentration, $[Cu^{2+}]$, as a function of the total copper concentration, $[Cu_T]$, expected during the course of the titration of the humic acid sample based on the constants obtained by the different analytical methods. Note that at 2×10^{-6} M total copper the difference in the ionic copper concentration calculated for the ISE and DPASV data is about 2 orders of magnitude. If one thinks in terms of metal toxicity effects ot the biota as a function of the ionic concentration of metal, this variation could, indeed, be significant.

Table V summarizes some characteristics of the techniques employed in this work. It is clear that of the methods available for the measurement of complexation parameters, the use of ion selective electrodes must be regarded as superior as their specificity limits the possibility for improper data evaluation.

TABLE V. Summary of characteristics of methods for complexation measurements.

METHOD	K' VALUE	[L]	M:L RATIO	OBSERVATIONS
DIALYSIS	YES	YES	YES	Results are function of the molecular cutoff used
MnO_2	YES	YES	YES	Improved by using radioisotopes
VOLTAMMETRY (ASV)	YES	total non-reducible ligands	YES	Results depend on the nature of the ligands
FLUORESCENCE	YES	YES	YES	Results depend on the nature of the ligands
ISE	YES	YES	YES	Wide titration range-highly selective

ACKNOWLEDGEMENTS

The autors would like to thank
Dr. C.M.G. van den Berg of the
Department of Oceanography, University
of Liverpool, for providing the MnO_2
suspension and to Dr. R.L. Beissinger of
the Department of Chemical Engineering,
Illinois Institute of Technology, for
use of the fluorometer. Wilson F. Jardim
is grateful to the Brazilian Government
for his supporting scholarship CNPq
200 110/80. We wish to thank the Pritzker
Department of Environmental Engineering,
Illinois Institute of Technology, and
the Botany Department, University of
Liverpool for the unlimited cooperation
in this joint research.

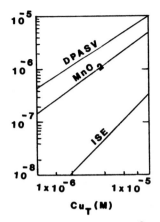

FIGURE 6. Predicted Cu^{2+} vatiation
as a function of the Cu_t concentration in the titration of
humic acid. The lines were obtained using the experimental
data presented in Table IV.

REFERENCES

-Allen, H.E., R.H. Hall and T.D. Brisbin, 1980. Metal speciation.
 Effects on aquatic toxicity. Environ. Sci. Technol. 14: 441-443.
-Anderson, D.M. and F.M.M. Morel, 1978. Copper sensitivity of Gonvaulax
 tamarensis. Limnol. Oceanogr. 23: 283-295.
-Beck, M.T., 1970. "Chemistry of Complex Equilibria". Van Nostrand
 Reinhold, London.
-Bender, M.E., W.R. Matson and R.A. Jordan, 1970. On the significance of
 metal complexing agents in secondary effluents. Environ. Sci. Technol.
 4: 520-521.
-Benes, P. and E, Steinnes, 1974. "In situ" dialysis for the determina-
 tion of the state of trace metals in natural waters. Water Res. 8:
 947-953.
-Blum, R. and H.M. Fogg, 1972. Metal buffers as standards in direct
 potentiometric determination of metal ion activities. J. Electroanal.
 Chem. 34: 485-488.
-Buffle, J., 1980. A critical comparison of studies of complex formation
 between Cu(II) and fulvic substances of natural waters. Anal. Chim. Acta
 118: 29-44.
-Buffle, J., P. Deladoey, F.L. Greter and W. Haerdi, 1980. Study of the
 complex formation of copper (II) by humic and fulvic substances. Anal.
 Chim. Acta 116: 255-274.
-Chau, Y.K., R. Gachter and K. Lum-Shue-Chan, 1974. Determination of the
 apparent complexing capacity of lake waters. J. Fish. Res. Bd. Can. 31:
 1515-1519.
-Christman, R.F. and E.T. Gjessing (eds.), 1983. "Aquatic and Terrestrial
 Humic Materials". Ann Arbor Science, Ann Arbor, Mich.

14

-Fogg, C.E. and R.F. Westlake, 1955. The importance of the extracellular products of algae in freshwater. Proc. Int. Assoc. Theor. Appl. Limnol. 12: 219-232.

-Gamble, D.S., A.W. Underdown and C.H. Langford, 1980. Copper (II) titration of fulvic acid ligand sites with theoretical, potentiometric, and spectrophotometric analysis. Anal. Chem. 52: 1901-1908.

-Gjessing, E.T., 1976. "Physical and Chemical Characteristics of Aquatic Humus". Ann Arbor Science, Ann Arbor, Mich.

-Hart, B.T., 1981. Trace metal complexing capacity of natural waters: A review. Environ. Technol. Letters 2: 95-110.

-Hart, B.T. and S.H.R. Davies, 1977. A new dialysis - ion exchange technique for determining the forms of trace metals in water. Aust. J. Mar. Freshwater Res. 28: 105-112.

-Jackson, G.A. and J.J. Morgan, 1978. Trace metal-chelator interactions and phytoplankton growth in seawater media: Theoretical analysis and comparison with reported observations. Limnol. Oceanogr. 23: 268-282.

-Jardim, W.F., 1982. unpublished results.

-McCrady, J.K. and G.A. Chapman, 1979. Determination of copper complexing capacity of natural river water, well water and artificially reconstituted water. Water Res. 13: 143-150.

-McKnight, D.M. and F.M.M. Morel, 1979. Release of weak and strong copper-complexing agents by algae. Limnol. Oceanogr. 24: 823-837.

-McKnight, D.M. and F.M.M. Morel, 1980. Copper complexation by sidero-phores from filamentous blue-green algae. Limnol. Oceanogr. 25: 62-71.

-Mantoura, R.F.C. and J.P. Riley, 1975. The use of gel filtration in the study of metal binding by humic acids and related compounds. Anal. Chim. Acta 78: 193-200.

-Murphy, T.P., D.R.S. Lean and C. Nalewajko, 1976. Blue-green algae: Their excretion of iron-selective chelators enables them to dominate other algae. Science 192: 900-902.

-Neubecker, T.A. and H.E. Allen, 1983. The measurement of complexation capacity and conditional stability constants for ligands in natural water. Water Res. 17: 1-14.

-Povoledo, D. and H.L. Golterman (eds.), 1975. "Humic Substances. Proceedings of International Meeting, The Netherlands, May 29-31, 1972". Wageningen Centre for Agricultural Publishing and Documentation, Wageningen, The Netherlands.

-Reuter, J.H. and E.M. Perdue, 1977. Importance of heavy metal-organic matter interactions in natural waters. Geochim. Cosmochim. Acta 41: 325-334.

-Rossotti, F.J.C. and H. Rossotti, 1961. The determination of stability constants. McGraw Hill, New York.

-Ruzic, I., 1982. Theoretical aspects of the direct titration of natural waters and its information yield for trace metal speciation. Anal. Chim. Acta 140: 99-113.

-Ryan, D.K. and J.H. Weber, 1982. Copper (II) complexing capacities of natural waters by fluorescence quenching. Environ. Sci. Technol. 16: 866-872.

-Saar, R.A. and J.H. Weber, 1980. Comparison of spectrofluorometry and ion-selective electrode potentiometry for determination of complexes between fulvic acid and heavy-metal ions. Anal. Chem. 52: 2095-2100.

-Saar, R.A. and J.H. Weber, 1982. Fulvic acid: Modifier of metal-ion chemistry. Environ. Sci. Technol. 16: 510A-517A.

-Sachdev, D.R., J.J. Ferris and N.L. Clesceri, 1976. Apparent molecular weight of organics in secondary effluents. J. Water Pollut. Control Fed. 48:570-579.

-Scatchard, G., J.S. Coleman and A.L. Shen, 1957. Physical chemistry of protein solutions. VII. The binding of some small anions to serum albumin. J. Am. Chem. Soc. 79: 12-20.

-Schnitzer, M. and S.U. Khan, 1972. "Humic Substances in the Environment". Dekker, New York.

-Shuman, M.S. and J.L. Cromer, 1979. Copper association with aquatic fulvic and humic acids. Estimation of conditional stability constants with a titrimetric anodic stripping voltammetry procedure. Environ. Sci. Technol., 13: 543-545.

-Shuman, M.S. and G.P. Woodward, 1977. Stability constants of copper-organic chelates in aquatic samples. Environ.Sci. Technol. 11: 809-813.

-Simpson, F.B. and J.B. Neilands, 1976. Siderochromes in cyanophyceae: Isolation and characterization of schizokinem from Anabaena sp. J. Phycol. 12: 44-48.

-Singer, P.C. (ed), 1973. Trace metals and metal-organic interactions in natural waters. Ann Arbor Science, Ann Arbor, Mich.

-Sondergaard, M. and Schierup, 1982. Release of extracellular organic carbon during a diatom bloom in Lake Mosso: Molecular weight fractionation. Freshwater Biology, 12: 313-320.

-Stevenson, F.J., 1982. "Humus Chemistry: Genesis, Composition, Reactions". Wiley, New York.

-Sueur, S., C.M.G. van den Berg and J.P. Riley, 1982. Measurements of the metal complexing ability and toxic effects of exudates of marine macro algae. Limnol. Oceanogr. (in press).

-Sunda, W.G. and R.R.L. Guillard, 1976. The relationship between cupric ion activity and the toxicity of copper to phytoplankton. J. Mar. Res. 34: 511-529.

-Sunda, W.G. and P.J. Hanson, 1979. Chemical speciation of copper in river water. Effect of total copper, pH, carbonate and dissolved organic matter. In: E.A. Jenne (Ed), Chemical modeling in aqueous systems. American Chemical Society Symp. Ser. 93: 147-180.

-Swallow, K.C., J.C. Westall, D.M. McKnight, N.M.L. Morel and F.M.M. Morel, 1978. Potentiometric determination of copper complexation by phytoplankton exudates. Limnol. Oceanogr. 23: 538-542.

-Truitt, R.E. and J.H. Weber, 1981. Determination of complexing capacity of fulvic acid for copper (II) and cadmium (II) by dialysis titration. Anal. Chem. 53: 337-342.

-Tuschall, J.R., Jr and P.L. Brezonik, 1981. Evaluation of the copper anodic stripping voltammetry complexometric titration for complexing capacities and conditional stability constants. Anal. Chem. 53: 1986-1989.

-van den Berg, C.M.G. and J.R. Kramer, 1979. Determination of complexing capacities of ligands in natural waters and conditional stability constants of the copper complexes by means of manganese dioxide. Anal. Chim. Acta 106: 113-120.

-van den Berg, C.M.G., 1982. Determination of copper complexation with natural organic ligands in seawater by equilibration with MnO_2 II. Experimental procedures and application to surface seawater. Mar. Chem. 11: 323-342.

-van den Berg, C.M.G., P.T.S.Wong and Y.K. Chau, 1979. Measurements of complexing materials excreted from algae and their ability to ameliorate copper toxicity. J. Fish Res. Bd. Can. 36: 901-905.

DETERMINATION OF COMPLEXING CAPACITIES AND CONDITIONAL STABILITY
CONSTANTS USING ION EXCHANGE AND LIGAND COMPETITION TECHNIQUES.

CONSTANT M. G. VAN DEN BERG.

1. INTRODUCTION

After the early reports on the effects of dissolved organic
material on the bioavailability of metal ions to aquatic organisms
(e.g. Goldberg, 1952, den Dooren, 1965, Neilands, 1967) we have come
a long way towards the development of sensitive chemical techniques
which provide a measure of the organic and inorganic metal fractions
in natural waters. A number of these techniques have been collected
in Table 1. Most techniques are based on the physical separation of
the organically complexed and free (inorganic) metal fractions.
Some methods achieve this separation by kinetic means, by adsorbing
the free metal onto a cation exchanger (chelex resin) or by adsorbing
the complexed metal fraction onto an anion exchanger (XAD2) or onto
a surface containing high molecular weight C-18 molecules (SEP-PAK).
It is then assumed that equilibria are not shifted when the sample is
passed through a column containing one of those adsorbents with a
contact time of only a few minutes. Usually these methods only
determine the complexed metal concentration as a fraction of the total
metal concentration, which is a measure of $K'_{ML}C_L$, the product of the
conditional stability constant, of the complex ML, with the ligand
concentration (see the section on theory). One technique, DPASV,
makes use of kinetic separation of organic complexes which are
polarographically reversible, while the free metal fraction is collected
quantitatively on the electrode surface, but evidence still has to be
provided to my knowledge that natural organic complexes do not
contribute significantly to the collected metal fraction.

Kramer, C.J.M. and Duinker, J.C. (eds.), Complexation of Trace Metals in Natural Waters. ISBN 90-247-2973-4
©1984 Martinus Nijhoff/Dr W. Junk Publishers, The Hague/Boston/Lancaster.
Printed in the Netherlands.

TABLE1   Comparison of literature reports of the organic speciation of copper in seawater.

organic fraction %	technique	Reference
30 - 50	chelex, radio-tracer DPASV	Batley and Florence (1976)
80	DPASV	Duinker and Kramer (1977)
50 - 75	ultrafiltration,DPASV	Hasle and Abdullah (1981)
97	XAD2 + ligand exchange	Hirose et al. (1982)
3 - 5	XAD2	Kremling et al. (1981)
50	SEP-PAK	Mills et al. (1982)
<50	filtration, DPASV	Nilsen and Lund (1982)
>80	XAD2	Sugimura et al. (1978)
6 - 30	bubbling of surfactants	Wallace (1982)
97 - 98	MnO_2 equilibration	van den Berg (1982, a,b)
99	CFSV of copper-catechol	van den Berg (1983, d)

Hirose et al. (1982) make use of the apparent differences in adsorption onto XAD2 of natural organic complex ions of copper and of complex ions with EDTA. The EDTA serves to keep a measurable quantity of metal ions in solution during passage through a XAD2-column. The total complexing capacity is obtained by passing sample aliquots containing added metal through the column. Natural ligands behaving similarly to EDTA would not be determined by this technique.

A large spread of results is apparent from a comparison of these techniques (Table 1). It is possible that this spread is caused at least in part by experimental and systematic inadequacies of these techniques, as most rely on kinetic separations. Very large organic metal fractions have been obtained with two techniques which measure in approximately equilibrium conditions. These two, the MnO_2 equilibration technique (van den Berg and Kramer, 1979 a,van den Berg, 1982, b) and the cathodic stripping voltammetric determination of complexes of catechol (van den Berg, 1983, d), will be discussed in detail, following the theory.

2. THEORY

2.1 Complexation

Some or all of the following equations form the basis of the reviewed studies of organic-metal interactions in water:

$$[ML] / ([M^{n+}][L']) = K'_{ML} \quad \text{where} \tag{1}$$

K'_{ML} = a conditional stability constant,

$[M^{n+}]$ = the concentration of the free metal ion,

$[L']$ = the concentration of ligand ions not complexed by M^{n+},

and

$[ML]$ = the concentration of complex ions of M and L.

Mass balances:

$$C_L = [L'] + [ML], \quad \text{and} \tag{2}$$

$$C_M = [M'] + [ML], \tag{3}$$

where C_L and C_M are the total concentration of L and M respectively, and $[M']$ = the concentration of M not complexed by L :

$$[M'] = \alpha_M [M^{n+}], \tag{4}$$

where all inorganic side-reactions of the metal ion are incorporated in α_M :

$$\alpha_M = 1 + \sum_j (K^*_{MX_j} [X^-]^j) + \sum_i (\beta^*_{M(OH)_i} / [H^+]^i)$$

where $[X^-]$ = the concentration of free ligand ions such as Cl^-, CO_3^{2-}, HPO_4^{2-}, and K^* and β^*_{MOH} are formation constants and acidity constants respectively as indicated by their subscripts, valid at the ionic strength of the sample.

There are two methods for data treatment. Both make use of the mass balances of the metal and of the ligand.

Method 1 : equations 1, 2, 3 and 4:

$$C_M = [M^{n+}] \alpha_M + [M^{n+}] K'_{ML} C_L / (K'_{ML} [M^{n+}] + 1). \tag{5}$$

It is possible to isolate $[M^{n+}]$ from this equation but one ends up with a quadratic equation (Shuman and Cromer, 1979). In case of more than one complexing ligand of type Lx the following equation can be derived:

$$C_M = [M^{n+}] \alpha_M + [M^{n+}] \sum_x (K'_{MLx} C_{Lx} / (K'_{MLx} [M^{n+}] + 1) \tag{6}$$

It is not possible to isolate $[M^{n+}]$ from this equation, so it is easier to use the equation as it stands (van den Berg, 1983,d). A plot of $[M^{n+}]$ as a function of C_M is shown in Fig. 1 for the case of

two complexing ligands. This plot is described by (6) although C_M is plotted along the X-axis. Such a plot is curved in case of one or more ligands. Values for K'_{MLx} and C_{Lx} can be calculated by means of a least squares non-linear curve fitting computer programme. Alternatively one can make use of approximately linear segments of the plot and calculate initial estimates for K'_{ML} and C_L:
At high metal concentrations the slope of the plot is equal to α_M and C_L is obtained from a linear extrapolation to the X-axis as (5) is then approximately:

$$C_M = [M^{n+}] \alpha_M + C_L \tag{7}$$

At low metal ion concentrations a value can be estimated for K'_{ML} from the slope, as then approximately

$$C_M = [M^{n+}] K'_{ML} C_L \tag{8}$$

The total ligand concentration, $\sum_x C_{Lx} m$ is obtained in case of more than one complexing ligand, and an average of the stability constants, weighted by the respective ligand concentrations.

Method 2 makes use of the mass balance of the ligand (2), and of (1), and a linear relationship between the free metal ion and the ratio of free over complexed metal is obtained (van den Berg and Kramer, 1979,a, van den Berg, 1982,a, Ruzic, 1982):

$$[M^{n+}]/[ML] = [M^{n+}]/C_L + 1/(K'_{ML} C_L) \tag{9}$$

The ligand concentration is obtained from the slope of a plot of $[M^{n+}]/[ML]$ as a function of $[M^{n+}]$, while K'_{ML} is obtained from the Y-axis intercept by linear extrapolation. In presence of more than one complexing ligand such a plot is curved and the following equation is valid (van den Berg 1982,a, and 1983,a):

$$[M^{n+}]/\sum_x [MLx] = [M^{n+}]/\sum_x C_{Lx} + \sum_x([MLx]/K'_{MLx})/(\sum_x[MLx]/\sum_x C_{Lx}) \tag{10}$$

A plot of $[M^{n+}]/\sum_x [MLx]$ as a function of $[M^{n+}]$ for the case of two complexing ligands is shown in Fig. 2. Values for K'_{MLx} and C_{Lx} can be calculated from a non-linear lease squares fit of the data to (10). Preliminary estimates can be obtained from approximately linear portions of the plot if at low metal concentrations only one complex is formed (as in Fig. 2). Equation (9) is used for this part of the data. Concentrations of ML2 can then be calculated at higher metal ion concentrations from the mass balance:

$$[ML2] = C_M - [M'] - [ML1] , \tag{11}$$

and an approximately linear plot can be made of $[M^{n+}]/[ML2]$ as a
function of $[M^{n+}]$, from which values for K'_{ML2} and C_{L2} can be
calculated. Concentrations of ML1 can then be re-calculated, from
the mass balance, at lower metal concentrations, and thus values
for K'_{ML1}, K'_{ML2}, C_{L1}, and C_{L2} are obtained by an iterative, linear
procedure.

2.2 Adsorption

Adsorption of metal ions onto manganese oxide can conveniently
be described by the Langmuir equation:

$$[M^{n+}]/_{ads} = [M^{n+}]/\Gamma_{max} + 1/(\Gamma_{max} B) \tag{12}$$

where Γ_{ads} = the concentration of adsorbed metal divided by the
manganese oxide concentration,

Γ_{max} = maximum or limiting value of Γ_{ads}, and

B = a constant representing the binding strength of
the oxide.

Free metal ion concentrations in equlibrium with the determined
amount of Γ_{ads} can be calculated from (12) after calibration of
B and Γ_{max}. It has been found, however, that these parameters are
constant only over a limited range of metal ion concentrations, and
that a greater adscrption capacity (= Γ_{max}) is obtained at increased
concentrations, indicating that two, or possibly a series of
adsorption sites are present, with decreasing magnitude of B
(van den Berg, 1982, b, Stroes, 1983). It is therefore necessary
to calibrate the oxide over the same range of Γ_{ads} as used during the
speciation measurement. Also, B has been found to be dependent on
solution parameters, such as pH and major cation composition. For
this reason Dempsey and Singer (1980) have added a term to the
Langmuir equation to correct for competition by calcium, while
Stroes (1983) developed the following, "implicit", equation which
incorporates the effect of the pH in simple, low ionic strength
electrolytes; as well as the effect of already adsorbed metal ions
on the residual surface charge (and therefore on B) of the oxide:

$$\Gamma_{ads} = \Gamma_{max} [M^{n+}]/(([H^+]^n e^{-(1-\Gamma_{ads}/\Gamma_{max})/B} + [M^{n+}]) \tag{13}$$

It may be possible to obtain a comprehensive equation by adding a
term for major cation competition. In practice, and for purposes
of metal complexation studies, it is most simple to either prepare

a calibration curve of adsorbed metal as a function of the free metal
ion concentration, or to fit any suitable, but non-descriptive,
polynomial through the data.

3.   THE MnO_2 EQUILIBRATION TECHNIQUE

The principle of this method is that inorganic, positively
charged metal ions adsorb on the surface of MnO_2, while neutral
or negatively charged, organic complex ions of those metals do not
adsorb. Small sample aliquots, of 5 ml or larger, are equilibrated
overnight with 30 μM MnO_2 which is added as a suspension. The
dissolved metal concentration is equal to $\Sigma[ML] + [M']$ and is
determined after filtration or centrifugation of the suspensions.
This measurement is repeated at increased metal concentrations as it
is necessary to approximately saturate the ligands in order to
determine the total ligand concentration. The dissolved metal
concentration is best determined by DPASV when seawater is
investigated, but atomic absorption spectrometry can be used more
easily for freshwater samples. The filtrates need to be acidified
to prevent adsorption of the metal ions onto container surfaces,
and in order to measure the total metal concentration by DPASV.
Such losses by container wall adsorption are no problem in presence
of MnO_2, as it acts as a free metal ion buffer: added metal quickly
forms a complex with organic ligands, and surplus metal adsorbs onto
the MnO_2 which represents a much greater surface area than the
container wall. Further exchange between the complexing ligands and
the MnO_2 takes place slowly, and it is necessary to allow an
equilibration period of at least four hours in seawater, and around
one hour in freshwater.

3.1 <u>Calibration of the MnO_2</u> It is normally necessary to calibrate the
MnO_2 in a UV-irradiated aliquot of every sample, unless the background
electrolyte composition and experimental pH are identical. Then, one
can either plot the data as a calibration curve of adsorbed metal as
a function of the free metal, or the inorganic metal, concentration
or the data can be fitted to an equation of the form of (12) or (13).
Normally a known complexing ligand (5 x 10^{-5}M glycine) is added to
the suspension before calibration, in order to increase the dissolved
metal concentration to measurable, and less easily contaminable,

levels, unless a radio-tracer is used. The adsorbed metal concentration
is determined by mass balance from the total and dissolved metal
concentrations.   It is important to calibrate the oxide over the
same range of adsorbed metal concentrations as obtained during sample
titration, because extrapolation to other metal concentrations is not
precise.

3.2 Possible interferences and detection limit   For this technique to
work it is necessary that only inorganic metal adsorbs, and that
neither organic complex ions, nor free organic ligands adsorb.
The author has attempted to find conditions in which organic ligands
can be made to adsorb, and from dissolved organic carbon measurements
it was found that a small fraction, some 25%, of dissolved glycine
adsorbs only at pH near 1. Similarly from voltammetric determinations
of electroactive dissolved organic compounds it appeared that such
compounds do not adsorb at natural pH, but that MnO_2 acts as an
efficient scavenger of dissolved organic material in seawater at
pH <2 (van den Berg, 1982, b). More recently, experiments with
radio-labelled NTA, glycine and aspartic acid in freshwater conditions
showed that these compounds did not adsorb onto MnO_2 (Stroes, 1983).
It is likely that this behaviour of MnO_2 is derived from its very
low pH of zero point of charge (pH_{ZPC}) which is at pH 2.2 - 3.3
(Morgan and Stumm, 1964, Murray, 1975, Gray et al., 1978). The
MnO_2 used in these experiments is prepared by a redox reaction of
permanganate and manganese ions at neutral pH (van den Berg, 1979),
and manganese in thus prepared MnO_2 has a high oxidation state of
+3.9. After ageing the particles are quite stable and resistant to
dissolution even at pH < 1. The surface of this MnO_2 bears a negative
charge at pH > pH_{ZPC}. Positively charged metal ions are therefore
attracted coulombically at neutral pH values, while dissolved organic
material of fulvic or marine nature tends to be repulsed because of
its predominantly negatively charged, reactive, carboxyl and
hydroxyl groups (e.g. Perdue, 1978). Recent experiments by Stroes
(1983) have indicated that the pH of the solution in which the MnO_2 is
first prepared, determines the reactivity of the solid to some extent.
It was found that MnO_2 prepared in alkaline conditions has a smaller
surface area than that prepared at neutral pH.   The MnO_2 prepared

at neutral pH slowly converses to cryptomelane (the rate of conversion depends on the potassium content) and becomes more filamentous over a period of years. The adsorption capacity is thereby reduced by a factor of 1.5. One should therefore regularly calibrate the MnO_2.

Some investigations have shown that similar MnO_2 can be reduced and is dissolved to some extent (0.8% per hour) by dissolved organic material in Sargasso seawater (Sunda et al., 1983), especially in presence of light. The concentrations of MnO_2 used in those experiments were about 10 to 20 times less than those used in the complexing capacity determinations, and the fraction (if any) which would dissolve is therefore quite small. In addition the MnO_2 used in the present experiments has been aged for a long period, and is therefore likely to be more resistant to dissolution. Experiments have shown that the dissolved manganese concentration was always very low (around $10^{-8}M$), also in freshwater samples containing large amounts of dissolved organic compounds (van den Berg, 1979), but it is possible that reduced Mn^{2+} immediately adsorbs onto the MnO_2, and is then not measurable in the filtrate. Stone (1983) demonstrated in a study with artificial organic substances that manganese oxide can be reduced and dissolved at very high concentrations (10 x greater than the oxide concentration) of for instance hydroquinone and catechol. However, he used a different type of manganese oxide of a much lower oxidation state, of average composition of $MnO_{1.6}$, instead of the $MnO_{1.95}$ used by van den Berg (1979) and by Stroes (1983).

More strongly reducing conditions are present in anoxic interstitial waters containing reduced inorganic and organic components. Tests in the laboratory of the author have indicated that in such conditions part of the MnO_2 dissolves. Aeration of the sample is necessary, by purging with air or oxygen for a considerable time, or pre-treatment of the sample with MnO_2 followed by filtration. The redox potential differs from the original, natural conditions after such treatment, and the complexing ability of the determined ligands is possibly not the same as before.

Another source of interference is contamination of the filtrate during separation of the MnO_2 from the solution, It is hard to avoid such contamination especially when small subsamples are used. This,

rather than the sensitivity of the hanging mercury drop electrode
(HMDE) employed in the DPASV measurements of the dissolved metal
concentrations, frequently determines the detection limit at around
10^{-8}M of complexing ligands.

3.3 Substitution of a different solid for MnO_2. The susceptibility of
MnO_2 to participate in natural redox processes limits its application
to oxidising solutions. Alternative solids need to have a low
pH_{ZPC}, and need to occur in a fine suspension for precise, low
concentration additions. Examples of such alternative solids could
be SiO_2, having a pH_{ZPC} of 3 (Yates and Healy, 1976), and possibly
TiO_2, having a pH_{ZPC} of 5.5 (Tewari and Lee, 1975).

3.4 Application and results of measurements with the MnO_2 method.
The MnO_2 method has been applied successfully to the determination of
complexing capacities for copper ions of freshwater samples and in
seawater. Typical results were ligand concentrations of $10^{-7} - 10^{-6}$M
in lakes, having conditional stability constants (log K'_{CuL}) of
7.2 - 9.5 (van den Berg and Kramer, 1979b), while in samples of the
Irish Sea ligand concentrations of 10^{-8} - 10^{-7} M were observed,
having values for log K'_{CuL} of 10.0 - 10.4 (van den Berg, 1983, e).
Recently the method was used in conjunction with radio-active Zn-65
to determine complexing capacities for zinc (van den Berg and
Dharmvanij, 1983). Thus the problem of contamination during the
filtration step is avoided while excellent sensitivity is obtained.
However, zinc cannot displace for instance copper from organic
complexes, so the observed apparent ligand concentrations were rather
low, 10^{-9} - 10^{-8} M. Experiments with cobalt (II) and lead (II) were
not successful because of the apparently irreversible adsorption of
cobalt, which is possibly oxidized to Co(III) (Murray and Dillard,
1979), and the combination of strong adsorption and weak organic
complex formation of lead.

4. CFSV OF METAL COMPLEXES

An interesting new polarographic technique has recently been
developed for the measurement of copper (II) in solution. It was
found that complex ions of copper with catechol (= ortho-dihydroxy-
benzene) of the type CuL_2^{2-} adsorb selectively onto the surface of
the HMDE, thus forming a film of adsorbed complexed ions (van den

Berg, 1983, c). Copper, as well as some other metals which are
preconcentrated in this film, is then reduced when the potential is
scanned from 0 V (versus the sat. AgCl/Ag reference electrode) in
negative direction, and a quantitative reduction (= cathodic) peak
is obtained at around - 0.25 V, using the differential pulse mode.
The technique was called cathodic film stripping voltammetry (CFSV),
because of similar nomenclature for determinations of for instance
Ni and Co, as reviewed by Brainina (1971, 1974). The technique is
very sensitive as the detection limit is of the order of 10^{-11}M Cu,
after a collection period of just 3 minutes (van den Berg, 1983, b).
In addition to the copper (II) peak at -0.2 V, the reduction peaks
of Fe (II) at - 0.4 V, and of U(VI) at -0.55 V (all potentials are
pH dependent) have now been identified (van den Berg and Huang, 1983,
a and b). Application of this technique to the study of organic
copper speciation is based on the different sensitivity (= ratio of
reduction current over metal concentration) observed in untreated and
in UV-irradiated seawater (SW and UV-SW in Fig. 3). In Fig. 3 the
peak height has been plotted as a function of the catechol concentration,
at constant copper concentration (from van den Berg, 1983, c).
An S-shaped curve is obtained reflecting an increase in the peak current
as a function of the catechol concentration until a maximum has been
reached. This curve is displaced to higher catechol concentrations in
untreated seawater, apparently as a result of ligand competition by
dissolved organic compounds for available metal ions. This titration
with catechol, and its displacement, is on its own a measure of copper
complexation. However, more information is obtained from a titration
with copper at constant catechol concentration intermediate of the two
optimal catechol concentrations, at a level below where all dissolved
copper is determined, but sufficiently high for good sensitivity.

The experimental procedures for the complexing capacity
titration are as follows (van den Berg, 1983, d) : 15 ml aliquots of
the untreated sample are pipetted into teflon cups which also serve
as voltammetric cells; 0.008 M HEPES (N-2-Hydroxyethylpiperazine-
N'-2-ethanesulphuric acid) buffer, pH 7.7, is added, and the total
copper concentration is varied between 2 x 10^{-9} and 10^{-7}M in 20 steps.
About three hours equilibration time is allowed, and the metal

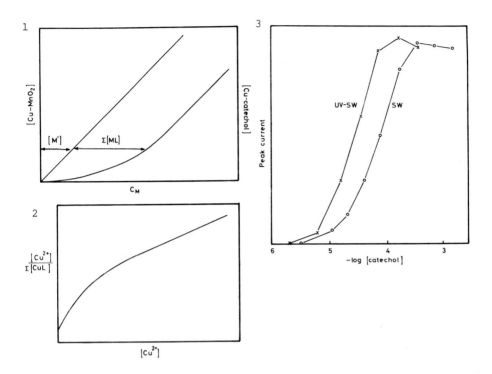

FIG. 1 Diagrammatic representation of the results of a measurement of the complexing capacity of seawater containing two complexing ligands by MnO_2 (left scale) and by cathodic stripping of catechol complex ions (right scale), using Eqn. 6.
FIG. 2 Representation of the data obtained in presence of two complexing ligands following Eqn. 10.
FIG. 3 Reduction current of complex ions of copper with catechol as a function of the catechol concentration in seawater before (SW) and after (UV-SW) UV-irradiation; data from van den Berg (1983, c).

additions are performed sequentially in order to ensure approximately equal equilibration time at all concentrations. Then each sample is purged with argon, and 2.5×10^{-5} M catechol is added; a new drop is made and the stirrer is started about one minute after the catechol addition. A scan is made after a collection time of three minutes. It is best not to add the catechol earlier as it degrades in presence of oxygen and because copper-catechol complex ions tend to adsorb onto the cell walls. Equilibrium is thought to be reached rapidly, as the catechol concentration is quite high. Tests have shown that the peak height is reproducible for about 30 minutes, after which it may decrease as a result of adsorption.

Titrations with copper were performed of 10^{-8} M EDTA in UV-irradiated seawater, and in untreated seawater originating from the surface of the South Atlantic. The data were fitted to the iterative linear regression of Eqn. 10, and ligand concentrations of 1.1 and 3.3 x 10^{-8}M were obtained, having conditional stability constants for complexes with copper (log K'_{CuL}) of 12.2 and 10.2 respectively (van den Berg, 1983, d). The fraction of copper that is present in organic complexes was calculated from these data and is given in Table 1. This fraction is large because of the high stability of the complexes.

The sensitivity of the determination of the dissolved inorganic copper concentration is calibrated by standard additions to an UV-irradiated aliquot of the sample. Only a few standard additions are necessary as the response is linear at least until 10^{-7}M copper. Any decrease in the peak height in the untreated sample is the result of a lowering of the free Cu^{2+} concentration, as the peak height is directly related to the concentration of Cu-catechol complex ions. The peak height can therefore be directly related to the free Cu^{2+} concentration by previous calibration at known dissolved copper concentrations, through calculation using known stability constants for the formation of copper catechol complex ions.

The advantage of this technique over the MnO_2 method is the greater simplicity: no filtrations are necessary, and the complexing capacity is determined from a single titration as calculation of the free copper ion concentrations is performed directly from the copper-catechol equilibria, rather than from a separate titration (as necessary for MnO_2) in an UV-irradiated sample aliquot. One can fairly comfortably finish a titration of a sample in one day, depending on the number of aliquots one wishes to analyse. The experimental time is limited by the rate with which one can do the voltammetric determinations.

The present experiments were done with copper. However, uranium (VI) and iron (III) (van den Berg and Huang, 1983,a and b) can also be determined, though the sensitivity is about ten times less than for copper. Nevertheless it may be very useful to apply this technique to the determination of free inorganic iron

concentrations, as iron is generally considered to be associated
with organic material. Similar measurements, but with different
complexing agents, have been reviewed for Co(II) and Ni(II) by
Brainina (1974). It may be possible to perform complexing capacity
determinations, similar to that developed for copper, with the ions
of other metal ions such as Fe, Ni, and Co.

4.1 Test of adsorption columns of the type SEP-PAK. In the course of
these experiments it was attempted to use SEP-PAK treatment to remove
organic complexing material from solution. These SEP-PAKs are used
by Mills et al. (1982) to collect the organic metal fraction of
metals in seawater. Test showed that the results of a titration of
thus treated seawater with catechol, followed by CFSV, as shown in
Fig. 3, were identical to that obtained with untreated seawater.
A complexing capacity titration with DPASV revealed that at most 5%
of the organic complexing ligands had been removed from the Atlantic
seawater sample. Apparently natural organic complexing ligands pass
rather well through these SEP-PAKs, which may have some bearing on
the determination of the organic metal fraction by the technique of
Mills et al. (1982). It is possible that the metal-organic complex
ions are adsorbed selectively, while free organic ligands pass through
the pre-packed columns. Their previously published organic metal
fractions were rather lower than those obtained with the MnO_2 and
CFSC techniques (Table 1), indicating that some metal-organic complexes
might pass through, but in more recent experiments, presented in this
volume, they report much greater organic copper fractions.

4.2 Possible interferences The collection is caused by adsorption of
metal-catechol complex ions onto the surface of the HMDE. The copper
peak is reduced in size as a result of competitive adsorption of

increased concentrations of other complex forming cations, such as
U(VI) and Fe(III). The concentrations of these ions therefore have
to be kept constant, and the sensitivity needs to be calibrated in
an UV-irradiated aliquot of the sample.

Surface active organic compounds can interfere similarly by
adsorbing and competing for a limited amount of surface area on the
drop. Measurements with Triton-X-100, used as a model non-ionic
surfactant, showed that the peak height for copper is indeed reduced,
but only at concentrations greater than normally found in natural

waters (van den Berg, 1983, b). It is possible, however, that in areas of high primary productivity the peak height is affected to some extent. In this case the sensitivity should be calibrated in the untreated sample, at high metal ion concentrations, where all complexing ligands are saturated. Such adsorption of surfactants probably affects all voltammetric determinations with the HMDE, anyway, also those using DPASV at low or neutral pH.

4.3 General comments

Both methods, the MnO_2 and the CFSV technique, are equilibrium techniques, as metal ions and ligands are allowed to equilibrate before, and remain in equilibrium during measurement. Interestingly the MnO_2 method to some degree reduces interference as a result of competition by complex forming cations for ligands, as the free concentrations of many dissolved cations are reduced by adsorption onto MnO_2. A similar effect is caused by catechol which is added to a concentration greater than the combined concentrations of all trace metal ions. The reaction time, available after addition of the catechol, is probably too short to allow re-allocation of metal ions over recently released ligand ions, but it may happen to a small extent.

The MnO_2 method allows the application of any suitable analytical technique for the determination of the dissolved metal fraction, such as atomic absorption spectrometry, DPASV, or scintillation counting of radio-tracers. The CFSV technique, on the other hand, is more sensitive, and less prone to metal contamination as a result of experimental procedures, and the practical detection limit is therefore about an order of magnitude less, below 10^{-9}M of complexing ligands. Its advantage over complexing capacity titrations by DPASV is the greater sensitivity of CFSV, while using a more simple and reproducible electrode (the HMDE). It is necessary to use a pre-plated mercury film electrode for the complexing capacity titrations by DPASV, which is less sensitive than when in-situ mercury plating is used.

REFERENCES

- Batley, G.E. and T.M. Florence, 1976. Determination of the chemical forms of dissolved cadmium, lead and copper in seawater. Mar. Chem. 4: 347-363.
- Brainina, Kh.Z., 1971. Film stripping voltammetry. Talanta, 18: 513-539.

- Brainina, Kh.Z., 1974. Stripping Voltammetry in Chemical Analysis. J. Wiley and Sons, New York.
- Dempsey, B.A. and P.C. Singer, 1980. The effects of calcium on the adsorption of zinc by MnOx(s) and Fe(OH)$_3$ (am). In : R. A. Baker (Ed), Contaminants and Sediments, Vol. 2, Ann. Arbor Sci., Michigan, p. 333-352.
- Den Dooren de Jong, L.E., 1965. Tolerance of Chlorella vulgaris for metallic and non-metallic ions. Ant. v. Leeuwenhoek, 31: 301-313.
- Duinker, J.C. and C.J.M. Kramer, 1977. An experimental study on the speciation of dissolved zinc, cadmium, lead and copper in river Rhine and North Sea water, by differential pulsed anodic stripping voltammetry. Mar. Chem. 5: 207-228.
- Goldberg, E.D., 1952. Iron assimilation by marine diatoms. Biol. Bull., 102: 243-248.
- Gray, M.J., M.A. Malati and M.W. Raphael, 1978. The point of zero charge of manganese dioxides. J. Electroanal. Chem. 89: 135-140.
- Hasle, J.R. and M.I. Abdullah, 1981. Analytical fractionation of dissolved copper, lead and cadmium in coastal seawater. Mar. Chem. 10: 487-503.
- Hirose, K., Y. Dokiya and Y. Sugimura, 1982. Determination of conditional stability constants of organic copper and zinc complexes dissolved in seawater using ligand exchange method with EDTA. Mar. Chem. 11: 343-354.
- Kremling, K., A. Wenck and C. Osterroht, 1981. Investigations on dissolved copper-organic substances in Baltic waters. Mar. Chem. 10: 209-219.
- Mills, G.L., A.K. Hanson, Jr., J.G. Quinn, W.R. Lammela and N.D. Chasteen, 1982. Chemical studies of copper-organic complexes isolated from estuarine waters using C$_{18}$ reverse-phase liquid chromatography. Mar. Chem. 11: 355-377.
- Morgan, J.J. and W. Stumm, 1964. Colloid chemical properties of manganese dioxide. J. Coll. Sci., 19: 347-359.
- Murray, J.W., 1975. The interaction of metal ions at the manganese dioxide-solution interface. Geoch. Cosmoch. Acta., 39: 505-519.
- Murray, J.W. and J.G. Dillard, 1979. The oxidation of Cobalt (II) adsorbed on manganese dioxide. Geoch. Cosmoch. Acta., 43: 781-787.
- Neilands, J.B., 1967. Hydroxamic acids in nature. Science, 156: 1443-1447.
- Nilsen, S.K. and W. Lund, 1982. The determination of weakly and strongly bound copper, lead and cadmium in Oslofjord samples. Mar. Chem. 11: 223-233.
- Perdue, E.M., 1978. Solution thermochemistry of humic substances - I. Acid-base equilibria of humic acid. Geoch. Cosmoch., Acta., 42: 1351-1358.
- Ruzic, I., 1982. Theoretical aspects of the direct titration of natural waters and its information yield for trace metal speciation. Anal. Chim. Acta. 14: 99-113.
- Shuman, M.S. and J.L. Cromer, 1979. Copper association with aquatic fulvic and humic acids. Estimation of conditional formation constants with a titrimetric anodic stripping voltammetric procedure. Env. Sci. Technol. 13: 543-545.
- Stone, A.T., 1983. The reduction and dissolution of Mn(III) and Mn(IV) oxides by organics. Thesis, Cal. Inst. Techol., Calif., Report No. AC-1-83.

- Stroes, S., 1983. Adsorption behaviour of delta-manganese dioxide in relation to its use as a resin in trace metal speciation studies. Thesis, McMaster University, Ontario, Canada.
- Sugimura, Y., Y. Suzuki and Y. Miyake, 1978. Chemical forms of minor metallic elements in the ocean. J. Oceanogr. Soc., Japan, 34: 93-96.
- Sunda, W.G., S.A. Huntsman and G.R. Harvey, 1983. Photoreduction of manganese oxides in seawater and its geochemical and biological implications. Nature, 301: 234-236.
- Tewari, P.H. and W. Lee, 1975. Adsorption of Co(II) at the oxide-water interface. J. Coll. Interf. Sci. 52: 77-88.
- van den Berg, C.M.G., 1979. Complexation of copper in natural waters. Thesis, McMaster University, Ontario, Canada.
- van den Berg, C.M.G., 1982, a. Determination of copper complexation with natural organic ligands in seawater by equilibration with MnO_2. I. Theory. Mar. Chem. 11: 307-322.
- van den Berg, C.M.G., 1982, b. II. Experimental procedures and application to surface seawater. Mar. Chem. 11: 323-342.
- van den Berg, C.M.G., 1983, a. Corrections to "Determination of copper complexation with natural organic ligands in seawater by equilibration with MnO_2. I. Theory". Mar. Chem. 13: 83-85.
- van den Berg, C.M.G., 1983, b. Determining trace concentrations of copper in water by cathodic film stripping voltammetry with adsorptive collection. Anal. Letters: submitted.
- van den Berg, C.M.G., 1983, c. Analysis of seawater by cathodic film stripping voltammetry preceded by adsorptive collection with the hanging mercury drop electrode. I. Copper. Anal. Chim. Acta: submitted.
- van den Berg, C.M.G., 1983, d. Determination of the complexing capacity and conditional stability constants of complexes of copper (II) with natural organic ligands in seawater by cathodic stripping voltammetry of copper-catechol complex ions. Mar. Chem. : submitted.
- van den Berg, C.M.G. 1983, e. Complexation of copper by organic material in surface water of the Irish Sea. Mar. Chem: submitted.
- van den Berg, C.M.G. and J.R. Kramer, 1979,a. Determination of complexing capacities and conditional stability constants for copper in natural waters using MnO_2. Anal. Chim. Acta. 106: 113-120.
- van den Berg, C.M.G. and J.R. Kramer, 1979, b. Conditional stability constants for copper ions with ligands in natural waters. In : E.A. Jenne (Ed.) Chemical Modelling in Aqueous Systems, A.C.S. Symposium Series, 93: 115-132.
- van den Berg, C.M.G. and S. Dharmvanij, 1983. Determination of complexing capacities of zinc using equilibration with MnO_2 and Zn-65. In preparation.
- van den Berg, C.M.G. and Z.Q. Huang, 1983,a. Analysis of seawater by cathodic film stripping voltammetry preceded by adsorptive collection with the hanging mercury drop electrode. II. Uranium (VI). Anal. Chim. Acta : submitted.
- van den Berg, C.M.G. and Z.Q. Huang, 1983,b. III. Iron (III). Anal. Chim. Acta : in preparation.
- Wallace, G.T., Jr., 1982. The association of copper, mercury and lead with surface-active organic matter in coastal seawater. Mar. Chem. 11: 379-394.
- Yates, D.E. and T.W. Healy, 1976. The structure of the silica/ electrolyte interface. J. Coll. Interf. Sci. 55 : 9-19.

THE USE OF ELECTROCHEMICAL TECHNIQUES TO MONITOR COMPLEXATION CAPACITY
TITRATIONS IN NATURAL WATERS

MARK S. VARNEY, DAVID R. TURNER, MICHAEL WHITFIELD and
RICHARD F.C. MANTOURA

1. INTRODUCTION

The chemical speciation of trace metals in natural waters provides the
key to an understanding of their biological availability and geochemical
reactivity. Complexation by natural organic matter is a major feature of
the speciation of certain trace metals and is widely studied by
complexation capacity titrations. The theory of such titrations is well-
documented (Shuman and Woodward 1973, 1977). It is assumed that the metal
ion whose concentration is measured forms a single 1:1 complex with the
organic matter

$$M + Y \rightleftharpoons MY \qquad\qquad (1)$$

It is further assumed that the technique used to monitor the titration
responds to the concentration of the free metal ion M or to some quantity
proportional to this concentration. The simplest method of data analysis
is to plot the response to the free metal ions against the total
concentration of metal added and to divide the resulting plot into two
linear portions (Figure 1). The complexation capacity (or the equivalent
site concentration) is given by the intersection point of the two straight
lines and the stability constant is derived from their slopes.

To follow such titrations it is necessary to measure the free
concentration of the added metal (usually copper) at each point in the
titration. A wide variety of techniques can be used for this measurement
including equilibration with chelating resins (Figura & McDuffie 1980),
dialysis equilibration (Guy & Chakrabarti 1976, Truitt & Weber 1981),
ultrafiltration (Smith 1976), gel filtration (Sugai & Healy 1978),
adsorption on a manganese dioxide suspension (van den Berg 1982) and
bioassay using marine bacteria (Davey et al. 1973, Gillespie & Vaccaro
1978, Sunda & Gillespie 1979, Sunda & Ferguson 1983). However the most

Kramer, C.J.M. and Duinker, J.C. (eds.), Complexation of Trace Metals in Natural Waters. ISBN 90-247-2973-4
©1984 Martinus Nijhoff/Dr W. Junk Publishers, The Hague/Boston/Lancaster.
Printed in the Netherlands.

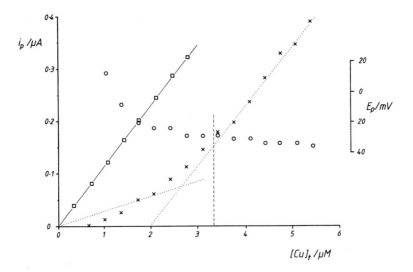

FIGURE 1. Plot of DPP peak currents (x) and peak potentials (o) for the titration of copper against 6.79 mg dm^{-3} fulvic acid and peak currents (□) for a blank titration at pH 6.80. The dotted lines indicate the two linear sections used for estimating the complexation capacity and the dashed vertical line indicates the computer-estimated complexation capacity.

popular technique is anodic stripping voltammetry (ASV) because it combines high sensitivity with ease of use (Chau et al. 1974, Duinker & Kramer 1977, Florence & Batley 1977, Baccini & Suter 1979, Bhat et al. 1981, Plavsic et al. 1982).

Unfortunately ASV does not simply measure the free metal ion concentration. The technique also responds to complexes which are sufficiently labile to dissociate during the timescale of measurement and it is subject to interferences associated with the adsorption of organic matter on the electrode surface. As a result there is considerable controversy concerning the use of ASV for complexation capacity measurements (Wilson et al. 1980, Tuschall & Brezonik 1981, Shuman et al. 1982, Bhat et al. 1982). The purpose of the present paper is to discuss some of the problems associated with the use of ASV for the measurement of complexation capacity and the stability constants for metal-organic matter interactions. The discussion will be illustrated with results from a recent study of metal-fulvic acid complexation in which the performances of a variety of electrochemical techniques were compared under similar chemical conditions (Varney 1982, Varney et al., in preparation).

2. EXPERIMENTAL DESIGN AND RESULTS'

Compleximetric titrations were carried out by following the response of a number of electrochemical techniques (Table 1) to the sequential addition of metal ions (Cu^{2+} or Pb^{2+}) to aged solutions of an aquatic fulvic acid (FA) in 0.1M $NaClO_4$. The pH was maintained constant during each titration by manual additions of perchloric acid or sodium hydroxide. All titrations were carried out at a pH close to 6.5 in order to avoid interference by metal hydrolysis while remaining as close as possible to the pH of natural waters. The equilibration time following each metal addition was held constant. An artificial medium was employed to minimise any effects due to inorganic complexation and to provide a uniform basis for the comparison of the response of the various electrochemical techniques. Furthermore the organic material used throughout was a single sample of aquatic fulvic acid (FA) extracted from a freshwater source. Detailed descriptions will be found elsewhere of the isolation and purification of the FA (Varney 1982), its acid-base chemistry (Varney et al. 1983) and of the metal-FA titrations (Varney et al., in preparation).

The electrochemical techniques employed enable the response of the complexation capacity measurement to variations in the timescale of the analysis to be assessed (Table 1). This timescale, which represents the time available for the dissociation of metal complexes in non-equilibrium systems (Davison 1978, Turner & Whitfield 1979), is determined largely by the hydrodynamic conditions close to the electrode and can also be expressed in terms of a diffusion layer thickness.

Rather than rely simply on the graphical procedure described in the INTRODUCTION, Ruzic (1982) developed a numerical procedure based on the linearisation of the equations describing the electrochemical response to the complexation and Bhat et al. (1981) described a non-linear least squares procedure. In the present study the titration data were fitted to a multi-site model of metal-FA binding by a non-linear least squares regression technique which used the degree of binding, α (= [free metal]/[total metal]) as the dependent variable. Each binding site is characterised by a stability constant Kj and a concentration Nj (Table

Table 1. Characteristics of electrochemical techniques used.

Technique	Timescale	Notes
Ion-selective electrode (ISE)	true equilibrium	Not sufficiently sensitive for work in natural samples. Useful reference point for non-equilibrium techniques
Linear scan ASV[a] (LSASV)	Dependent on diffusion layer thickness during deposition (100 ms at hanging mercury drop electrode)	Preconcentration of metal during electrodeposition. Metal accumulated is measured during oxidative stripping with a linear voltage scan.
Differential pulse[a] ASV (DPASV)	As for LSASV	As for LSASV but pulsed scan used to achieve extra sensitivity by redepositing and stripping metal several times
Differential pulse polarography (DPP)	Equivalent to pulse duration (47 ms in this study)	Direct measure of current associated with reduction of trace metal

[a] Using a hanging mercury drop electrode

2). The 1-site model corresponds to the simple model (equation 1) with N_1 being equal to the complexation capacity.

The values of the binding strengths obtained by ISE are in reasonable agreement with previously published work for copper and lead (e.g. Buffle et al. 1977, Saar & Weber 1980), and are consistent with the relative strengths of bidentate chelation in 'model' compounds (e.g. Cu-citrate, $\log K_1 = 5.2$, $\log K_2 = 8.4$; Martell & Smith 1976). They are also compatible with the strengths of the metal complexes with o-carboxyphenate and o-carboxybenzoate groupings which have been demonstrated to complex copper ions in soil fulvic acids (Schnitzer & Khan 1972) and are thought to be involved in similar interactions with aquatic fulvic acids (Gamble et al. 1970, Manning & Ramamoorthy 1973, Stevenson, 1976).

3. DISCUSSION

The results summarised in Table 2 relate to titrations carried out under similar chemical conditions using the same organic material. If the complexation capacity model were valid for the metal-FA binding and if each technique simply measured the free metal ion concentration (or a proportional quantity) then we should obtain the same results from all four techniques for each metal. In fact three striking differences are observed: (i) the non-equilibrium techniques fail to detect any binding of lead by FA, (ii) the ASV techniques seriously underestimate the concentration of copper binding sites, and (iii) whereas the ISE titrations indicate the presence of two binding sites (with a poor fit to a 1-site model), the other techniques indicate only a single binding site. These differences will be considered in terms of the characteristics of the individual measuring techniques (Table 1).

Table 2. Comparison of the parameters obtained from titrations of copper and lead against fulvic acid in 0.1M NaClO$_4$

Metal	technique	binding model	log $K_j{}^a$	$N_j{}^a$ (meq g^{-1})	F-ratio	pH	[FA] (mg dm^{-3})
Cu	ISE	1-site	6.63	0.427	205	6.22	29.1
		2-site	8.12	0.098	485		
			5.79	0.528			
	DPP[b]	1-site	6.81	0.490	2042	6.80	6.79
	LSASV[b]	1-site	6.58	0.180	1149	6.37	13.04
	DPASV[b]	1-site	7.23	0.185	2316	6.41	13.04
Pb[c]	ISE	1-site	5.36	0.823	176	6.40	72.75
		2-site	6.78	0.117	1329		
			4.19	2.421			

[a] Mean values from a series of experiments. For statistical analyses see Varney (1982) and Varney et al. (in preparation). Similar metal/FA concentration ratios were used in all of the titrations.

[b] 2-site model failed to converge.

[c] No evidence of binding from peak current measurements using DPP, LSASV or DPASV.

3.1. Lability of complexes

The non-equilibrium electrochemical techniques (DPP and ASV) respond not only to the free metal ion but also to metal bound in complexes which are able to dissociate over the timescale of the measurement (Davison 1978, Turner & Whitfield 1979). Such dissociation is known to be significant in complexation capacity titrations (Bhat et al. 1981), and has been exploited in electrochemical estimations of the dissociation rate constants of Cu-organic complexes in a natural water (Shuman & Michael 1978). The influence of this kinetic component on the response of the voltammetric techniques is clearly illustrated by the comparative experiments.

If we express the values of the binding site concentration (N_1) for the kinetic (non-equilibrium) techniques as a fraction of the total site concentration for the same metal from potentiometric measurements (N_1 + N_2), we obtain the sequence:

| 0% | = | 0% | = | 0% | < | 29% | < | 30% | < | 78% |
| Pb(LSASV) | | Pb(DPASV) | | Pb(DPP) | | Cu(LSASV) | | Cu(DPASV) | | Cu(DPP) |

How far are these observations consistent with the complexation patterns obtained from potentiometric data? Davison (1978) has suggested that, for labile behaviour in DPP measurements, the timescale of the experiment (47 ms in this instance) should exceed a characteristic time (t) given by

$$t = 0.01 \cdot K_j^2 \cdot [FA] \cdot N_j / \pi \, k_1 \qquad (2)$$

where k_1 is the rate constant for the dissociation of the complex. Assuming that $k_1 = 10^9 \, mol^{-1} \, dm^3 \, s^{-1}$ (Davison 1978) we find (Table 3) that for lead both sites identified by the potentiometric titration (Table 2) should be labile. This is consistent with the observations. For copper, however, equation (2) suggests that DPP should only sense site-1 (equivalent to 20% of the sites identified by the ISE titrations). The concentration of sites apparently sensed is equivalent to nearly 80% of the total. The contribution of a complex to the electrochemical response does not, however, necessarily imply that the complex is labile. A non-labile complex may also be reduced irreversibly at the electrode surface (Turner & Whitfield 1979, Florence 1982). These two possibilities may be distinguished by varying the timescale of the measurement (Shuman & Michael 1978, van Leuwen 1979) and/or the deposition potential (Turner & Whitfield 1979).

Table 3. Lability criteria for non-equilibrium techniques derived from the two-site potentiometric model (Davison 1978)[a]

Technique	Parameter	Metal	site 1	site 2
DPP	t/s (equation 2)	Cu	$\underline{0.04}$[b]	4.4×10^{-6}
		Pb	7.4×10^{-4}	1.0×10^{-7}
LSASV	$K/s^{\frac{1}{2}}$ (equation 3)	Cu	$\underline{5.15}$	0.05
		Pb	0.22	1.6×10^{-3}
DPASV	$K/s^{\frac{1}{2}}$ (equation 3)	Cu	$\underline{5.15}$	0.05
		Pb	0.38	3.5×10^{-3}

[a] Stability constants and ligand concentrations taken from Table 2. For the lead experiments the fulvic acid concentrations were as follows: DPP, 54.5 mg dm^{-3}; LSASV, 13.04 mg dm^{-3}; DPASV, 36.12 mg dm^{-3}.

[b] Non-labile complexes are underlined.

For the ASV measurements the equivalent criterion for labile behaviour (Davison 1978) is that

$$K = K_j^{1.5} . [FA] . N_j / k_1 (1 + K_j . [FA] . N_j))^{\frac{1}{2}} < 2.85 \qquad (3)$$

Substituting in the appropriate values we find once more (Table 3) that, in accord with the observations, both of the lead sites should be labile with respect to ASV. Only site-1 is non-labile for copper and, on this occasion, the percentage of copper sites sensed by the ASV measurements (30% of the total identified by the ISE experiments) is in reasonable agreement with the percentage anticipated by the lability criteria (site-1, 20%, Table 3). The experimental results and the calculations both indicate that the copper complexes are least labile with respect to the technique with the shortest timescale (DPP). There is little difference between the lability of copper with respect to the two ASV techniques since the timescale is controlled by the deposition step which is common to both.

3.2. Adsorption

Interference in electrochemical measurements by adsorption of organic matter on the electrode is well documented (see for example Batley & Florence 1976, van Leeuwen 1979, Greter et al. 1979, Buffle & Greter 1979, Wilson et al. 1980, Sagberg & Lund 1982). Adsorption of organic matter (and its metal complexes) on the electrode surface may affect the electrode processes by: (i) physical blocking of the electrode surface,

(ii) complexation of metal by the high local organic matter concentration at the electrode surface and (iii) deposition of metal by irreversible reduction of adsorbed complexes. Physical blocking (process (i)) would not have a serious effect on complexation capacity titrations since it would merely imply a reduction in the effective area of the electrode. Earlier studies have provided evidence of more active interference by adsorbed organic matter by processes (ii) and (iii) (Buffle & Greter 1979, Greter et al. 1979). We can obtain further evidence of the significance of such interferences by a study of the shifts in peak potential observed during the titrations (Table 4).

Table 4. Peak potential shifts (ΔE mV) for the various titrations calculated from the two-site potentiometric model[a]

Metal	Procedure	Both sites	Most labile site	Observed
Cu	DPP	58	15[b]	58
	LSASV	66	21	0
	DPASV	66	21	190
Pb	DPP	48	14	38
	LSASV	31	5	75
	DPASV	43	11	85

[a] Calculated from equation 4 using the data shown in Table 2.

[b] The values underlined are those which are consistent with the lability criteria in Table 3.

The shifts in potential are related to complex formation and dissociation during the DPP measurement and during the deposition and/or stripping processes in ASV. Here there are important distinctions to be made between the techniques. In the case of DPP the ligand is in excess over the metal at the start of a titration, while at the end the metal is in excess and the effect of the ligand can be discounted. This means that, assuming reversible reduction throughout, the shift in peak potential during the titration should be given by the equation (Crow 1969)

$$\Delta E = \frac{RT}{2F} \ln (1 + \Sigma_j K_j N_j) \qquad (4)$$

Only labile complexes will contribute to this potential shift. We can therefore calculate a potential shift from the values of K_j and N_j derived from the potentiometric measurements (Table 2) and the FA concentrations used in the DPP titrations.

For lead the lability criteria (Table 3) suggest that both sites identified by the potentiometric titrations should be labile. The peak potential shift calculated on this basis is similar to the observed value (Table 4). The potential shift calculated for copper on the assumption that only the second site is labile (Table 3) is, however, only one quarter of the observed value. This is consistent with the indications from the lability calculations that adsorption processes are making an important contribution to the response of DPP in the copper-fulvic acid titrations.

The peak potentials observed in the ASV techniques are governed by processes occurring during stripping rather than during deposition. While metal is being stripped from the electrode, very high local metal concentrations can build up at the electrode surface so that we can expect the metal to be in excess of the bulk ligand concentration throughout the titration and therefore no shift in peak potential should be observed. However the experimental potential shifts shown in Table 4 are large (except for Cu in LSASV) and do not follow a clear pattern. Shifts are similar for lead in both DPASV and LSASV, while copper shows a large variation. With the exception of copper in LSASV the calculated potential shifts are once more much lower than the observed values (Table 4). This provides clear evidence that adsorption is affecting the ASV measurements for both metals, and that in the case of copper the effect is strongly dependent on the nature of the stripping technique.

The differences revealed by the potential shift data can be understood in a qualitative sense by recalling the details of the techniques in question. In DPP the electrode is exposed to the sample for a very short period ($\sim$ 1s) before measurement and the metal crosses the solution/ electrode interface only once. In ASV the electrode is exposed to the sample for several minutes before measurement and the metal has to cross the solution/electrode interface twice (LSASV) or many times (DPASV). It is therefore to be expected that interferences from adsorbed organic matter will increase in the order

DPP < LSASV < DPASV

which is in broad agreement with the effects observed (Table 4).

3.3. Derivation of complexation capacity and stability constants

Any description of FA binding in terms of a small number of sites is necessarily an approximation since the organic material used is in effect a mixture of different compounds offering a wide variety of sites for metal binding. The analysis of the data from the titration experiments is therefore a statistical exercise involving the correct selection of the dependent variable and the use of statistical tests to determine the most probable binding model. The non-linear least squares method used here provides clear statistical tests of the suitability of the various binding models for all of the detection techniques employed (Table 2). Comparison with the graphical method of estimation of complexation capacity for the non-equilibrium techniques (Fig. 1) shows that differences of the order of twenty to thirty percent can arise (Varney 1982, Varney et al. in preparation).

The observations of only a single binding site in the DPP and ASV titrations is in agreement with the results of many other workers (see for example Chau et al. 1974, Duinker & Kramer 1977, Bhat et al. 1981, Plavsic et al. 1982). It is, however, in sharp contrast to the results obtained here on the same material, under similar conditions from the potentiometric titrations. In this case the single-site model does not provide a good fit and a two-site model has to be invoked to provide an adequate representation of the data (Table 2). No significant improvement was obtained by the application of a three-site model (Varney 1982). The striking difference between the potentiometric measurements and the non-equilibrium techniques may in part be a reflection of the higher precision of potentiometric titrations. The stability constants for the one-site fittings to the ISE, DPP and LSASV data for copper are all closely similar (Table 2). However, the single site fitting to the non-equilibrium data may also, in part, be an artefact arising from a combination of kinetic and adsorption effects. The K_1 value obtained for the Cu-FA complex from the DPASV titrations, which are most prone to adsorption artefacts, is significantly higher than the values obtained from the other techniques (Table 2).

4. SUMMARY AND CONCLUSIONS

The preceding discussion has highlighted the importance of complex lability and electrode adsorption in determining the results of complexation capacity titrations monitored by DPP and ASV. These observations are also relevant to the problem of estimating the free metal ion concentration in a natural sample, which in many cases corresponds to the biological availability of the metal (Turner 1983). The sensitivities of the four electrochemical techniques used increase in the order

$$ISE < DPP < LSASV < DPASV$$

Unfortunately a general conclusion of this paper is that the importance of interferences also increases in the same order, suggesting that one can have unequivocal measurements or sensitive measurements but not both. We must therefore seek to understand in more detail the processes giving rise to the interferences. Specifically we need to improve our models of metal-organic binding and to develop adequate models to describe the kinetics of complex formation and dissociation. We must also try to understand more clearly the adsorption of organic matter on the mercury electrode and its effect on electrode processes involving trace metals. In this context it is particularly important to assess the effective timescale of the procedure being used and, where possible, to make observations over a range of timescales. At the same time we should not overlook possibilities which may be provided by new techniques such as manganese dioxide adsorption (van den Berg 1982) and Cu(II)/Cu(I) amperometry (Waite & Morel 1983) where these artefacts might be less obtrusive.

NOTES AND ACKNOWLEDGEMENTS

The work was funded in part by a Natural Environment Research Council CASE studentship to Mark Varney. The present paper is designated contribution No. 674 by the Institute of Marine Environmental Research.

REFERENCES

- Baccini, P., Suter, S., 1979. Chemical speciation and biological availability of copper in lake water. Schweiz. Z. Hydrol., 41: 191-314.
- Batley, G.E., Florence, T.M., 1976. The effect of dissolved organics on the stripping voltammetry of seawater. J. Electroanal. Chem., 72: 121-126.
- van den Berg, C.M.G., 1982. Determination of copper complexation with natural organic ligands by equilibration with MnO_2. Mar. Chem., 11: 307-342.
- Bhat, G.A., Saar, R.A., Smart, R.B., Weber, J.H., 1981. Titration of soil-derived fulvic acid by copper II and measurement of free copper II by anodic stripping voltammetry and copper II selective electrodes. Anal. Chem., 53: 2275-2280.
- Bhat, G.A., Weber, J.H., Tuschall, J.R., Brezonik, P.L., 1982. Exchange of comments on evaluation of the copper anodic stripping voltammetry complexometric titration for complexation capacities and conditional stability constants. Anal. Chem., 54: 2116-2117.
- Buffle, J., Greter, F.-L., Haerdi, W., 1977. Measurement of complexation properties of humic and fulvic acids in natural waters with lead and copper ion-selective electrodes. Anal. Chem. 49: 216-222.
- Buffle, J., Greter, F.-L., 1979. Voltammetric study of humic and fulvic substances. Part II. Mechanism of reactions of the Pb-fulvic complexes on the mercury electrodes. J. Electroanal. Chem. 101: 231-251.
- Chau, Y.K., Gachter, R., Lum-Shue-Chan, K., 1974. Determination of the apparent complexing capacity of lake waters. J. Fish. Res. Bd. Can., 31: 1515-1519.
- Crow, D.R., 1969. Polarography of metal complexes. Academic Press, London, 171 pp.
- Davey, E.W., Morgan, M.J., Erickson, S.J., 1973. A biological measurement of copper complexation capacity of seawater. Limnol. Ocean. 18: 993-997.
- Davison, W., 1978. Defining the electroanalytically measured species in a natural water sample. J. Electroanal. Chem. 87: 395-404.
- Duinker, J.C., Kramer, C.J.M., 1977. An experimental study on the speciation of dissolved zinc, cadmium, lead and copper in River Rhine and North Sea water by differential pulse anodic stripping voltammetry. Mar. Chem. 5: 207-228.
- Figura, P., McDuffie, B., 1980. Determination of labilities of soluble trace metal species in aqueous environmental samples by anodic stripping voltammetry and chelex resin and batch methods. Anal. Chem., 52: 1433-1439.
- Florence, T.M., 1982. Development of physico-chemical speciation procedures to investigate the toxicity of copper, lead, cadmium and zinc towards aquatic biota. Anal. Chem. Acta 141: 73-94.
- Florence, T.M., Batley, G.E., 1977. Determination of copper in sea water by anodic stripping voltammetry. J. Electroanal. Chem. 75: 791-798.
- Gamble, D.S., Schnitzer, M., Hoffman, P., 1970. Cu^{2+} fulvic acid chelation equilibrium in 0.1 m KCl at 25°C. Can. J. Chem. 48: 3187-3204.
- Gillespie, P.A., Vaccaro, R.F., 1978. A bacterial bioassay for measuring the copper-chelation capacity of seawater. Limnol. Oceanogr. 23: 543-548.
- Greter, F.-L., Buffle, J., Haerdi, W., 1979. Voltammetric study of humic and fulvic substances. Part I. Study of the factors influencing the measurement of their complexing properties with lead. J. Electroanal. Chem. 101: 211-229.

- Guy, R.D., Chakrabarti, C.L., 1976. Studies of metal organic interactions in model systems pertaining to natural waters. Can. J. Chem. 54: 2600-2611.
- van Leeuwen, H.P., 1979. Complications in the interpretation of pulse polarographic data on complexation of heavy metals with natural polyelectrolytes. Anal. Chem. 51: 1322-1323.
- Manning, P.G., Ramamoorthy, S., 1973. Equilibrium studies of metal-ion complexes of interest to natural waters. VII. J. Inorg. Nucl. Chem. 35: 1577-81.
- Martell, A.E., Smith, R.M., 1976. Critical stability constants Vol. 3. Plenum Press, New York.
- Plavsic, M., Kraznaric, D., Branica, M., 1982. Determination of the apparent complexing capacity of seawater by anodic stripping voltammetry. Mar. Chem. 11: 17-31.
- Ruzic, I., 1982. Theoretical aspects of the direct titration of natural waters and its information yield for trace metal speciation. Anal. Chim. Acta 140: 99-113.
- Saar, R.A., Weber, J.H., 1980. Lead II complexation by fulvic acid: how it differs from fulvic acid complexation of copper II and cadmium II. Geochim. Cosmochim. Acta 44: 1381-1384.
- Sagberg, P., Lund, W., 1982. Trace metal analysis by anodic stripping voltammetry. Effect of surface-active substances. Talanta 29: 457-460.
- Schnitzer, M., Khan, S.U., 1972. Humic substances in the environment. Marcel Dekker, New York.
- Shuman, M.S., Michael, L.C., 1978. Application of the rotated disc electrode to measurement of copper complex dissociation rate constants in marine coastal samples. Environ. Sci. Technol. 12: 1069-1072.
- Shuman, M.S., Brezonik, P.L., Tuschall, J.R., 1982. Exchange of comments on evaluation of the copper anodic stripping voltammetry complexometric titration for complexing capacities and conditional stability constants. Anal. Chem. 54: 998-1000.
- Shuman, M.S., Woodward, G.P., 1973. Chemical constants of metal complexes from a complexometric titration followed with anodic stripping voltammetry. Anal. Chem. 45: 2032-2035.
- Shuman, M.S., Woodward, G.P., 1977. Stability constants of copper-organic chelates in aquatic samples. Environ. Sci. Tech. 11: 809-813.
- Smith, R.G., 1976. Evaluation of combined applications of ultrafiltration and complexation capacity techniques to natural waters. Anal. Chem. 48: 468-9.
- Stevenson, F.J., 1976. Stability constants of Cu, Pb and Cd complexes with humic acids. Soil Sci. Soc. Am. J. 40: 665-672.
- Sugai, S.F., Healy, M.L., 1978. Voltammetric studies of the organic association of copper and lead in two Canadian inlets. Mar. Chem. 6: 291-308.
- Sunda, W.G., Ferguson, 1983. Sensitivity of natural bacterial communities to additions of copper and to cupric ion activity. A bioassay of copper complexation in seawater. In "Trace Metals in Seawater", Wong, C.S., Boyle, E., Bruland, K.W., Burton, J.D., Goldberg, E.D., Eds. Plenum, New York, 871-891.
- Sunda, W.G., Gillespie, P.A., 1979. The response of a marine bacterium to cupric ion and its use to estimate cupric ion activity in seawater. J.Mar. Res. 37: 761-777.
- Truitt, R.E., Weber, J.H., 1981. Determination of complexing capacity of fulvic acid for copper (II) and cadmium (II) by dialysis titration. Anal. Chem. 53: 337-342.

- Turner, D.R., 1983. Relationships between biological availability and chemical measurements, in "Metal Ions in Biological Systems", Sigel, H., Ed. Volume 18, Marcel Dekker, New York, in press.
- Turner, D.R., Whitfield, M., 1979. The reversible electrodeposition of trace metal ions from multi-ligand systems. Part II. Calculations on the electrochemical availability of lead at trace levels in seawater. J. Electroanal. Chem. 103: 61-79.
- Tuschall, J.R., Brezonik, P.L., 1981. Evaluation of the copper anodic stripping voltammetry complexometric titration for complexing capacities and conditional stability constants. Anal. Chem. 53: 1986-1989.
- Varney, M.S., Mantoura, R.F.C., Whitfield, M., Turner, D.R., Riley, J.P., 1983. Potentiometric and conformational studies of the acid base properties of fulvic acid extracted from natural waters. In: NATO Conference Proceedings "Trace metals in sea water". pp. 751-772.
- Varney, M., 1982. Organometallic interactions of fulvic acid extracted from natural waters. Ph.D. thesis, University of Liverpool, 286 pp.
- Waite, T.D., Morel, F.M.M., 1983. Characterisation of complexing agents in natural waters by copper (II)/copper (I) amperometry. Anal. Chem. 55: 1268-1274.
- Wilson, S.A., Huth, T.C., Arndt, R.A., Skogerboe, R.K., 1980. Voltammetric techniques for determination of metal binding by fulvic acid. Anal. Chem. 52: 1515-1518.

CHROMATOGRAPHIC SEPARATION FOR THE MEASUREMENT OF STRONG AND MODERATELY
STRONG COMPLEXING CAPACITY IN LAKE WATERS

RICHARD J. STOLZBERG

1. INTRODUCTION

The technique I am going to describe for measuring complexing ability
is unique in that it is capable of rapidly measuring the quantities of
both strong and intermediate strength ligands. As such it should be a
useful tool for investigating changes in metal binding ability that
accompany, preceed, or follow changes in primary productivity of lakes.

Hart (1981) and Neubecker and Allen (1983) have reviewed the wide
variety of techniques for measuring complexing capacity (CC) in natural
water. There is disagreement over the meaning or validity of the numbers
generated, and I do not wish to suggest that this technique is
necessarily any less operational or more valid than any other. This
Chelex chromatography method is rapid compared to other techniques, and
it can be done using a wide variety of metals, including copper, nickel,
cobalt(II), and zinc. In addition, the method differs from most others
in that complexed, rather than free, metal is measured. The results I
will present in this paper are primarily laboratory data with well
characterized ligands that show that the technique works and why it
works. In addition, I will show some lake water results that suggest
useful and exciting results that may come from applying the technique
during times of rapid changes in lake water chemistry.

2. PROCEDURE

2.1 Method

Complexing capacity (CC) measurements in lake water were made as
described previously (Stolzberg and Rosin, 1977). In brief, a lake water
sample was filtered (0.45 μm), 0.01 M tris was added, and the pH was
adjusted to 7.5. Duplicate 10 mL aliquots were taken and were spiked
with 13 μmol dm^{-3} Cu or 14 μmol dm^{-3} Ni. The sample was equilibrated for

Kramer, C.J.M. and Duinker, J.C. (eds.), Complexation of Trace Metals in Natural Waters. ISBN 90-247-2973-4
© *1984 Martinus Nijhoff/Dr W. Junk Publishers, The Hague/Boston/Lancaster.*
Printed in the Netherlands.

1 hr and was chromatographed on a 3.0 cm long, 0.6 cm i.d. column of NaChelex at a nominal flow rate of 2.7 mL min^{-1}. The column effluent was collected and analyzed for Cu or Ni by atomic absorption spectrometry using standard conditions. The molar concentration of Cu is taken as strong CC, and the molar concentration of Ni is taken as the sum of moderate plus strong CC. Moderate CC is calculated by difference.

Column experiments with well characterized ligands were done similarly. Filtration was omitted. Calibration curves were made by preparing solutions of 0,2,4,6,8 and 10 µmol dm^{-3} of each of 15 well characterized ligands (Table 1). The solutions were spiked and chromatographed as described above. A plot of concentration of spike metal in the effluent vs. concentration of ligand in the solution constituted the calibration curve. A slope of 1.00 corresponds to complete ligand detection.

TABLE 1. Ligands used in the study.

Number	Abbreviation	log K'_{NiL}[a]	Ligand
1	DTPA	17.0	Diethylenetrinitrilopentaacetic acid
2	EDTA	16.7	Ethylenedinitrilotetraacetic acid
3	CDTA	16.2	trans-1,2-cyclohexylenedinitrilo-tetraacetic acid
4	HEDTA	15.4	N-(2-hydroxyethyl)ethylenedinitrilo-N,N',N'-triacetic acid
5	TTHA	15.3	Triethylenetetranitrilohexaacetic acid
6	EDDA	12.1	Ethylenediimodiacetic acid
7	EDHPA	11.8	Ethylenediiminobis[(2-hydroxyphenyl)acetic acid]
8	EGTA	10.9	Ethylenebis(oxyethylenenitrilo)-tetraacetic acid
9	TRIEN	10.5	Triethylenetetramine
10	NTA	10.0	Nitrilotriacetic acid
11	HIDA	8.5	N-(2-hydroxyethyl)iminodiacetic acid
12	IDA	6.9	Iminodiacetic acid
13	HIS	5.5	Histidine
14	CYS	2.7	Cysteine
15	GLY	1.9	Glycine

[a] pH 7.5, I = 0.01

Batch experiments were done in 500 mL flasks with magnetic stirring in such a way as to be similar to those described by Figura and McDuffie

(1979). A 400 mL portion of solution containing 10^{-4} mol dm^{-3} of test ligand and 10^{-5} mol dm^{-3} each of Cu, Ni, Co, and Zn was taken for each run. After initial sampling, 1.00 g of air dried (15 min in Buchner funnel with suction) Chelex was added to the flask. Samples were withdrawn and filtered immediately (medium porosity glass) after 10 min, 90 min, and 7 days.

2.2. Materials.

The NaChelex (Bio-Rad Laboratories) was prepared as described in a previous paper (Stolzberg and Rosin, 1977). Stock metal solutions (1000 µg dm^{-3}) were prepared by methods suggested by Smith and Parsons (1973). Stock ligand solutions (0.1 mol dm^{-3}) were prepared by weight. Strong and moderately strong ligands (numbers 1 through 10, Table 1) were standardized by titration with Cu, using a copper ion selective electrode for endpoint detection.

Near-surface water was collected from Smith Lake and Ballaine Lake, near Fairbanks, Alaska. Smith Lake has been characterized (Alexander and Barsdate, 1971).

2.3. Calculations

Stability constants used were the critical values compiled by Martell and Smith (1974, 1975). Conditional stability constants (Ringbom, 1963) were calculated at pH 7.5 and at an ionic strength of 0.01 using the extended Debye-Huckel equation.

3. RESULTS AND DISCUSSION

3.1. Calibration Curves for Well-Characterized Ligands

Calibration curve slopes for ligands 1 through 5 were close to 1.0 for Cu, Ni, Co, and Zn. Results varied slightly between metals as shown in Table 2 for two typical ligands. Intercepts were generally indistinguishable from zero. The slope for NiDTPA was 1.62 and for TTHA (all metals) 1.92 to 1.96. All of these ligands are completely detectable, primarily due to their large stability constant, and they should be considered "strong". High values for TTHA and for NiDTPA are due to the formation of stable dinuclear complexes, M_2L.

TABLE 2. Calibration curve slopes and intercepts for EDTA and HEDTA.

Metal	EDTA Slope	EDTA Intercept[a]	HEDTA Slope	HEDTA Intercept
Cu	1.00	0.0	0.99	0.1
Ni	0.95	-0.2	0.95	-0.2
Co	0.95	-0.2	1.00	-0.3
Zn	0.91	0.0	0.92	-0.1

[a] Intercept values are in μmol dm^{-3}

Calibration curve slopes and intercepts for ligands 11 through 15 were indistinguishable from zero for the four metals. These ligands are undetectable under normal analysis conditions, primarily because the stability constants are small and the reaction

$$ML(aq) + Resin(s) \rightleftharpoons L(aq) + MResin(s) \qquad (1)$$

proceeds completely to the right. They should be considered "weak".

Calibration curve slopes for ligands 6 through 10 are highly variable. They range from 0.0 to 1.02. In general, the Ni slopes are larger than the others. Intercepts are indistinguishable from zero. These are the "intermediate" strength ligands.

The simplest explanation is that detectability depends on the size of the conditional stability constant. Strong ligands are detectable, weak ligands are not detectable, intermediate strength ligands are partially detectable. This is true for strong and weak ligands, but it is not necessarily true for intermediate strength ligands as shown in Figure 1 for Cu and Ni. There is no obvious correlation between log K$_{ML}'$ and detectability for intermediate strength ligands. Similar results are observed for Co and Zn.

The rate of approach to equilibrium (i.e. the kinetics of the system) is the factor which controls detectability of intermediate strength ligands when reaction 1 lies to the right at equilibrium. Batch experiments for all five ligands are consistent with this interpretation. Results with NTA in Table 3 show the correlation. Results with EGTA and EDDA are very similar. Less than 5% of the TETREN is detected, and batch uptake is 85% to 99% complete in 10 minutes. Results with EDHPA are more difficult to categorize because critical stability constant data for Cu and Co(II) are not available. In addition, the reaction between CuEDHPA and Chelex is less than 10% complete and the reaction between CoEDHPA and Chelex is only approximately 40% complete after 1 week. Nonetheless, nothing in the EDHPA data is inconsistent with the above statement of the importance of kinetics.

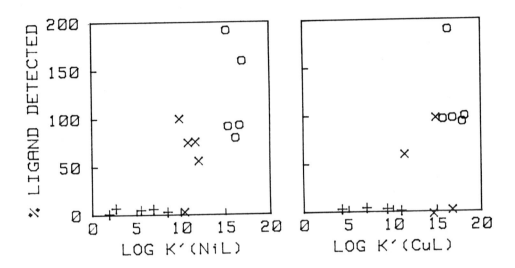

FIGURE 1. Percent ligand detected vs. log K'_{ML} for Ni (left) and Cu (right). Legend: o ligands 1-5, x ligands 6-10, + ligands 11-15.

TABLE 3. Role of K'_{MNTA} and rate of reaction between MNTA and NaChelex on NTA detectability.

Metal	log K'_{MNTA}	% NTA Detected	% M Remaining After		
			10 min	90 min	7 days
Cu	11.4	62	84	15	5
Ni	10.0	102	99	97	41
Co	8.8	94	96	48	4
Zn	9.1	82	92	31	3

That Ni is capable of detecting these ligands more completely than the other metals is due to the inherent slow rate of ligand exchange exhibited by Ni. For a given ligand, which will have similar values of K_{CuL} and K_{NiL}, the rate constant for dissociation of the metal complex is typically 10^5 times slower for Ni than for Cu (Wilkins, 1974). As a result, a given ligand may not be detected when using copper as spike metal because reaction 1 proceeds to completion during the few seconds of contact on the column. This same ligand may be detected using nickel as spike metal because the slow *kinetics* of ML dissociation in reaction 1 prevents more than a few percent conversion to products (L(aq) + MR(s)) during the few seconds of contact on the column, even though the position of *equilibrium* does favor near-complete conversion to products. Thus we expect that a Ni spike will detect weaker ligands than will a Cu spike. Based on this we can measure the concentration of "strong" ligands with Cu, and "strong" plus "intermediate" strength ligands with Ni. The concentration of "intermediate" strength ligands is calculated by difference.

There may be some confusion in the difference in time scales between batch (minutes to hours) and column (seconds) experiments. This difference is due to the extreme inefficiency of mass transfer in the batch experiments.

We can make a generalization about ligand detectability in these well characterized systems. A ligand is detectable when:

a) It forms a strong complex so that this exchange reaction lies far to the left

$$ML(aq) + Resin(s) \rightleftarrows L(aq) + MResin(s) \tag{1}$$

or

b) The above exchange reaction is slow proceeding from left to right.

3.2. Results for Lake Waters

The value of Cu complexing capacity is nearly always (98% of the time) smaller than the value of Ni complexing capacity of the lake waters studied. Values of Zn and Co complexing capacities are smaller yet. Typical results are shown in Table 4 for Ballaine Lake. Data in the last entry would be interpreted as 4.9 μmol dm^{-3} strong complexing capacity and 0.7 μmol dm^{-3} moderately strong complexing capacity.

TABLE 4. Ballaine Lake complexing capacity and optical absorbance.

Date	Cu	Ni	Zn	Co	Abs[a]
10 May	3.7	3.9	0.8	0.9	1.62
21 May	3.8	5.1	0.8	1.0	1.60
9 June	3.9	5.9	0.7	1.1	1.63
24 July	4.2	5.4	ND[b]	ND	1.56
9 Aug	4.2	5.3	0.7	1.1	1.51
6 Sept	3.5	4.2	0.7	1.1	1.43
21 Oct	4.8	5.6	ND	ND	1.89
23 Dec	4.4	4.9	0.7	0.9	1.99
18 Apr	4.9	5.6	0.6	ND	2.50

[a] Optical absorbance at 250 nm, 1 cm cell path
[b] ND, not determined

A more closely spaced sequence of data from Smith Lake shows rapid variation of moderately strong and strong complexing capacity during breakup in Spring, 1983 (submitted for publication). This time period corresponds to the historical algal bloom period (Alexander and Barsdate, 1971). We do not have biological data during this time period for 1983, and we cannot ascribe changes in measured CC to a particular cause. Nonetheless, this method does detect rapid fluctuations in the metal complexing capabilities when the chemical and physical properties of the lake are changing rapidly due to a large influx of runoff water and rapid changes in insolation. We plan to do simultaneous CC and primary productivity measurements during Spring, 1984, to determine if a correlation exists.

REFERENCES

- Alexander, V. and Barsdate, R.J., 1971. Physical Limnology, Chemistry and Plant Productivity of a Taiga Lake. Int. Revue ges. Hydrobiol. 56: 825-872.
- Figura, P. and McDuffie, B., 1979. Use of Chelex Resin for Determination of Labile Trace Metal Fractions in Aqueous Ligand Media and Comparison of the Method with Anodic Stripping Voltammetry. Anal. Chem. 51: 120-125.
- Hart, B., 1981. Trace Metal Complexing Capacity of Natural Waters: A Review. Environ. Technol. Lett. 2: 95-110.
- Martell, A.E. and Smith, R.M., 1974. Critical Stability Constants Volume 1: Amino Acids. Plenum Press, New York.
- Martell, A.E. and Smith, R.M., 1975. Critical Stability Constants Volume 2: Amines. Plenum Press, New York.

-Neubecker, T.A. and Allen, H.E., 1983. The Measurement of Complexation Capacity and Conditional Stability Constants for Ligands in Natural Waters. Water Res. 17, 1-14.

-Ringbom, A., 1963. Complexation in Analytical Chemistry. Interscience, New York.

-Smith, B.W. and Parsons, M.L., 1973. Preparation of Standard Solutions: Critically Selected Compounds. J. Chem. Educ. 10: 679-681.

-Stolzberg, R.J. and Rosin, D., 1977. Chromatographic Measurement of Submicromolar Strong Complexing Capacity in Phytoplankton Media. Anal. Chem. 49: 226-230.

-Wilkins, R.G., 1974. The Study of Kinetics and Mechanism of Reactions of Transition Metal Complexes. Allyn and Bacon, Boston, p388.

AN INVESTIGATION OF METAL-ORGANICS IN SEAWATER USING HPLC WITH ATOMIC
FLUORESCENCE DETECTION

D.J. MACKEY

1. INTRODUCTION

In the past decade, a number of elegant experiments have
demonstrated that for some marine organisms, the bioavailability of
copper, zinc and iron is related to the free ion activity rather than
the total concentration of trace metal (Sunda and Guillard, 1976;
Anderson et al., 1978; Anderson and Morel, 1982). This does not
necessarily imply that the free metal ion is the only form taken up by
these organisms but it does mean that studies of chemical speciation are
of vital importance in understanding the role of essential or toxic
trace metals in determining the fertility of natural waters. While the
inorganic speciation is fairly well characterized, virtually nothing is
known about the chemical nature and co-ordinating strength of naturally
occurring organic ligands. In this paper I discuss one approach to
investigating trace metal organic complexes in seawater.

The total concentration of most biologically important trace metals
in seawater is in the range 10^{-10}M to 10^{-8}M and hence the concentration
of any individual metal organic complex must be considerably lower.
There are no chemical techniques which can characterize or identify
individual compounds at these concentrations in seawater and any
investigation of such compounds must therefore consist of three distinct
phases. Firstly, we must have some method of extracting and
concentrating the compounds from seawater and in this paper I will only
consider adsorption on hydrophobic resins. Secondly, we must
fractionate the compounds on the basis of as many different chemical and
physical properties as possible and since the compounds may be
involatile, the most useful technique is high performance liquid
chromatography (HPLC). Thirdly, we need to know which fractions contain
trace metals as most organic matter extracted from seawater is unlikely
to be co-ordinated to trace metals.

Kramer, C.J.M. and Duinker, J.C. (eds.), Complexation of Trace Metals in Natural Waters. ISBN 90-247-2973-4
© 1984 Martinus Nijhoff/Dr W. Junk Publishers, The Hague/Boston/Lancaster.
Printed in the Netherlands.

2. PROCEDURE

2.1. Detector

Let us first consider the choice of detector. Since we wish to look at a range of biologically important metals, the detector should be capable of simultaneously detecting a number of metals, be compatible with the eluent from the liquid chromatograph and be sensitive enough to detect metals at part per billion (10^9) concentrations. The technique used here is based on atomic fluorescence (AF) using conventional hollow cathode lamps. An adjustable nebulizer and a cylindrical burner are used and multielement capability is achieved by modulating the individual lamps at different frequencies and using lock in amplifiers to separate the component signals from the photomultiplier. The detector has been described in detail in the literature (Mackey, 1982a, 1982b, 1982c). Typical detection limits for the metals described in this paper are 1 µg ℓ^{-1} (Mg), 5 µg ℓ^{-1} (Zn,Cd), 15 µg ℓ^{-1} (Ni,Mn, Cr) and 25 µg ℓ^{-1} (Cu,Fe).

2.2. Adsorbents

In Table I are listed some possible types of organic ligands that could occur in seawater. Metal complexes of these, and other compounds, would differ considerably in chemical and physical properties. The ideal technique for extracting such compounds from seawater should be capable of quantitatively removing them regardless of differences in molecular weight, polarity or hydrophilicity of the molecule or the thermodynamic stability of the metal organic complexes. Inorganic species such as the aquated cations or carbonato complexes should be unaffected by the procedure which should also be able to preconcentrate the species of interest by a few orders of magnitude without altering the chemical equilibria in the original sample. No such technique is presently known and in choosing a specific procedure it is essential to be aware of its limitations. This is particularly important when working at the ultra trace levels that occur in seawater.

Three basic types of hydrophobic resin have been used to extract organic compounds from air or natural waters. These are based on cross linked polystyrene (XAD-2, Porapak-Q, Chromosorb 101), organic groups chemically bonded to porous silica (SEP-PAK C18) and finally polymers

Table 1.   Possible organic ligands in seawater

---

1. Humic and Fulvic Acids
2. Organic compounds formed by exudation and autolysis,
    e.g. proteins, tetrapyrroles
3. Specific ion complexing agents, e.g. siderophores.

---

containing functional groups that increase the polarity but also are
potential donor sites to metal cations (XAD-7, Tenax). The latter group
have quite large cation exchange capacities and will not be considered
further. Since metal-organic compounds tend to be more hydrophobic than
the parent organic molecules, the first two types of resin would seem to
be ideally suited to isolating metal-organics from seawater. However,
the compounds removed from seawater by adsorption onto these resins
depend on such factors as the hydrophilicity of the organic compound,
the nature of the resin, the pH of the seawater, the pore size of the
resin and the flow rate. Moreover, a wide range of such resins have
been shown to have a cation exchange capacity that is small but highly
significant when studying metals at the part per billion level (Mackey,
1982c). These resins are capable of adsorbing inorganic trace metals
from aqueous solution and these metals can be desorbed by strong eluants
such as methanolic hydrochloric acid or methanolic ammonia. However, if
such strong solvents are not used then it is likely that all the
adsorbed metal-organics will not be eluted from the resin (Mackey,
1982d).

The following experiments illustrate some of the limitations of
hydrophobic resins. Milli-Q grade water was pumped through a column of
XAD-2 resin and the effluent connected to the multichannel AF
detector. A spike containing 5 µg each of iron, zinc, copper,
magnesium, manganese and nickel was then injected into the system. When
the response of the AF detector returned to the baseline another spike
was injected and the process repeated a total of five times. The column
was then eluted with methanol, acetonitrile and 10^{-3}M EDTA. The
procedure was repeated but the elutions with organic solvents were
omitted. Finally, an attempt was made to eliminate potential donor
sites by methylating the resin with dimethylsulphate. The results are
shown in Fig. 1 and the conclusions can be summarized as follows:

58

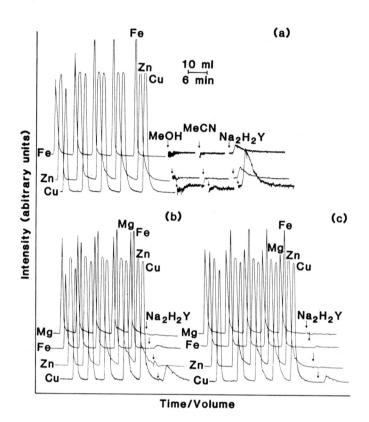

FIGURE 1. a) Iron, zinc and copper fluorescence signals from the
effluent from a column of XAD-2 after five successive 500 μℓ injections
of stock solution containing 5 μg of iron, copper, zinc, magnesium,
manganese and nickel. The column was then eluted successively with
methanol (MeOH), acetonitrile (MeCN) and 10^{-3}M disodium dihydrogen
ethylenediaminetetracetate (Na_2H_2Y). b) Same as a), except that the
magnesium fluorescence was also monitored and the MeOH and MeCN elutions
were eliminated. c) Same as b), except that the resin had been
methylated with dimethylsulphate. The signals were recorded
simultaneously but have been offset for clarity. (Reproduced with
permission from Mackey, 1982a).

a) XAD-2 can remove inorganic trace metal ions from aqueous solution;

b) this must occur through impurity donor sites on the resin that cannot

be eliminated by methylation; c) metal EDTA complexes are not retained

by the resin; d) the resin can adsorb trace metals from organic

solvents; and e) methanol or acetonitrile do not elute adsorbed trace

metal cations.

Similar conclusions pertain to a wide range of adsorbents I have
investigated and none are suitable for quantitative studies of trace
metal speciation. The polystyrene resins have the added disadvantage
that some readily polarizable solutes may interact strongly with the
benzene rings in the resin and may not be eluted by simple organic
solvents. This irreversible adsorption via charge transfer interactions
cannot occur for reverse phase C18 adsorbents as used in the SEP-PAK
cartridges. While these cartridges also adsorb aquated trace metal
cations, this adsorption should be due entirely to free silanol groups
on the surface rather than a number of unknown sites as occurs on the
polymeric resins. There are also obvious advantages in using the same
material for adsorbing organics and for separating them
chromatographically. For these reasons, SEP-PAK cartridges were
subsequently used to extract metal organic compounds from seawater.

2.3. Sample preparation

In a typical experiment, 5-8 litres of seawater were drawn via a
peristaltic pump through a 47 μm diameter furnace fired GFC filter and a
prewashed SEP-PAK cartridge. Seasalts were removed with 50 ml of Milli-
Q water and the adsorbed material eluted from the cartridge with 5 ml of
methanol. The methanol was evaporated to dryness in a stream of
filtered high purity nitrogen. The residue was dissolved in 1.5 ml of
Milli-Q water, filtered and injected into the liquid chromatograph using
water as the mobile phase. Two stock solutions were also prepared by
combining the extracts from ca. 50 ℓ of seawater, evaporating the
methanol as before and dissolving the residue in about 8 ml of Milli-Q
water.

Samples have been collected from the entrance to Port Hacking
estuary in New South Wales, Australia (34° 04.5'S, 151° 08.8'E). This
is a comparatively unpolluted marine dominated estuary and analyses of
the stock solutions gave average metal-organic concentrations in the
original seawater of 1.0 and 3.6 nM (Cu) 440 and 630 pM (Zn) and 480 and
660 pM (Fe).

Samples were also collected from about 18 locations on the North
West Shelf of Australia in a region centred on 17°S, 121°E. The
continental shelf is wide with strong tidal currents and the area is

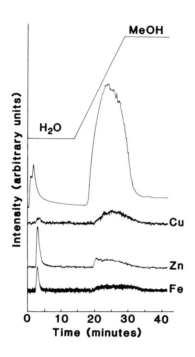

FIGURE 2. Chromatogram of the material extracted from a SEP-PAK cartridge after the passage of 6 ℓ of seawater. Organics were detected at 254 nm (0.4 AUFS) and trace metals were detected using an atomic fluorescence detector. A 5 mm i.d. RAD-PAK C18 column was used and the flow rate was 1.7 mℓ min.$^{-1}$. The eluant changed from water to methanol according to the profile shown. (Reproduced with permission from Mackey, 1983)

remote from any riverine or anthropogenic influx of trace metals. The average values of metal-organic compounds isolated from the original seawater were 540 ± 140 pM (Cu), 130 ± 63 pM (Ni), 8.8 ± 7.6 pM (Cd), 93 ± 100 pM (Zn), 190 ± 150 pM (Fe) and 8.9 ± 6.9 pM (Mn).

2.4. Chromatography

If one of the seawater extracts is chromatographed on a C18 reverse phase column using a solvent gradient from water to methanol then the compounds should be retained at the head of the column and be eluted in order of decreasing polarity as the solvent composition changes with the least polar fraction being completely eluted by pure methanol. This does not occur using a Waters RAD-PAK C18 column and a substantial fraction of the trace metal ions are eluted from the column by water

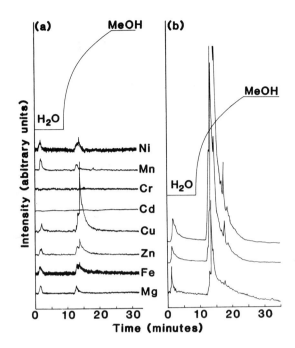

FIGURE 3. Chromatograms of a stock solution of material extracted from
seawater by SEP-PAK cartridges. a) Atomic fluorescence detector
responses from two separate 1000 μℓ loadings; b) response of
molecular fluorescence detector at loadings and sensitivities
of 1000 μℓ, 1 μA (top), 400 μℓ, 1 μA (middle), 25 μℓ, 0.1/0.2 μA
(bottom - the sensitivity was changed after the first peak). Excitation
was at 360 nm and emission was detected above 470 nm. An 8 mm i.d. RAD-
PAK C18 column was used. Other conditions as in Fig. 2 (Reproduced
with permission from Mackey, 1983).

(Fig 2). A smaller fraction of the organic material (defined in terms

of absorption at 254 nm) is also not retained by the column with water

as the eluent and it is thought that this is due to very polar compounds

that are only partially removed from seawater by the SEP-PAK

cartridges. The amount of trace metal that is organically bound may

therefore be considerably underestimated by the SEP-PAK technique.

From Fig. 2 it can be seen that as the solvent is changed linearly

from water to methanol, the trace metals are eluted in broad bands

covering a wide range of polarities. Zinc tends to be eluted before

62

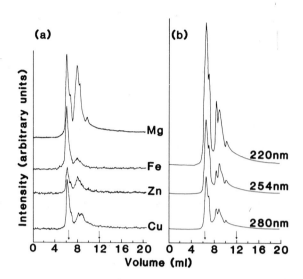

FIGURE 4. Chromatograms of a stock solution of material extracted from seawater by SEP-PAK cartridges. a) Atomic fluorescence detector responses; b) response of UV detector at three different wavelengths (2 AUFS). A Waters I-125 column was used with water as the mobile phase at a flow rate of 1.7 $m\ell$ min^{-1}. The exclusion and total permeation volumes are marked by arrows.

iron and copper implying greater polarity for the zinc complexes and this is in agreement with the comparatively large initial peak due to very polar compounds eluted by water. If a convex gradient is used, the compounds are eluted more rapidly and this increases the response of the AF detector albeit with a loss in resolution during portions of the chromatogram. In Fig. 3 are shown the chromatograms obtained from a number of injections of a standard solution such that 1000 $\mu\ell$ corresponds to the material extracted from 8 ℓ of Port Hacking seawater. Metal-organic compounds were observed for Ni, Mn, Cu, Zn, Fe and Mg but not for Cd and Cr. Once again there is a tendency for those metals which form weaker complexes (eg. Mg, Mn) to be eluted before Cu while intermediate behaviour is observed for the other trace metals. The shape of the Cu band is almost identical to that obtained from the molecular fluorescence detector implying that the trace metals are uniformly distributed amongst the fluorescent organic compounds.

One of the advantages of HPLC is that it is possible to fractionate compounds on the basis of a wide range of chemical and physical properties by varying the nature of the chromatographic support and the composition of the solvent. In Fig. 4 are shown the chromatograms obtained using a gel permeation column designed for fractionating proteins by molecular size. If it is assumed that there is no interaction between the solute molecules and the chromatographic support, then all compounds should be eluted between the arrows shown. If the metal-organics behaved like proteins, then compounds with a molecular weight greater than about 30,000 Daltons should elute at 6.3 mℓ while those with a molecular weight less than 1000 Daltons should elute at 12.0 mℓ. Intermediate molecular weight compounds should elute between these extremes.

The response of the UV detector was reproducible on repeated loadings of a stock solution but there were large increases in the response of the magnesium detector and variations in the relative intensities of the two copper signals. The latter variations are thought to be due to copper ions exchanging between the different molecular weight fractions via co-ordination by the diol groups on the chromatographic support. The increase in the magnesium signal is attributed to free silanol groups on the support removing magnesium from organic complexes. The magnesium signal increases as more silanol groups become co-ordinated to magnesium ions. The signals due to magnesium in Fig. 4 are much greater than those observed in Fig. 3 although prolonged use of the RAD-PAK C18 cartridge leads to increases in the intensity of the magnesium signal. This is consistent with the presence of more free silanol groups on the C18 resin than on the C3 diol resin as the smaller C3 groups would provide less steric hindrance and lead to a more complete surface coverage by the silylating agent.

The relative intensities of the peaks from the UV detector are virtually independent of wave length and the various molecular weight fractions must have very similar UV spectra in the 220-280 nm range. Moreover, the magnesium signal closely follows the UV signal suggesting that magnesium is indiscriminately complexed by organic compounds.

3. DISCUSSION

The preliminary results described here suggest that trace metals in seawater are complexed by organic molecules covering a wide range of polarities and molecular sizes. There are small differences in the relative affinity of the various fractions for trace metals and the amounts of trace metal complexed are in general agreement with expectations based on the Irving-Williams series. No evidence has yet been found for the presence of specific metal organic complexes such as tetrapyrroles or siderophores although the characteristic fluorescence or absorption spectra of these compounds should make it easy to search for them in seawater and perhaps put an upper limit on their natural concentration.

The chromatographic resolution of the system could be enhanced by using longer columns, columns in series, slower flow rates or more gradual changes in solvent composition. These modifications would decrease the rate of mass transfer to the AF detector and lead to a corresponding decrease in sensitivity. In principle, this could be overcome by loading the chromatograph with the extract from a larger volume of seawater. In practice, not only is this undesirable due to the increased time needed to process the seawater, but there are also problems due to the high UV absorbance of the extracted organics as well as potential solubility problems with very high concentration factors.

One way out of this dilemma is by improving the sensitivity of the AF detector and this is possible using demountable cathode lamps that are now commercially available. These lamps have up to 100 times the spectral output of conventional hollow cathode lamps and since the intensity of the fluorescence emission is directly proportional to the intensity of the excitation source, the detection limit for the AF detector will be improved by the same factor. There is no reason why the AF detector cannot be coupled to a gas chromatograph and the combination of atomic fluorescence with liquid and/or gas chromatography promises to be a very powerful tool in investigating the speciation of metals and metalloids in the natural environment.

REFERENCES

- Sunda W, Guillard RRL. 1976. The relationship between cupric ion
  activity and the toxicity of copper to phytoplankton. J. Mar. Res.,
  34, 511-529.
- Anderson MA, Morel FMM, Guillard RRL. 1978. Growth limitation of a
  coastal diatom by low zinc ion activity. Nature, 276, 70-71.
- Anderson MA, Morel FMM. 1982. The influence of aqueous iron
  chemistry on the uptake of iron by the coastal diatom *Thalassiosira
  weisflogii*. Limnol. Oceanogr., 27, 789-813.
- Mackey DJ. 1982a. The adsorption of simple trace metal cations on
  Amberlite XAD-1 and XAD-2. A study using a multichannel non-
  dispersive atomic fluorescence detector with quantitation by batch
  measurements. J. Chromatogr., 236, 81-95.
- Mackey DJ. 1982b. Amberlite XAD-2 and XAD-4 as cation exchange
  resins of low capacity. An investigation using an atomic
  fluorescence detector directly coupled to a liquid chromatograph.
  J. Chromatogr., 237, 79-88.
- Mackey DJ. 1982c. Cation exchange behaviour of a range of adsorbents
  and chromatographic supports with regard to their suitability for
  investigating trace metal speciation in natural waters.
  J. Chromatogr., 242, 275-287.
- Mackey DJ. 1982d. An investigation of the suitability of Amberlite
  XAD-1 resin for studying trace metal speciation in seawater. Mar.
  Chem., 11, 169-181.
- Mackey DJ. 1983. Metal organic complexes in seawater - an
  investigation of naturally occurring complexes of Cu, Zn, Fe, Mg,
  Ni, Cr, Mn and Cd using high performance liquid chromatography with
  atomic fluorescence detection. Mar. Chem., 13, 169-180.

DETERMINATION OF THE COMPLEXATION CAPACITY OF NATURAL WATERS USING METAL
SOLUBILIZATION TECHNIQUES

P.G.C. CAMPBELL and A. TESSIER

1. INTRODUCTION

The speciation of trace metals in the aquatic environment has a determining
influence on their geochemical mobility, chemical reactivity and biological
availability. Inorganic ligands of importance in determining trace metal
speciation are relatively few in number and their concentrations are readily
determined. In contrast, a diverse spectrum of organic compounds of potential
complexing ability is known to exist in natural waters, for example: amino
acids, proteins, monosaccharides, polysaccharides, porphyrins, hydroxamic
acids, fulvic acids, humic acids.

The recognition of the impracticability of identifying and quantifying all
the organic ligands present in a given water sample has led to the development
of the "complexation capacity" (CC) concept. As the term implies, complexation capacity is a measure of the ability of a water sample to complex or mask
trace metals with respect to conventional chemical and biological metal-
sensing techniques. Use of the complexation capacity concept implies that for
a particular metal it is possible to group together the various ligands
present and represent their overall behaviour by means of certain "average"
properties.

$$M + L \underset{k_d}{\overset{k_f}{\rightleftharpoons}} ML_n$$

Examples of such properties are the stoichiometric constant, n; the conditional formation constant, $*K$; the total concentration of ligand capable of
binding metal M, $[L]_T$; the rate constants for formation / dissociation of ML,
k_f and k_d.

Since the early 1970's many experimental procedures, varying greatly in
manipulative complexity, have been proposed for measuring the complexation
capacity of natural waters. These can conveniently be grouped into chemical

Kramer, C.J.M. and Duinker, J.C. (eds.), Complexation of Trace Metals in Natural Waters. ISBN 90-247-2973-4
© 1984 Martinus Nijhoff/Dr W. Junk Publishers, The Hague/Boston/Lancaster.
Printed in the Netherlands.

and biological methods (Table 1). Many of these techniques are critically discussed elsewhere in this volume; in the present chapter we consider only those approaches belonging to the general class of "solubilization methods".

Table 1: Classification of methods for determining the complexation capacity of natural waters.

Chemical:	complexometric titration
	• anodic stripping voltammetry
	• potentiometry
	• dialysis
	• fluorescence
	equilibrium competition
	• ligand exchange
	• ion exchange
	• solubilization
Biological:	bio-assay
	• unicellular algae
	• bacteria

2. SOLUBILIZATION METHODS FOR DETERMINING COMPLEXATION CAPACITY

2.1 Historical perspective

As a class, the solubilization methods are intrinsically very simple. Metal-containing solid phases are equilibrated with the water sample of interest, and the concentration of dissolved metal is determined. Blank values are determined in the absence of any organic complexation (e.g. in synthetic inorganic media or in photo-oxidized samples). Any increase in metal concentration above this background value is then attributed to solubilization or complexation involving organic species.

Most of the solid phases that have been used in such solubilization procedures have contained copper(II), a metal with a known strong affinity for many organic ligands. The use of copper in this regard was pioneered by Pope and Stevens (1939), who introduced a procedure based on the solubilization of copper(II) phosphate for determining the total concentration of amino acids in protein hydrolysates. Several improvements to this procedure were proposed

in the 1950's for use in biochemical determinations, but the method
subsequently fell into disuse, presumably as a result of the introduction of
more specific methods for determining amino acids.

The first application of solubilization techniques in an environmental
context, involving not Cu(II) but rather Fe(III), would appear to be that of
Shapiro (1964), who determined the "iron-holding capacity" of yellow organic
acids isolated from pond water. However no attempt was made to apply the
technique directly to unconcentrated natural waters. The first such application was that of Kunkel and Manahan (1973), who resurrected the Cu(II)
solubilization method and suggested its use for determining strong trace metal
chelating agents in aqueous environmental samples. In subsequent years a
number of applications of this procedure to natural waters have appeared and,
in addition, several other solubilization methods have been proposed. In the
remainder of this review particular emphasis will be placed on the Cu(II)
solubilization methods, these being the most-studied of the procedures in this
class, but mention will also be made of other solubilization methods that have
been employed in an environmental context.

2.2   Cu(II) solubilization method: copper hydroxide

2.2.1   <u>Method description</u>. Acording to the experimental procedure
introduced by Kunkel and Manahan (1973) for measuring complexation capacity,
the water sample is filtered (0.45 μm) and excess copper is added in dissolved
form (e.g., $CuSO_4$). After adjustment of the pH to 10 by addition of Na_2CO_3
solution, the samples are heated at 95-100°C for 1 h to favour chemical
equilibrium and promote aggregation of the Cu(II) hydroxide precipitate. The
cooled samples are diluted to volume, filtered (0.45 μm) to remove the
precipitate, and then analyzed for total dissolved Cu by atomic absorption
spectrophotometry. The chemical equilibria involved under these experimental
conditions are summarized in Table 2 (equations 1-2, 4-10). At pH 10 the
calculated concentration of inorganic Cu(II) is about 2.4×10^{-7} $mol.dm^{-3}$.
Any increase in copper concentration above this low value is attributed to
solubilization or complexation involving species other than hydroxo or
carbonato ligands. As indicated in the first section of Table 3, the
experimentally observed blank values fall in the range $3-5 \times 10^{-7}$ $mol.dm^{-3}$, in
reasonable agreement with the predicted value.

Table 2: Inorganic equilibria involved in the Cu(II) solubilization
methods.

	reaction	log K (I=0)[a]
(1)	$Cu^{+2} + 2\ OH^- = Cu(OH)_2(s)$	20.35
(2)	$2\ Cu^{+2} + 2\ OH^- + CO_3^{-2} = Cu_2(OH)_2CO_3(s)$	31.99
(3)	$3\ Cu^{+2} + 2\ PO_4^{-3} = Cu_3(PO_4)_2(s)$	37.7*
(4)	$Cu^{+2} + OH^- = Cu(OH)^+$	6.0
(5)	$Cu^{+2} + 2\ OH^- = Cu(OH)_2^{\ °}$	13.7
(6)	$Cu^{+2} + 3\ OH^- = Cu(OH)_3^-$	15.2
(7)	$Cu^{+2} + 4\ OH^- = Cu(OH)_4^{-2}$	16.1
(8)	$2\ Cu^{+2} + 2\ OH^- = Cu_2(OH)_2^{+2}$	17.0
(9)	$Cu^{+2} + CO_3^{-2} = CuCO_3^{\ °}$	6.77
(10)	$Cu^{+2} + 2\ CO_3^{-2} = Cu(CO_3)_2^{-2}$	10.01
(11)	$Cu^{+2} + Cl^- = CuCl^+$	0.6*
(12)	$Cu^{+2} + 2\ Cl^- = CuCl_2^{\ °}$	0.4*
(13)	$Cu^{+2} + H^+ + PO_4^{-3} = CuHPO_4^{\ °}$	16.6*

(a) Thermodynamic constants from Stumm and Morgan (1981) or directly from
the MINEQL data bank (Westall et al., 1976); the latter values are
indicated by an asterisk.

Table 3: Blank values obtained with various Cu(II) solubilization methods.

insoluble salt	pH	blank (10^{-6} mol.dm^{-3})	reference
hydroxide	10	0.5	Montgomery and Echevarria (1975)
	10	0.3 - 0.5	Campbell et al. (1977)
	10	0.3	Elder and Horne (1978)
phosphate	9	20 - 30	Borchers (1959)
	7	2	Frimmel et al. (1980)
	7	1.5 - 3	Huber (1980)

2.2.2 <u>Method validation</u>. Quantitative recovery of strong synthetic chelating agents (e.g.: ethylenediamine tetraacetate, EDTA; nitrilotriacetate, NTA) has been demonstrated in the concentration range 10^{-6} to 10^{-4} mol.dm^{-3}, both for laboratory standards and for spikes added to wastewater effluents (Kunkel and Manahan, 1973; Manahan and Smith, 1973). Elder and Horne (1978) reported similar findings for the recuperation of standard additions of EDTA to lake water (92 - 96% recovery). Working in an estuarine environment, however, Montgomery and Echevarria (1975) experienced some difficulty in reproducing these results and reported a relationship of the type CC = 0.8 $[$EDTA$]$ + 0.4 for samples spiked with known quantities of EDTA. These results suggest that recovery of added ligand may be incomplete in the presence of high concentrations of Ca and Mg. With synthetic solutions prepared in the laboratory (10^{-6} and 10^{-5} mol EDTA.dm^{-3}), the Cu(OH)$_2$ solubilization method is highly reproducible (relative standard deviation 4-5%); with subsamples drawn from a natural water sample, however, the relative standard deviation tends to be somewhat higher, normally in the range 10-20% (Montgomery and Echevarria, 1975; Campbell et al., 1977). The detection limit is about 3 x 10^{-7} mol.dm^{-3}.

The sensitivity of the Cu(OH)$_2$ method to monomeric biogenic ligands has not been verified experimentally, but for three representative amino acids (aspartic acid, glycine, lysine), as well as for salicylic acid, the calculated complexation capacity values are far below the detection limit of 3 x 10^{-7} mol.dm^{-3} (Campbell et al., 1977). We recently repeated this calculation for the group of ligands shown in Table 4. This suite of ligands corresponds closely to the Mixture Model proposed by Sposito (1981) to simulate the metal complexing behaviour of a soil fulvic acid solution at a concentration of 10^{-5} kg C.dm^{-3} (i.e., 10 mg C.L^{-1}); EDTA was included in the calculation for comparison purposes. The calculated complexetion capacities for the individual ligands at pH 10 vary from 10^{-6} to $10^{-13\ 8}$ mol.dm^{-3} (Table 4); with the exception of citrate and the synthetic ligand EDTA, these levels are again below the operational detection limit. It follows that monomeric biogenic ligands, similar in nature and complexation potential to those studied here, will <u>not</u> be detected by the Cu(OH)$_2$ solubilization technique.

Table 4:   Calculated complexation capacities for representative monomeric
           ligands in the presence of solid copper(II) hydroxide.

ligand	$[L]_T^{(a)}$	CC at pH 10[a]	$[\dfrac{CC\ pH\ 7}{CC\ pH\ 10}]$
phthalate	4.79	1.6×10^{-8}	9.9×10^{5}
salicylate	2.40	1.1×10^{-5}	9.1×10^{5}
arginine	1.45	5.2×10^{-4}	940
maleate	4.79	6.1×10^{-8}	580
valine	1.95	1.3×10^{-3}	420
ornithine	1.95	5.4×10^{-6}	18
citrate	3.23	2.6	1.2
lysine	1.95	1.0×10^{-5}	1.2×10^{-4}
EDTA	1.00	1.0	1.0

(a)   Concentrations expressed as 10^{-6} mol.dm^{-3} (i.e., μmol.L^{-1}).

The calculations also indicate that the complexation capacity at pH 10 will
normally be lower than that which would be expected at natural pH closer to 7;
for all but one of the biogenic ligands studied, the CC (pH 7): CC (pH 10)
ratios were greater than 1 (range 990000 - 1.2).   Since this effect depends
upon the acid-base properties of the ligand, and the ligand concentration, the
same trend will not necessarily be observed for all ligands.   Nevertheless,
these calculations clearly demonstrate the potential sensitivity of the
complexation capacity to pH changes in the critical pH range 7-10.

The sensitivity of the $Cu(OH)_2$ solubilization method to several polymeric
ligands of natural origin has been tested experimentally.   Of the four ligands
studied (gelatin, casein, fulvic acid, humic acid) only the latter two
maintained appreciable quantities of Cu in "solution" under the conditions of
the experiment (Campbell et al., 1977).   High CC: dissolved organic carbon
(DOC) ratios were observed (2.5 - 3.3 mol Cu per kg C), suggesting a
considerable solubilization potential for the humic materials.   However an
ultrafiltration experiment, designed to verify the physical dimensions of the

"dissolved" copper species passing through the standard 0.45 μm membrane
filter after treatment of the water sample, revealed that the major portion
(~ 96%) of the copper in the initial filtrate was colloïdal in nature rather
than truly dissolved. The function of the fulvic / humic acid is not to form
dissolved Cu complexes that are stable under the experimental conditions, but
rather to allow colloïdal Cu(II) to persist in the treated sample, presumably
by stabilizing colloïdal copper hydroxide. Shapiro (1964) proposed just such
an interaction, or peptization, to explain the ability of yellow organic acids
to prevent the precipitation of Fe(III) and Cu(II) in the pH range 5-11. This
stabilization of colloïdal copper by fulvic / humic acids is a potentially
damaging artefact in the original experimental procedure as it does not
correspond to a true complexation capacity available under realistic
environmental conditions.

2.2.3 <u>Method application.</u> The $Cu(OH)_2$ solubilization method has been
applied to a variety of natural and waste waters. Representative complexation
capacity values have been compiled in Tables 5, 6 and 7, and where possible
the CC:DOC ratios are indicated. For natural waters (Table 5) these ratios
are similar in magnitude to the average value of 0.17 mol/kg C suggested by
Mantoura (1981) for Cu complexation in inland waters. Note that the CC values
presented in these tables were obtained without a final ultrafiltration step
and thus represent an overestimation of the complexation capacity, since
presumably both complexed and peptized copper were present in the filtrate.
Indeed, for several river water samples (4 of the 6 tested) the inclusion of a
final ultrafiltration step (0.018 μm) resulted in a significant reduction (70
- 86%) in the apparent complexation capacity (Campbell et al. 1977); the two
samples on which ultrafiltration had little or no effect were collected
downstream from large lakes, at sampling points where the water quality was
largely determined by the characteristics of the epilimnetic waters upstream.
The proportion of allochthonous organic matter (including humic acids) would
be expected to be lower at these points than elsewhere in the watershed; low
concentrations of surfactive fulvic / humic materials would explain the
absence of any appreciable colloïdal Cu in these sample filtrates.

For these same river samples it also was demonstrated that the experimental
conditions required by the $Cu(OH)_2$ solubilization method favour removal of
dissolved organic matter, presumably by co-precipitation. Significant losses

Table 5: Compilation of complexation capacity determinations by solubilization measurements - surface waters.

solid phase	sample	$CC^{(a)}$ $(10^{-6}$ mol.dm$^{-3})$	$CC:DOC^{(a)}$ (mol/kg C)	reference
Cu(OH)$_2$	Bee Fork Creek Missouri	23	-	Kunkel and Manahan (1973)
	Courtois Creek Missouri	15	-	Kunkel and Manahan (1973)
	Yamaska River Quebec	0.3 - 10 (1.8)	0.01 - 0.34 (0.14)	Campbell et al. (1977)
	St-François River Quebec	0.3 - 19 (2.6)	0.02 - 1.4 (0.22)	Campbell et al. (1977)
	Rio Guanajibo Puerto Rico	0.5 - 24 (4.0)	-	Montgomery and Echevarria (1975)
	Lake Tahoe Nevada	<1	-	Elder et al. (1976)
	Lake Perris California	0.2 - 7.2 (1.4)	-	Elder and Horne (1978)
Cu$_3$(PO$_4$)$_2$	Lake Tjeukemeer Netherlands	1.6 - 17 (7.7)	-	De Haan et al. (1981)
PbCO$_3$	Lake water	5	-	Lautenbacher and Baker (1974)

(a) Mean values given in parentheses.

(30-50%) of DOC were observed for all but one of the samples tested. Such entrapment of dissolved organic matter by co-precipitation has been observed by several workers (Jeffrey and Hood, 1958; Chapman and Rae, 1967). In the present context such losses would seem to compromise the use of this method to determine the complexation capacity of natural waters, as a significant fraction of the very material being measured may be lost during analysis. Indeed, it has been suggested that those organic solutes with metal-complexing properties should be preferentially removed by such co-precipitation reactions (Chapman and Rae, 1967)!

Table 6: Compilation of complexation capacity determinations by solubilization measurements - wastewaters.

solid phase	sample	CC (10^{-6} mol.dm^{-3})	CC:DOC (mol/kg C)	reference
	Raw sewage Columbia, Mo	53	-	Kunkel and Manahan (1973)
	Primary effluent Columbia, Mo	47	-	Kunkel and Manahan (1973)
$Cu(OH)_2$	Secondary effluent Columbia, Mo	14	-	Kunkel and Manahan (1973)
	Spent wash, Whiskey distillery Northern Ireland	19700 - 51100	-	Quinn et al. (1982)
$CaCO_3$	Sanitary landfill leachate	147000	33	Avnimelech and Raveh (1982)

Table 7: Compilation of complexation capacity determinations by solubilization measurements - dissolved organic matter isolated from natural waters.

solid phase	sample	pH	CC:DOC (mol/kg C)	reference
$Fe(OH)_3$	Lindsey Pond Connecticut	7-9	4 - 8[a]	Shapiro (1964)
CuS	Sea waters Atlantic Ocean	7	0.6 - 1.0	Kerr and Quinn (1980)
$Cu(OH)_2$	Lake water Laurentide Park Quebec	10	3.3	Campbell et al. (1977)
$Cu(OH)_2$	Bog lakewater Germany	7	0.8	Frimmel (pers. communication, 1983)

(a) Original results (mol/kg organic acids) transformed assuming an organic carbon content of 40%.

2.3 Cu(II) solubilization method: copper phosphate

2.3.1 Method description. In this variation of the Cu(II) solubilization method, the water sample is filtered, buffered and then stirred with excess copper added as solid $Cu_3(PO_4)_2$. After a short reaction time (5-60 min) the suspended Cu(II) is removed by filtration or centrifugation and the total concentration of Cu in the filtrate or supernatant is determined by a suitable sensitive technique. The chemical equilibria involved are similar to those prevailing in the $Cu(OH)_2$ procedure (Table 2), but the actual experimental conditions are quite different (T, pH and $[CO_3^{-2}]$ lower; $[Cl^-]$ and $[PO_4^{-3}]$ much higher).

Two versions of this general procedure have been proposed. In the first, adapted by De Haan and co-workers (1981) from the alkaline cupric salt procedure described by Borchers (1959), the copper phosphate is suspended in borate buffer at pH 9.1, aged several days, mixed with the water sample and then stirred for 15 min. Thermodynamic calculations indicate that, despite the presence of relatively high phosphate concentrations, the solid phase present at equilibrium at pH 9.1 should be cupric hydroxide rather than the cupric phosphate originally added in solid form; the calculated concentration of dissolved inorganic Cu(II) at this pH is about 4×10^{-7} mol.dm^{-3}. Because of the short contact times used and the low reaction temperature, it is highly unlikely that thermodynamic equilibrium will be achieved by the end of the solubilization step. Unlike the situation prevailing with the $Cu(OH)_2$ procedure, this $Cu_3(PO_4)_2$ system will thus be under kinetic rather than thermodynamic control. Not surprisingly, then, the experimentally observed blank values (Table 3) exceed the calculated value by a considerable margin and indeed are markedly higher than those obtained with the $Cu(OH)_2$ procedure.

In the second version of the $Cu_3(PO_4)_2$ method (Huber, 1980; Frimmel et al. 1980) the copper phosphate suspension is prepared by partial neutralization of an acid solution of $CuCl_2/Na_2HPO_4$ and aged two weeks. The water sample is adjusted to pH 7.2 with a phosphate buffer, mixed with the copper phosphate suspension (final pH 7.0) and stirred for 60 min. Under these conditions the thermodynamically stable solid phase is indeed $Cu_3(PO_4)_2$ and the calculated concentration of dissolved inorganic Cu(II) is about 2×10^{-6} mol.dm^{-3}. The experimentally observed blank values fall in the 1 to 3×10^{-6} mol.dm^{-3} range (Table 3).

2.3.2 <u>Method validation</u>. Only in the case of the second version of the $Cu_3(PO_4)_2$ procedure has the analytical chemistry been explored in any detail. For moderately concentrated laboratory standards (2×10^{-5} to 10^{-4} mol.dm^{-3}), stoichiometric recovery of NTA and EDTA has been reported (Frimmel et al., 1980). For such synthetic solutions the analytical standard deviation was about $1-3 \times 10^{-6}$ mol.dm^{-3}. Under similar conditions the recoveries of citrate, tartarate and glycine were somewhat lower: 87, 85 and < 10% respectively. The low recovery of glycine is consistent with the thermodynamic calculations reported earlier for similar monomeric ligands (section 2.2.2; Table 4). A marked effect of water hardness was observed in the case of tartarate; at calcium concentrations $>10^{-4}$ mol.dm^{-3} the recovery of tartarate dropped to 15-20%. No comparable recovery data could be found for standard additions to natural water samples.

2.3.3 <u>Method application</u>. The only direct application of the $Cu_3(PO_4)_2$ solubilization method to natural waters would appear to be that of De Haan and co-workers (1981), who studied the seasonal variations of copper binding capacity in an alkaline humic lake (Tjeukemeer, The Netherlands; see Table 5). On occasion these workers used ultrafiltration as well as centrifugation to remove the suspended Cu(II) precipitate; the relative contributions of complexation and peptization to Cu solubilization at pH 9.1 in Tjeukemeer water were almost equal.

An indirect application has been attempted by Frimmel (personal communication, 1983), who first isolated a humic acid from bog water and then determined its complexation capacity in solution at pH 7 (Table 7). Appreciable losses of dissolved organic carbon (< 20%) were noted during the solubilization step, as was also observed at pH 10 with the Cu(OH)$_2$ method (section 2.2.3).

2.4 Other Cu(II) solubilization methods

Two additional Cu containing phases have been proposed as insoluble substrates from which copper could be solubilized and measured: copper sulphide (Kerr and Quinn, 1980), and copper-loaded ion-exchange resins (Manahan and Jones, 1973; M. Florence, personal communication, 1983). Little or no method development work has been reported for these solid phases, however, and such basic analytical parameters as sensitivity, precision and accuracy are largely unavailable. Critical comparison with the $Cu(OH)_2$ or $Cu_3(PO_4)_2$ methods is clearly not feasible.

Only in the case of the copper(II) sulphide method have environmental samples been tested (Kerr and Quinn, 1980). Dissolved organic matter was isolated by activated-charcoal chromatography from coastal and open-ocean surface waters, as well as from seawater of intermediate depths, and the Cu-solubilizing capacities of highly concentrated reconstituted solutions were determined at pH 7 in N_2-purged solutions (Table 7).

2.5 Miscellaneous solubilization methods

Although most of the solid phases used in solubilization procedures for measuring complexation capacity have contained copper(II), a number of other metals have been employed in specific studies ($Fe(OH)_3$, $PbCO_3$, $CaCO_3$). As alluded to earlier, Shapiro (1964) determined the capacity of yellow organic acids isolated from pond water to prevent the precipitation of Fe(III) hydroxide at ambient temperature in the presence of hydrogen peroxide over th pH range 7-9 (Table 7). Filtration experiments showed that the solubilized iron was largely colloïdal in nature, and Shapiro suggested that the iron hydroxide precipitate had been peptized as colloïdal ferric hydroxide as a result of adsorption of the organic acids onto the surface of the particles.

Solubilization of lead carbonate has been suggested as a means of determining NTA concentrations in the 10^{-5} to 10^{-4} mol.dm^{-3} concentration range (Lautenbacher and Baker, 1974). Samples were shaken with powdered PbCO at room temperature for 15 min at pH ~ 7, filtered and analyzed for dissolved Pb by atomic absorption spectrophotometry. The proposed method was designed for use in laboratory toxicological studies and was not recommended for field determination at ambient NTA levels. Recovery of added NTA was indeed lower in natural waters (~ 65%) than for the laboratory standards prepared in distilled water (~ 95%), presumably due at least in part to competition between the major cations (Ca, Mg) and Pb for the NTA. In a similar approach Avnimelech and Raveh (1982) have used the solubilization of solid calcium carbonate as an indication of the presence of organic ligands in sanitary landfill leachates (Table 6).

3. CONCLUSIONS

Of the various solubilization techniques discussed in the preceding
section, only the copper hydroxide and copper phosphate methods have yet been
studied in sufficient detail to warrant consideration as means of determining
the complexation capacities of natural waters. Both these methods have as an
advantage their conceptual and experimental simplicity. In common with the
other solubilization methods, however, these two techniques yield only one of
the complexation capacity parameters, $[L]$, the concentration of ligand
available to bind the added metal (Cu). In addition to this inherent
disadvantage, the $Cu(OH)_2$ method has a number of specific deficiencies: (1)
the required use of an alkaline reaction medium (pH 10), usually very
different from the original sample; (2) the need to heat the solution for an
extended period, with the attendant danger of altering the chemical nature of
the complexing agents originally present; (3) the inherent insensitivity of
the method to a variety of weakly complexing ligands; (4) the dual response o:
the method, in the absence of an ultrafiltration step, to both complexed and
peptized copper(II); (5) the appreciable loss of dissolved organic matter by
co-precipitation with copper(II) hydroxide. The $Cu_3(PO_4)_2$ methods, both the
pH 7 (phosphate buffer) and pH 9 (borate buffer) variations, share many of
these same deficiencies (notably (3), (4) and (5) above) and, in addition,
suffer from high blank values and an inadequate analytical sensitivity.
Clearly these considerations seriously compromise the determination of the
complexation capacity of natural waters by the copper(II) hydroxide or
copper(II) phosphate methods.

ACKNOWLEDGEMENTS

The authors thank H. De Haan (Tjeukemeer Laboratory, The Netherlands),
J.F. Elder (US Geological Survey, Tallahasee, Florida, USA), F.H. Frimmel
(Inst. Wasserchem. und Chem. Balneolog., Munich, FRG), W. Huber (BASF AG,
Ludwigshafen, FRG), J.R Montgomery (Harbor Branch Inst., Ft. Pierce, Florida,
USA) and J.P. Quinn (New Univ. Ulster, Coleraine, N. Ireland) for helpful
comments and discussion.

REFERENCES

- Avnimelech, Y. and A. Raveh, 1982. Decomposition of chelates leached from waste disposal sites. J. Environ. Qual. 11: 69-72.
- Borchers, R., 1959. Spectrophotometric determination of amino acids. Akaline copper salt method using cuprizone, biscyclohexanoneoxalyldihydrazone. Anal. Chem. 31: 1179-1180.
- Campbell, P.G.C., M. Bisson, R. Gagné and A. Tessier, 1977. Critical evaluation of the copper(II) solubilization method for the determination of the complexation capacity of natural waters. Anal. Chem. 49: 2358-2363.
- Chapman, G. and A.C. Rae, 1967. Isolation of organic solutes from sea water by coprecipitation. Nature 214: 627-628.
- De Haan, H., T. de Boer and H.L. Hoogveld, 1981. Metal binding capacity in relation to hydrology and algal periodicity in Tjeukemeer, The Netherlands. Arch. Hydrobiol. 92: 11-23.
- Elder, J.F. and A.J. Horne, 1978. Ephemeral cyanophycean blooms and their relationships to micronutrient chemistry in a southern California reservoir. Sanitary Eng. Res. Lab., Univ. California, Berkeley, UCB-SERL Report No. 78-1.
- Elder, J.F., K.E. Osborn and C.R. Goldman, 1976. Iron transport in a lake Tahoe tributary and its potential influence upon phytoplankton growth. Water Res. 10: 783-787.
- Frimmel, F.H., F. Dietz, J. Elbert, W. Huber, E. Keck, H.-T. Kempf, W.P. Meier, J.K. Reichert, U. Schöttler, M. Schorer and J. Wernet, 1980. Zur summarischen Bestimmung der Komplexbildungsfähigkeit II. Umsetzungen des Kupferphosphatbodenkörpers mit definierten Liganden. Z. Wasser Abwasser Forsch. 13: 12-15.
- Huber, W., 1980. Zur summarischen Bestimmung der Komplexbildungsfähigkeit. Die Kupferphosphat-Methode. Z. Wasser Abwasser Forsch. 13: 8-11.
- Jeffrey, L.M. and D.W. Hood, 1958. Organic matter in sea water: an evaluation of various methods for isolation. J. Mar. Res. 17: 247-271.
- Kerr, R.A. and J.G. Quinn, 1980. Chemical comparison of dissolved organic matter isolated from different oceanic environments. Mar. Chem. 8: 217-229.
- Kunkel, R. and S.E. Manahan, 1973. Atomic absorption analysis of strong heavy metal chelating agents in water and waste water. Anal. Chem. 45: 1465-1468.
- Lautenbacher, H.W. and H.W. Baker, 1974. Determination of NTA in aquatic toxicity studies: a compleximetric method using atomic absorption spectroscopy. Bull. Environ.Contam. Toxicol. 11: 57-63.
- Manahan, S.E. and D.R. Jones, 1973. Atomic absorption detector for liquid-liquid chromatography. Anal. Lett. 6: 745-753.
- Manahan, S.E. and M.J. Smith, 1973. The importance of chelating agents in natural waters and wastewaters. Water Sewage Works 120: 102-106.
- Mantoura, R.F.C., 1981. Organo-metallic interactions in natural waters. In: E.K. Duursma and R. Dawson (eds.), Marine organic chemistry: evolution, composition, interactions and chemistry of organic matter in seawater. Elsevier Scientific Publishing Co., Amsterdam, Chapt. 7, pp. 179-223.
- Montgomery, J.R. and J.E. Echevarria, 1975. Organically complexed copper, zinc and chelating agents in the rivers of western Puerto Rico. In: F.G. Howell, J.B. Gentry and M.H. Smith (Eds.), Mineral cycling in southeastern ecosystems. ERDA symposium series 36: 423-434.

- Pope, C.J. and M.F. Stevens, 1939. The determination of amino-nitrogen using a copper method. Biochem. J. 33: 1070-1077.
- Quinn, J.P., T.W. Barker and R. Marchant, 1982. Soluble complexes of copper and zinc in whiskey distillery spent wash. J. Inst. Brew. 88: 95-97.
- Shapiro, J., 1964. Effect of yellow organic acids on iron and other metals in water. J. Am. Water Works Assoc. 56: 1062-1082.
- Sposito, G., 1981. Trace metals in contaminated waters. Environ. Sci. Technol. 15: 396-403.
- Stumm, W. and J.J. Morgan, 1981. Aquatic chemistry. An introduction emphasizing chemical equilibria in natural waters. Second edition, J. Wiley and Sons, New York, 780 pp.
- Westall, J.C., J.L. Zachary and F.M.M. Morel, 1976. MINEQL, a computer program for the calculation of the chemical equilibrium composition of aqueous systems. Massachusetts Institute of Technology, Dept. Civil Eng., Tech. Report No. 18, 91 pp.

ANALYTICAL METHODS FOR MEASUREMENT AND INTERPRETATION OF METAL BINDING BY
AQUATIC HUMUS AND MODEL COMPOUNDS

JOHN R. TUSCHALL, JR. AND PATRICK L. BREZONIK

1. INTRODUCTION

Progress in understanding and measuring the role of natural organic
matter as a complexing agent for heavy metals has been hindered by two
analytical difficulties. The first obstacle lies in obtaining reliable data
for metal complexation with organic matter over a useful range of metal
levels, and the other lies in reporting the results in a simple and accurate
fashion. This paper addresses both of these aspects, first by demonstrating
the reliability of binding data for identical solutions among four dis-
tinctly different methods, and second by treating these data using a
mathematical model that is based conceptually on electrostatic interactions
as a cause of variable strength binding. Natural water organic matter and
synthetic polymers were used to evaluate these methods.

2. EVALUATION OF METHODS

Water samples were evaluated by each of five analytical methods under
identical conditions so that direct comparisons could be made. Three of the
methods - ion selective electrode (ISE), anodic stripping voltammetry (ASV),
and fluorescence quenching (FQ) - were developed by others and had been
applied previously to natural-water complexation studies. The other two
methods --continuous-flow ultrafiltration (CUF) and competing-ligand
differential spectroscopy (CLDS) - were developed by us in this study. In
each case, copper(II) (Cu^{2+}) was added incrementally to samples maintained
at pH 6.25 in 0.1 M KNO_3; measurements were made throughout the Cu^{2+}
titrations by each of the five analytical methods. Complete experimental
details are presented in Tuschall and Brezonik (1983; 1983a).

Natural water samples were collected from two hardwood swamps,
Waldo Swamp and Basin Swamp, in northern Florida, U.S.A. Water in these
swamps was soft, highly colored, and slightly acidic (pH ~6). Dissolved
organic carbon levels were 50 and 25 mg C/L in Waldo and Basin Swamps. In
addition, three model peptides (homopolyamino acids) - polyalanine (mol. wt.

Kramer, C.J.M. and Duinker, J.C. (eds.), Complexation of Trace Metals in Natural Waters. ISBN 90-247-2973-4
©1984 Martinus Nijhoff/Dr W. Junk Publishers, The Hague/Boston/Lancaster.
Printed in the Netherlands.

3,900), polyarginine (mol. wt. 13,900), and polyaspartic acid (mol. wt. 5,400) (Miles Biochemicals) - were used in the comparative study. Of the three, only polyaspartic acid complexed Cu^{2+} to any measurable extent.

The two parameters used to compare methods are maximum Cu^{2+} binding capacity, C_L, and conditional stability constant, β'. Values of C_L determined on the two swamp waters and polyaspartic acid by the five analytical methods are listed in Table 1. For each sample, agreement among the procedures was quite good, except for the ASV method which yielded C_L values consistently lower than those from the other methods. The cause of the lower ASV results is discussed below.

Direct comparisons of β' values are difficult because single values are seldom obtained over the desired range of total metal concentrations and no satisfactory method of accurately calculating multiple β' values is available. Scatchard analysis has been applied to natural water studies, and frequently results have been interpreted in terms of two distinct binding sites, one strong and one weak. This approach, which is an improvement on the idea of a single β', remains a simplification of reality and can be criticized as "curve-fitting." Our results did not fit the two-component Scatchard model, and consequently, for comparison purposes only, we divided Scatchard plots (Scatchard 1949) into three section of $\bar{v}$ and fitted experimental data to straight lines. Although the slopes of these segments are only an approximation of true β' values, they should allow of valid comparisons among methods (Tuschall and Brezonik, 1983a).

For any given range of $\bar{v}$, values of β' are similar among the methods except for ASV, which produced β' values that are 15-300 times lower than corresponding average values (Table 2). The disparate values for ASV

TABLE 1. Binding Capacities for Surface Waters and Polyaspartic Acid.

Method of Analysis	Waldo Swamp	Basin Swamp	Polyaspartic Acid $(3.7 \cdot 10^{-5}$ M$)$
ASV	$8.5 \cdot 10^{-6}$	$5.8 \cdot 10^{-6}$	$<10^{-7}$
ISE	$7.9 \cdot 10^{-5}$	$1.4 \cdot 10^{-5}$	$5.0 \cdot 10^{-4}$
FQ	$7.4 \cdot 10^{-5}$	$2.1 \cdot 10^{-5}$	---
CUF-CUF	$4.3 \cdot 10^{-5}$	$1.2 \cdot 10^{-5}$	$3.0 \cdot 10^{-5}$
CLDS	$6.8 \cdot 10^{-5}$	$1.3 \cdot 10^{-5}$	$4.2 \cdot 10^{-4}$
σ	$2.9 \cdot 10^{-5}$	$5.42 \cdot 10^{-6}$	$1.0 \cdot 10^{-4}$

TABLE 2. Log β' Values for Surface Waters and Polyaspartic Acid With Copper

| v̄ range | Analytical Technique | | | | | |
	ASV	ISE	FQ	CUF	CLDS	X̄
Waldo Swamp						
0-0.02	5.25	8.11	ND	ND	ND	7.81
0-021-0.2	5.36	6.72	6.80	6.54	6.61	6.60
0.21-0.8	ND	5.34	5.17	5.70	5.48	5.47
Basin Swamp						
0-0.025	6.04	7.82	ND	ND	ND	7.53
0.026-0.125	5.30	6.85	5.70	6.67	6.72	6.55
0.126-0.4	ND	5.26	4.87	5.56	5.54	5.38
PAA						
0-0.25	ND	8.50	NA	ND	ND	8.50
0.26-4	ND	7.04	NA	7.50	7.31	7.32
4.1-15	ND	5.44	NA	5.36	5.18	5.34

ND = not detected   NA = not analyzed

suggest noncompliance with one of the criteria necessary for the ASV method
to produce accurate results, namely that the metal-organic complex is
reduced at the Hg electrode (Shuman and Woodward, 1973). We examined this
possibility by performing cyclic voltammetry on samples containing added
metal.

Figure 1 shows the expected 30 mV difference between the reduction peak
(+0.01V SCE) and the oxidation peak (+0.04 V vs. SCE) for Cu^{2+} (10^{-5} M) in
background electrolytes. In the presence of organic matter (Waldo Swamp
water), the reduction peak of copper shifted to -0.11 vs. SCE, and a
decrease in current was observed (Fig. 1). This peak is likely due to the
direct reduction of complexed copper, and the low and rounded nature of the
reduction peak indicates that the copper complex diffused at a lower rate
than Cu^{2+}, hence causing less reduction of metal and a decrease in peak
current (i_p). The broader nature of the peak also suggests that reduction
was irreversible. The copper oxidation peak is a shoulder on the rising
mercury oxidation peak, which is shifted cathodically in the presence of
organic matter. This cathodic shift in mercury oxidation, which is caused
by complexation of mercury by soluble organic ligands, subsides as copper
titrant is added. The varying nature of the mercury oxidation potential
makes accurate measurements of i_p difficult.

Nevertheless, the copper-organic complex examined here is readily
reducible at a potential 120 mV cathodic of the reduction potential of Cu^{2+}.

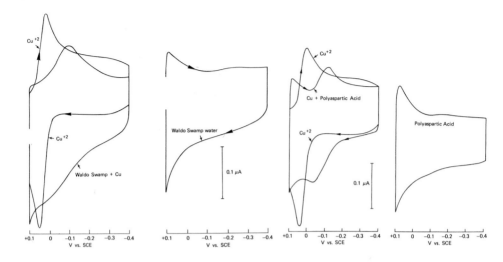

FIGURE 1. Cyclic voltaammograms of 10^{-5} M Cu^{2+}, Waldo Swamp water plus 10^{-5} M Cu^{2+}, and Waldo Swamp water alone. Scan rates were 50 mV·sec^{-1} and solutions contained 0.1 M KNO_3 at pH 6.25.

FIGURE 2. Cyclic voltammograms of 10^{-5} M Cu^{2+} and polyaspartic acid ($1.8 \cdot 10^{-5}$ and $3.6 \cdot 10^{-5}$ M). Conditions as described in Fig. 1.

This difference in potentials is not great enough to allow plating of Cu^{2+} without reducing the complexed metal, and consequently, the concentration of free Cu^{2+} cannot be measured accurately. A second experiment indicated that organic matter slowly adsorbed to the electrode: because the magnitude of the reduction peak increased with increased time of waiting after forming the mercury drop, both with and without an applied potential. No such effect was observed for solutions containing only copper or only organic matter. Buffle et al. (1976) observed the same increase in reduction current for lead with increasing waiting times, but they also reported a cathodic shift in reduction potential. We did not observe a shift in reduction potential with varying waiting times. Experiments with Cu^{2+} and Basin Swamp water produced cyclic voltammograms and adsorption behavior similar to those described above for Waldo Swamp.

The cyclic voltammograms of cadmium ($5 \cdot 10^{-6}$ M) in background electrolytes displayed expected reversible behavior. For cadmium in the Waldo Swamp water both the reduction and oxidation peaks were shifted 30 mV in the cathodic direction, but the Cd reduction current was greater than in the electrolytes alone. This increase appears to be caused by adsorption of organically complexed cadmium to the electrode surface. In another experiment, the reduction current increased with increasing waiting time after drop formation (both with and without an applied potential) in a manner similar to that observed with copper and the Waldo Swamp water. Other techniques confirmed the presence of a cadmium-organic matter complex in Waldo Swamp water (Tuschall 1981). The above experiments show that, like copper, the cadmium complex adsorbed to the working electrode and was ASV-labile.

The cyclic voltammograms of copper (10^{-5} M) and polyaspartic acid (PAA) ($1.8 \cdot 10^{-5}$ M) produced a reduction peak that was lower, broader, and 120 mV more cathodic than that for a solution of Cu^{2+} containing no organics (Fig. 2). However, the oxidation peak for copper with PAA was only 50 mV cathodic of the oxidation of Cu^{2+}. For a reversible two-electron process, the peak potentials for reduction and oxidation in cyclic voltammetry should differ by 30 mV (Greter, et al. 1979) but the difference observed here was 100 mV. Consequently, the electrochemical reversibility of the copper-PAA complex is in question. The decrease in peak current for the solution containing PAA (Fig. 2) indicates that some step in the reduction process (probably diffusion to the electrode surface) is slower for the copper-PAA acid complex than for Cu^{2+}.

The adsorption phenomenon observed for the swamp samples was not detected with PAA. No change in potential or current occurred for various waiting times after drop formation. This result is corroborated by the observation that discarded drops of mercury remained discrete in the bottom of the cell for the Waldo Swamp samples (due to a change in surface tension), whereas the mercury drops coalesced in solutions of PAA and solutions containing only inorganic species.

Further evidence that the copper-PAA complex is reducible was obtained by doubling the PAA concentration of a solution initially containing $1.8 \cdot 10^{-5}$ M PAA and 10^{-5} M Cu^{2+}. The cyclic voltammograms before and after the increase were identical. This suggests that the reduction peak was due to the direct reduction of a copper-PAA complex. If complexation of copper were nearly complete, increases in PAA would not appreciably affect the concentration of the copper complex. However, if the reduction peak were due to the reduction of Cu^{2+}, then an increase in polymer concentration

would decrease the peak further. Taken together, these experiments indicate
that the copper-PAA complex exists in appreciable concentrations and that
the complex is reducible at a potential near that of Cu^{2+}.

Cyclic voltammetric scans of copper with polyarginine or polyalanine
resembled scans for Cu^{2+} in organic-free medium, indicating that no
complexation or adsorption to the Hg electrode occurred. No copper
complexation by these two polymers was observed with any of the other
methods.

3. CALCULATION OF INTRINSIC STABILITY CONSTANTS

The concordance of metal speciation data obtained using four distinct
analytical methods was encouraging, but expression of the data in an
accurate and useful manner remains problematic. Conditional stability
constants can be incorporated into equilibrium calculations (e.g. computer
programs) if the constant can be expressed simply. Values of β' are
accurate only in a narrow range of metal concentrations because β' values
appear to vary several fold throughout a metal titration. More elaborate
attempts at estimating β' suffer either from being too complicated or not
adequate to describe β' over an entire range of metal concentrations for a
variety of environmental samples. Perdue and Lytle (1983) have proposed
characterizing stability constants for metal-humic complexes in terms of a
normal distribution function. They obtained good fits of complexation data
to this model. There is merit to this proposal for aquatic humus because it
most likely does have a large population of similar, but not equivalent,
binding sites. On the other hand, this approach is essentially empirical
and does not consider some factors such as electrostatic effects, which are
likely to be important in metal binding with natural water organics and
which can be treated in a more rigorous fashion.

The mathematical description of electrostatic interactions was formulated by Tanford (1961) and since has been applied to metal and proton
binding with a variety of proteins and synthetic polymers (see King 1965).
The equations that describe proton affinity for identical charged sites on a
polymer are

$$pK' = pH - \log(\alpha/1 - \alpha) = pK_{INT}h + 0.868 \, \omega n\alpha \qquad (1)$$

where pK' is the apparent dissociation constant, $pK_{INT}h$ is the intrinsic
dissociation constant, α is the degree of ionization, ω is the electrostatic
interaction term, and n is the total number of ionic sites (Tanford 1961).

In the case of a metal ion competing with a proton for a site on a polymer, the following equation is solved for the intrinsic metal-binding constant, $\beta_{INT}m$, which represents binding to a polymer that is (hypothetically) devoid of charge:

$$\frac{\theta_m}{1 - \theta_m} = \frac{\beta_{int}m(M)\exp(-2\omega z_m Z^*)}{1 + (H^+)(K_{int}h)^{-1}\exp(-2\omega Z^*)} \tag{2}$$

where θ_m is the fraction of bonding sites occupied by metal ions, (M) is the concentration of uncomplexed metal, z_m is the charge on the free metal, and Z^* is the charge on the macromolecule. The latter term was evaluated by the equation: $Z^* = z_m n_m \theta_m - \alpha n$, where n_m is the ultimate number of sites occupied by metal.

In deriving and applying equations 1 and 2, it must be assumed that (1) all ionized sites are identical, (2) protons and metal ions compete for sites in a similar fashion, (3) the polymer is spherical in shape, and (4) the change is randomly distributed among sites (King 1965). Although these conditions have yet to be demonstrated for aquatic organics, it is reasonable to assume compliance to these strictures.

Several researchers reported good fit of their data to equation 1 for potentiometric titrations of humics, suggesting that they behave like polyelectrolytes with similar binding sites (Posner 1964; Wilson and Kinney 1977; Dempsey and O'Melia 1983). Wilson and Kinney (1977) applied their results for variable pH titrations of aquatic humics at a single level of zinc to equations 1-2 and produced a nearly constant value for β_{INT}. We applied Tanford's model to variable metal titrations conducted at a fixed pH to examine the usefulness of the electrostatic model in this case.

We initiated our examination with polyaspartic acid (PAA), which has identical, repeating sites that bind Cu^{2+}. To evaluate $K_{INT}h$, the polymer was titrated potentiometrically under conditions (0.1M KNO_3) consistent with those used for the complexometric copper titrations. Nitric acid or KOH were added by an automatic titrator (Metrohm E535-536) and pH was monitored continuously. Throughout the titration, the pH and amounts of acid or base added to the polymer were recorded and used to determine the degree of ionization (α) of the polymer. Thus, according to equation 1, a plot of α vs. pH - log ($\alpha/1 - \alpha$) should produce a line with slope ωn and intercept $pK_{INT}h$. Our results (Fig. 3) were linear for $\alpha > 0.25$, and extrapolation of these linear points produced $pK_{INT}h = 3.25$ and $\omega = 0.042$. The nonlinear

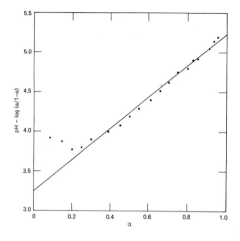

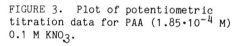

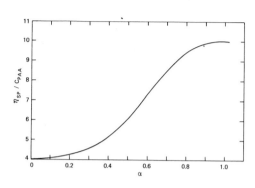

FIGURE 3. Plot of potentiometric titration data for PAA $(1.85 \cdot 10^{-4}$ M) 0.1 M KNO_3.

FIGURE 4. Variation of η_{sp}/C_{PAA} (specific viscosity/concentration of PAA) with α, the degree of ionization, for polyaspartic acid (mol. wt. 7000) (Adapted from Berger and Katchalski 1951).

region below α = 0.25 can be ascribed to a change in polymer conformation caused by hydrogen bonding at low pH. Specific viscosity measurements on PAA (Fig. 4) made by Berger and Katchalski (1951) support this hypothesis in that the polymer was less viscous (more tightly coiled) at low pH than at high pH. Thus, at low pH (α < 0.25) equation 1 does not accurately model the electrostatic interactions for PAA. However, the parameters derived from Fig. 3 should be accurate for our purpose because the complexometric titration of PAA with Cu^{2+} was performed at pH 6.25 (α = 0.9), which is well above the observed region of non-theoretical behavior.

Our value of 3.25 for $pK_{INT}h$ of PAA (5400 mol. wt.; μ = 0.1) agrees well with the value of 3.53 obtained from electrophoretic measurements for PAA (4130 mol. wt.; μ = 0.5) by Katchalsky, et al. (1954). Both values agree well with pK_1 = 3.37 (Kokufuta et al. 1977) for N-acetylaspartic acid. This compound should have an acid dissociation constant similar to the intrinsic value for PAA because of its N-acetylated amino group.

We solved equation 2 for for $\beta_{INT}m$ at various concentrations of Cu^{2+} that were measured in the copper ISE titration. The resulting $\beta_{INT}m$ values (Table 3) ranged from 3.90 to 3.34 for corresponding levels of total copper that varied over 100-fold. Although the intrinsic constants are relatively

TABLE 3.   Intrinsic Binding Constants for Copper with PAA.

pCu_T	LOG $\beta_{INT}M$	pCu_T	LOG $\beta_{INT}M$
5.57	3.90	4.09	3.67
5.46	3.88	3.95	3.53
5.35	3.81	3.84	3.45
5.11	3.80	3.72	3.34
4.71	3.80	3.57	3.34
4.46	3.78	3.46	3.50
4.28	3.72		

close, less variation was expected for a homopolymer such as PAA. Reasons
for the variation are discussed below.

4.   OBSERVATIONS ON GRAPHICAL TREATMENTS OF BINDING DATA

Although Scatchard plots have been used frequently to display
metal-binding data, attainment of the endpoint of a complexometric titration
is not obvious from such a plot. Klotz (1974; 1982) has suggested an
alternative plot of $\bar{v}$ vs. -log ($metal_{free}$) to verify the completeness of a
titration. For an ideal system in which every site has the same affinity
for binding, a Klotz plot will produce a symmetric, sigmoidal curve with an
inflection point at 1/2 of the ultimate value of $\bar{v}$. When our results for a
copper titration of PAA were plotted in the manner suggested by Klotz
(Fig. 5), an asymmetric curve was produced with the inflection point near
the ultimate value of $\bar{v}$. Curves of similar shape were obtained for copper
binding with naturally occurring organic matter (Shuman et al. 1983;
Tuschall and Brezonik 1983a).

We attempted to reproduce this asymmetric curve using a hypothetical
system containing metal with various ligands that form variable-strength
complexes. Results (using MINEQL) for a copper titration of 11 ligands
($9.1 \cdot 10^{-6}$ M each) that complexed copper with log β' ranging from 3.5 to 7.0
showed a symmetric, sigmoidal curve with the inflection point at 1/2 the
ultimate value of $\bar{v}$ (Fig. 6). Varying the distribution or magnitude of β'
for the complexes did not alter the symmetric nature (i.e. inflection point)
of the curves.

In light of these observations, two explanations for the asymmetrical
shape of the experimental curves are feasible: (1) The binding sites have
not been complete satisfied by the added metal and the apparent endpoint was
caused by inaccurate measurements or by precipitation or flocculation of

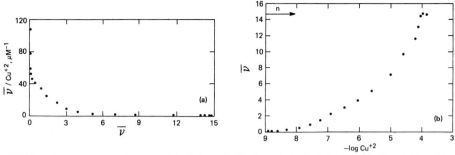

FIGURE 5. Scatchard plot (left) and Klotz plot (right) for the titration of polyaspartic acid with copper.

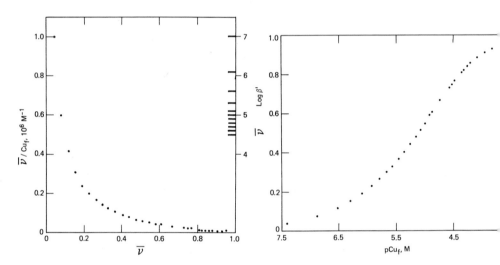

FIGURE 6. Scatchard plot (left) and Klotz plot (right) for the hypothetical titration of multiple ligands with Cu^{2+}. The apparent stability of the complexes, β', ranged from 10^7 to $10^{4.5}$.

components, or (2) the conformation of the organic ligand changed during the titration with metal, thereby causing apparently stronger binding sites to be exposed for complexation. The latter possibility was shown by computer simulation to produce a curve similar to that obtained empirically. If, as stated in the former case, titrations are not being measured to completion, the validity of C_L and other parameters that incorporate C_L (e.g. most β' values) is questionable.

5. DISCUSSION AND CONCLUSIONS

Complexometric titrations of aquatic humus and polyaspartic acid displayed similar results for four of five separate analytical methods used. The ASV method produced consistently lower C_L and β' values and further investigations using cyclic voltammetry indicated that the reducible nature of the copper-organic matter complex was causing inaccurate results. However, reduction of the complex occurred at a slower rate than for Cu^{2+} either because of electrochemical irreversibility or because complexed copper diffused to the Hg electrode at a lower rate, and thus deceptively typical titration curves were observed.

It is difficult to estimate precisely the effect of complexation on rates of Cu^{2+} diffusion to Hg electrode surfaces. Diffusion coefficients (D_O) of solutes decrease slowly and non-linearly with increasing molecular weight. However, other factors also will affect diffusion of copper complexes to Hg electrodes, most importantly electrostatic factors (net charge on complex; electrode potential). Adequate information is not available to quantify these factors in our samples.

Nonetheless, these results are disconcerting because any practitioner of the ASV method who does not ensure that the metal-organic complex is non-reducible may generate erroneous C_L and β' values. The developers of the ASV method plainly state that the non-reducible nature of the complex is a necessary condition for accuracy of the method (Shuman and Woodward 1973; 1977), but convincing demonstrations of this condition for natural organic matter remain to be disclosed.

The data for Cu^{2+} titrations of PAA were applied to a mathematical model that is predicated on electrostatic phenomena and was found to conform well to the model. However, the variability of β_{INT} values suggests that further refinements in the model are necessary before applying it to natural organic matter. The variability of results is likely caused by (1) conformational changes in the polymer during complexation, (2) non-random distribution of charged sites, or (3) the non-spherical shape of the polymer. Once refined, the model could be applied to titrations of natural water samples to eliminate apparent metal-binding variability caused by electrostatic interactions. Then these data could be treated further by other methods (e.g. Gaussian distribution model) to account for any non-identical binding sites.

REFERENCES

-Berger, A., and E. Katchalski. 1951. Poly-L-Aspartic acid. J. Am. Chem. Soc. 73:4084-4088.

-Buffle, J., F. L. Greter, G. Nembrini, J. Paul, and W. Haerdi. 1976. Capabilities of voltammetric techniques for water quality control problems. Z. Anal. Chem. 282:339-350.

-Dempsey, B. A. and C. R. O'Melia. 1983. Proton and calcium complexation of four fulvic acid fractions. In: R. F. Christman and E. T. Gjessing (Eds), Aquatic and Terrestrial Humic Materials. Ann Arbor Science, p. 239-274.

-Greter, F. L., J. Buffle, and W. Haerdi. 1979. Voltammetric study of humic and fulvic substances, I. Study of the factors influencing the measurement their complexing properties with lead. J. Electroanal. Chem. 101:211-229.

-Katchalsky, A., N. Shavit and H. Eisenberg. 1954. Dissociation of weak pol meric acids and bases. J. Polymer Sci., 13:69-84.

-King, E. J. 1965. Equilibrium properties of electrolyte solutions, Volume Acid-Base equilibria. Pergamon Press, NY, p. 218-247.

-Klotz, I. M. 1974. Protein interactions with small molecules. Acc. Chem. Res. 7:162-168.

-Klotz, I. M. 1982. Number of receptor sites from Scatchard Graphs: Facts a fantacies. Science 217:1247-1249.

-Kokufuta, E., S. Suzuki, and K. Harada. 1977. Potentiometric titration behavior of polyaspartic acid prepared by thermal polycondensation. Bio Systems 9:211-214.

-Perdue, E. M. and C. R. Lytle. 1983. A critical examination of metal-ligan complexation models: Application to defined multi-ligand mixtures. In: R. F. Christman and E. T. Gjessing (Eds)., Aquatic and Terrestrial Humic Materials, Ann Arbor Science, p. 295-314.

-Posner, A. M. 1964. Titration curves of humic acid. Trans. 8th Intern. Congr. Soil Sci., Bucharest, Romania, 3:161-174.

-Scatchard, G. 1949. The attractions of proteins for small molecules and ions. Ann. NY. Acad. Sci. 51:660-672.

-Shuman, M. S. and G. P. Woodward, Jr. 1973. Chemical constants of metal complexes from a complexometric titration followed with anodic stripping voltammetry. Anal. Chem. 45:2032-2035.

-Shuman, M. S. and G. P. Woodward, Jr. 1977. Stability constants of copper-organic chelates in aquatic samples. Environ. Sci. Technol. 11:809-313.

-Shuman, M. S., B. J. Collins, P. J. Fitzgerald, and D. L. Olson. 1983. Distribution of stability constants and dissociation rate constants among bind sites on estuarine copper-organic complexes: Rotated disk electrode studies and an affinity spectrum analysis of ion selective electrode and photometric data. In: R. F. Christman and E. T. Gjessing (Eds.), Aquatic and Terrestri Humic Materials, Ann Arbor Science, p.349-370.

-Tanford, C. 1961. Physical chemistry of macromolecules. Wiley, NY, 572 p

-Tuschall, J. R., Jr. 1981. Heavy metal complexation with naturally occurri organic ligands in wetland ecosystems. Ph.D. disser., Univ. of Fla., 212 p

-Tuschall, J. R., Jr., and P. L. Brezonik. 1983. Complexation of heavy meta by aquatic humus: A comparative study of five analytical methods. In: R. F. Christman and E. T. Gjessing (Eds.), Aquatic and Terrestrial Humic Materials. Ann Arbor Science, p. 275-294.

-Tuschall, J. R., Jr., and P. L. Brezonik. 1983a. Application of continuous-flow ultrafiltration and competing ligand/differential spectrophotometry for measurement of heavy metal complexation by dissolved organic matter. Analytica. Chimica Acta, 149:47-58.

-Wilson, D. E., and P. Kinney. 1977. Effects of polymeric charge variations on the proton-metal ion equilibria of humic materials. Limnol. Oceanogr. 22:281-289.

POTENTIALITIES OF VOLTAMMETRY FOR THE STUDY OF PHYSICOCHEMICAL
ASPECTS OF HEAVY METAL COMPLEXATION IN NATURAL WATERS

H.W. NÜRNBERG

1. INTRODUCTION

Physicochemical interactions of heavy metals present in natural
waters with interfaces of suspended particles, sediments and
aquatic organisms depend strongly on the chemical forms (species)
in which the dissolved metals exist in the respective natural
water type (Stumm and Morgan, 1981). The resulting tendencies and
potentialities of a heavy metal to accumulate at suspended particles has e.g. a decisive influence on the concentration regulation of heavy metals in the water column down to the bottom sediments (Breder et al., 1982; Sigg et al., 1982). In this context
also determinations of the distribution of the total heavy metal
content in the water column between the dissolved and suspended
matter phase can be considered, taking a broad view, as belonging
to the field of speciation studies, although in the future more
detailed studies on the chemical states in which metals exist at
or in the various types of suspended matter are very much needed.
As the interaction of metals with biological interfaces depends
as well strongly on the metal speciation in the dissolved phase
it has pronounced significance for the metal uptake by organisms
and thus ultimately even for the ecotoxic effects in aquatic biota
and the levels of toxic heavy metals in sea food extending in this
way its impact even to man.

From the methodological viewpoint it has to be realized, that
heavy metals exist in all natural waters at the trace or even
ultra trace level. The total contents cover a range of about
10^{-3} to several hundred $\mu g/l$ depending on the water type and
its heavy metal pollution degree. However, the heavy metal levels
in the dissolved phase remain in the lower part of this range

Kramer, C.J.M. and Duinker, J.C. (eds.), Complexation of Trace Metals in Natural Waters. ISBN 90-247-2973-4
©1984 Martinus Nijhoff/Dr W. Junk Publishers, The Hague/Boston/Lancaster.
Printed in the Netherlands.

and hardly exceed o.5/ug/l. Consequently substance specific meth
of high determination sensitivity are needed, if the investigations are to be carried out at realistic concentration levels.
The latter are usually particularly low in the sea, often decreasing in the sequence estuary>coastal water>open sea (Mart et
al., 1982).

The dissolved phase is generally of great significance in
aquatic ecosystems, as this phase provides the medium in which
predominantly the transfer of the heavy metals to the other
components of the aquatic ecosystem (suspended matter, sediments
and organisms) takes place. The dissolved overall concentration
of a heavy metal Me is generally distributed in a given natural
water type, as function of the present types and concentrations
of the prevailing inorganic and organic ligands, between labile
MeX_j and nonlabile MeL_m having rather high stability constants β,
according the following general complexation scheme:

(1) $\quad \Sigma MeX_j \rightleftharpoons \Sigma jX^- + Me^{n+} + \Sigma mL \rightleftharpoons \Sigma MeL_m$

Therefore, in many natural water types, e.g. sea water, only
small amount of the overall dissolved metal concentrations exist
as free hydrated cations Me^{n+}. Basic parameters and features
of the natural water types, i.e. salinity, pH, oxygen level,
concentration and composition of dissolved organic matter (DOM),
will obviously strongly influence the species distribution of a
given heavy metal in the dissolved phase. On the contrary the
heavy metal concentrations will in practice not influence the
basic parameters or the actions of the major chemical constituen
of a natural water type, because the heavy metals are always pre
sent only at low trace levels. As consequence of this high dilut
of heavy metal concentrations also mutual influences of the vari
heavy metal traces on their speciation distributions in the
dissolved phase remain negligible.

Among the methodological approaches for heavy metal speciatio
in the dissolved phase advanced modes of voltammetry, particular
differential pulse voltammetry (DPV) and differential pulse stri
ping voltammetry (DPSV), provide until now one of the most eluci

dating approaches (Florence and Batley, 198o; Nürnberg 1983 a; Nürnberg and Valenta, 1983). Not all heavy metals are well accessible to voltammetry, but it just so happens, that this is the case for those heavy metals and metalloids of prime ecotoxic significance, e.g. Cd, Pb, Cu, Zn, Hg, As(III) etc. (Nürnberg, 1982; 1984). In the ecotoxicological context also the oxidation states in which a metal exists in the considered natural water can be a speciation problem of interest. A representative example is As(III) and As(V) of which only the highly toxic As(III) is recorded by voltammetry (Bodewig et al., 1982).

It has to be borne in mind that one indispensable task of a speciation study is also an accurate determination of the overall level of the heavy metal traces in the dissolved and particulate matter phases. For this analytical task voltammetry provides one of the most sensitive, reliable and convenient approaches in aquatic trace metal chemistry.

The salient aspects of the potentialities provided by voltammetry for speciation studies of exploratory and diagnostic nature as well as for detailed studies on defined species and on speciation aspects of general significance and validity will be treated subsequently. For more detailed and comprehensive treatments of the subject the reader is referred to the cited references (Nürnberg and Raspor, 1981; Nürnberg and Valenta, 1983; Nürnberg 1983 a, b; Valenta 1983).

2. EXPLORATORY AND DIAGNOSTIC STUDIES

2.1. Trace Analytical Measurements

Certain general speciation informations typical for the studied water type can be obtained already by performing the analytical procedure for the determination of the overall concentrations of heavy metals in the respective water sample under various normalized conditions (viz.fig.1).

At first the dissolved heavy metal amounts are separated from the heavy metals bound to suspended particles by filtration through a membrane filter with o.45/um pore size. Analysing subsequently the filtrate and the suspended material one obtains the distribution of the total heavy metal content between both phases in

98

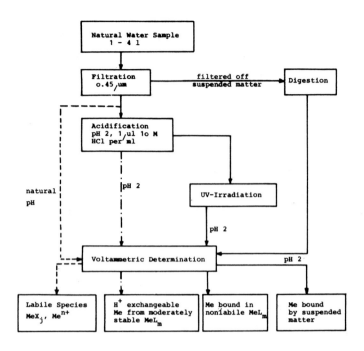

FIGURE 1. Flow chart of a speciation-orientated analytical procedure for the voltammetric determination of heavy metals Me in natural waters. (---) natural pH, (-.-.) acidified to pH 2, (——)
UV-irradiated, pH 2. For the determination of the total dissolved
Me-concentration the procedure goes always via acidification and
at non-negligible DOM-concentrations also via UV-irradiation. To
master contamination risks ultra trace levels of Me in oceanic waters are determined if possible without prior filtration if, as
frequently, suspended matter amount is negligible. The total of
4 categories corresponds to the total heavy metal level in the water column at the sampled depth in ng/l

the water column at the respective depth and location of sampling.
The heavy metal concentrations in the filtrate are regarded as
dissolved. Although this is strictly spoken an operational definition, it is frequently not a bad approximation of the reality
bearing in mind that small colloidal particles may pass the filter
and their heavy metal content will be counted virtually as dissolved.

The filtrate can be divided into three aliquots. In the first
the heavy metal levels are determined at natural pH. In this manner
the heavy metal amount present as rather labile complex species
MeX_j and free hydrated cations Me^{n+} will be obtained. These MeX_j

species are in the first place complexes with the usually in natural waters significant inorganic ligands Cl^-, OH^-, CO_3^{2-}, HCO_3^-, SO_4^{2-} and to a small extent also certain organic ligands as most amino acids. All these labile heavy metal species yield a reversible voltammetric response as the as central ion contained heavy metal ion Me(II or III) undergoes a sufficiently fast electrode process. It has, however, to be recognized, that at the natural pH, usually around 8, of a natural water sample the determination conditions are not optimal and therefore not the ultimate possible determination sensitivity can be reached.

In the second aliquot the pH is adjusted to an optimal value of about 2 by addition of a small amount of 1o M HCl, E.Merck, suprapur grade. Then the ultimate determination sensitivity of DPSV down to about 1 ng/l can be attained. If the studied natural water contains a heavy metal fraction in more stable species at H^+-exchangeable sites this additional heavy metal amount will be determined. Also most colloidal species will release their heavy metal content upon this acidification. By comparison with the determined heavy metal levels at natural pH the heavy metal content contained in those moderately stable species can be evaluated.

A number of natural water types, i.e. many rivers and lakes, frequently estuaries and a number of coastal water areas, but with respect to the in vast regions of the oceans rather low biological productivity much more rarely oceanic samples, contain higher DOM levels. Then the water will contain sufficiently high concentrations of organic ligands forming nonlabile heavy metal complexes MeL_m with rather high stability constants β_m. The metal content bound in those complexes is usually not accessible to DPSV.

The reason is, that if they undergo an electrode process at all within the potential range available for the most frequently applied mercury electrode the required overvoltage is so large that hydrogen evolution would interfere. Therefore, the heavy metal content bound in those nonlabile complexes MeL_m has to be made accessible to voltammetric determination by prior decomposition of these complexes. This is achieved by subjecting the third aliquot of the filtrate after acidification to UV-irradiation from a mercury arc lamp (15o W, 238-265 mm wave length) for 1-2 h. (Mart et al., 198o)

In this manner photolytic decomposition of the nonlabile complexes is achieved, without any contamination risk. Comparison
the determined heavy metal concentration gives that heavy metal
amount present in the dissolved phase in nonlabile complexes Me.

In this manner the distribution of the heavy metal concentrations over three species categories of increasing stability can
be determined in the studied natural water type. The heavy meta
content in the suspended particle phase is also determined afte
digestion. This is performed either by a wet digestion with
o.2-1.o ml conc. HNO_3 and o.1-o.5 ml $HClO_4$, avoiding by these
small amounts safety risks (Ostapczuk et al., 1983) or preferentially by low temperature ashing (LTA) in a microwave induced
oxygen plasma at 15o$^{\circ}$C (Mart et al., 198o).

By these digestion procedures the organic particles, organic
films on clay and silica particles and the carbonates are diges
i.e. those components of the suspended matter taking up heavy
metals from the dissolved phase.

The usual voltammetric mode applied for heavy metal determin
tions in natural waters is with respect to the sensitivity requ
ments differential pulse anodic stripping voltammetry (DPASV)
(Nürnberg, 1982). The most frequently applied working electrode
a mercury electrode. Ultratrace levels below typically 1oo ng/l
quire a thin mercury film electrode (MFE) on a glassy carbon su
strate (Mart et al., 198o), while for higher levels the hanging
mercury drop electrode (HMDE) suffices. Hg or As(III) require a
specially treated gold electrode which is also very suitable fo
Cu (Sipos et al., 1979; 198o a; Bodewig et al., 1982). Heavy me
not capable to form amalgams, as Ni and Co, are very sensitivel
determinable down to 1 ng/l at the HMDE by adsorption different
pulse voltammetry (ADPV). Here efficient preconcentration is
achieved by adsorption of a suitable organic heavy metal comple
the electrode formed in the case of Ni and Co by addition of di
methyl glyoxime (Pilhar et al., 1981; Ostapczuk et al., 1983).
course, this ADPV approach is only suitable for the discussed a
lytical determinations, as the deliberate addition of an organi
complexator changes the natural speciation pattern.

2.2 Complexation Capacity

The total concentration of dissolved ligands $\Sigma[L]$ capable to form nonlabile complexes MeL_m with heavy metals can be determined by titration of the filtrate of a water sample at natural pH with a standard solution of a suitable heavy metal. Usually Cu and Pb are applied, as they tend to form readily nonlabile complexes with DOM components. The in this manner determined complexation capacity is an important diagnostic parameter of a natural water type. As in important natural water types, particularly in sea water, the DOM levels are rather low, a highly sensitive method as DPASV is conveniently applicable.

Small portions of a rather concentrated standard solution of the titrant heavy metal Me are added to the filtrate. After a sufficient equilibration time (up to 3o min or more) the fraction of Me remaining uncomplexed is recorded via its reversible DPASV response (viz.fig.2). The uncomplexed fraction of Me may actually consist of labile complexes MeX_j formed with inorganic ligands present in the natural water, e.g. Cl^-, OH^-, CO_3^{2-}, but producing a common reversible response of Me. In this manner in the titration plot the points of the initial branch with the lower slope are obtained at first. Once the total ligand concentration $\Sigma[L]$ forming nonlabile complexes MeL_m is consumed by the titrant metal any further addition of Me will only produce a corresponding increase of the reversible response which consequently will increase more rapidly. The corresponding branch of the titration plot is therefore steeper than the initial branch before the equivalence point, which corresponds to the intersection point of both branches. As the change of the slope of the titration plot is rather gradual in the vicinity of the equivalence point its accurate determination requires a sufficient number of experimental points in the linear sections of both branches to enable a meaningful extrapolation to the accurate intersection. From the equivalence point follows the complexation capacity of the studied natural water for the as titrant applied heavy metal.

In this manner characteristic relative scales of different natural water types for the capability of binding heavy metals in nonlabile complex species can be established. It has to be emphasi-

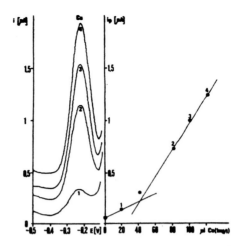

.FIGURE 2. Voltammetric determination of the complexation capacity
for Cu of southern Pacific sea water with DPASV at HMDE (56oo m
depth), pH 7.75. The resulting complexation capacity is 3.6 ug Cu

zed, however, that these complexation capacities for heavy metals
reflect usually not the actual complexation degree by nonlabile
complexes MeL$_m$ of the actual heavy metal trace concentrations in
the studied natural water but rather the ultimate complexation po
tential, if a sufficient excess of the respective heavy metal is
adjusted. In sea water and also in hard fresh waters a substantia
amount of the available complexation sites of DOM components will
be occupied by Ca and Mg being present in the water in many order
of magnitude higher concentrations than the heavy metal traces. A
though they usually form complexes with larger stability constant
than the alkaline earth metal ions, the natural trace concentrati
of the heavy metals are only able to displace Ca and Mg to a certain extent from the complexation sites.

A substantial number of complexation capacity values in variou
regions of the sea with differing biological productivity, in
a number of fresh water systems and in substrate solutions of
bioassays of plankton are reported in the recent literature
(Duinker and Kramer, 1977; Florence and Batley, 198o; Neubecker
and Allen, 1983; Nürnberg, 198o; Shuman and Woodward 1977).

3. STUDIES ON GENERAL PHYSICOCHEMICAL PARAMETERS OF SPECIATION AND
   CHARACTERIZATION OF DEFINED HEAVY METAL COMPLEXES

This field is a very important and successful application area
of voltammetry. Studies have been carried out particularly for Cd,
Pb and Zn. The results have deepened considerably the understanding
of the common physicochemical effects and properties and the specific influences of the major chemical constituents for dissolved
heavy metal speciation in various natural water types. Based on
the established findings important prognostic conclusions on the
speciation by yet to be investigated ligand types have become
possible. The high determination sensitivity of DPASV or LSASV permits to carry out such investigations at or close to the realistic
overall dissolved heavy metal concentrations in the low trace range.
The experimental approaches differ for the study of labile complexes
MeX_j and nonlabile complexes MeL_m.

3.1. Labile Complexes

Labile complexes MeX_j produce a reversible voltammetric response.
This reversible behaviour is a decisive prerequisite for the
following evaluation strategy.

A reversible peak potential E_p, or in conventional dc-polaro-
graphy the half wave potential $E_{1/2}$, will be shifted with increasing ligand concentration [X] to more negative values, because labile complexes of the corresponding consecutive series MeX_j are
formed. All this complexes undergo a reversible electrode process
and therefore the following Nernstian expression is valid for the
potential shift:

$$(2) \quad \Delta E_p \text{ or } \Delta E_{1/2} = -\frac{RT}{nF} \ln \sum_{j=0}^{j=N} \beta_j [X]^j$$

R is the gas constant, T the temperature in K, n (having usually
the value 2) is the number of electrons transferred in the ele-
mentary step of the electrode process and F the Faraday
($955oo$ C $mole^{-1}$). β_j are the stoichiometric values of the overall
stability constants for the complexes MeX_j with the ligand number
j according to eq.(3). Charges have been omitted as usual for
simplicity.

$$(3) \quad \beta_j = \frac{[MeX_j]}{[Me][X]^j}$$

These β_j-values refer to the respective medium and contain therefore the activity coefficients accounting for the general salt effects due to the ionic strength.

From the measurement of the potential shift as function of t concentration of a certain ligand [X] follow according to eq.(2 the stability constants β_j and the ligand numbers j of the corr ponding consecutive complex series MeX_j of the investigated hea metal Me in the studied water type. Precise measurements of the potential shifts are the proviso for accurate β_j-values.

At more elevated overall heavy metal levels of $10^{-6}M$ to 10^{-7} conveniently DPP can be applied to follow the shift ΔE_p. At eve higher levels in the range of $10^{-3}M$ even very precise potentio-metric measurements become possible (Simoes-Goncalves and Valen 1982; Simoes-Goncalces et al., 1983).

In aquatic trace metal chemistry it has been, however, regar as desirable to measure close the ultratrace overall levels of heavy metals actually existing in natural waters, particularly in order to be sure that at higher concentrations insignificant side effects play not an important role on the speciation by th studied ligand type. Then the highly sensitive LSASV has to be applied to record the so called pseudo-polarograms (Branica et 1977). This approach permits to work at very low overall heavy tal levels of $10^{-9}M$ which are at or at least close to the actua levels in natural waters.

The pseudo-polarogram approach consists in the pointwise cor struction of the hypothetical, for sensitivity reasons not dire measurable dc-polarogram, from LSASV or DPASV measurements. For this purpose (viz.fig.3) at each adjusted ligand concentration a series of plating potentials (e.g. ~10) is adjusted within th potential range of the electrode process of the studied complex MeX_j. Plotting the values of the resulting ASV-peak heights vs. the corresponding plating potentials the pseudo-polarogram is constructed pointwise. Repeating this procedure over at least two or more orders of magnitude of the ligand concentration [X]

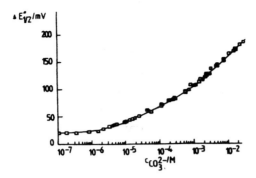

FIGURE 3. Relationship between pseudo-polarogram half-wave potential shift $\Delta E_{1/2}^*$ and logarithm of ligand concentration [X]. Investigation of labile Pb-carbonato complex formation in sea water. Dissolved Pb(II)-concentration adjusted to 6×10^{-9}M.

the resulting dependence of the half wave potential shift $\Delta E_{1/2}^*$ is established for the studied ligand X and the studied heavy metal Me in the investigated water type (viz.fig.4). All ASV-parameters, e.g. plating time, stirring rate during plating, etc. have to be kept constant. Therefore, and also with respect to convenience, automation of the measurements for the recording of pseudo-polarograms is advantageous (Sipos et al., 1980b).

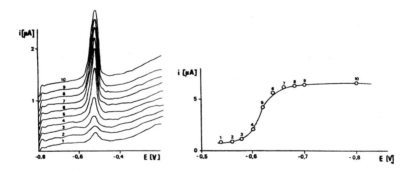

Figure 4. Construction of a pseudo-polarogram from ASV peaks (left side) recorded after application of plating potentials corresponding to the points incidated in the pseudo-polarogram (right side)

From these relationships of $\Delta E_{1/2}^*$ or ΔE_p vs. lg [X] the stability constants β_j and the ligand numbers j of the consecutive complexes MeX_j can be evaluated by the procedure of DeFord and Hume (1951), which is based on validity of eq.(2). If these complexation parameters β_j and j have been established in the studied natural water type for the prevailing heavy metal species MeX_j the distribution of the overall dissolved concentration of the respective heavy metal Me over the prevailing complex species MeX_j can be evaluated provided the DOM level is sufficiently low, that contributions by ligands L forming nonlabile species MeL_m can be neglect

This is frequently the case in large areas of the open sea where as consequence of low phytoplankton productivity the DOM levels remain very low. Consequently for substantial parts of the open sea the discussed heavy metal speciation by labile complexes MeX_j with the inorganic anionic macroconstituents Cl^-, OH^-, CO_3^{2-}, HCO_3^- and SO_4^{2-} prevails. Different conditions exist in anoxic bottom waters where S^{2-} becomes a very efficient and strong ligand. For the aforementioned open sea areas the determined inorganic speciation pattern of Pb and Cd (viz. Table 1) has general validity. It reflects significant differences in the speciation tendencies for both ecotoxic heavy metals. Both metals are in terms of fundamental coordination chemistry border cases between the hard and soft type while Co, Ni, Zn and Cu are typical hard metals and Hg is a really soft metal. Hard metals tend to com plex the inorganic anions by electrostatic forces only and form consequently with them rather labile complexes in contrast to the strong complexes formed with those inorganic anions by a typical soft metal, as Hg, due to predominant covalent bonding of the inorganic ligands X.

The precision of the evaluated species distribution depends strongly on the precision of the β_j-values. This can be improved if it is taken into account, that high salinity media, as sea water, are a rather concentrated aqueous multielectrolyte in which ion pairing between the alkali and alkaline earth cations and the anionic macroconstituents acting as inorganic ligands X for the heavy metal traces, exist (Sipos et al., 1980c)

3.2. Nonlabile Complexes with Organic Ligands

Corresponding ligands are predominatly DOM-components. In certain polluted waters also anthropogenic complexators, e.g. NTA and EDTA can contribute. A fundamental difference to the anionic macro constituents of inorganic nature is that the organic ligands are

TABLE 1. Inorganic labile species (MeX$_j$) distribution for Cd and Pb in the regions of the sea with negligible DOM levels

Species MeX$_j$	Percentage %	β_1
[CdCl]$^+$	1o	7.o
[CdCl$_2$]	87	146
other [CdX$_j$] including Cd^{2+}	3	–
[PbCO$_3$]	43	1.8×10^6
[Pb(CO$_3$)$_2$]$^{2-}$	3.7	9.1×10^9
[PbOH]$^+$	3o	7.9×10^6
[Pb(OH)$_2$]	o.5	6.3×10^{10}
[PbCl]$^+$	8.6	1o.3
[PbCl$_2$]	12.6	32
other [PbX$_j$] including Pb^{2+}	1.8	–

The data for the Cd-distribution are based on a recent more precise redetermination of β_1 (Simoes-Goncalves et al. 1981) instead of previous older data. Nevertheless, the general conclusion on inorganic Cd-speciation in the sea remain rather similar.

comparison only available at a restricted level. This holds particularly for sea water where the DOM-level hardly exceeds 1 mg/l and is often lower while the labile complexes MeX$_j$ forming inorganic macroconstituents X are present in substantially higher concentrations. In estuarine, river and lake water significantly higher DOM-levels often occur. Although the stability constants β_m of the rather strong heavy metal complexes MeL$_m$ forming DOM components are much larger as those of the labile MeX$_j$, there exists a certain competition of the inorganic ligands X for the heavy metals Me in the sea and hard fresh waters. Even more counteracting to the formation of organic complexes MeL$_m$ is, particularly in the sea, the fact, that the in comparison to the heavy metal traces 6 to 8 orders of magnitude in excess existing concentrations of 1o^{-2}M Ca(II) and 5.36 x 1o^{-2}M Mg(II) compete heavily for the restricted level of organic ligands L, because the stability constants of the alkaline earth complexes of the similar MeL$_m$-type are almost smaller than

those of the corresponding heavy metal complexes with L but by
no means negligible. This situation exists in a somewhat moderat
manner also in hard fresh water systems where the situation for
heavy metal complexation is already somewhat more favourable,
due to the smaller alkaline earth level and generally the smalle
ionic strength concomitant with a higher DOM-level. Most favourable for the formation of nonlabile MeL_m-species is the situati
in soft fresh waters with negligible ionic strength and if at th
same time elevated DOM-levels, e.g. in form of dissolved humics,
exist.

The magnitude of the specific competition effects, in the fir
place by the alkaline earth ions, for the organic ligands L and
in the second place the competition of the inorganic anionic macroconstituents and thirdly the general salt effect correspondin
to the existing ionic strength have a strong counteracting influ
ce on the complexation of heavy metal traces by organic ligands
This is particularly the case in high salinity waters, as the se
where in addition the concentration of organic ligands L is limi

A characteristic important speciation parameter of a natural
ter type are therefore the ligand concentrations [L] required to
attain a certain complexation degree, e.g. 2o % or 5o %, for the
respective heavy metal present at its typical trace level. These
ligand concentrations have to be determined, separately for the
relevant ligands or at least ligand types L, empirically for the
heavy metals of interest in the studied natural water type. This
quires that one has an idea of the nature of the predominant lig
or ligand types L to be expected in the considered natural water

Then the ligand concentration [L] required for a particular l
gand L to attain a complexation degree of p% for the investigate
heavy metal trace in the studied water type can be determined by
the following voltammetric titration procedure (Raspor et al.,
1978; 1981; Nürnberg and Raspor, 1981).

The typical dissolved overall concentration of the studied he
metal Me amounts to or will be if necessary adjusted in the 10^{-9}
range in the investigated natural water type. Suspended matter
has been filtered off before and if necessary the natural DOM
level is diminished to zero by prior UV-treatment of the water.

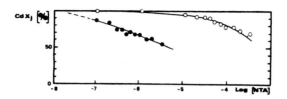

IGURE 5. Determination of ligand concentration [L] required to ob-
ain complexation degrees for Cd above 1o % with NTA. ● o.59 M NaCl
o.11 M NaClO4, pH 8.5, borate buffer, model solution of main sa-
inity components in sea water; o Adriatic sea water, pH 7.9;
verall Cd(II)-concentration 3 x 1o^{-9}M.

pplying with respect to this rather realistic ultra trace heavy
etal level DPASV at first the reversible response corresponding
o 1oo % of the dissolved heavy metal, present as labile inorganic
pecies MeX$_j$, is determined before increasing amounts of the
espective studied organic ligand L are added. With increasing
evels of L increasing amounts of Me are complexed as nonlabile
pecies MeL$_m$ and the reversible response, due to Σ[MeX$_j$] decreases
orrespondingly. This decrease is an indicator for the complexation
egree of Me as MeL$_m$ according to:

4)   1oo % Σ[MeX$_j$] - y % Σ[MeX$_j$] = p % [MeL$_m$]

An example for those titration plots obtained for a given heavy
etal with a defined organic ligand L in given aquatic media shows
ig.5. If this procedure is appropriately applied, the theoretically
xisting slight perturbation of the complex equilibrium for MeL$_m$
emains negligible in practice. An indisputable test is, that from
ach point of those titration curves the relevant values of the
tability constants of MeL$_m$ can be evaluated, provided the parameters
f all significant complex equilibria of MeX$_j$-species involved are
nown and can be taken into account (Nürnberg and Raspor, 1981). The
xample in fig.5 emphasizes the signficance of the above mentioned
ompetition effects by the alkaline earth ions, if one compares the
igand concentrations [L] required to decrease Σ[CdX$_j$] by y % in
.59 M NaCl, adjusted by addition of o.11 M NaClO$_4$ to sea water
onic strength but containing no Ca(II) and Mg(II), with the analogous relationship obtained in genuine Adriatic sea water.

Extended previous investigations of the complexation of Cd, Pb
and Zn by the defined anthropogenic ligands NTA and EDTA in sea w
ter and lake water with this procedure have clarified the aforeme
tioned general aspects of the speciation of heavy metals by ligan
L forming nonlabile complexes MeL_m in natural waters. To obtain s
porting evidence also relevant model solutions of important compo
have been studied, e.g. $10^{-2}M$ Ca(II) in a with NaCl + $NaClO_4$ + bo
or carbonate buffer to sea water adapted medium (Nürnberg and Ras
1981; Raspor et al., 1978; 1981).

From these studies with NTA and EDTA prognostic conclusions on
the significance of DOM-components for the speciation of those
heavy metals in sea water and lake water arised. Thus, it was
predicted, based on known stoichiometric stability constants
(Mantoura et al. 1978) and concentration levels of complexing
DOM-components, that dissolved humics and amino acids hardly
have any significance for the speciation of Cd, Pb and Zn in sea
water or hard lake water. The reason is, that with respect to the
moderate stability constants of the corresponding MeL_m-species
and the magnitude of the counteracting competition effects, particularly by Ca(II) and Mg(II), the available ligand concentrations [L] remain too low. The dissolved humic and fulvic acid
concentration exceeds e.g. in sea water rarely o.1 mg/l. Recently
these predictions have been verified by direct investigations of
dissolved humics (Raspor et al. 1984) and amino acids (Valenta
et al. 1984) in sea water and fresh water and the levels of [L]
required for complexation degrees above 1o %, which exceed substantially the naturally available levels of both ligand types,
have been determined. It should, however, be emphasized, that
these findings refer to the dissolved levels of these natural
ligands L in the studied types of natural waters. Actually there
are also frequently present fine suspended clay particles coated
with a film of adsorbed humic substances. Due to the high interfacial concentration of humics, such particles can act as efficient
heavy metal scavengers. This is a topic needing more detailed
studies.

3.3. Kinetics and mechanism of nonlabile complex formation
with organic ligands

Significant evidence of general validity has been also obtained on the formation kinetics and the operating mechanism from studies with Cd, Pb and Zn and EDTA in sea water and lake water. For this purpose equal concentrations of the heavy metal Me and the ligand L have been adjusted to establish 2nd order kinetics. Obviously a very strong ligand as EDTA had to be applied with respect to the ultra trace levels to be used to establish realistic conditions. After addition of the reactants to the studied natural water the time function of the decrease of the reversible voltammetric response corresponding to $\Sigma[MeX_j]$ is followed until the complex equilibrium of MeL_m is attained. Fig.6 shows an example for Pb(II). From those time functions the conditional formation rate constant k_f, referring to the respective medium, can be evaluated (Raspor et al., 1977; 1980; 1981).

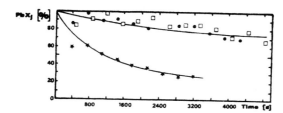

FIGURE 6. Time function of unchelated $\Sigma[PbX_j]$ percentage in study of formation kinetics of PbEDTA. Initial concentration of Pb(II) and Na_3HEDTA (1:1) $2 \times 10^{-8}M$. Media: * $5.37 \times 10^{-2}M$ $MgCl_2$; o $1 \times 10^{-2}M$ $CaCl_2$, model solutions of alkaline earth ion concentrations in sea water at pH 8; □ sea water.

For the complexation of Cd, Pb and Zn with the strong ligand EDTA the k_f-values fall in sea water and lake water into the range from 3×10^2 to 5×10^3 1 $mole^{-1}s^{-1}$ being for Cd and Zn about one order of magnitude lower than for the a higher complexing tendency having Pb. The determined k_f-values are at any rate by many orders of magnitude smaller than rate constants of 10^9 to 10^{10} 1 $mole^{-1}s^{-1}$ to be expected, if the complex formation mechanism of MeL_m species would be in the studied natural waters the simple recombination mechanism according to eq.(5), where the rate determining step would be the diffusional approach of the reactants to the critical distance (Eigen 1963).

(5) $\quad MeX_j + L \xrightarrow{\;k_f\;} MeL + jX$

As the experimental findings reveal (viz. e.g. fig.6) the actually in natural waters, containing non-negligible amounts of
Ca(II), operative formation mechanism of nonlabile heavy metal complexes MeL_m with organic ligands L is a ligand exchange mechanism.
In this mechanism usually, due to its excess, the complexation
sites occupying Ca(II) and Mg(II) are displaced by the heavy
metal Me, because the formed heavy metal MeL_m-species has a higher
stability than the corresponding species where Me is Ca(II) or
Mg(II). In such a ligand exchange mechanism

(6) $\quad \begin{matrix} CaL \\ \updownarrow \\ MgL \end{matrix} + MeX_j \xrightarrow{\;k_f\;} MeL + Ca(II), Mg(II) + jX$

the ligand exchange and not the diffusional reactant approach is
the rate determining step with the much smaller formation rate constants k_f observed. Only in media with negligible alkaline earth
concentrations the above by eq.(5) represented simple recombination
mechanism is responsible for the formation of nonlabile heavy metal species MeL_m.

4. CONCLUDING REMARKS

The heavy metal speciation is an important and experimentally
rather demanding field of aquatic trace metal chemistry. Many
problems are still to be solved. The prudent application of advanced
voltammetric methods has significantly contributed to clarify already a number of essential general aspects and has provided a certain degree of understanding on the physicochemical effects governing heavy metal speciation. This applies at present quantitatively
mainly to the dissolved phase. Certainly voltammetry will remain
also in the future a very promising approach for speciation research
in natural waters.

REFERENCES

- Bodewig FG, Valenta P and Nürnberg HW. 1982. Trace determination of As(III) and As(V) in natural waters by differential pulse anodic stripping voltammetry. Fresenius Z.Anal.Chem. 311: 187-191
- Branica M, Novak DH and Bubic S. 1977. Application of anodic stripping voltammetry to determination of the state of complexation of traces of metal ions at low concentration levels. Croat. Chem. Acta 49: 231-251
- Breder R, Flucht R and Nürnberg HW. 1982. A comparative study on the toxic trace metal situation in Tyrrhenian estuaries. Thalassia Jugosl. 18: 135-171
- DeFord DD and Hume DN. 1951 . The determination of consecutive formation constants of complex ions from polarographic data. J. Am. Chem. Soc. 73: 5321-5321
- Duinker JC and Kramer CJM. 1977. An experimental study on the speciation of dissolved zinc, cadmium, lead and copper in river Rhine and North Sea Water by differential pulse anodic stripping voltammetry. Mar. Chem. 5: 2o7-228
- Eigen M. 1963. Ionen-Ladungsübertragungsreaktionen in Lösung. Ber. Bunsenges. Phys. Chem. 67: 753-762
Florence TM and Batley GE. 198o. Chemical speciation in natural waters. A review. CRC Crit. Revs. Anal. Chem. 9: 219-296
- Mantoura RFM, Dickson A and Riley JP. 1978. The complexation of metals with humic materials in natural waters. Estuar. Coast. Mar. Sci. 6: 387-4o8
- Mart L, Nürnberg HW and Valenta P. 198o. Voltammetric ultratrace analysis with a multicell system for clean bench working. Fresenius Z. Anal. Chem. 3oo: 35o-362
- Mart L; Rützel H; Klahre P, Sipos L, Platzek U, Valenta P and Nürnberg HW. 1982. Comparative studies on the distribution of heavy metals in the oceans and coastal waters. Sci. Tot. Environm. 26: 1-17
- Neubecker A and Allen HE. 1983. The measurement of complexation capacity and conditional stability constants for ligands in natural waters. Water Res. 17: 1-14
- Nürnberg HW. 198o. Features of voltammetric investigations on trace metal speciation in sea water and inland waters. Thalassia Jugosl. 16: 95-11o
- Nürnberg HW. 1982. Voltammetric trace analysis in ecological chemistry of toxic metals. Pure Appl. Chem. 54: 853-878
- Nürnberg HW. 1983 a. Investigations on heavy metal speciation in natural waters by voltammetric procedures. Fresenius Z. Anal. Chem. 316: 557-565
- Nürnberg HW. 1983 b. Voltammetric studies on trace metal speciation in natural waters, Part II. Application and conclusions for chemical oceanography and chemical limnology In: GG Leppard (Ed.) Trace element speciation in surface waters and its ecological implications. Plenum Publ. Corp. New York - London, pp. 211-23o
- Nürnberg HW. 1984. Trace analytical procedures with modern voltammetric determination methods for the investigation and monitoring of ecotoxic heavy metals in natural waters and atmospheric precipitates. Sci. Tot. Environm. 35: to appear

- Nürnberg HW and Raspor B. 1981. Applications of voltammetry in studies of the speciation of heavy metals by organic chelators in the sea. Environm. Technol. Letters 2: 457-484
- Nürnberg HW and Valenta P. 1983. Potentialities and applicatic of voltammetry in chemical speciation of trace metals in the sea. In: CS Wong, E Boyle, KW Bruland, D Burton and ED Goldber (Eds.), Trace metals in sea water. Plenum Press, New York - Lc don, pp. 671-697
- Ostapczuk P, Valenta P, Stoeppler M and Nürnberg HW. 1983. Vol tammetric determination of nickel and cobalt in body fluids ar other biological materials. In: SS Brown and J Savory (Eds.) Chemical toxicology and clinical chemistry of metals, Academic Press London-New York, pp. 61-64
- Pilhar B, Valenta P and Nürnberg HW. 1981. New high-performanc analytical procedure for the voltammetric determination of nickel in routine analysis of waters, biological materials and food. Fresenius Z. Anal. Chem. 3o7: 337-346
- Raspor B, Valenta P, Nürnberg HW and Branica M. 1977. Polaro graphic studies on the kinetics and mechanism of Cd(II)-chelat formation with EDTA in sea water. Thalassia Jugosl. 13: 79-91
- Raspor B, Valenta P, Nürnberg HW and Branica M. 1978. The chel tion of cadmium with NTA in sea water as a model for the typi cal behaviour of trace heavy metal chelates in natural waters. Sci. Tot. Environm. 9: 87-1o9
- Raspor B, Nürnberg HW, Valenta P and Branica M. 198o. Kinetics and mechanism of trace metal chelation in sea water. J. Electr anal. Chem. 115: 293-3o8
- Raspor B, Nürnberg HW, Valenta P and Branica M. 1981. Voltamme tric studies on the stability of Zn(II) chelates with NTA and EDTA and the kinetics of their formation in Lake Ontario water Limnol. Oceanogr. 26: 54-66
- Raspor B, Nürnberg, HW, Valenta P and Branica M. 1984. Signifi cance of dissolved humic substances for heavy metal speciation natural waters. this book, pp.317-328
- Sigg L, Sturm M, Stumm W, Mart L and Nürnberg HW. 1982. Schwer metalle im Bodensee - Mechanismen der Konzentrationsregulierunc Naturwissenschaften 69: 546-547
- Simoes-Goncalves, MLS and Valenta P. 1982. Voltammetric and po tentiometric investigations on the complexation of Zn(II) by glycine in sea water. J.Electroanal. Chem. 132: 357-375
- Simoes-Goncalves MLS, Valenta P and Nürnberg HW. 1983. Voltamme tric and potentiometric investigations on the complexation of Cd(II) by glycine in sea water. J. Electroanal. Chem. 149: 249-262
- Simoes-Gonvalces MLS, Vaz MCTA and Frausto da Silva JJR. 1981. Stability constants of chloro-complexes of Cd(II) in sea water medium. Talanta 28: 237-24o
- Sipos L, Golimowski J, Valenta P and Nürnberg HW. 1979. New vol tammetric procedure for the simultaneous determination of coppe and mercury in environmental samples. Fresenius Z.Anal.Chem. 298: 1-8
- Sipos L, Nürnberg HW, Valenta P and Branica M. 198o. The reliab determination of mercury traces in sea water by subtractive differential pulse voltammetry at the twin gold electrode. Anal Chim. Acta 115: 25-42

- Sipos L, Raspor B, Nürnberg HW and Pytkowicz RM. 1980. Inter-
action of metal complexes with coulombic ion pairs in aqueous
media of high salinity. Mar. Chem. 9: 37-47
- Sipos L, Valenta P, Nürnberg HW and Branica M. 1980. Voltammetric determination of the stability constants of the predomi-
nant labile lead complexes in sea water. In: M Branica and
Z Konrad (Eds.) Lead in the marine environment. Pergamon Press,
Oxford, pp. 61-76
- Stumm W and Morgan JJ. 1981. Aquatic Chemistry, 2nd Ed.,
J. Wiley, New York, 780 pp.
- Shuman MS and Woodward GP. 1977. Stability constants of copper
organic chelates in aquatic systems. Environm. Sci. Technol.
11: 809-813
- Valenta P. 1983. Voltammetric studies on trace metal speciation in natural waters. Part I. Methods. In: GG Leppard, Ed.,
Trace element speciation in surface waters and its ecological
implications. Plenum Publ. Corp. New York-London, pp. 49-69
- Valenta P, Simoes-Goncalves MLS and Sugawara M. 1984. Voltammetric studies on the speciation of Cd and Zn by amino acids in
sea water. this book, pp. 357-366

DIRECT DETERMINATION OF METAL COMPLEXATION

MARKO BRANICA and GINA BRANICA

ABSTRACT

   Interactions between trace metals and organic ligands in the natural
aquatic system as well as in the polluted waters influence the metal species
distribution the result of which are various geochemical pathways. Electrochemical methods are very valuable for trace metal speciation. However,
their application must be performed cautiously because of very specific
sensitivity of the electrode reaction to the different groups of heavy
metal species. At the same time possible redistribution of species at the
electrode surface with respect to the bulk of the solution must be taken
into account.

   A new approach to determination of trace metal complexation in natural
waters, based on the voltammetric analysis of the inert metal complex, is
proposed. From the dependence of the inert metal complex wave height on
the added quantity of metal-ion titrant, one can evaluate the complexing
capacity and apparent stability constant of the formed metal complex.
Because the measurement directly correspond to the formed inert metal
complex many drawbacks of indirect methods, based on diminished signal of
the 'free metal' wave, are avoided.

   The additional physico-chemical characterization of the formed metal
complex, from the 'pseudo-polarographic' analysis of the corresponding
separated wave can be obtained, also. The use of different metals as
titrant of natural water samples will be discussed. The applicability of
the new proposed method on the voltammetric titration with cadmium
of the ligands mixture of NTA and EDTA in sea water will be demonstrated.

Kramer, C.J.M. and Duinker, J.C. (eds.), Complexation of Trace Metals in Natural Waters. ISBN 90-247-2973-4
©1984 Martinus Nijhoff/Dr W. Junk Publishers, The Hague/Boston/Lancaster.
Printed in the Netherlands.

ANODIC STRIPPING VOLTAMMETRY OF COPPER IN ESTUARINE MEDIA

L.A. NELSON and R.F.C. MANTOURA

1. INTRODUCTION

Anodic stripping voltammetry (ASV) has been applied often in the past decade for the measurement of trace metals and their speciation in natural waters. However, there remain ambiquities in the interpretation of the results due to two fundamental uncertainties associated with the ASV assay. In the first instance, the species measured in an ASV analysis has so far been only operationally defined. There is no surety on the precise metal species being reduced at the mercury electrode. Secondly, voltammetric measurements are sensitive to natural organic material adsorbed on the working electrode but the extent of this interference in natural waters has not as yet been comprehensively examined. Accordingly, in this paper we address ourselves to the first problem and how it relates to copper speciation measurements in estuarine waters using differential pulse ASV (DPASV).

2. ELECTROCHEMISTRY OF COPPER IN CHLORIDE SOLUTIONS

Prior to embarking on a voltammetric study of copper speciation in estuarine waters one should consider the electrochemistry of copper in chloride media.

In the DC polarography of $Cu(II)$ in chloride solutions (>0.02 M, $[Cl^-]$) $Cu(II)$ is reduced in two one electron reversible waves via a $CuCl_2^-$ intermediate (v. Stackelberg and v. Freyhold, 1940). The potential separation of the two waves is directly related to the stability of the $CuCl_2^-$ species which can be evaluated from the equilibrium constant for the reaction:

$$Cu(II) + Cu(O) + 4 Cl^- = 2 CuCl_2^-$$

Kramer, C.J.M. and Duinker, J.C. (eds.), Complexation of Trace Metals in Natural Waters. ISBN 90-247-2973-4
© *1984 Martinus Nijhoff/Dr W. Junk Publishers, The Hague/Boston/Lancaster.*
Printed in the Netherlands.

$$K = \frac{[CuCl_2{}^-]^2}{[Cu(II)]\ [Cu(0)]\ [Cl^-]^4}$$

$$= (\beta_{CuCl_2}{}^-)^2 \cdot \frac{[Cu(I)]}{[Cu(II)]\ [Cu(0)]} = (10^{5.78})^2\ 10^{-4.24}$$

$$= 10^{7.32}$$

where $\beta_{CuCl_2}{}^-$ is the cumulative stability constant for the species, $CuCl_2{}^-$. It is readily seen that the stability of $CuCl_2{}^-$ is a function of chloride concentration and can be expressed:

$$K[Cl^-]^4 = \frac{[CuCl_2]^2}{[Cu(II)]} = x_0. \tag{1}$$

Where x_0 is the conditional Cu(I) stability constant for Cu equilibria in chloride media. The dependence of the half-wave potential ($E_{\frac{1}{2}}$) of the two reduction steps of Cu(II) on chloride concentration has been previously illustrated very clearly and reproduced in Fig. I.

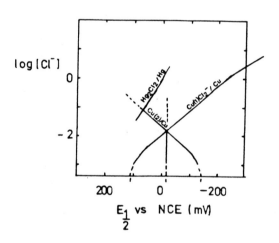

FIGURE 1. The polarographic half-wave potentials of copper in chloride solution (v. Stackelberg and v. Freyhold, 1940).

It is to be noted that the calomel reduction wave masks the initial reduction step of Cu(II) in chloride solutions above 0.1 M, [Cl$^-$].

ASV differs from the DC polarographic case in that only the electron step immediately prior to Cu amalgamation will be detected and this will determine the electrochemical response.

The question now to be addressed, therefore, is how the final electron step is related to the speciation of Cu(II) in chloride solution with added Cu(II) complexing organic ligand. In the first instance the effect of this will be to reduce the value of the conditional CuCl$_2$$^-$ stability constant. Thus equation (1) can be modified:

$$x_1 = \frac{x_o}{\beta_{Cu:L} \cdot [L]} \tag{2}$$

Where β is the conditional stability constant for a 1:1 organic ligand - Cu(II) complex, [L] is the concentration of ligand and x_1 is the conditional stability constant for Cu(I) in organic ligand-chloride media. Consider the situation where the Cu(II) complex equilibrium is mobile such that the complex is reduced at a mercury electrode by a reversible two electron process. In DC polarography, at low values of x_1 the electrodeposition of Cu will be determined by the reduction of the Cu(II) complex; however, at higher values of x_1 the reduction of CuCl$_2$$^-$ will control the Cu electrodeposition. The point at which the final reduction step assumes chloride control is given by:

$$E_{\frac{1}{2}}{}_{CuCl_2^- \rightarrow Cu(Hg)} = E_{\frac{1}{2}}{}_{Cu:L \rightarrow Cu(Hg)}$$

or when $x_1 = 1$ (Nelson and Mantoura, 1983a).

For ASV the situation is different since the mean $E_{\frac{1}{2}}$ for a reversible reduction is more cathodic than the DC polarographic $E_{\frac{1}{2}}$ due to the accumulation of reduced depolariser in the amalgam (Zirino and Kounaves, 1977). Also the cathodic shift for a one electron wave is twice that for a two electron wave. The effect of this is to stabilise CuCl$_2$$^-$ relative to Cu:L in the diffusion layer. As a result in organic ligand-chloride media, the electrodeposition of Cu assumes chloride control at $x_1 = 10^{-2.2}$

for a typical electrolysis time of 5 mins and HMDE drop radius
0.0265 cm (Nelson and Mantoura, 1983a).

3. Cu(II) REDUCTION IN GLYCINE-CHLORIDE MEDIA

In order to substantiate the theoretical considerations in t⟩
above section, an experiment was performed in synthetic organic
ligand chloride media described below.

Solutions were prepared with varying concentrations of [Cl⁻]
KCl. KNO₃ was used as a base electrolyte to adjust the media t⟨
constant ionic strength. Glycine [Gl] was added as organic
ligand. Glycine forms a 1:2 complex with Cu(II), $CuGl_2$, which ⟩
been well characterised by polarography (Li and Doody, 1952).
$CuGl_2$ is reduced reversibly at a dropping mercury electrode by ⟨
two electron process. The $E_{\frac{1}{2}}$ of the polarographic wave varies
a classical manner with the concentration of free ligand in
accordance with the v. Stackelberg-Lingane equation (Crow, 1969
It is particularly advantageous that the concentration of free
ligand in solution can be varied by altering the pH. Thus:

$$\text{Log } [Gl^-] = pH - pK + \log [Gl] \tag{3}$$

where [Gl⁻] is the concentration of free ligand
and [Gl] is the concentration of undissociated glycine. The pK
glycine is 9.69.

A small quantity of $NaHCO_3$ was added to establish a carbona⟩
alkalinity thus the pH of the solution could be varied by alter⟩
the partial pressure of CO_2 in equilibrium with the medium. Cu
voltammetry was studied in two solutions 0.19M [Cl⁻] and 0.34M
[Cl⁻] respectively (ionic strength 0.34) with 0.01M glycine and
10 µg/L Cu. Solutions at 6 pH values were studied. Stripping
polarography (pseudopolarography) was used to investigate the
electrochemistry of Cu deposition during DPASV. The apparatus
used and experimental procedure have been described elsewhere
(Nelson and Mantoura, 1983a, b). The stripping polarographic l⟨
plots are displayed in Fig. 2.

It is observed that at pH 8.90 and 8.25 a reversible two
electron reduction characteristic of $CuGl_2$ is observed. However
at lower pH values, the electrodeposition is controlled by a one

electron reversible $CuCl_2^-$ reduction.

Indeed, the x_1 value where the electrodeposition changes from glycine to chloride control lies between $10^{-2.8}$ and $10^{-2.0}$. This estimate is in good agreement with the theoretical prediction of $x_1 = 10^{-2.2}$ (see previous section). Interestingly, when the Cu(II)-glycinate dissociation is kinetically hindered, a higher value of x_1 (10^{-1} - 10^{-2}) is required before reversible reduction is observed.

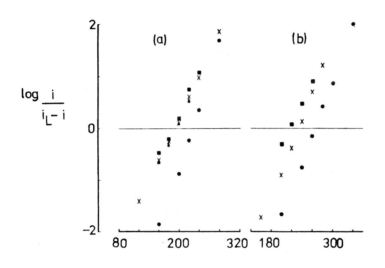

E applied $(-mV)$

FIGURE 2.   Stripping polarography of copper in glycine-chloride media.   Cu, 10μg/L 0.01M glycine.   Log. analysis plots.

    (a)   0.19M [Cl$^-$] - log x_1 (b)   0.34M [Cl$^-$] - log x_1

       ● pH 8.25   3.8   ● pH 8.92   4.1

       X pH 7.31   2.0   X pH 8.26   2.8

       ■ pH 6.62   0.6   ■ pH 7.36   1.0

       ▲ pH 3.85 (+ 25 μL, HCl)

4. IRREVERSIBLE REDUCTION OF Cu(II) SPECIES IN CHLORIDE MEDIA

How is a Cu(II) complex irreversible reduction affected by chloride? This is shown schematically in Fig. 3. In cases 2 and 3 the reduced copper establishes equilibrium with $CuCl_2^-$ in the diffusion layer and the electrodeposition comes under chloride control. This is a mixed process where one depolariser is generated from another:-

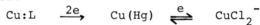

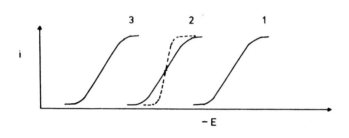

FIGURE 3. The influence of the $CuCl_2^-/Cu(Hg)$ redox process on copper electrodeposition following the irreversible reduction of Cu(II) species.

Also, addition of chloride will affect an irreversible Cu(II) complex reduction by increasing the probability of the mechanism:

$$Cu:L \xrightarrow{e} CuCl_2^- \xrightarrow{e} Cu(Hg)$$

due to the enhanced stability of $CuCl_2^-$.

In order to substantiate the above predictions regarding irreversible systems, Cu(II) electrodeposition during DPASV was studied by stripping polarography. An estuarine water sample, salinity 0.7‰ was chosen as a medium for study with 15 µg/L added Cu equilibriated for 12 hours, and various additions of chloride as KCl. Ionic strength was maintained constant at 0.55 with KNO_3

In Fig. 4 it may be seen that the stripping polarographic wave
for the aliquot with no added chloride is irreversible and
undefined showing direct reduction of Cu-organic complex (see
later section). With added chloride the electrodeposition assumes
chloride control and at 0.33M [Cl⁻] and above, the stripping
polarographic slope approaches 60 mV at cathodic potentials.

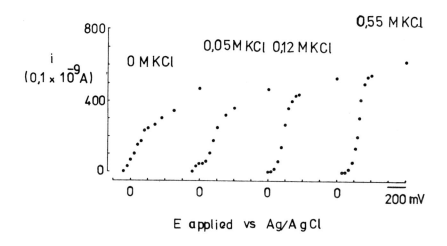

FIGURE 4. Stripping polarographic assays of estuarine waters
(Salinity, 0.7°/oo; pH, 7.60). Cu, 15 µg/L. Ionic strength 0.55,
adjusted with KNO_3. Additions of [Cl⁻] as KCl. 1 mg/L gelatin.

5. COPPER SPECIES REDUCTION AT THE HMDE IN ESTUARINE WATER

Stripping polarograms for ambient Cu reduction in estuarine
waters at different pH values are shown in Fig. 5. At natural pH,
reduction of Cu(II) takes place in an irreversible manner.
Decreasing pH, protonates free ligand thus increasing the
stability of the $CuCl_2^-$ state and its importance in the electrodeposition process. Accordingly, the presence of a limiting
current becomes more apparent with decrease in pH. A wave is
defined which shows a trend to reversible character and anodic
shift in $E_{\frac{1}{2}}$. At pH ∿2.5, reversible reduction is occurring
indicated by a log plot slope of 60 mV per decade showing a one

126

electron transfer. This same stripping polarographic response is
seen in a sample at natural pH with sufficient added Cu to remove
organic ligand competition to $CuCl_2^-$ electroreduction. At low pH,
liberation of Cu(II) from complexes is indicated by the marked
increase in wave height.

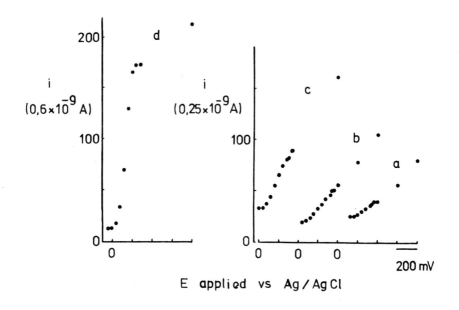

FIGURE 5.   Stripping polarographic assays of estuarine water
salinity, 5°/oo; DOC, 4.60 mg/L ambient Cu.   1 mg/L gelatin.

pH   a.   7.65
     b.   7.14
     c.   6.20
     d.   2.41 (+ 25 µL HCl).

For a series of acidified estuarine waters of varying salinity
$E_{\frac{1}{2}}$ for the electroreduction shows a linear relationship with log
$[Cl^-]$.   The same is observed for estuarine media at natural pH
where the organic ligand competition to Cu electrodeposition has
been removed by addition of Cu (Fig. 6).   This plot has a slope c
120 mV per decade indicating the reduction of a Cu(I) complex of
co-ordination number 2.   The coincidence of $E_{\frac{1}{2}}$ values of a

chloride dosed sample on this plot confirms the Cu(I) species to be CuCl$_2^-$. Moreover, the plot is identical to that of v. Stackelberg and v. Freyhold (1940) when corrections are made for the cathodic shift of the stripping polarographic wave.

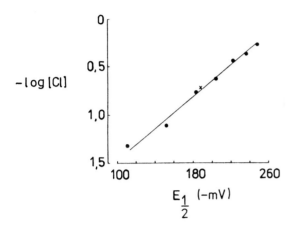

FIGURE 6.   Estuarine waters + added Cu (sufficient to remove organic ligand competition to Cu electrodeposition) E$_\frac{1}{2}$ versus - log [Cl$^-$] for Cu reduction.   1 mg/L gelatin.

●   Estuarine waters.

X   Estuarine water, salinity, 3.1^o/oo dosed to 0.19M, [Cl$^-$] with KCl.

6.   VOLTAMMETRY OF Cu IN UV IRRADIATED WATER

Stripping polarographic assays of UV irradiated water (Fig. 7) shows that the form of the copper reduction has changed markedly from the natural sample (e.g. curve a in Fig. 5). Noticeably the wave height has increased and the current reaches a constant limiting value. The conclusion from this therefore is that the ligands affecting the electrodeposition of copper from estuarine waters are of an organic nature and these are altered or removed on UV-irradiation.

Nonetheless, reduction of Cu in UV irradiated water at natural pH is irreversible. We attribute this to residual organic material and by-products of organic oxidation following UV irradiation influencing the electrodeposition of copper (Nelson and Mantoura, 1983b).

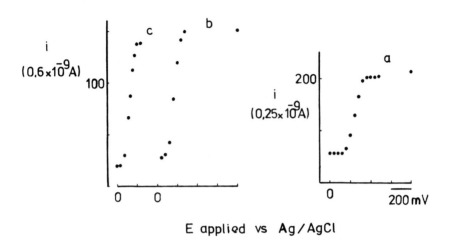

E applied vs Ag/AgCl

FIGURE 7. Stripping polarographic assays of UV irradiated estuarine waters (salinity, 5°/oo) 1 mg/L gelatin.

> pH a. 7.65
> b. 6.25
> c. 2.50 (+ 25 µL HCl)

7. VOLTAMMETRIC TITRATIONS

Voltammetric titrations have been based on the premise that where metal is present with an excess of ligand only free metal equilibrium with the complex reacts at the electrode. Thus on sequential addition of metal to the sample, there comes a point where the ligand complexing sites are filled and with further titrant addition the full voltammetric sensitivity for the metal is realised. However, as we have found under the conditions used in this study, Cu complexes with organic material show a wide range of reaction at the Hg electrode. Indeed, during successive addition of Cu to an estuarine water sample, the free ligand concentration is reduced and the importance of the $CuCl_2^-$ state in the electrodeposition increases until reversible $CuCl_2^-$ reduction is observed. This will occur before all ligand sites are filled. These predictions are confirmed in Fig. 8 which shows an example of such a titration together with associated stripping polarograms.

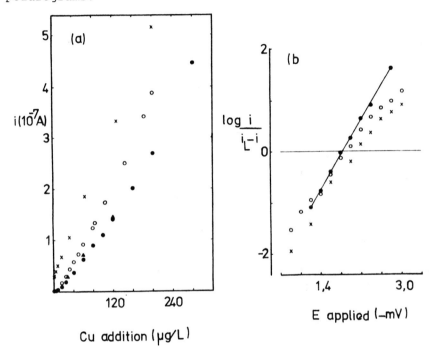

FIGURE 8. Voltammetric titration. Estuarine water, (salinity, 11°/oo; DOC, 4.79 mg/L). 1 mg/L gelatin.

(a) Titration.

X   pH 2.80 (+ 25 μL HCl).

O   pH 7.58 sequential titrant addition.

●   pH 7.58 12 hours equilibriation of each
         titrant addition.

(b) Stripping polarographic assays of titrant
     additions. Log analysis plots. pH 7.58.

X   8 μg/L. Cu.

O   40 μg/L. Cu.

●   140 μg/L. Cu.

8. CONCLUSIONS

The mechanism of Cu species reduction in estuarine waters can be summarised as follows.

At ambient Cu levels and natural pH, the copper (II)-organic species are reduced by an adsorption mechanism (Nelson and Mantoura, 1983b). By reducing pH or adding Cu, the concentration of free ligand is decreased and the importance of $CuCl_2^-$ in the electrodeposition increases both before and after amalgamation. At natural pH in the absence of organic ligand competition, the predominant forms of Cu are as carbonate species (Turner et al., 1981) and these are reduced via $CuCl_2^-$ (Nelson and Mantoura, 198

ACKNOWLEDGEMENTS

This work forms part of the Estuarine Ecology Programme of th Institute for Marine Environmental Research, a component of the Natural Environment Research Council, and was partly supported b the Department of the Environment under Contract No. DGR 480/605

REFERENCES

Crow, D.R., 1969. Polarography of Metal Complexes. Academic Press, London, 203 pp.

Li, N.C. and E. Doody, 1952. Metal-amino acid complexes. II. Polarographic and potentiometric studies on complex formation between copper (II) and amino acid ion. Journal of the American Chemical Society, 74, 4184-4188.

Nelson, A. and R.F.C. Mantoura. 1983a. Voltammetry of copper species in estuarine waters. Part I. Electrochemistry of coppe species in chloride media. Journal of Electroanalytical and Interfacial Chemistry, submitted.

Nelson, A. and R.F.C. Mantoura, 1983b. Voltammetry of copper species in estuarine waters. Part II. Copper species reduction at the HMDE in estuarine waters. Journal of Electroanalytical and Interfacial Chemistry, submitted.

Stackelberg, M.v. and H.V. Freyhold, 1940. Polarographische Untersuchungen an Komplexen in wäszriger Lösung. Zeitschrift fü Elektrochemie, 46, 3: 120-129.

Turner, D.R., M. Whitfield and A.G. Dickson, 1981. The equilibrium speciation of dissolved components in freshwater and seawater at $25^{\circ}C$ and 1 atm pressure. Geochimica et Cosmochimica Acta. 45: 855-881.

Zirino, A. and S.P. Kounaves, 1977. Anodic stripping peak currents: Electrolysis potential relationships for reversible systems. Analytical Chemistry, 49, 1: 56-59.

KINETICS OF COMPLEXATION AND DETERMINATION OF COMPLEXATION
PARAMETERS IN NATURAL WATERS

I. RŮŽIĆ

1. INTRODUCTION

Trace metals play an important role in the environment. Their behaviour in the environment depends, however, on their speciation. Trace
metals are present in natural waters as ions or labile (mostly inorganic)
complexes, nonlabile (mostly organic) complexes or chelated forms and
adsorbed or incorporated in colloidal particles. They can be characterized by different degrees of hydration and oxydation states. In the natural environment all these different forms of trace metals can be in equilibrium or at least in a steady state (Florence and Batley, 1980; Stumm
and Morgan, 1981).

With convenient physicochemical or biological techniques it is possible to determine individual species and parameters describing their
stability and lability. Sometimes an even minimum treatment of natural
water samples can shift the state of a trace metal in a desired direction
and a comparative study of treated and untreated samples can be undertaken. Such a procedure can be of a great help in attempting to characterize the species of trace metals in natural waters. From the point of
view of the analytical technique used the trace metal content can be divided into components which are detectable with or without the pretreat-
ment of natural water samples and which are not detectable under such
conditions. Such different components can sometimes be identified as some
of the known physicochemical forms of trace metals.

A series of different investigations has been reported in the literature including the use of the theoretical models (Jenne, 1979), laboratory experiments with artificial (model) ligands and colloidal particles
and with ligands and particles of natural origin (Singer, 1973; Mantoura
et al., 1978; Duinker, 1980). Because natural water systems contain a
very large number of interacting components it is very difficult to pre-

Kramer, C.J.M. and Duinker, J.C. (eds.), Complexation of Trace Metals in Natural Waters. ISBN 90-247-2973-4
© 1984 Martinus Nijhoff/Dr W. Junk Publishers, The Hague/Boston/Lancaster.
Printed in the Netherlands.

dict the actual chemical state of trace metals in such an environment on the basis of the available thermodynamical data.

The interaction of trace metals with ligands and colloidal particles has been often studied by the direct titration of natural water samples (treated or untreated). On the basis of such experiments with natural water samples concepts of trace metal complexing capacity of natural water and corresponding conditional (integral or differential) equilibrium (staiblity or formation) constants have been developed (Hart, 1981; Neubecker and Allen, 1983).

It is reasonable to suppose, however, that at trace levels of concentration equilibrium may exist or alternatively, reactions may be kinetically controlled. Therefore an adequate speed of reaction will not be always attained. While in the common cases of the titration analysis (Ashworth, 1965) there is a strong need to use reactants which can ensure a complete reaction (without the observable influence of the kinetics) and the formation of very stable products (the shift of equilibrium mainly in one direction); titration of natural waters brings before us sometimes completely different priorities. Kinetics of complexation and lower stabilities of complexes formed in natural waters represent a useful source of information in some cases and the corresponding equilibrium and kinetic parameters can be determined simultaneously.

The speed of a complexation reaction depends on the electronic structure of the trace metal and the properties of the ligand. In addition, the speed of the reaction will also depend on the mechanism of complexation and dissociation. Two possible mechanisms can be taken into account and these are dissociation and displacement mechanisms. The latter is usually a case of a significantly slower reaction. The formation of polynuclear or mixed complexes is usually also very slow (Ringbom, 1963).

When a very fast complexation is followed by the formation of very stable complexes the interpretation of direct titration curves is relatively simple. In such a case, however, only the complexation capacity can be determined. Information about the stability or the kinetics of formation of such complexes will not be available from such a simple experiment. When the formation of weaker complexes occurs, the information on their conditional stability can be obtained, but the determination of complexing capacity will not be simple and special methods for the interpretation of titration curves should be used. In cases where complexation kinetics play a role, the interpretation of experimental results will be even more complex.

The purpose of this paper is to present a critical review of the methods used in the interpretation of the direct titration of natural waters with trace metals as well as to indicate some of the problems encountered and some solutions for these problems. The cases, in which the effects of complexation kinetics and low stability of complexes are important, will be discussed in detail. The influence of complexation kinetics can appear from two principal reasons:

1.   Distrubed equilibrium in the bulk of the sample after the addition of the trace metal during titration (Maljković and Branica, 1971; Figura and McDuffie, 1980);

2.   Distrubed equilibrium near the solid-liquid interface if such one is used in the process of measurement during titration (Shuman and Michael, 1978; Shuman and Cromer, 1979).

These two cases will be treated separately.

2. THE THEORY OF TRACE METAL BINDING IN NATURAL WATERS

In this section of the paper we will discuss the problems of interpretation of experimental results obtained by the direct titration of natural waters with trace metals. First, we will discuss the application of the model of a single 1:1 complex formation. This model is based on the following assumptions:

1.   There is no interference between different trace metals with respect to complex formation.

2.   The mixture of ligands present in natural waters produce a series of complexes with the trace metal used in titration, which have comparable stabilities and speed of complex formation.

3.   The complexes formed will be 1:1 complexes, i.e., one metal ion will be bound to a single ligand molecule.

After this we will procede with more complex models.

2.1. The theory of a single 1:1 complex formation. Let us assume that in the natural water the trace metal M is present in a total concentration C_M and the corresponding ligand L in a total concentration C_L. After the addition of trace metal during titration the trace metal concentration will increase for Δ_M and the total trace metal concentration, after the addition, will be $M_T = C_M + \Delta_M$. Using the known expression for the 1:1

complex formation the following equation for the titration curve can be obtained (Ružić, 1982):

$$\Delta_M = C_L/(1 + 1/K\,[M\,]) - C_M + [M\,] \tag{1}$$

where [M] is the concentration of the free (uncomplexed) metal ion which is detectable by a chosen analytical technique, and K is the corresponding conditional equilibrium constant. In natural waters [M] represent the trace metal ion concentration including all labile complexes formed mostly with inorganic ligands which are in much larger excess in comparison with the trace metal and organic ligand concentrations. Under such conditions all labile forms of the trace metal will act similarly to the free ionic trace metal. It is also assumed that during titration, pH and other conditions which may influence complex formation in any way, are constant.

If the complex formed is very stable ($K \gg 1$) then eqn. (1) simplifies to the following expression:

$$[M\,] = C_M + \Delta_M - C_L = M_T - C_L \quad \text{for} \quad M_T > C_L \tag{2}$$

and

$$M = 0 \qquad\qquad\qquad \text{for} \quad M_T < C_L$$

The titration diagram intersects the Δ_M axis at $\Delta_M = C_L - C_M$ and if C_M is available (from titration of the mineralized sample) the complexing capacity of natural waters can be obtained from such an intercept.

If the complex formed is not very stable ($0.1/C_L < K < 1000/C_L$) then the location of the equivalence point is not simple. Several methods have been reported in the literature for the interpretation of such titration curves. We can divide these methods in four groups:

i.   Approximative methods which are based on the interpretation of individual segments of titration curve (Shuman and Woodward, 1973, 1977; Duinker and Kramer, 1977; Stolzberg, 1981). It has been shown that all these methods can produce significant errors in the complexing capacity and conditional stability determination (Rosenthal et al., 1971; Coenegracht and Duisenberg, 1976; Ružić, 1982).

ii. Nonlinear methods which are based on solving nonlinear equation in order to calculate free metal concentration (Shuman and Woodward, 1973; Stolzberg, 1977; Buffle et al., 1977). These methods are more accurate but inconvenient for routine work.

iii. Methods based on the model of continuous changes of stabilities. In

these methods a transformation of data is used in order to obtain, a continuous distribution function of stabilities (Gamble, 1970,1972,1973;
Gamble et al., 1976,1980). From such a distribution function dominant
values of conditional stability constants and the concentration of corresponding ligands should be obtained. The concept of this method is developed from the fact that the single 1:1 complex formation model is not
always valid. Forcing such a simple model on experimental data has for a
consequence a continuous change of stabilities at different additions of
the trace metal during the titration (Saar and Weber, 1979). Such a behaviour can also be explained by the model of relatively small number of
discreate values of staiblity constants. Taking into account the quality
of data obtained by routine measurements, the optimum number of different
ligand groups which can be detected from such experiments is two. In addition, the model of continuous distribution of stabilities requires also
complex mathematical analyses without gaining any increase in data quality. Whatever the results, such amethod produces again only a set of
conditional parameters but not necessarily always the same as those which
can be obtained by more simple and equally justified methods.

iv. Linearization methods which are based on forcing the data to a single
1:1 complex formation model. Similar methods have been used for the determination of stability constants if concentrations of reactants are
known (Rossotti and Rossotti, 1961; Brady and Pagenkopf, 1978). If data
agree reasonably well with such a model then both complexation parameters,
K and C_L, can be obtained (Scatchard, 1949,1957; van den Berg, 1979,1982;
Ružić, 1980,1982; Lee, 1983). If there is a discrepancy from this simple
model, then the data should be interpreted by two 1:1 complex formation
models, or 2:1, 1:2 and any other more complex model if there is a strong
evidence that such a model can be used (Ružić, 1982,1983). For this purpose original data are transformed into a diagram which should have a
form of a stright line if the single 1:1 complex formation model is applicable. If this is not the case, a departure from linearity should occur in such a diagram.

In principle there are two different linearization methods:

1. The Scatchard method (Scatchard, 1949; Scatchard et al., 1957) is
based on plotting the ratio of bound and free metal concentration
$(M_T - [M])/[M]$ vs. bound metal concentration $M_T - [M]$. If the single 1:1
complex formation is applicable then:

136

$$(M_T - [M])/[M] = K(C_L - M_T + [M])$$ (3)

2. The method proposed by van den Berg (1979,1982), Ružić (1980,1982) and Lee (1983) is based on plotting the ratio of free and bound metal concentration $[M]/(M_T - [M])$ vs. free metal concentration $[M]$. If the model of a single 1:1 complex formation is applicable then:

$$[M]/(M_T - [M]) = ([M] + 1/K)/C_L$$ (4)

In principle, there are no significant differences between these two methods. The only difference could be in the form of nonlinearity when a single 1:1 complex formation model is not valid (see Fig. 1).

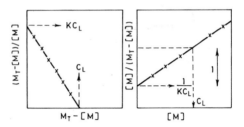

FIGURE 1: Comparison of the special plots for the interpretation of titration curves as proposed by Scatchard (left); as proposed by van den Berg, Ružić and Lee (right). Both for the case of a single 1:1 complex formation.

Very often a good agreement between experimental results and the single 1:1 complex formation model is obtained (van den Berg, 1979,1982; Plavšić et al., 1980,1982; Ružić, 1982; Lee, 1983).

2.2. The influence of complexation kinetics for a single 1:1 complex formation. Assuming a model of a single 1:1 complex formation is valid the following equation could be used to describe the complexation kinetics:

$$d[M]/dt = -k_f[M][L] -k_r(M_T-[M]) = -k_f ([M]+\alpha)^2 -\beta^2$$ (5)

where $\alpha = (C_L - M_T + 1/K)/2$ and $\beta = (\alpha^2 + M_T/K)^{1/2}$, and k_f, k_r are the forward and the reverse rate constants for complexation reaction. This equation has been solved recently by Ružić and Nikolić (1982) (see Fig. 2). When the $k_f C_L \Delta t$ is between 0.1 and 5.0 the equilibrium and kinetic data can be obtained simultaneously from the transformed titration curve (Ružić and Nikolić, 1982). In such a case Scatchard's plot is not convenient for the determination of equilibrium parameters.

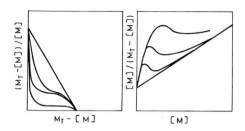

FIGURE 2: The influence of complexation kinetics on the plots as proposed
by Scatchard (left) and van den Berg, Ružić and Lee (right).

2.3. The theory of two 1:1 complex formations. If the model of a single 1:1
complex formation is not valid a departure from linearity should be observed in a transformed diagram. In such a case the basic assumption used
in a single 1:1 complex formation model are obviously not valid. This
means that there may be some interference between several metals competing
for the same ligand for complex formation, or the assemblage of ligands
present in natural water produces a series of complexes of significantly
different stabilities and also complexes formed of perhaps other than the
1:1 type. If the formation of 1:1 complexes is assumed, then from the
analysis of the titration curve different cases can sometimes be distinguished one from the other (Ružić, 1982).

2.3.1. The case of two 1:1 complexes between one metal and two ligands.
Let us assume that the assemblage of ligands present in the natural water
forms two groups of complexes with the trace metal with significantly different stabilities. Total complexing capacity of the natural water will
be a sum of the total concentrations of individual type of ligands C_{L1} +
C_{L2}. It has been shown that in such a case the transformed diagram can be
described by the following equations:

$$[M]/(M_T - [M]) = ([M] + 1/K)/(C_{L1} + C_{L2}) \tag{6}$$

where

$$\overline{K} = ([M] + 1/K_2^*)/([M]/K_1^* + 1/K_1K_2)$$
$$K_1^* = (C_{L1} + C_{L2})/(C_{L1}/K_1 + C_{L2}/K_2)$$
$$K_2^* = (C_{L1} + C_{L2})/(C_{L1}/K_2 + C_{L2}/K_1)$$

This means that if we force a single 1:1 complex formation model the result
will be an overall value of the stability constant $\overline{K}$ which will be continuously changing during the titration depending on the free metal concentration (or on the addition of a trace metal as well). This, however, does

not mean that complexes are formed in the natural water with a continuous
distribution of real conditional stability constants. It is evident from
eqn. (6) that even only two different values for stability constants are
sufficient to produce the same effect. Therefore, the model proposed by
Gamble (1970, 1972, 1973) and Gamble et al. (1976, 1980) not at all more
suitable for the interpretation of experimental data. It could, however
be more complex with almost the same degree of fitting the data as in the
case of a simple model of two 1:1 complex formations.

At a very large addition of a trace metal, K will attain a constant
value K_1^*. Such a value can be obtained by extrapolation of transformed
diagram from the region of high to the region of low M values. As proposed by Ružić (1982) the original diagram can be subtracted from the extrapolated one and the inverse value of the resulting difference can be
again plotted vs. M . If the model of two 1:1 complex formation is valid,
the result of such a procedure should be a stright line. From the coefficients of the extrapolated and resulting straight lines individual stability constants and corresponding concentrations can be calculated
(Ružić, 1982). A similar procedure has not yet been developed for the
Scatchard's type of plot. A similar analysis has been reported by van den
Berg (1982); individual segments of the titration curve were used rather
than the whole diagram, resulting in less precise calculations.

The constant value of K at large additions of the trace metal (equal
to K_1^*) is controlled more by a weaker complex, supposing that the concentrations of individual ligands are comparable. Such an effect has been observed experimentally when an aritificial ligand forming very strong complexes was added to the natural water samples (Krznarić, 1983).

2.3.2. The case of two 1:1 complexes between two metals and one ligand.
Let us assume that at least two trace metals compete for the complex formation with a certain type of L ligands present in the natural water. One
trace metal (M_1) is used in the titration of the natural water and the
other may not be known to us (M_2). The transformed diagram can then be
described by the following equation (Ružić, 1982):

$$[M_1]/(M_{T1} - [M_1]) = ([M_1] - 1/\overline{K} /C_1 \tag{7}$$

where

$$\overline{K} = K_1 / \left\{ 1 + C_{M2}K_2/(1 + \frac{K_2}{K_1} \frac{M_{T1} - [M_1]}{[M_1]}) \right\}$$

Again the overall stability constant K reuslting from forcing a single

1:1 complex formation model will be continuously changing with the $[M_1]$ value, but not exactly in the same way as in the case of the model of two complex formation between one metal and two ligands. At large addition of the trace metal K will attain a constant value of $K_1/(1 + C_{M2}K_2)$. The use of the same procedure of extrapolation and subtraction of the extrapolated diagram from the original one as in the previous case results in an inverse value of the difference which is linearly dependent on $[M_1]/(M_{T1} - [M_1])$ ratio (Ružić 1982), (see Fig. 3).

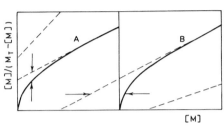

FIGURE 3: Special plot proposed for two 1:1 complex formations: one metal with two ligands (A); two metals with one ligand (B).

Again, the analogous analysis for Scatchard's plot has not yet been reported.

In the case of very strong complexes the expression for an overall stability constant reduces to the simpler form:

$$\bar{K} = (K_1 + K_2 \frac{M_{T2} - [M_1]}{[M_1]})/C_{M2}K_2 \qquad (8)$$

This means that the effect of stabilities of complexes formed on titration curve will not disappear except if K_2 K_1 and then:

$$[M] = M_{T1} - C_L + C_{M2} \qquad \text{for } M_T > C_L - C_{M2} \qquad (9)$$

and

$$[M] = 0 \qquad \text{for } M_T < C_L - C_{M2}$$

Because the value of C_{M2} generally will not be known, the observed complexing capacity of the system will be smaller than the total concentration of the ligand C_L. Theoretically, there is a possibility to check wheather the complexing capacity observed corresponds to the total concentration of ligands that can bind metal M_1. For this purpose we suggest the use of back titration with an artificial ligand forming very strong complexes (see Fig. 4). If the back titration curve is completely symmetrical to the usual titration curve with the trace metal, then the interference of other metals is not significant. However, significant asymmetry of the back titration curve with respect to the usual titration curve indicates that a part of the ligand is probably bound to another unknown trace metal, and that an underestimated value of the complexing capacity of the system has been obtained. Such experiments have not been performed yet and we believe that they could be very helpful in attempting to obtain more adequate complexation paraemters.

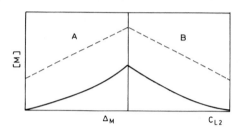

FIGURE 4: Usual titration curves for a trace metal (A). Curves for back titration with an artificial ligand L_2 (B). Original sample fule line; mineralized sample dotted line.

2.3.3. <u>Other 1:1 complex formation models.</u> Ligands sometimes can be present in natural waters in two conformational forms (one of which can be the bidentate form), or each ligand molecule can have two different sites for complex formation. If only 1:1 complexes are formed then the system will behave as a single 1:1 complex model (Ružić, 1983). The Langmuirian type of adsorption behaves in the same way as a single 1:1 complex formation. Therefore the Langmuirian type of adsorption cannot be distinguished from the 1:1 complex formation solely on the basis of direct titration curves. The overall stability constant for 1:1 complex formation in the case of more sites available for complex formation is equal to the sum of stability constants for individual sites.

2.4. <u>Theory of other than 1:1 complex formation.</u> In most cases with different than 1:1 complex formation, the analysis proposed for 1:1 complex formation should fail to produce adequate results. It is interesting, however, that if both 1:1 and 2:1 complexes are formed with the same ligand they behave analogously to two complexes formed with two independent ligands forming 1:1 complexes. The meaning of overall parameters is different but the form of the corresponding titration curves is the same (Ružić, 1983). This means that on the basis of direct titration curves alone we cannot distinguish two different ligands forming 1:1 complexes from a single ligand with two different sites available for metal binding. Such analogy does not exist in the case of 1:2 complexes (at least if titration is performed with a metal).

3. THE INFLUENCE OF COMPLEXATION KINETICS THROUGH THE INTERFACE USED IN THE PROCESS OF MEASUREMENT

If the solid-liquid interface is involved in the process of measurement, a mass transport from the bulk of the sample towards the interface will have a significant effect on the results measured. In the course of measurement several processes will take place:

1. A heterogeneous process at the interface (charge transfer, sorption or a new phase formation).

2. Mass transport will be induced by such heterogeneous processes resulting in the formation of a concentration profile (mass transport layer owing to diffusion or dispersion).

3. Complexation reaction may be displaced from the equilibrium inside a part of the mass transport layer (the so called reaction layer). Such a reaction layer depends on the relative difference between the rate of mass transport and the rate of complex formation.

The process of measurement will always be connected in a certain way with the flux of material near the interface. In the case of charge transfer this will be the electric current which is directly proportional to the surface flux at the electrode-solution interface, or electric charge which is proportional to the time integral of such a flux (Delahay, 1954 Ernst et al., 1975; Whitfield and Jagner, 1981). In the case of sorption the time integral of the surface flux is equal to the surface concentration of the sorbed molecules. This quantity can be measured from the changes of the properties of the interfacial double layer (capacity currents, Jehring, 1969) or, for example, from the changes in the surface tension (Guidelli and Moncelli, 1978).

The theory of mass transport controlled processes is based on the solution of the corresponding differential equations describing this mass transport. The general solution can be described by the following expression:

$$C = C^* (1 - aI - bJ) \tag{10}$$

$$I = \int F(\tau) \, G(t - \tau) \, d\tau$$

$$J = \int F(\tau) \, G(t - \tau) \, e^{-H(t-\tau)} \, d$$

where a and b are the constants containing diffusion coefficients of the species involved in the complexation reaction and the corresponding stability constant. $G(t)$ and $H(t)$ are parameters describing the regime of mass transport and complexation kinetics, respectively. The surface flux $F(t)$ can be described with the corresponding rate equation which can be combined with eqn. (10) to produce the final integral equation (Ružić, 1977) describing the surface flux $F(t)$:

$$F(t) = L - \Lambda I - MJ \tag{11}$$

where L, Λ, and M represent certain combinations of the parameters a, b and the corresponding heterogeneous rate constants. The solution of integral eqn. (11) can be used for the interpretation of experimental results. However, analytical solution is rarely available and if so, then for very simple special cases only.

In the case of a very fast complexation the integral describing complexation kinetics can be simplified (Koutecky, 1953; Brdička, 1954):

$$J \sim F(t) / \sqrt{k(1 + K)D} \tag{12}$$

and

$$F(t) \sim (L - \Lambda \cdot I)/(1 - M/\sqrt{k(1 + K)D}) = L^* - \Lambda^* I$$

where k, K are the rate constant and the stability constant for complexation reaction respectively. In such a case the reaction layer is very small and the surface flux has a form equivalent to the case when the effect of complexation kinetics cannot be observed. However, the information about complexation kinetics is contained in an overall kinetic term $\Lambda^* = 1/(1/\Lambda + M/\sqrt{k(1 + K)D})$. If the heterogeneous reaction at the interface is much faster than the complexation reaction, then the overall rate parameter Λ^* will be controlled by homogeneous kinetics. This is a case of the so-called steady state homogeneous kinetics under the conditions of the so-called heterogeneous equivalent (Ružić and Feldberg, 1974; Ružić, 1983).

When there is a very significant convective contribution to the mass transport both integrals I and J may be simplified:

$$I \sim F(t) \, \delta /D \qquad \text{and} \qquad J \sim F(t) \, \varepsilon /D \tag{13}$$

and

$$F(t) \sim L/(1 + \frac{\Delta\delta + M\varepsilon}{D}) = (LD/\Delta\,\delta_{,})\,\frac{\Delta_{.}^{*}\,\delta\,/D}{1+\Delta^{*}\,\delta/D}$$

where δ is the mass transport layer and $\varepsilon = \sqrt{D/k(1 + K)}$ is the reaction layer. The first part of such an approximative solution describes complexation equilibrium ($LD/\Delta\,\delta_{,}$). The second part is a simple function of the overall kinetic parameter ($\Delta_{.}^{*}\,\delta\,/D$). In the case of an extremly fast complexation the second part of the solution approaches to unity and the information on complexation kinetics is lost. All the solutions used so far in studying the trace metal complexation are analogous (Turner and Whitfield, 1979) to the approximative solution (13).

Approximative solutions (12) and (13) are valid only for extreme conditions and should not be used without necessary precaution. For example, these approximative solutions fail to describe the exact form of pseudopolarogram (Brown and Kowalski, 1979; Shuman and Cromer, 1979; Lovrić and Branica, 1980). However, in the last few decades a large number of papers have been dedicated to the theory of mass transport controlled heterogeneous processes, especially in the field of electroanalytical chemistry (Galus, 1976; Bond, 1980; Bard and Faulkner, 1980; Delahay, 1954; Heyrovsky and Küta, 1965). Today there are no mathematical limitations for a rigorous description of complexation kinetics under the experimental conditions used so far in this kind of research. A very powerful method is the so-called digital simulation which is based on time and space incrementisation and finite difference approximation (Mattson et al., 1972; Britz, 1981). The main effects of complexation kinetics on the results measured can be summarized as follows:

1. Decrease of the surface flux $F(t)$, i.e., the change of availability of trace metals for heterogeneous process.

2. Coupling of complexation with heterogeneous kinetics. Therefore more rigorous theoretical predictions should be used for interpretation of complexation kinetics in the future.

ACKNOWLEDGEMENT: The author is grateful to Drs. Duinker and Kramer whose experimental work initiated the part of this work connected with complexation kinetics. This work has been supported by Republic Council for Scientific Research of Croatia. The author is also grateful to The Netherlands Institute for Sea Research at Texel for their hospitality during the time when the first draft of this work has been compiled.

REFERENCES

- Ashworth, A.R.F., 1965. Titrimetric organic analysis. In: P.J. Elving and I.M. Kolthoff (Eds.), Chemical Analysis Interscience Publ., Vol. 15, Part 1, Section 1.
- Bard, A.J. and L.R. Faulkner, 1980. Electrochemical methods. Fundamentals and applications. John Wiley & Sons.
- Van den Berg, C.M.G., 1979. Complexation of copper in natural waters. Determination of complexing capacities and conditional stability constants in natural waters by MnO_2 and the applications for phytoplankton toxicity. Ph.D. Thesis, McMaster Univ., Ontario.
- Van den Berg, C.M.G., 1982. Determination of copper complexation with natural organic ligands in sea water by equilibrium with MnO_2. I. Theory. Marine Chemistry, Vol. 11, pp. 307-322.
- Van den Berg, C.M.G. and J.R. Kramer, 1979. Determination of complexing capacity and conditional stability constants for copper in natural water using MnO_2. Anal. Chim. Acta, Vol. 106, pp. 113-120.
- Bond, A.J., 1980. Modern polarographic methods in analytical chemistry. Marcel Dekker Inc.
- Brady, B. and G.K. Pagenkopf, 1978. Cadmium complexation by soil fulvic acid. Canad. J. Chem., Vol. 56, pp. 2331.
- Brdicka, R., 1954. Evaluation of the rate constants of reactions involved in polarographic electrode processes. Coll. Czech. Chem. Comm., Vol. 19, Suppl. p. 41.
- Britz, D., 1981. Digital simulation in electrochemistry. In: Lecture notes in chemistry. Springer-Verlag, No. 23.
- Brown, S.D. and Kowalsky, B.R., 1979. Pseudopolarographic determination of metal complex stability constants in dilute solution by rapid scan anodic stripping voltammetry. Anal. Chem., Vol. 51, p. 2133.
- Buffle, J., F. Greter and W. Haerdi, 1977. Measurement of complexation properties of humic and fulvic acids in natural waters with lead and copper ion-selective electrodes. Anal. Chem., Vol. 49, p. 216.
- Coenergracht, P.M.J. and A.J.M. Duisenberg, 1976. End-point construction and systematic titration error in linear titration curves-complexation reactions. Anal. Chim. Acta, Vol. 78, p. 183.
- Delahay, P., 1956. New instrumental methods in electrochemistry. Interscience Publ.
- Duinker, J.C., 1980. Suspended matter in estuaries: adsorption desorption processes. In: Olausson, E. and I. Cato (Eds.), Chemistry and Biochemistry of Estuaries. John Wiley & Sons, pp. 121-151.
- Duinker, J.C., C.J.M. Kramer, 1977. An experimental study on the speciation of dissolved zinc, cadmium, lead and copper in river Rhine and North Sea water by differential pulse anodic stripping voltammetry. Marine Chem., Vol. 5, pp. 207-228.
- Ernst, R., H.E. Allen and K.H. Mancy, 1975. Characterization of trace metal species and measurement of trace metal stability constants by electrochemical techniques. Water Res., Vol. 9, pp. 969-979.
- Figura, P. and McDuffie, 1980. Determination of labilities of soluble trace metal species in aqueous environmental samples by anodic stripping voltammetry and chelex column and batch methods. Anal. Chem., Vol. 52, p. 1433.
- Florence, T.M. and G.E. Batley, 1980. Chemical speciation in natural waters. CRC critical reviews in Anal. Chem., pp. 219-296.
- Galus, Z., 1976. Fundamentals of electrochemical analysis, Polish Sci. Publ. and Ellis Horwood Ltd.

- Gamble, D.S., 1970. Titration curves of fulvic acid: the analytical chemistry of a weak acid polyelectrolyte. Canad. J. Chem., Vol. 48, p. 2662.
- Gamble, D.S., 1972. Potentiometric titration of fulvic acid: Equivalence point calculation and acid functional groups. Canad. J. Chem., Vol. 50, p. 2680.
- Gamble, D.S., 1973. Na^+ and K^+ binding by fulvic acid. Canad. J. Chem., Vol. 51, p. 3217.
- Gamble, D.S., C.H. Langford and J.R.K. Tong, 1976. The structure and equilibria of a manganese (II) complex of fulvic acid studied by ion exchange and nuclear magnetic resonance. Canad. J. Chem., Vol. 54, pp. 1239-1245.
- Gamble, D.S., A.W. Underdown and C.H. Langford, 1980. Copper (II) titration of fulvic acid ligand sites with theoretical, potentiometric and spectrophotometric analysis. Anal. Chem., Vol. 52, pp. 1901-1908.
- Guidelli, R. and M.R. Moncelli, 1978. The influence of diffusion controlled adsorption upon interfacial tension measurements. J. Electroanal. Chem., Vol. 89, pp. 261-270.
- Hart, B.T., 1981. Trace metal complexing capacity of natural waters: A review. Environ. Tech. Letters, Vol. 2, pp. 95-110.
- Heyrovsky, J. and J. Küta, 1965. Principles of polarography. Czechoslovak Acad. Sci., Prague.
- Jehring, H., 1969. Untersuchungen mit der Wechselstrompolarographie. IV. Erniedrigung des momentanen und mitteleren Kapazitätsstromes durch Adsorption ohne und mit Diffusionshemmung. J. Electroanal. Chem., Vol. 20, pp. 33-46.
- Jenne, E.A. (ed.), 1979. Chemical modeling in aqueous systems. ACS Symp. Ser. No. 93. American Chemical Society, Washington, D.C.
- Koutecky, J., 1953. Theorie langsamer Elektrodenreaktionen in der Polarographie und polarographischen Verhalten eines Systems bei welchem der Depolarisator durch eine schnelle chemische Reaktion aus einem elektroaktiven Stoff ensteht. Coll. Czech. Chem. Comm., Vol. 18, p. 597.
- Krznarić, D., 1983. The influence of surfactants upon the measurements of copper and cadmium speciation in model sea water by differential pulse anodic stripping voltammetry. Submitted for publication.
- Lee, J., 1983. Complexation analysis of fresh waters by equilibrium diafiltration. Water Res., Vol. 17, pp. 501-510.
- Lovrić, M. and M. Branica, 1980. Application of ASV for trace metal speciation. Croat. Chem. Acta, Vol. 53, pp. 477-483, 485-501 and 503-506.
- Maljković, D. and M. Branica, 1971. Polarography of seawater. II. Complex formation of cadmium with EDTA. Limnol. Oceanogr., Vol. 16, pp. 779-785.
- Mantoura, R.F.C., A. Dickson and J.P. Riley, 1978. The complexation of metals with humic materials in natural waters. Estuar. Coast. Mar. Sci., Vol. 6, pp. 387-408.
- Mattson, J.S., H.B. Mark Jr. and H.C. MacDonald Jr, 1972. Electrochemistry, calculation, simulation and instrumentations. Marcel Dekker, Vol. 2.
- Neubecker, T.A. and H.E. Allen, 1983. The measurement of complexing capacity and conditional stability constants for ligands in natural waters. Water Res., Vol. 17, pp. 1-14.
- Plavsić, M., S. Kozar, D. Krznarić, H. Bilinski and M. Branica, 1980. The influence of organics on the adsorption of copper (II) on γ-Al_2O_3 in sea water. Marine Chem., Vol. 9, pp. 175-182.

- Plavsić, M., D. Krznarić and M. Branica, 1982. Determination of the apparent copper complexing capacity of sea water by anodic stripping voltammetry. Marine Chem., Vol. 11, pp. 17-31.
- Ringbom, A.J., 1963. Complexation in analytical chemistry. Interscience, New York.
- Rosenthal, D., G.L. Jones Jr. and R. Megargle, 1971. Feasibility and error in linear extrapolation titration procedures. Anal. Chim. Acta, Vol. 53, p. 141.
- Rossottoi, F.J.C. and H. Rossotti, 1961. The determination of stability constants and other equilibrium constants in solution. McGraw Hill Inc.
- Ruzić, I., 1977. Theory of pulse polarography and related chronoamperometric and chronocoulometric techniques. I. Influence of mass transport regime and heterogeneous kinetics on current-potential curves. J. Electroanal. Chem., Vol. 75, p. 24-44.
- Ruzić, I., 1980. Interpretation of titration curves for nonlabile complex formation with trace metals. Thalassia Jugoslavica, Vol. 16, p. 325.
- Ruzić, I., 1982. Theoretical aspects of the direct titration of natural waters and its information yield for trace metal speciation. Anal. Chim. Acta, Vol. 140, p. 99-113.
- Ruzić, I., 1983. Digital simulation of very rapid coupled chemical reactions by electrode reaction. J. Electroanal. Chem., Vol. 144, pp. 433-436.
- Ruzić, I. and S.W. Feldberg, 1974. The heterogeneous equivalent: a method for digital simulation of electrochemical systems with compact reaction layers. J. Electroanal. Chem., Vol. 50, pp. 153-162.
- Ruzić, I. and S. Nikolić, 1982. The influence of kinetics on the direct titration curves of natural water systems. Theoretical considerations. Anal. Chim. Acta, Vol. 140, pp. 331-334.
- Saar, R.A. and J.H. Weber, 1979. Complexation of cadmium(II) with water- and soil-derived fulvic acids: Effect of pH and fulvic acid concentration. Canad. J. Chem., Vol. 57, p. 1263.
- Scatchard, G., 1949. The attractions of proteins for small molecules and ions. Ann. N.J. Acad. Sci., Vol. 51, pp. 660-672.
- Scatchard, G., J.S. Coleman and A.L. Shen, 1957. Physical chemistry of protein solutions. VII. The binding of some small anions to serum albuminum. J. Am. Chem. Soc., Vol. 79, pp. 12-20.
- Shuman, M.S. and J.L. Cromer, 1979. Copper association with aquatic fulvic and humic acid. Estimation of conditional formation constants with a titrimetric anodic stripping voltammetry procedure. Environ. Sci. Tech., Vol. 13, p. 543.
- Shuman, M.S. and J.L. Cromer, 1979a. Pseudopolarograms: applied potential-anodic stripping peak current relationships. Anal. Chem., Vol. 51, p. 1548.
- Shuman, M.S. and L.C. Michael, 1978. Reversibility of copper in dilute aqueous carbonate and its significance to anodic stripping voltammetry of copper in natural waters. Anal. Chem., Vol. 50, p. 2104.
- Shuman, M.S. and L.C. Michael, 1978a. Application of the rotated disk electrode to measurement of copper complex dissociation rate constants in marine coastal samples. Environ. Sci. Tech., Vol. 12, p. 1609.
- Shuman, M.S. and G.P. Woodward, 1973. Chemical constants of metal complexes from a complexometric titration followed with anodic stripping voltammetry. Anal. Chem., Vol. 45, p. 2032.
- Shuman, M.S. and G.P. Woodward, 1977. Stability constants of copper-organic chelates in aquatic samples. Environ. Sci. Tech., Vol. 11, pp. 809-813.

- Singer, P.C. (Ed.), 1973. Trace metals and metal organic interactions in natural waters. Ann. Arbor Science Publ.
- Stolzberg, R.J., 1977. Potential inaccuracy in trace metal speciation measurements by differential pulse polarography. Anal. Chim. Acta, Vol. 92, p. 193.
- Stolzberg, R.J., 1981. Uncertainty in calculated values of measurements by differential pulse polarography. Anal. Chem., Vol. 53, pp. 1286-1291.
- Stumm, W. and J.J. Morgan, 1981. Aquatic chemistry. Wiley Intersci., New York, 2nd ed.
- Turner, D.R. and M. Whitfield, 1979. Reversible electrodeposition of trace metal ions from Multy-ligand systems. Part I. Theory. J. Electroanal. Chem., Vol. 103, pp. 43-60.
- Whitfield, M. and D. Jagner, 1981. Marine electrochemistry. Practical introduction. John Wiley & Sons.

CHEMICAL MODELS, COMPUTER PROGRAMS AND METAL COMPLEXATION IN NATURAL WATERS

DARRELL KIRK NORDSTROM AND JAMES W. BALL

1.   INTRODUCTION

It has been about 22 years since the first publication of a paper on the speciation of a natural water (Garrels and Thompson, 1962) and about 15 years since the appearance of papers in which computers were used to model the chemistry of water-mineral reactions (Helgeson, et al., 1969, 1970). The use of computers to solve chemical equilibrium problems for aqueous systems has expanded enormously and "chemical modeling" is now a common phrase describing the application of physico-chemical principles to the interpretation of natural hydrogeochemical systems (Jenne, 1979). Nordstrom, et al. (1979) reviewed computerized chemical models for aqueous systems but in the last five years there have been significant advances in electrolyte theory and evaluated thermodynamic data which have found applications in chemical modeling. The number of computer programs has nearly doubled since the 1979 review paper and an attempt is made in this paper to inventory these programs, review the shortcomings of some of the traditional methods of calculating speciation and suggest avenues for improvement in the chemical models which could be incorporated into the programs.

2.   MATHEMATICAL AND THERMODYNAMICAL NATURE OF THE CHEMICAL EQUILIBRIUM PROBLEM

The simplest way of describing the chemical equilibrium problem is to say that it is the attempt to find the most stable state of a system for a given set of pressure, temperature and compositional constraints. This definition doesn't mean that pressure, temperature and composition have to be fixed at constant values but it does mean that they have to be defined. In thermodynamic terms, a chemical equilibrium calculation attempts to find the minimum value for the Gibbs free energy of a defined system. The calculation itself can be carried out in one of two ways: by minimizing a free energy function

Kramer, C.J.M. and Duinker, J.C. (eds.), Complexation of Trace Metals in Natural Waters. ISBN 90-247-2973-4
© 1984 Martinus Nijhoff/Dr W. Junk Publishers, The Hague/Boston/Lancaster.
Printed in the Netherlands.

or by solving a set of non-linear equations consisting of equilibrium constants and mass balance constraints. These have become known as the "free
energy minimization" method and the "equilibrium constant" approach. In each
method the reaction affinity (expressed as free energies or mass action
equilibrium constants) must be solved within mass balance constraints for all
species and components within the system. The two methods are thermodynamically equivalent but they can lead to different mathematical statements of
the problem, which in turn suggest different numerical algorithms for their
solution. Even this difference, however, is a superficial one because at
least one computer program (SOLGASWATER) can accept either free energies or
equilibrium constants for the database. The major disadvantage of using a
free energy database is that these values are not nearly as reliable as
directly measured equilibrium constants and the chemical modeler must evaluat
the data very critically. The process of critical evaluation is a lengthy
and tedious procedure for all the species that occur in a natural water and,
therefore, the equilibrium constant method is generally prefered.

The numerical algorithms can be conveniently characterized into (a) pure
iteration, (b) Newton-Raphson iteration and (c) integration of ordinary
differential equations (Van Zeggeren and Storey, 1970). Wigley (1977) and
Nordstrom, et al. (1979) have described two forms of pure iteration or
successive approximation used in water chemistry calculations: the brute
force approach of simple back-substitution and the continued fraction
approach. The Newton-Raphson method is perhaps the most widely used numerica
method of approximation although it frequently does not converge as quickly
as pure iteration techniques (Ingri, et al., 1967; Nordstrom, et al., 1979).
Complicated reaction path calculations are handled by techniques that solve
ordinary differential equations (Helgeson, et al., 1970; Wolery, 1979).

Further information on the mathematics and thermodynamics of the chemica
equilibrium problem formulation can be found in many books, papers and report
The simplest introduction to homogeneous phase calculations can be found in
Wigley (1977) and is recommended for teaching purposes. Two books, one by
Van Zeggeren and Storey (1970) and the other by Smith and Missen (1982) are
highly recommended. The additional reviews by Zeleznik and Gordon (1968) and
Wolery (1979) are also recommended. An excellent recent review by Rubin
(1983) provides the mathematical formalism needed to solve the chemistry of
reacting systems during fluid flow for both equilibrium and non-equilibrium
conditions. These six references cover virtually all the mathematical,
thermodynamical and computational aspects of the chemical equilibrium problem.

3. COMPUTERIZED CHEMICAL MODELS FOR NATURAL WATERS

More than 50 computer programs that calculate chemical equilibrium in natural waters or similar aqueous systems have appeared in the literature since 1965. This number represents an increase of nearly 100% since the 1979 review. These programs are given in Table 1 along with an indication of their application. The application annotations do not mean that other programs lacking the annotation cannot do the prescribed job, but only that they have not been applied in such a manner. Obviously any of these programs can be made to include trace elements but only a few of them have actually been used for such purposes. Several of these programs deserve further description, especially those that were not mentioned in the 1979 review paper. References for these programs are found in Table 1.

All the programs listed in Table 1 perform distribution-of-species calculations in a single aqueous phase. In addition, many can do "mass transfer" calculations and some can do "mass transport" calculations. "Mass transfer" is here defined as the transfer of mass between two or more phases such as the precipitation or dissolution of a soluble mineral. "Mass transport" is here defined as solute movement during fluid flow such as the movement of dissolved trace metals in a flowing river. Combining mass transport with chemical equilibrium models is a very new subject in which only a limited amount of work has been done. Chapman et al. (1982) have combined a physical model for fluid flow appropriate to a river (based on a convective-dispersion equation) with the chemical model MINEQL to calculate chemical changes during short term neutralization of an acid mine drainage. The resulting program, RIVEQL, was fairly successful in predicting the precipitation of trace metals compared to the actual field data. Grove and Wood (1979) modeled water quality changes during artificial recharge of an aquifer with a program that combined one-dimensional fluid flow with gypsum solubility equilibrium and ion exchange.

TABLE 1. AQUEOUS CHEMICAL EQUILIBRIUM COMPUTER PROGRAMS

Program	Application	Reference
AION	B, TE	Van Luik and Jurinak (1978, 1979)
ADSORP	A, TE	Bourg (1982)
AQ/SALT	B	Bassett, et al. (1981)
ASAME	TE	Matisoff, et al. (1980)
CHEMTRN	F, M	Miller and Benson (1983)
COMICS		Perrin and Sayce (1967)

DISSOL	M	Droubi (1976); Droubi, et al. (1976)
ECES		Zemaitis and Rafal (1975)
EHMSYS		Proeseler and Wagner (1982)
EQ 3/6	M, TE, T, R	Wolery (1979, 1983)
EQBRAT		Detar (1969)
EQUIL		Bos and Meershoek (1972)
EQUIL	M, TE	Fritz (1975)
EQUIL		I and Nancollas (1972)
EVAPOR	M	Droubi (1976); Droubi, et al. (1976)
FASTPATH	M, TE, T	Herrick (see Mercer, et al., 1982)
GEOCHEM	A, M, TE	Mattigod and Sposito (1979); Sposito and Mattigod (1980)
GIBBS		Gautem and Seider (1979); Seider, et al.(1980
HALTAFALL	M, TE	Ingri, et al. (1967)
HITEQ	T	Iglesias and Weres (1981)
IONPAIR		Thrailkill (1970)
KATKLE1		van Breemen (1973)
MICROQL	A, M	Westall (1979)
MINEQL	A, M	Westall, et al. (1976)
MINTEQ	A, M, TE	Felmy, et al. (1983)
MIRE		Holdren and Bricker (1977)
MIX2		Plummer, et al. (1975)
NOPAIR		Thrailkill (1970)
PATHI	M, TE, T	Helgeson, et al. (1970)
PHREEQE	M, R	Parkhurst, et al. (1980)
REDEQL	A, M, TE	Morel and Morgan (1972)
REDEQL 2	A, M, TE	McDuff and Morel (1973)
REDEQL.EPAK	A, M, TE	Ingle, et al. (1979)
RIVEQL	A, F, M, TE	Chapman, et al. (1982)
SEAWAT		Lafon (1969)
SENECA		Ma and Shipman (1972)
SENECA2	T	Kerrisk (1981)
SIAS		Fardy and Sylva (1978)
SOLGASWATER	M, TE, R	Eriksson (1979)
SOLMNEQ	TE, T	Kharaka and Barnes (1973)
SOLVEQ	M, TE, T, R	Reed (1982)
THERMAL	T	Fritz (1981)

WATCH1	T	Arnorsson, et al. (1982)
WATCHEM		Barnes and Clarke (1969)
WATEQ		Truesdell and Jones (1974)
WATEQF		Plummer, et al. (1976)
WATEQFC		Runnells and Lindberg (1981)
WATEQ2	TE	Ball, et al. (1979, 1980)
WATEQ3	TE	Ball, et al. (1981)
WATSPEC		Wigley (1977)
--		Crerar (1975)
--	F	Grove and Wood (1979)
--	B	Harvie and Weare (1980)
--		Karpov, et al. (1972, 1973)
--	M, T	Ryzhenko, et al. (1981)
--		Vallochi, et al. (1981)
--	B, T	Wood (1975, 1976)

Application symbols: A = adsorption; B = brines; F = fluid flow (mass transport); M = mass transfer; TE = trace elements; T = high temperature geothermal chemistry; R = redox balance

Miller and Benson (1983) have developed a program called CHEMTRN to model simple equilibria reactions with fluid flow in a porous medium. They have used Newton-Raphson iteration to solve a set of differential equations for fluid transport and algebraic equations for chemical equilibria. Ion exchange equilibria have been combined with fluid flow equations to simulate exchange equilibria in a groundwater system by Vallocchi, et al. (1981).

An additional degree of complexity can be induced by considering reaction progress wherein states of partial equilibrium are attained during the path towards complete equilibrium. A simplified version of these reaction path models is offered by PHREEQE which uses a continued fraction scheme to calculate species distribution for everything except pH, pE and mass transfer which are calculated by a modified Newton-Raphson iteration. PHREEQE has been developed for the interpretation of groundwater chemistry and, when used in conjunction with its companion program BALANCE (Parkhurst, et al., 1982) provides a valuable tool in the understanding of geochemical processes in groundwater systems (Plummer, et al., 1983). The first reaction path model came out of the pioneering efforts of Helgeson and his colleagues who developed the PATHI program. Improvements in the efficiency of this type of program have been made by Herrick (see Mercer, et al., 1982) which led to the develo ment of FASTPATH and by Wolery (1979) who developed EQ3/6 and by Reed (1982) who developed SOLVEQ. No attempt has yet been made to combine a reaction path program with a complete fluid flow model in 3 dimensions and the result ing computations may be exhorbitantly expensive as well as being limited by computer memory. Numerical and chemical simplifications will continue to provide an important role in the development of such programs.

Adsorption reactions are incorporated into the REDEQL series including MINEQL and MICROQL. For modelers who wish to use microcomputers, MICROQL is particularly convenient for this purpose. It is written in BASIC and can be quickly and easily adapted to small computers such as a Tektronix 4052.*   A new adsorption program developed by Bourg (1982) shows promise in the inter- pretation of trace metal behavior in riverine systems (Bourg and Mouvet, thi volume).

Geothermal waters can be modeled by several programs including SOLMNEQ, HITEQ, THERMAL and WATCH1. WATCH1 has been applied to the geothermal waters of Iceland by Arnorsson and colleagues. HITEQ has been used to simulate the geothermal chemistry of the Cierro Prieto brines.

Probably the most challenging natural water to model is a brine. Wood (1975, 1976) first modeled successfully brine solutions with a computer program that used Harned's rule with the Scatchard specific ion interaction model. Van Luik and Jurinak (1978, 1979) have adopted the cluster integral expansion theory for calculating the chemical equilibrium of major and trace elements in the Great Salt Lake, Utah. The most successful approach has been derived from the application of the Pitzer equations (Pitzer, 1979) by Harvie and Weare (1980). AQ/SALT developed by Bassett, et al. (1981) also uses the Pitzer equations. Further discussions concerning the problems of brine cal- culations and activity coefficients will be taken up later in this paper.

Several programs have incorporated the speciation of uranium into the chemical model. F. J. Pearson wrote a uranium speciation program for the Tektronix 4050-series computers that is compatible with the WATEQF program. Ball, et al. (1981) incorporated uranium speciation into WATEQ2 and, along with a few other minor modifications, renamed the expanded version WATEQ3. Runnells and Lindberg (1981) included uranium speciation into the WATEQF

---

*The use of brand names is for identification purposes only and does not imply endorsement by the U.S. Geological Survey.

program which led to WATEQFC and this program was used as a guide to uranium ore exploration.

Increasing use of the free energy minimization technique is apparent, e.g., SOLGASWATER, GIBBS, EHMSYS, the program developed by Shvarov (1981) and Ryzhenko, et al. (1981) and the program developed by Harvie and Weare (1980).

Two questions commonly arise: (1) Why are there so many programs? and (2) Which is the best program? It is easiest to answer the last question first. There is no "best" program.* Although there are a great many computational algorithms, there is no major difference as to which one is used. As each program has somewhat different objectives and can do slightly different computations (i.e., some do adsorption calculations, some do variable temperatures calculations and some do reaction path calculations) each researcher should use a program according to his own objectives. Usually the type of computer facilities available and the compatibility of the program is more decisive than other factors.

It is a bit more difficult to answer the question as to why there are so many programs. Certainly no one program can do all the desirable calculations in a perfectly general way. Programs such as EQ 3/6 come close to this, but there is still the law of diminishing returns: the more general and comprehensive the computer program, the more difficult it is to use and the more costly. Furthermore, simple 2- or 3-component systems would be a waste of time for a program such as EQ 3/6 to compute. Therefore, the large number of programs represents the wide variety of user needs. However, there is still a considerable amount of obvious duplication. Many programs serve the same purpose. In addition, some of the more powerful programs seem to be under-utilized. Several programs are not readily available or not well documented or, if documented, not easy to use. Even these explanations are not fully satisfactory and it must be concluded that researchers enjoy programming.

4. LIMITATIONS OF THE CHEMICAL MODELS

4.1 Testing vs. simulation and the equilibrium assumption

Perhaps the most fundamental assumption that forms the basis for chemical equilibrium calculations on natural waters is that equilibrium exists in the system chosen for study. For most aqueous phase reactions such as ion pairing

---

*This statement is not entirely true. A FORTRAN program written for refining stability constants from potentiometric data is called BEST (Motekaitis & Martell, 1982).

of inorganic species, this assumption is quite safe. The assumption is not applicable to all mineral solubilities. Some minerals are known to reach equilibrium saturation rapidly whereas others do not attain equilibrium at lc temperatures. Therefore, saturation index computations can only indicate the tendency for the precipitation or dissolution of a mineral. Such computations are always useful as a reference base for interpreting water quality data but they cannot be taken as proof that precipitation or dissolution is taking place. The equilibrium assumption is used implicitly in all "simulation" studies. When water-mineral reactions are simulated for a given set of conditions it is always assumed that at least partial equilibrium holds. This assumption is probably a good one at high temperatures but not at low temperatures. For low temperature geochemical processes, a great deal of information on dissolution kinetics, precipitation kinetics, organic reaction and biological processes is required.

4.2 The ion association (IA) model and the specific ion interaction (SI) model

There are currently two different approaches to solving the problem of non-ideality in aqueous solutions, i.e. the problem of activity coefficient calculations. The IA model assumes that only interactions of oppositely charged ions are of importance and it describes the short range interactions by ion pair or complex equilibrium reactions and the long-range interactions by the electrostatic Debye-Hückel theory. Alternatively, the SI model describes the short-range interactions by a difference function which is specific for each solute. This function is simply the difference between the Debye-Hückel value and the measured mean activity coefficient for a solute. The SI difference function or "interaction coefficient" is a function of the composition and of ionic strength of the solution. The interaction coefficie can take a variety of forms from a simple constant term to an expanded virial form as developed by Pitzer(1979). The severe limitation of the IA model can easily be demonstrated by the following argument. According to the definitic of the mean activity coefficient:

$$\gamma_{\pm} = [(\gamma_{+})^{\nu+} (\gamma_{-})^{\nu-}]^{1/\nu}$$

where γ_{+} is the free cation activity coefficient, γ_{-} is the free anion activi coefficient, γ_{+} is the stoichiometric amount of the cation in the salt and γ_{-} is the stoichiometric amount of the anion in the salt. Very strong electro-lytes such as NaCl and KCl are considered to be fully dissociated. Therefore the following ratio should have the value of unity at all concentrations:

$$\frac{\gamma_{\pm KCl} \quad \gamma_{\pm NaBr}}{\gamma_{\pm NaCl} \quad \gamma_{\pm KBr}}$$

as long as each free ion activity coefficent is not dependent on the particular medium of the salt, i.e. γ_{Cl^-} is the same·whether measured in a KCl or a NaCl or a RbCl solution. The ratio shown above is plotted from the data of Hamer and Wu (1972) and shown in Figure 1. The deviations from unity are quite significant, approaching 10% at 3 molal. This example demonstrates clearly that activity coefficients are a function of the specific ionic composition and that for ionic strengths or compositions above about 0.5 molal they must be taken into account. Although freshwaters and estuaries can be modeled without worrying about these activity coefficient problems, there is another very important area where the SI method can be especially helpful. A large number of ion pair or complex reactions have been measured only at high ionic strengths and a reliable method of extrapolating them to infinite dilution is needed. Some form of SI equation must be used to carry out this extrapolation and only then can the data be available for chemical modeling purposes.

Several types of SI models have been proposed in the literature and three of these deserve special attention for chemical modeling. The most successful approach has been the virial equations of Pitzer and his colleagues (Pitzer, 1979). Pitzer has been able to reproduce most of the available solute data for activity and osmotic coefficients to within the experimental precision over a large range of ionic strength. This semi-empirical, semi-theoretical approach has been developed and applied to simple brine solutions by Harvie and Weare (1980). Their results have been extremely successful. They have been able to reproduce complex solubility data for highly soluble salts based on two or three component solubilities. Unfortunately, not all the solute data needed to model natural waters by the Pitzer method are available, pH and pE conventions need to be established, strong complexing has to be taken into account, and a considerable amount of solute and solubility data have to be critically evaluated.

Alternatively, a hybrid model that might combine the advantages of the large database of equilibrium constants in the IA method with the more reliable activity coefficient expressions in the SI method could be a better way to solve the problem. A hybrid model was suggested by Whitfield (1975) and discussed in a general manner in Whitfield (1979), in which the major free ion activity coefficients were calculated by the Pitzer equations, ion pair constants were used for strong interactions and a simplified Debye–Hückel equation was used for the activity coefficient of the ion pair. Another hybrid model proposed by Biedermann (1975) takes advantage of the simpler Guggenheim–Scatchard SI equation for calculating activity coefficients. This approach has the advantage that the interaction coefficient can be related in a simple manner to the size and charge of the ion or the ion pair (Biedermann, et al., 1982). Both of these approaches need to be further examined as, for example, Millero and Schreiber (1982) have done with Whitfield's hybrid model.

4.3. Model validation: an example with gypsum solubility.

The biggest weakness of chemical models, whether computerized or not, is the reliability of the thermodynamic data base. The problem of activity coefficients is also inextricably bound up with equilibrium constants. Choosing slightly different values for an equilibrium constant and/or an activity coefficient can greatly change the outcome of a calculation and therefore the interpretation of the chemistry of a natural water. It is also necessary to validate the calculations by comparing the simulation of complex electrolyte solution with actual measured values. For example, a chemical model should be able to reproduce activity coefficients and solubility data for a wide variety of single and mixed electrolyte data. This type of model validation can help to point out the limitations of the chemical model, especially the range of validity for certain kinds of solutes. The solubilities of gypsum in NaCl and $NaClO_4$ solutions provide good examples for model validation. Kerrisk (1981) made a comparison of 4 programs by testing their ability to simulate the solubilities at ionic strengths of 0–4 molal and at 25–300°C. The comparisons were generally poor and differences in activity coefficients and the thermodynamic data base were blamed for the discrepencies. We have modified some of the variables in the WATEQ3 program to demonstrate how sensitive the gypsum solubility simulations are to various parameters.

Gypsum solubility has been measured numerous times. The solubility in NaCl solutions has been taken from the work of Marshall and Slusher (1966) which agrees well with most other measurements. The solubility has been plotted in Figure 2. In addition to the $NaSO_4^-$ ion pair, the $CaSO_4^\circ$ ion pair has been considered and kept constant with a stability constant of $10^{2.3}$ throughout the computations. When the Davies equation is used for γ

of $NaSO_4^-$ and $K=10^{0.72}$ for the ion pair formation constant the correct
shape of the solubility curve is maintained (Line A) and a fairly close fit
is obtained if a $NaSO_4^-$ ion pair constant of $10^{1.1}$ is combined with the
Davies equation (Line B). However, if this same constant is used with a
Debye-Hückel expression then the simulated solubility is much too high (line
D). The best fit is obtained when the ion pair constant is kept to $10^{0.82}$
(Reardon, 1975) and the interaction coefficient in the Truesdell-Jones activity
coefficient for Ca^{2+} is changed to 0.109 (Line C). Unfortunately, this
adjustment also makes the Ca^{2+} ion activity coefficient less compatible
with mean activity coefficient data on $CaCl_2$. Further adjustments of the
simulation could be made by changing the activity coefficient for SO_4^{2-} and
for $CaSO_4^o$ but there is no guarantee that the result will be consistent
with other measured data containing some of the same species. Only simultaneous regression on all available data can achieve that degree of consistency.
There is one further point that makes a simultaneous fitting worthless with
the IA method alone. The solubility of gypsum in $NaClO_4$ solutions will be
simulated the same as in NaCl solutions with the IA approach because the
same ion pairs and activity coefficient expressions occur. However, the
measured solubility in $NaClO_4$ medium is considerably lower than in NaCl
medium, as shown in Figure 2. The simple Guggenheim-Scatchard SI model or
the Pitzer SI model can both simulate these solubilities very well where the
IA model fails. The data for the $NaClO_4$ solubilities is from Kalyanaraman,
et al. (1973). These authors have also tried to apply an IA model to their
results with no success. The conclusions from these calculations is that
(1) activity coefficients and equilibrium constants must be evaluated and
used in a consistent manner, and (2) at ionic strengths of seawater and
higher, the SI method must be used for reliable calculations.

As a final note on the hazards of chemical modeling I should like to
emphasize that there is a second sense in which model verification is urgently
needed. I refer to the application of the equilibrium assumption mentioned
earlier. This assumption must be tested, for example, by comparing reliable
speciation analyses with the calculated speciation or by comparing a saturation state calculation with the actual occurrence and precipitation/dissolu-
tion kinetics of a mineral in natural waters. The underlying assumption
about equilibrium and the inevitability of natural systems reaching that
state may prove to be the greatest uncertainty in model calculations.

160

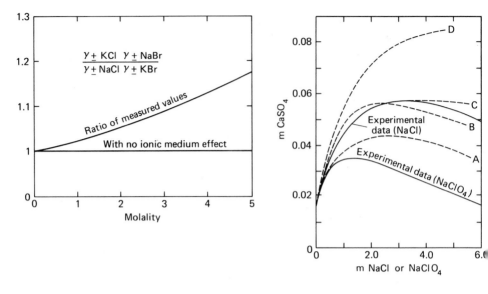

Figure 1. Plot of the $\dfrac{\gamma_{+KCl}\ \ \gamma_{+NaBr}}{\gamma_{+NaCl}\ \ \gamma_{+KBr}}$ function vs. $\overline{\text{molality}}$.

Figure 2. Computer simulations of gypsum solubility in NaCl or NaClO₄ at 25°C.

REFERENCES

-Arnorsson,S., Sigurdsson,S. and Svavarsson,H., 1982. The chemistry of geo-thermal waters in Iceland. I. Calculation of aqueous speciation from 0 to 270°C, Geochim. Cosmochim. Acta 46: 1513-1532.
-Ball,J.W., Jenne,E.A. and Nordstrom,D.K., 1979. WATEQ2 - A computerized chemical model for trace and major element specation and mineral equilibria of natural waters, In: E.A. Jenne (Ed), Chemical modeling in aqueous systems Amer. Chem. Soc. Symp. Series 93: 815-835.
-Ball,J.W., Nordstrom,D.K. and Jenne,E.A., 1980. Additonal and revised thermochemical data and computer code for WATEQ2 - A computerized chemical model for trace and major element speciation and mineral equilibria of natural waters. U.S. Geol. Survey Water Resour. Invest. Rept. 78-116, 109 pp.
-Ball,J.W., Jenne,E.A. and Cantrell,M.W., 1981. WATEQ3 - A geochemical model with uranium added. U.S. Geol. Survey Open-File Rept. 81-1183, 81 pp.
-Barnes,I. and Clarke,F.E., 1969. Chemical properties of groundwater and their encrustation effect on wells. U.S.Geol. Survey Prof. Paper 498-D, 58
-Bassett,R.L., Bentley,M.E., Duncan,E.A. and Griffin,J.A., 1981. Predicting the reaction state of brines in proposed regions of nuclear waste isolation In: J.G. Moore (Ed), Scientific basis for nuclear waste management 3, Plenum Press: 27-34.
-Biedermann,G., 1975. Ionic media. Dahlem workshop on the nature of seawater 339-362.
-Biedermann,G., Bruno,J., Ferri,D., Grenthe,I., Salvatore,F. and Spahiu,K., 1982. Modelling of the migration of lanthanoids and actinoids in ground water; the medium dependence of equilibrium constants. In: W. Lutze (Ed)

Scientific basis for radioactive waste management – V:791–800.
-Bos,M. and Meershoek, H.Q.J., 1972. A computer program for the calculation of equilibrium concentrations in complex systems. Anal. Chim. Acta 61:185–194.
-Bourg,A.C.M., 1982. ADSORP, A chemical equilibria computer program accounting for adsorption processes in aquatic systems. Environ. Tech. Letters 3:305–310.
-Chapman,B., James,R.O., Jung,R.F., and Washington,H.G., 1982. Modelling the transport of reacting chemical contaminants in natural waters, Aust. J. Mar. Freshwater Res. 33:617–628.
-Clement,A., and Fritz,B., 1983. THERMAL and EQUIL(T): Two simulation models for the interactions between natural waters and minerals, at variable temperature. Inst. Geol. Strasbourg Tech. Note 15.
-Crerar,D., 1975. A method for computing multicomponent chemical equilibria based on equilibrium constants. Geochim. Cosmochim. Acta 39:1375–1384.
-Detar,D.F., 1969. Computer programs for chemistry. II: 260 pp. W.A. Benjamin.
-Droubi,A., 1976. Geochemistry of salts and solution conncentrations upon evaporation. Thermodynamic simulation model. Application to the salt deposits of Tchad. Ph.D. Thesis Univ. Louis Pasteur, Strasbourg, France: 177 pp.
-Droubi,A., Fritz,B. and Tardy,Y., 1976. Equilibrium between minerals and solutions; Calculation programs applied to the prediction of soil salinity and optimal doses for irrigation, Cah. ORSTOM, ser. Pedol. XIV: 13–38.
-Eriksson,G., 1979. An algorithm for the computation of aqueous multicomponent, multiphase equilibria. Anal. Chim. Acta 112:375–383.
-Fardy,J.J. and Sylva,R.N., 1978. SIAS, a computer program for the generalized calculation of speciation in mixed metal-ligand aqueous systems, AAEC/E445, Lucas Heights, Australia: 20 pp.
-Felmy,A.R., Girvin,D.C. and Jenne,E.A., 1983. MINTEQ: A computer program for calculating aqueous geochemical equilibria, EPA Rept. (in press).
-Fisher,F.H. and Fox,A.P., 1975. Monosodium sulfate (-1) ion pairs in aqueous solutions at pressures up to 2000 atm., J. Solution Chem. 4: 225–236.
-Fritz,B., 1975. Thermodynamic study and simulation of the reactions between minerals and solutions. Application to the alteration geochemistry of continental waters. Doct. Eng. Thesis, University of Strasbourg, France: Sci. Geol. Mem. 41: 152 pp.
-Fritz,B., 1981. Thermodynamic study and modeling of hydrothermal and diagenetic reactions. Ph.D. Thesis, University of Strasbourg, France: Sci. Geol. Mem. 65: 197 pp.
-Garrels,R.M. and Thompson,M.E., 1962. A chemical model for seawater at 25°C and one atmosphere total pressure. Am. J. Sci. 260: 57–66.
-Gautem,R. and Seider,W.D., 1979. Computation of phase and chemical equilibrium: Part I. Local and constrained minima in Gibbs free energy. Part II. Phasesplitting. Part III. Electrolytic solutions. Am. Inst. Chem. Eng. J. 25:991–1015.
-Grove,D.B. and Wood,W.W., 1979. Prediction and field verification of subsurfacewater quality changes during artificial recharge, Lubbock, Texas. Ground water 17:250–257.
-Hamer,W.J. and Wu,Y.C., 1972. Osmotic coefficients and mean activity coefficients of uniunivalent electrolytes in water at 25°C, J. Phys. Chem. Ref. Data 1:1047–1099.
-Harvie,C.E. and Weare,J.H., 1980. The prediction of mineral solubilities in natural waters: the $Na-K-Mg-Ca-Cl-SO_4-H_2O$ system from zero to high concentration at 25°C. Geochim. Cosmochim. Acta 44: 981–997.
-Helgeson,H.C., Garrels,R.M. and MacKenzie,F.T., 1969. Evaluation of irreversible reactions in geochemical processes involving minerals and aqueous solutions: II. Applications. Geochim. Cosmochim. Acta 33: 455–481.

-Helgeson,H.C., Brown,T.H., Nigrini,A. and Jones,T.A., 1970. Calculation of mass transfer in geochemical processes involving aqueous solutions. Geochim. Cosmochim. Acta 34: 569-592.

-Holdren,G.R.,Jr. and Bricker,P.O., 1977. Distribution and control of dissolved iron and manganese in the interstitial waters of the Chesapeake Bay, In: H. Drucker and R.E. Wildung (Eds), Biological implications of metals in the environment. ERDA Symp. Series 42: 178-196.

I.T.P. and Nancollas,G.H., 1972. EQUIL - A computational method for the calculation of solution equilibria. Anal. Chem. 44: 1940-1950.

-Iglesias,E.R. and Weres,O., 1981. Theoretical studies of Cerro Prieto brines' chemical equilibria. Geothermics 10: 239-244.

-Ingle,S.E., Keniston,J.A. and Schults,D.W., 1979. REDEQL.EPAK, aqueous chemical equilibrium program. Corvallis Environmental Research Laboratory, Corvallis, Oregon.

-Ingri,N., Kakolowicz,W., Sillen,L.G. and Warnquist,B., 1967. High-speed computers as a supplement of graphical methods - V. HALTAFALL, a general program for calculating the composition of equilibrium mixtures. Talanta 14: 1261-1286.

-Jenne,E.A. (Ed), 1979. Chemical modelling in aqueous systems: speciation, sorption, solubility, and kinetics. Am. Chem. Soc. Symp. Series 93: 914 pp.

-Kalyanaraman,R., Yeatts,L.B. and Marshall,W.L., 1973. High-temperature Deby Huckel correlated solubilities of calcium sulfate in aqueous sodium perchlor solutions. Solubility of calcium sulfate and association equilibriums in $CaSO_4 + Na_2SO_4 + NaClO_4 + H_2O$ at 273-623 K. J. Chem. Thermodyn. 5: 891-909.

-Karpov,I.K. and Kaz'min,L.A., 1972. Calculation of geochemical equilibria i heterogeneous multicomponent systems. Geochem. Int. 9; 252-262.

-Karpov,L.K., Kaz'min,L.A. and Kashik,S.A., 1973. Optimal programming for computer calculation of irreversible evolution in geochemical systems. Geochem. Int. 10: 464-470

-Kerrisk,J.F., 1981. Chemical equilibrium calculations for aqueous geotherm brines. Los Alamos, New Mexico. Rept. LA-8851-MS: 21 pp.

-Kharaka,Y.K. and Barnes,I., 1973. SOLMEQ: Solution-mineral equilibrium computations. National Tech. Infor. Serv. Tech. Rept. PB214-899: 82 pp.

-Lafon,G.M., 1969. Some quantitative aspects of the chemical evolution of th oceans. Ph.D. Thesis, Northwestern University, Illinois: 136 pp.

-Ma,Y.H. and Shipman,C.W., 1972. On the computation of complex equilibria. Am. Inst. Chem. Eng. J. 18: 299-304.

-Marshall,W.L. and Slusher,R., 1966. Thermodynamics of calcium sulfate dihydrate in aqueous sodium chloride solutions, 0-100°C. J. Phys. Chem. 70: 4015-4027.

-Matisoff,G., Lindsay,A.H., Matis,S. and Soster,F.M., 1980. Trace metal mineral equilibriums in Lake Erie sediments. J. Great Lakes Res. 6:353-366.

-Mattigod,S.V. and Sposito,G., 1979. Chemical modelling of trace metal equilibrium in contaminated soil solutions using the computer program GEOCHI In: E.A. Jenne (Ed), Chemical modeling in aqueous systems. Amer. Chem. Soc. Symp. Series 93: 837-856.

-McDuff,R.E. and Morel,F.M., 1973. Description and use of the chemical equilibrium program REDEQL2. Keck Lab. Tech. Rept. EQ-73-02, California Institute of Technology, California: 75 pp.

-Mercer,J.W., Faust,C.R., Miller,W.J. and Pearson, F.J.,Jr., 1982. Review of simulation techniques for aquifer thermal energy storage (ATES), In: V.T. Chow (Ed) Advances in Hydroscience, Vol 13: 1-129.

-Miller,D. and Benson,L., 1983. Simulation of solute transport in a chemical reactive heterogeneous system: model development and application, Water Resour. Res. 19: 381-391.

-Millero,F.J. and Schreiber,D.R., 1982. Use of the ion pairing model to estimate activity coefficients of the ionic components of natural waters. Am. J. Sci. 282: 1508-1540.

-Morel,F. and Morgan,J.J., 1972. A numerical method for computing equilibria in aqueous systems. Env. Sci. Tech. 6: 58-67.

-Motekaitis,R.J. and Martell,A.E., 1982. BEST - a new program for rigorous calculation of equilibrium parameters of complex multicomponent systems. Can. J. Chem. 60: 2403-2409.

-Nordstrom,D.K., Plummer,L.N., Wigley,T.M.L., Wolery,T.J., Ball, J.W., Jenne,E.A., Bassett,R.L., Crerar,D.A., Florence,T.M., Fritz,B., Hoffman,M., Holdren,G.R.,Jr., Lafon,G.M., Mattigod,S.V., McDuff,R.E., Morel,F., Reddy,M.M., Sposito,G. and Thrailkill,J., 1979. A comparison of computerized chemical models for equilibrium calculations in aqueous systems, In: E.A. Jenne (Ed) Chemical modeling in aqueous systems. Amer. Chem. Soc. Symp. Series 93: 857-892.

-Parkhurst,D.L., Thorstenson,D.C. and Plummer,L.N., 1980. PHREEQE - a computer program for geochemical calculations. U.S. Geol. Survey WRI 80-96: 210 pp.

-Parkhurst,D.L., Plummer,L.N. and Thorstenson,D.C., 1982. BALANCE - a computer program for calculating mass transfer for geochemical reactions in ground water. U.S. Geol. Survey WRI 82-14: 29 pp.

-Perrin,D.D. and Sayce,I.G., 1967. Computer calculation of equilibrium concentrations in mixtures of metal ions and complexing species. Talanta 14: 833-842.

-Pitzer,K.S., 1979. Theory: ion interaction approach, In: R.M. Pytkowicz (Ed) Activity coefficients in electrolyte solutions, Vol 1, CRC Press: 157-208.

-Plummer,L.N., Parkhurst,D.L. and Kosiur,D.R., 1975. MIX2: a computer program for modeling chemical reactions in natural waters. U.S. Geol. Survey WRI 75-61: 68 pp.

-Plummer,L.N., Jones,B.F. and Truesdell,A.H., 1976. WATEQF - A FORTRAN IV version of WATEQ, a computer program for calculating chemical equilibrium of natural waters. U.S. Geol. Survey WRI 76-13: 61 pp.

-Plummer,L.N., Parkhurst,D.L. and Thorstenson,D.C., 1983. Development of reaction models for groundwater systems. Geochim. Cosmochim. Acta 47: 665-685.

-Proeseler,M. and Wagner,W., 1982. Calculation of complex equilibriums in heterogeneous multicomponent systems. Z. Phys. Chem. 263: 561-569.

-Reardon,E.J., 1975. Dissociation constants of some monovalent sulfate ion pairs at 25° from stoichiometric activity coefficiennts. J. Phys. Chem. 79: 422-425.

-Reed,M.H., 1982. Calculation of multicomponent chemical equilibria and reaction processes involving minerals, gases and an aqueous phase. Geochim. Cosmochim. Acta 46: 513-528.

-Rubin,J., 1983. Transport of reacting solutes in porous media: relation between mathematical nature of problem formulation and chemical nature of reactions. Water Resour. Res. 19: 1231-1252.

-Runnells,D.D. and Lindberg,R.L., 1981. Hydrogeochemical exploration for uranium ore deposits: use of the computer model WATEQFC. J. Geochem. Explor. 15: 37-50.

-Ryzhenko,B.N., Mel'nikova,G.L. and Shvarov,Y.V., 1981. Computer modeling of formation of the formation of the chemical composition of natural solutions during interaction in the water-rock system. Geochem. Int. 18: 94-108.

-Seider,W.D., Gautem,R. and White,C.W., III, 1980. Computation of phase and chemical equilibrium, In: R.G. Squires and G.V. Reklaitis (Eds) Computer applications to chemical engineering, Amer. Chem. Soc. Symp. Series 124: 115-134.

-Shvarov,Y.V., 1981. General equilibrium criterion for the isobaric-isothermic model of a chemical system. Geokhimiya 7: 981-988.

-Smith,W.R. and Missen,R.W., 1982. Chemical reaction equilibrium analysis. Wiley-Interscience: 364 pp.

-Sposito,G. and Mattigod,S.V., 1980. GEOCHEM: A computer program for the cal culation of chemical equilibria in soil solutions and other natural water systems. U.C.R., Dept. Soil Environ. Sci. Rept.: 92 pp.

-Thrailkill,J., 1970. Solution geochemistry of the water of limestone terrains. Univ. Kentucky Water Resour. Inst. Res. Rept. 19: 125 pp.

-Truesdell,A.H. and Jones,B.F., 1974. WATEQ, a computer program for calculat ing chemical equilibria of natural waters. t. U.S. Geol. Survey J. Res. 2: 233-274.

-Van Breemen,N., 1973. Calculation of ionic activities in natural waters. Geochim. Cosmochim. Acta 37:101-107.

-Van Luik, A.E. and Jurinak, J.J., 1978. A chemical model of heavy metals in the Great Salt Lake. Utah State University Res. Rept. 34: 155 pp.

-Van Luik,A. E. and Jurinak,J.J., 1979. Equilibrium chemistry of heavy metals in concentrated electrolyte solution, In: E.A. Jenne (Ed) Chemical modeling in aqueous systems, Amer. Chem Soc. Symp. Series 93: 683-710.

-Van Zeggeren; F. and Storey, S.H., 1970. The computation of chemical equili bria. Cambridge Univ. Press: 176 pp.

-Westall,J., 1979. MICROQL: I. A chemical equilibrium program in BASIC II. Computation of adsorption equilibria in BASIC. Swiss Inst. Tech., EAWAG: 77

-Westall,J., Zachary,J.L. and Morel,F.M.M., 1976. MINEQL, a computer program for the calculation of chemical equilibrium composition of aqueous systems. Mass. Inst. Tech. Dept. Civil Eng. Tech. Note 18: 91 pp.

-Whitfield,M., 1975. An improved specific interaction model for seawater at 25°C and 1 atmosphere pressure. Mar. Chem. 3: 197-205.

-Whitfield, M., 1979. Activity coefficients in natural waters, In: R.M. Pytkowicz (Ed) Activity coefficients in electrolyte solutions. Vol II: 153-299.

-Wigley,T.M.L., 1977. WATSPEC: a computer program for determining the equilibrium speciation of aqueous solutions. Brit. Geomorph. Res. Group Tech.Bull. 20: 48 pp.

-Wolery,T.J., 1979. Calculation of chemical equilibrium between aqueous solution and minerals: the EQ 3/6 software package. Livermore, California UCRL-52658: 41 pp.

-Wolery,T.J., 1983. EQ3NR, a computer program for geochemical aqueous speciation-solubility calculations User's Guide and Documentation:UCRL-5341

-Wood,J.R., 1975. Thermodynamics of brine-salt equilibria - I. The systems NaCl-KCl-MgCl$_2$-CaCl$_2$-H$_2$0 at 25°C. Geochim. Cosmochim. Acta 39: 1147-1163.

-Wood,J.R., 1976. Thermodynamics of brine-salt equilibria-II. The system Na KCl-H$_2$0 from 0 to 200°C. Geochim. Cosmochim Acta 40:1211-1220.

-Vallochi,A.J., Street,R.L. and Roberts,P.V., 1981. Transport of ion-exchang ing solutes in groundwater: chromatographic theory and field simulation. Water Resour. Res. 17: 1517-1527.

-Zeleznik,F.J. and Gordon,S., 1968. Calculation of complex chemical equili bria. Ind. Eng. Chem. 60: 27-57.

-Zemaitis,J.F. and Rafal,M., 1975. ECES - a computer system to predict the equilibrium composition of electrolyte solutions. 68th Ann. Mtg. Am. Inst. Chem. Eng., Louisiana.

METAL-ORGANIC BINDING: A COMPARISON OF MODELS

S.E. CABANISS, M.S. SHUMAN AND B.J. COLLINS

1.   INTRODUCTION

A model for metal-DOM speciation in oceans and other
environments must reflect the complex nature of DOM.   The
simplest possible model assumes a single binding site, 1:1
stoichiometry and no interactions between sites but is
inappropriate for most samples of DOM (Gamble, et al., 1980;
Perdue and Lytle, 1983a).   More complex models proposed for
metal-DOM binding include assumptions of electrostatic
interactions between sites (Wilson and Kinney, 1977), variable
stoichiometry (Buffle, et al., 1977), a continuous, Gaussian
distribution of binding constants among sites (Perdue and Lytle,
1983a) and a continuous distribution of binding constants of
unknown form (Gamble, et al., 1980; Shuman, et al., 1983).   Data
analysis methods proposed for other binding systems (Scatchard,
1949; Gordon, 1979) can also be used to analyze metal-DOM
equilibria.

The complexity of the DOM and its low concentration lead to
analytical and theoretical problems.   A model may fit very well a
wide range of experimental data, but unless it fits well at
environmental conditions, which are usually those of low $\bar{v}$ (bound
metal concentration/total ligand concentration) it is of
questionable use for metal-DOM speciation.   Extrapolations from
high $\bar{v}$ to low $\bar{v}$ are unreliable (Saar and Weber, 1979; Gamble, et
al., 1980); thus, the availability of good experimental binding
data at environmental levels and below is extremely important.
Even a model which fits low $\bar{v}$ data well, though useful, will not
substitute for sensitive analytical techniques.

The most general, rigorous model of metal-DOM binding is

Kramer, C.J.M. and Duinker, J.C. (eds.), Complexation of Trace Metals in Natural Waters. ISBN 90-247-2973-4
© 1984 Martinus Nijhoff/Dr W. Junk Publishers, The Hague/Boston/Lancaster.
Printed in the Netherlands.

conceptually straightforward, but very complex in practice. Ea
different binding site requires a set of acidity constants, a
series of complex formation constants, β_j, and terms correcting
them for interactions between sites. Each possible grouping of
sites which could bind to a single metal ion requires a
mixed-ligand stability constant. The number of parameters
necessary to describe a five or six component system is quite
large (Byrne, 1983), and greatly exceeds the number of data
points in typical titrations (15-40). Clearly this type of mode
is unsuitable for data analysis.

Models in current use represent a compromise between rigor
and practicality. Proton competition terms are unnecessary if
measurements at constant pH are used to obtain conditional
constants. Although not rigorously correct, certain assumption
about stoichiometry, interactions and the number of sites
simplify the data analysis considerably. Models using some of
these assumptions with three or more parameters often fit the
data much better than the simple single-site model. This
improved fit has been cited as supporting evidence for these
models (Buffle, et al., 1977; Wilson and Kinney, 1977).
Goodness-of-fit criteria alone cannot justify these assumptions
however; several models with contradictory assumptions can
curve-fit metal-DOM binding data (see below).

Any comparison of models must include the following importa
considerations:

1. Chemical correctness - What are the assumptions and are the
   reasonable? Can the model be used to think intuitively abo
   metal-DOM chemistry?

2. Accuracy - Does the model fit the data within experimental
   error? Are the important data points at low saturation wel
   described?

3. Simplicity - Is the data analysis readily reproducible by
   other workers? Are the resulting parameters useable with
   equilibrium speciation programs, e.g. REDEQL, MINEQL, etc.?

The purpose of this paper is to review critically the model
that are presently used with metal-DOM binding data and to

recommend a data analysis method that is easy to use, realistic and accurate.

2. REVIEW OF MODELS IN CURRENT USE

The simplest possible model makes three convenient assumptions - 1) all metal binding sites on DOM have the same binding constant K, 2) sites do not interact with each other through electrostatic, conformational, or other changes and 3) only 1:1 complexes are formed. This model is appropriate for many laboratory systems containing a single, pure chelator and can occasionally fit metal-DOM data over a narrow range of pM values, but the assumptions cannot be justified and it often fails to fit metal-DOM data over a wide range of pM values.

Several models which do not assume a single binding K have been applied to metal-DOM systems. The most common postulates two distinct binding sites and uses graphical data analysis similar to that suggested by Scatchard (1949) for systems of small molecules or ions binding to proteins. The binding equation

$$\bar{v} = K\ [M]/(n + K[M]) \qquad (1)$$

(where n = number of sites per molecule, K = the association constant, [M] = the concentration of free metal and $\bar{v}$ = metal bound/total binding sites) can be rearranged to

$$\bar{v}/[M] = K\ (n-\bar{v}) \qquad (2)$$

If $\bar{v}/[M]$ plotted versus $\bar{v}$ is linear, a single binding K can be used to fit the data. More commonly, the plot is curved and the limiting slopes and intercepts are used to determine the K and concentration of each of two types of sites. The use of two sites instead of one often greatly improves the fit to the data, and the graphical technique is easy to use.

Two site models have several problems. Limiting slopes are often difficult to determine, since the slope is often a function of the number of points used to fit the lines. This subjectivity can be avoided by using a least squares curve fit of the entire data set. The choice of two sites is somewhat arbitrary, since increasing the number of sites improves the least squares fit.

The values of K and site concentration found using this model must be regarded as fitting parameters, rather than representing two distinct classes of sites (Perdue and Lytle, 1983a).

A model postulating a small number (n<10) of discrete sites is conceptually similar to the two-site model, since it consider only 1:1 complexes and requires no specific interactions between sites. However, this model can describe systems with interactin sites. Simms (1926) showed that a polyprotic acid can be rigorously represented by a series of "titration" constants. This analysis also applies to metal-ligand binding if only 1:1 complexes are formed. A model with n sites can thus describe up to n independent binding sites or n interacting sites.

Data analysis for a model with n discrete sites presents unique problems. Graphical analysis after Scatchard (1949) is inappropriate when six or more parameters must be determined (Klotz and Hunston, 1971). Least squares methods often fail to converge without good initial estimates. Increasing the model r generally improves the fit regardless of the number of sites actually present. A program developed by Gordon (1982) for acid-base titrations of mixtures provides both good starting values in the least-square algorithm and criteria for choosing r We are currently modifying this program to handle metal-DOM data

A more radical challenge to the single-site assumption postulates a continuous distribution of binding Ks. The probability that any given site has a binding log K between a a b is

$$P\ (a,b)\ =\ \int_a^b N(k)\ d\ \log\ K \qquad (3)$$

and the binding curve can be described by

$$\bar{v}\ =\ \int_{-\infty}^{\infty} N(K)\ \frac{[M]K}{1+[M]K}\ d\ \log\ K \qquad (4)$$

where N(K) is the distribution function of log K. This approach is well suited to mixtures with a large number of slightly different sites. Like the n discrete site model, a continuous distribution can treat data for systems with interacting sites.

Two continuous distribution models are current in the literature. Karush and Sonenberg (1949) used a Gaussian distribution of log K to model binding on serum albumen; Posner

(1966) used this model for proton-humic acid equilibria. Perdue and Lytle (1983a) have proposed its use for metal-DOM equilibria as well. The Gaussian distribution of log K is expressed

$$P(a,b) = \frac{1}{\sigma\sqrt{2\pi}} \int_a^b e^{-(\mu-\log K)^2/2\sigma^2} d \log K. \quad (5)$$

The distribution can be completely described by the mean, μ, and the standard deviation, σ, both easily understood parameters. Data analysis by non-linear least squares can easily incorporate bimodal and trimodal distributions (Perdue and Lytle, 1983b).

The a priori choice of a Gaussian or any other distribution is somewhat arbitrary. A model without any a priori assumptions about the shape of the distribution has been used to model biochemical (Hunston, 1975; Thakur, et al., 1980) and metal-DOM equilibria (Shuman, et al., 1983). The integral in equation 4 can be solved by transform techniques or finite difference methods (Ferry, 1981) to obtain N(K), the affinity spectrum (Fig. 1). This is more general than the Gaussian distribution, but the affinity spectrum cannot be easily summarized in a few parameters. Neither continuous model produces output parameters compatible with common equilibrium speciation programs.

The assumption that sites do not interact has also been challenged. Metal binding to macromolecules can cause coagulation, conformational changes and changes in local and overall electron density. The initial binding of small molecules to biological macromolecules can enhance or inhibit further binding (Tanford, 1961; Marshall, 1978). Electrostatic effects are likely to dominate the site interaction when the small molecule is charged and the bonding is ionic (e.g., H+). The apparent binding K_{app} for a given saturation level can be expressed

$$K_{app} = K'_{int} e^{-2Zzw} = K_{int} e^{-\omega\bar{v}} \quad (6)$$

(Z = average charge on the macromolecule, z = charge on the ion, w = electrostatic factor, $\omega = 2wz$, $K_{int} = K_{app}$ at $\bar{v} = 0$. See Tanford, 1961 for a fuller description).

This model has been proposed by Wilson and Kinney (1977) for use with proton-DOM and metal-DOM equilibria. It requires that

the acid-base behavior of the DOM be known to study metal
binding. This model can easily be expanded to include two or
more sites at the cost of increasing complexity. The output K_{in}
and ω are easily understood as the intrinsic binding K and a
measure of the strength of the interaction, but they are not
compatible with commonly available equilibrium speciation
programs. The theoretical basis of this model is most
appropriate for systems of large macromolecules and ionic bondin
(Tanford, 1961). Consequently, applying it to metal-DOM systems
with small molecules (< 10 sites per molecule) or metal ions
which may not bind ionically (e.g., $Fe3+$, $Cu2+$) is questionable.
    The assumption that binding to DOM has a 1:1 (metal:site)
stoichiometry is very reasonable for proton-DOM binding but is

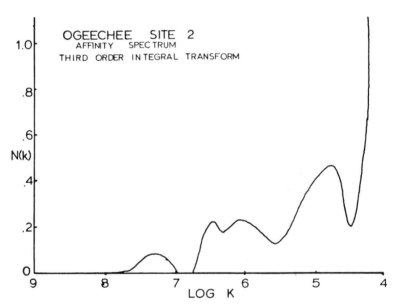

Figure 1.   Affinity Spectrum of Ogeechee site 2
             binding data. Peak size is proportional
             to site concentration.

unrealistic for binding of metals with high coordination numbers
(C.N. > 2, Cu^{2+}, etc.). The metal ion: binding site ratio must
be used to denote stoichiometry, since several functional groups
on a single molecule may bind to a single metal ion. For
example, a DOM molecule with two carboxyl groups and two phenol:

groups might bind a single metal ion to all four groups at low metal levels, and bind one metal ion to each carboxyl-phenolic pair at higher metal levels. This is analogous to the formation of 1:2 and 1:1 copper-salicylic acid complexes, and can be expressed

$$\beta_1 = \frac{(ML)}{(M)(L)} \qquad (7a)$$

$$\beta_2 = \frac{(ML_2)}{(ML)(L)} \qquad (7b)$$

Where β_1 and β_2 are the first and second formation constants. Changes in stoichiometry need not require two or more DOM molecules, but can occur among different sites on one molecule.

Buffle, et al., (1977) used a model postulating 1:1 and 1:2 complexes and only one type of site. Assuming that only the 1:1 complex is formed at high MT/LT ratios (MT = total metal concentration), a plot of the linear portion of the equation

$$\frac{\alpha}{(1-\alpha)M_T} = \frac{1}{L_T} + \frac{1}{L_T \beta_1 (M)} \qquad (8)$$

($\alpha = MT/[M]$) is used to obtain LT, and β_1 and β_2 are found by non-linear least squares. These three parameters may be used with MINEQL, REDEQL, etc.

A rigorous treatment of 1:j stoichiometries must consider the formation of mixed ligand complexes. Buffle (1980) specifically considers the possibility of copper-hydroxide-DOM and copper-carbonate-DOM complex formation. Byrne (1983) points out that the formation of mixed-ligand complexes is statistically favored in complex media, and offers an approximate method of modelling these equilibria when all components and pure-ligand formation constants are known. MacCarthy and Smith (1979) consider the case of a complex mixture of ligands with metal:site ratios from 1:1 to 1:4. They conclude that the system is too complex to fully resolve, and advocate a stability surface treatment. The variable stoichiometry model of Buffle, et al., (1977) is clearly an oversimplification of a difficult problem, but it is the only model which allows 1:j complexes (where j is not equal to 1) and attempts to resolve binding β_j from the experimental data.

3. COMPARISON OF MODEL PREDICTIONS

Eight of the models discussed above were used to analyze th
experimental results of a copper-into-ligand titration of
estuarine DOM. The DOM was taken from the Ogeechee estuary,
Georgia, and then concentrated, size fractionated and desalted
ultrafiltration and diafiltration (Fitzgerald, 1982). The
1,000-50,000 molecular weight fraction was diluted to 19.95 mg
c/liter, buffered to pH 6.00 by .001 M carbonate and adjusted t
ionic strength .1 by KNO_3. The [Cu 2+] was measured with an
Orion cupric ion selective electrode and a Fisher digital
voltmeter (Shuman, et al., 1983). The endpoint was determined
a modified Gran's plot procedure (Collins, 1983). Results for
one sample are discussed, although analysis of other data from
the Ogeechee and the data of Gamble, et al., (1980) gave simila
results.

Data analysis for the electrostatic model, the variable
stoichiometry model, and the one, two and three site models use
the SAS procedure NLIN, a Marquardt non-linear least-squares
fitting routine (SAS Institute, 1980). The two site graphical
analysis and the Gaussian distribution models used Basic progra
on an Apple II microcomputer. The affinity spectrum analysis w
performed by a Fortran program using a Stieltjes transform
technique (Collins, 1983).

Three indices of the model's numerical predictive power are
shown in Table 1; twenty-nine data points with $\bar{v}$ up to 0.8 were
used. The residual sum of squares (RSS) of $\bar{v}$ is the simplest
index of overall fit and was minimized for all models except th
two-site graphical analysis and the Gaussian distribution. The
tendency of the RSS to emphasize small percentage errors at hig
saturation is balanced by the greater number of data points at
low saturation. The RSS for $\bar{v} < 0.1$ tabulation indicates how
well the models fit in the environmentally important low $\bar{v}$ data
The weighted residual sum of squares (WRSS) normalizes the squa
error to $\bar{v}$; the Gaussian model analysis minimized WRSS (Perdue
and Lytle, 1983b).

A good fit by RSS does not imply that the low saturation da
are well described. The electrostatic, variable stoichiometry,

Table 1. Model Predictions: Error Analysis

Model	RSS	RSS $(\bar{v}<.1)$	WRSS
Simple Single Site	.0919	.0162	.641
Two Site Graphical	.0540	.00486	.330
Two Site Least Squares	.00102	.000324	.0167
Three Site	.000756	.0000441	.00217
Gaussian Distribution	.0164	.000199	.0370
Affinity Spectrum	.0122	.000108	.0234
Electrostatic Interaction	.0165	.000549	.0555
Variable Stoichiometry	.0203	.00144	.0873

$$RSS = (\bar{v}-\bar{v}_{calc})^2 \qquad\qquad WRSS = \frac{(\bar{v}-\bar{v}_{cal})^2}{\bar{v}}$$

Gaussian distribution and affinity spectrum models all have RSS within a factor of two, and the RSS of the two site (least squares) model is a factor of ten lower. However, the RSS at $\bar{v}$ <.1 conditions indicates that the two continuous distribution models fit the data much better than the other three in this range.

The large errors in the single site and the two site (graphical) models confirm that they are inadequate to describe the data. Except for these two, all the models can approximately fit the data (see Figure 2). Exact fitting to within experimental error is achieved over most of the $\bar{v}$ range, and data sets covering a narrow range of $\bar{v}$ may be fitted by any of them. Goodness of fit is not a valid criterion for judging the chemical correctness of a model, since these models based on contradictory assumptions all have "good" fits.

The best fit to this data is obtained from the three site model, followed by the two continuous models. The only model without a large relative error (>25%) in the low $\bar{v}$ region is the three site model. For other data sets analyzed, the same ranking is observed, although sets with fewer data points in the low $\bar{v}$ region have a smaller gap between the error values.

174

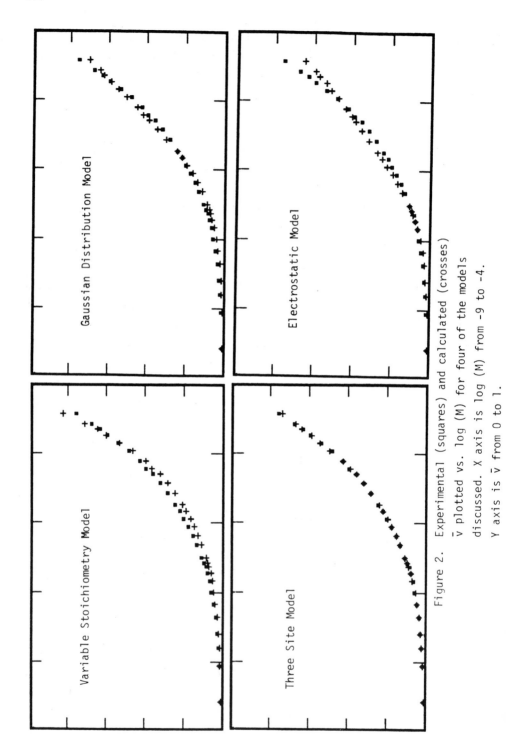

Figure 2. Experimental (squares) and calculated (crosses) v̄ plotted vs. log (M) for four of the models discussed. X axis is log (M) from -9 to -4. Y axis is v̄ from 0 to 1.

The five best fitting models indicate a large concentration of sites in the region log K = 4.6 -4.9 (Table 2) and four of these indicate a small concentration of very strong binding sites (log K > 7.5).

4. CRITIQUE

Models can be compared on theoretical and practical grounds, according to the criteria discussed above - chemical correctness, numerical accuracy, and simplicity.

1). The electrostatic interaction model reflects an important aspect of metal-DOM equilibria. Nonetheless, it is inadequate on theoretical grounds because it ignores the diverse nature of DOM binding sites, it ignores the possibility of changing stoichiometry, and it is inappropriate for binding of metals like Cu and Fe. (Dempsey and O'Melia (1983) found that although this model could describe acid-base data, it was unable to fit calcium-DOM data.) Furthermore, models postulating a large number of sites may rigorously represent interactions (Simms, 1926), so that those models are more general than this one. The electrostatic model does not fit the data especially well and is incompatible with MINEQL, etc.

2). The variable stoichiometry model emphasizes the possibility of 1:j complexes, including mixed ligand complexes. It is the only model to account for changing stoichiometry, which cannot be neglected in any rigorous model. The data analysis is straightforward and produces easily understood parameters compatible with MINEQL, etc. However, this model ignores the diverse nature of DOM binding sites and does not allow for interactions between sites. The fit to the data is clearly inadequate at low $\bar{v}$.

3). The n discrete site model stresses the differences in binding sites on DOM and is simple to use. It allows for limited site diversity and interaction. The fit to the data is excellent for n > 2, and the output parameters are easily understood and compatible with MINEQL, etc. Theoretical objections are that it ignores the possibility of 1:j complexes and underestimates the diversity of the binding sites. A practical objection is the

Table 2. Model Predictions: Output Parameters

Model	Parameters
Simple Single Site	Log K = 5.01
Two Site Graphical	$\log K_1$ = 7.86; C_1 = 0.303; $\log K_2$ = 4.91; C_2 = .9697
Two Site Least Squares	$\log K_1$ = 6.44; C_1 = .194; $\log K_2$ = 4.74; C_2 = .806
Three Site	$\log K_1$ = 8.25; C_1 = .0125; $\log K_2$ = 6.32; C_2 = .192; $\log K_3$ = 4.73, C_3 .795
Gaussian Distribution	μ = 4.94; σ = .94
Affinity Spectrum	peaks in log K at 7.7, 7.3, 6.6, 6.1, 4.9, 4.2
Electrostatic Interaction	$\log K_{int1}$ = 6.03; ω = 3.11; C_1 = .483; $\log K_{int2}$ = 4.616; ω = 0; C_2 .517
Variable Stoichiometry	$\log \beta_1$ = 4.17; $\log \beta_2$ = 3.71

lack of a data analysis method with criteria for determining n and a method for ensuring rapid convergence of the least squares algorithm.

4). Continuous distribution models emphasize the diversity o DOM binding sites. The Gaussian distribution model sacrifices generality for simplicity. This model can account for some interactions and provides a good fit to the data. The output parameters are easily understood, although incompatible with common equilibrium speciation programs. Two theoretical problem are that 1:j complex formation is ignored and the a priori assumption of a Gaussian distribution is somewhat arbitrary.

The affinity spectrum model makes no assumptions about the shape of the ligand distribution, so that it is more general tha the Gaussian model and better able to fit the data. It can account for any type of interaction between sites. The principa

theoretical drawback is that it does not consider 1:j complex
formation. Practical difficulties are that output parameters are
an array of N(K) values, incompatible with equilibrium speciation
programs, and its limited resolution prevents a simple
description of a single ligand binding.

5. RECOMMENDATIONS

The description of metal-DOM speciation requires both a
theoretical, conceptual model and a numerically accurate
predictive model. Goodness-of-fit criteria cannot validate a
theoretical model, since several models with contradictory
assumptions can fit binding data. None of the models discussed
above is rigorously correct, and hence none is a satisfactory
conceptual model. The continuous binding K distribution models
are the most general, accounting for site diversity and
interactions between sites, but neglecting changes in
stoichiometry due to formation of 1:j complexes. A good
conceptual model should account for site diversity, site
interactions and variable stoichiometry.

Potentially the most useful predictive model is the n
discrete site model. This model accounts for limited site
diversity and interactions, and the output parameters are
compatible with commonly available equilibrium speciation
programs. Despite its approximations, the n discrete site model
fits the data better than any other considered. Data analysis
using an affinity spectrum technique to find starting values for
a least squares algorithm speeds convergence (Gordon, 1982) and
provides a graphical, qualitative description of the binding
(Hunston, 1975; Shuman, et al., 1983). This deconvolution of the
affinity spectrum into discrete sites is highly accurate and
represents the DOM as a series of easily manipulated titration
constants and concentrations.

REFERENCES

-Buffle J. 1980. A critical comparison of studies of complex formation between copper (II) and fulvic substances of natural waters, Anal. Chim. Acta 118:29-44.

-Buffle J, Greter F, Haerdi W. 1977. Measurement of complexation properties of humic and fulvic acids in natural waters with lead and copper ion-selective electrodes, Anal. Chem. 49: 216-222.

-Byrne RH. 1983. Trace metal complexation in high ligand variety media, Mar. Chem. 12:15-24.

-Collins BJ. 1983. Affinity spectrum method for metal-organic binding in natural waters, MSPH thesis, Univ. North Carolina at Chapel Hill.

-Dempsey BA, O'Melia CR. 1983. Proton and calcium complexation of four fulvic acid fractions. In: Aquatic and Terrestrial Humic Materials, R.F. Christman and E.T. Gjessing, eds., Ann Arbor Science, Ann Arbor, MI, 239-273.

-Ferry JD. 1981. Viscoelastic properties of polymers, Third edition, Wiley-Interscience, New York, NY.

-Fitzgerald PJ. 1982. Dissociation rate constants of copper-estuarine organic complexes estimated with a rotating disk electrode technique, MSPH thesis, Univ. North Carolina at Chapel Hill.

-Gamble DS, Underdown AW, Langford CH. 1980. Copper (II) titration of fulvic acid ligand sites with theoretical, potentiometric and spectrophotometric analysis, Anal. Chem. 52: 1901-1908.

-Gordon WE. 1982. Data analysis for acid-base titration of an unknown solution Anal. Chem. 54:1595-1601.

-Gordon WE. 1979. Component discrimination in acid-base titration, J. Phys. Chem. 83:1365-1377.

-Hunston DL. 1975. Two techniques for evaluating small molecule-macromolecule binding in complex systems, Anal. Biochem. 63:99-109.

-Karush F, Sonenberg M. 1949. Interaction of homologous alkyl sulfates with bovine serum albumin, J. Am. Chem. Soc. 71:1369-1376.

-Klotz IM, Hunston DL. 1971. Properties of graphical representations of multiple classes of binding sites, Biochem. 10:3065-3069.

-MacCarthy P, Smith GC. 1979. Stability surface concept. In: Chemical Modeling in Aqueous Systems, E.A. Jenne, ed., ACS Symposium Series No. 93, ACS, Washington, DC., 201-222.

-Marshall AG. 1978. Biophysical Chemistry, John Wiley and Sons, New York, NY 812 pp.

-Perdue EM, Lytle CR. 1983a. A critical examination of metal-ligand complexation models: Application to defined multiligand mixtures. In: Aquatic and Terrestrial Humic Materials, R.F. Christman and E.T. Gjessing, eds. Ann Arbor Science, Ann Arbor, MI, 295-313.

-Perdue EM, Lytle CR. 1983b. Distribution model for binding of protons and metal ions by humic substances, Environ. Sci. Technol., in press.

- Posner AM. 1966. The humic acids extracted by various reagents from a soil. Part I. Yield, inorganic components and titration curves, J. Soil Sci. 17:65-78.
- Saar RA, Weber JH. 1979. Complexation of cadmium (II) with water and soil-derived fulvic acids: Effect of pH and fulvic acid concentration, Can. J. Chem. 57:1263-1268.
- SAS Institute. 1979. SAS Users Guide. SAS Institute, Cary, North Carolina, 317-329.
- Scatchard G. 1949. The attraction of proteins for small molecules and ions, Ann. N.Y. Acad. Sci. 51:660-672.
- Shuman MS, Collins BJ, Fitzgerald PJ, Olson DL. 1983. Distribution of stability constants and dissociation rate constants among binding sites on estuarine copper-organic complexes. In: Aquatic and Terrestrial Humic Materials, R.F. Christman and E.T. Gjessing, eds., Ann Arbor Science, Ann Arbor, MI, 349-370.
- Simms HS. 1926. Dissociation of polyvalent substances. I. Relation of constants to titration data, J. Am. Chem. Soc. 48:1239-1250.
- Tanford C. 1961. Physical Chemistry of Macromolecules, John Wiley and Sons, Inc., New York, NY.
- Thakur AK, Munson PJ, Hunston DL, Rodbard R. 1980. Characterization of ligand-binding systems by continuous affinity distributions of arbitrary shape, Anal. Biochem. 103:240-254.
- Wilson DE, Kinney P. 1977. Effects of polymeric charge variations on the proton-metal equilibria of humic materials, Limnol. Oceanog. 22:281-289.

THERMODYNAMIC AND ANALYTICAL UNCERTAINTIES IN TRACE METAL SPECIATION
CALCULATIONS

RICHARD W. ZUEHLKE AND ROBERT H. BYRNE

1. INTRODUCTION

The validity of chemical speciation models should be assessed in the
context of field measurements. Such an assessment must also include
detailed consideration of the cumulative uncertainties inherent in the
process of constructing chemical models. In this note we examine the
effect of uncertainties in trace metal complexation constants and pH on the
predicted fraction of free metal ion. Only mononuclear inorganic complexes
are included. Analytical uncertainties (other than in pH determinations)
are not treated.

Using the familiar formulation for speciation models:

$$T(M) = [M^{n+}] \cdot \left\{ 1 + \sum_{i,j} \beta_{ij} \cdot [L_i]^j \right\}$$
$$= [M^{n+}] \cdot \Sigma \tag{1}$$

where β_{ij} is a formation constant for complexation of M^{n+} with ligand
L_i, we may derive two equations relating uncertainties in $[M^{n+}]$ and
β_{ij}.

$$dpM = \frac{[L_i]^j}{2.303 \cdot \Sigma} \cdot d\beta_{ij} \tag{2}$$

and by integrating (2):

$$\frac{\Delta [M^{n+}]}{[M^{n+}]_o} = - \frac{[L_i]^j}{\Sigma} \cdot \Delta \beta_{ij} \tag{3}$$

where pM is defined as $-\log [M^{n+}]$. Equations (2) and (3) are extensions
of those derived by Stolzberg (1981) whose work was confined to a 1-1

Kramer, C.J.M. and Duinker, J.C. (eds.), Complexation of Trace Metals in Natural Waters. ISBN 90-247-2973-4
©*1984 Martinus Nijhoff/Dr W. Junk Publishers, The Hague/Boston/Lancaster.*
Printed in the Netherlands.

complex between a given trace metal and a given ligand. Here, $d\beta_{ij}$ or $\Delta\beta_{ij}$ may be viewed in either of two ways:

1. Uncertainty in β_{ij} which contributes to the uncertainty in pM.
2. The minimum value of a heretofore omitted complex which would make a given contribution to pM.

We now explore the application of these equations, through a propagation-of-errors analysis, to three trace metal systems in seawater:

1. Cd -- bound almost entirely by the conservative ligand, Cl^-.
2. Cu -- bound almost entirely by pH-sensitive ligands.
3. Pb -- a case intermediate between Cd and Cu.

A compilation of overall formation constants and uncertainties for Cd, Cu and Pb is shown in Table I. Values in the table represent, wherever possible, studies carried out at $35^\circ/oo$ salinity; data requiring activity coefficient estimates and conversion from one salinity to another were not used if directly appropriate experimental values were available. Where several sources for the same constant were available, the relative standard deviation (RSD) was based on the mean and resulting standard deviation; if only a single source was used, the RSD was based on its data set.

TABLE I. Formation constants and uncertainties for Cd, Pb and Cu at $35^\circ/oo$ salinity, 25°C and 1 atm.

Species	Cd	Pb	Cu
$-OH$	-10.19 (?)	-7.89 (40.2%)	-8.02 (12.4%)
$-(OH)_2$	-21.67 (?)	-17.29 (36.6%)	-16.7 (10.8%)
$-Cl$	1.35 (7.2%)	0.89 (2.7%)	-.37 (16.7%)
$-Cl_2$	1.70 (25.9%)	1.16 (4.8%)	
$-Cl_3$	1.48 (25.9%)	1.07 (11.2%)	
$-Cl_4$	1.36 (75%)		
$-F$	.46 (?)		.90 (?)
$-SO_4$	1.05 (?)	1.42 (?)	.38 (?)
$-HCO_3$	**	*	2.75 (43.9%)
$-CO_3$	2.95 (?)	6.07 (41.5%)	6.29 (15.7%)
$-(CO_3)_2$		9.32 (106.8%)	8.47 (47.0%)
$-B(OH)_4$	**	*	*

NOTE: Form gives logarithm of formation constant with a percent relative standard deviation in parentheses.

* A missing, but probably significant species. See text.
** A missing, but probably minor species. See text.

2. APPLICATIONS

2.1. Cadmium

Formation constants and associated uncertainties are shown in Table I.

A derived plot of pCd-log T (Cd) through a salinity gradient is shown in Figure 1; the envelope displayed shows the relative importance of uncertainties in the 1-1, 1-2, 1-3, 1-4 chloride formation constants as they change with chloride concentrations. At a salinity of 35°/oo, this corresponds to an uncertainty of $\pm$ 0.06 pCd unit.

2.2. Copper

The effects of the uncertainties of Table I on copper speciation are shown in Figure 2a. Current knowledge of copper's equilibrium chemistry will not allow prediction of pCu in seawater to better than $\pm$ 0.06 unit. If uncertainties in pH, ligand concentration, total copper and organic binding are added, the potential error in pCu rises dramatically.

The slope of Figure 2a is presented in Figure 2b and emphasizes that Cu's pH dependence reaches a maximum at the pH of seawater. Any uncertainty in pH (caused by liquid junction problems, improper selection of calibration buffers, or use of poorly defined pH scales) can be directly translated into a pCu error by use of the slope value. Thus, a pH measurement of 0.05 unit translates to a pCu error of 0.04 unit.

2.3. Lead

Uncertainties in Pb speciation are also shown in Figures 2a and 2b; the similarity to the case of Cu is apparent.

The effect of omitting a potentially important species such as the mixed ligand complex, $Pb(OH)(CO_3)^-$, is shown in Figure 3. Mixed complexes in Equation 3 may be accounted for in the following modification:

$$\frac{[M^{n+}]}{[M^{n+}]_o} = - \frac{[L_a] \cdot [L_b] \cdot \Delta\beta_{ab}}{\Sigma + [L_a] \cdot [L_b] \cdot \Delta\beta_{ab}} \tag{4}$$

where $\underline{a}$ and $\underline{b}$ refer to OH^- and CO_3^{2-}, and $\Delta\beta_{ab}$ is now interpreted as the new mixed ligand constant (i.e., it is incremented from zero).

The new formation constant is approximated by (see Byrne 1983)):

$$\beta_{ab} \text{ (or } \Delta\beta_{ab}) = 2 \cdot \{\beta_{Pb(OH)_2} \cdot \beta_{Pb(CO_3)_2}\}^{1/2} \tag{5}$$

Its importance is observed at high pH, where its omission is expected to cause about a 10% error in Σ, or a 0.04 unit error in pPb.

2.4. Omitted Species

Equation 2 can be used to identify potentially important complexes which have been omitted from speciation models. One might question, for

184

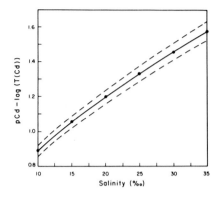

FIGURE 1. Uncertainties inCd^{2+} speciation through a salinity gradient.
[envelope shows relative importance of 1-1, 1-2, 1-3, and 1-4 chloride
species and their associated formation constant uncertainties as they
change with chloride concentrations].

(a)   (b)

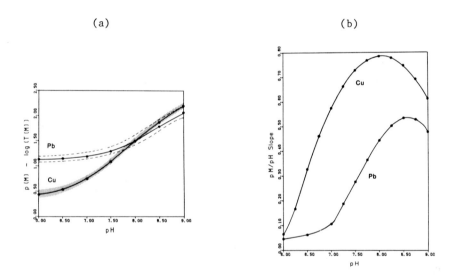

FIGURE 2(a). Values of pM for Cu and Pb as a function of pH. Uncertainty
envelopes computed from data in Table I.
(b). Slope of (a) plotted against pH.

instance, whether borate complexes will affect pCu by 0.05 or more unit.
From equation 2, given that Σ for Cu at pH 8 is about 31 and that the free
borate concentration is 3.6×10^{-5}M, the minimum required formation
constant is about 1×10^{5}.

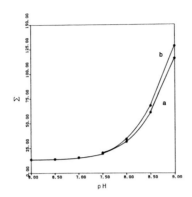

FIGURE 3. Effect of (a) omitting or (b) including the mixed ligand Pb(OH)(CO$_3$)$^-$ complex on the speciation of Pb^{2+}.

Such a preliminary value has indeed been found by one of us (R.W.Z., unpublished results) and by C. van den Berg (personal communication, 1983). Similar reasoning suggests that borate complexation of Pb^{2+} may also be significant.

The results of these considerations are included in Table I.

ACKNOWLEDGEMENTS

This work was supported by the U.S. National Science Foundation through Grants OCE 81-17850 and OCE 81-10162.

REFERENCES

-Byrne, R.H., 1983. Trace metal complexation in high ligand variety natural media. Mar. Chem. 12: 15-24.

-Stolzberg, R.J., 1981. Uncertainty in calculated values of uncomplexed metal ion concentration. Anal. Chem. 53: 1286-1291.

THE ACTIVITY OF METAL IONS AT HIGH IONIC STRENGTHS

FRANK J. MILLERO

1. INTRODUCTION

The effect of chemical composition on the activity of metal ions in natural waters can be determined by using ionic interaction models (Whitfield, 1979). The two most popular models used by various workers in recent years are the ion pairing model (Dickson and Whitfield, 1981; Millero and Schreiber, 1982) and the specific interaction model as formulated by Pitzer (Pitzer, 1973; Whitfield, 1975; Harvie and Weare, 1980; Krumgalz and Millero, 1982; Millero, 1983a,b). The specific interaction model yields reliable activity estimates for the major ionic components, while the ion pairing model yields reliable estimates for the minor ionic components (Whitfield, 1979; Millero and Schreiber, 1982). The failure of the equations of Pitzer is not due to deficiencies in the model, but a lack of parameters for the interactions of Mg^{2+} and Ca^{2+} ions with minor anions (OH^-, HCO_3^-, CO_3^{2-}, etc.).

Recently, we have attempted to derive Pitzer parameters from acid ionization (Millero, 1983a,b; Thurmond and Millero, 1982; Millero and Thurmond, 1983) and solubility data (Millero et al., 1983). We have also attempted to combine the ion pairing and Pitzer equations (Whitfield,1975) to treat the ionic interactions of trace metals in ionic media (Millero and Byrne, 1983). In this paper we will briefly review the highlights of our recent studies.

2. PITZER'S EQUATIONS

The activity (a_i) and total concentration, $[i]_T$, of an ionic component (i) are related by

$$a_i = [i]_T \, \gamma_T \, (i)$$

(1)

Kramer, C.J.M. and Duinker, J.C. (eds.), Complexation of Trace Metals in Natural Waters. ISBN 90-247-2973-4
© *1984 Martinus Nijhoff/Dr W. Junk Publishers, The Hague/Boston/Lancaster.*
Printed in the Netherlands.

The total or stoichiometric activity coefficient (γ_T) is related to the non-ideal behavior of the solute due to ionic interactions. Most of the early models used to estimate activity coefficients of ions in natural waters were based on extensions of the Debye-Hückel theory. The value of γ_i for an ionic solute of charge Z_i is given by

$$\ln\gamma_i = Z_i^2 A I^{1/2}/(1 + Ca_i I^{1/2}) + B_i I \tag{2}$$

where A and C are Debye-Hückel constants (Millero, 1974), a_i (the ion size parameter) and B_i are adjustable parameters. This equation works well in dilute solutions, but is inaccurate at high ionic strengths. Ionic solution theory (Mayer, 1950) suggests that the value of γ_i should be given by

$$\ln\gamma_i = D.H. + \sum_j B_{ij} m_j + \sum_{jk} C_{ijk} m_{jmk} \tag{3}$$

where D.H. is the Debye-Hückel term, B_{ij} and C_{ijk} are, respectively, second and third virial coefficients which are functions of ionic strength. Pitzer (1973) has developed a simple functional form for B_{ij} and found that the mean activity coefficients for electrolytes in binary solution can be represented by (for a 1-1 electrolyte)

$$\ln\gamma\pm(MX) = Z_M Z_X f + m (\beta^0 + f^1\beta^1) + 1.5m^2 C^\phi \tag{4}$$

where $f = -0.392 [I^{1/2}/(1 + 1.2I^{1/2}) + (2/1.2)\ln(1 + 1.2I^{1/2})]$, $f^1 = (1/2I) [1 - \exp(-2I^{1/2}) (1 + 2I^{1/2} - 2I)]$ and I is the ionic strength. Tabulations of the adjustable parameters β^0, β^1 and C^ϕ are given by Pitzer and Mayorga (1973, 1974) and Pitzer (1979).

For mixed electrolyte solutions, the activity coefficients of cation M and anion X are given by

$$\ln\gamma_M = Z_M^2 f + 2\sum_a m_a (B_{Ma} + EC_{Ma}) +$$
$$Z_M^2 \sum_c \sum_a m_c m_a B_{ca}^1 + Z_M \sum_c \sum_a m_c m_a C_{ca} \tag{5}$$

$$\ln\gamma_X = Z_X^2 f + 2\Sigma m_c (B_{cX} + EC_{cX}) +$$

$$Z_X^2 \Sigma\Sigma m_c m_a B_{Ca}^1 + Z_X \Sigma\Sigma m_c m_a C_{ca} \tag{6}$$

where m_i is the molality of cations (c) or anions (a) in the mixed solution and $E = 1/2\Sigma_i |Z_i|$, the equivalent molality. The second and third virial coefficients are given by

$$B_{MX} = \beta_{MX}^0 + (\beta_{MX}^1/2I)[1 - (1 + 2I^{1/2})\exp(-2I^{1/2})] \tag{7}$$

$$B_{MX}^1 = (\beta_{MX}^1/2I^2)[-1 + (1 + 2I^{1/2} + 2I)\exp(-2I^{1/2})] \tag{8}$$

$$C_{MX} = c_{MX}^\phi/(2|Z_M Z_X|^{1/2}) \tag{9}$$

To account for the interactions of like charged ions ($Na^+ - K^+$) and triple ions ($Na^+ - K^+$, Cl^-), the full equations of Pitzer are given by (Pitzer and Kim, 1974; Harvie and Weare, 1980)

$$\ln\gamma_M = \text{eq. } 5 + \Sigma m_c \{2\Theta_{Mc} + \Sigma m_a \Psi_{Mca}\} +$$

$$\Sigma\Sigma m_a m_a^1 \Psi_{aa^1M} \tag{10}$$

$$\ln\gamma_X = \text{eq. } 6 + \Sigma m_a \{2\Theta_{Xa} + \Sigma m_c \Psi_{Xac}\} +$$

$$\Sigma\Sigma m_c m_c^1 \Psi_{cc^1X} \tag{11}$$

The term Θ_{ij} is related to the interactions of ions i and j (e.g. Na^+-K^+ or $Cl^- - Br^-$). The Ψ_{ijk} term is related to the triple ion interactions of two similarly charged ions with an ion of opposite charge. These terms are determined from the free energies of ternary mixtures with a common ion (NaCl + KCl) and are tabulated elsewhere (Pitzer, 1979).

For solutions that contain electrolytes of different charge types ($Na^+ - Mg^{2+}$), higher order electrostatic terms ($^E\Theta_{ij}$ and $^E\Theta_{ij}^1 = \partial^E\Theta_{ij}/\partial I$) have been incorporated into the equations of Pitzer (Pitzer, 1975). These unsymmetrical or higher order electrostatic terms for mixing two cations or anions can be estimated from the approximate equations of Pitzer (1975). When these terms are used, the values of Θ_{ij} and Ψ_{ijk} derived with these terms must also be used (Pitzer, 1975; Harvie and Weare, 1980).

The details of using these equations, for seawater have been given elsewhere (Millero, 1983a). The equations have been applied to natural waters by a number of workers (Whitfield, 1975; Harvie and Weare, 1980; Krumgalz and Millero, 1982; Millero, 1983a, b).

3. THE IONIZATION OF ACIDS IN IONIC MEDIA

Much of our knowledge of the interactions of ions in natural waters comes from measurements of the ionization of weak acids in various ionic media (Millero, 1983b)

$$HA \rightarrow H^+ + A^- \tag{12}$$

From the measured stoichiometric ionization constants

$$K_{HA}^* = [H] \ [A] \ / \ [HA] \tag{13}$$

in various ionic media, it is possible to study the interactions of various cations (M) with weak acid anions (A). From the values of pK_{HA}^* in NaCl and NaCl plus Mg^{2+}, one can estimate the values for the formation of MgA^+ ion pairs

$$K_{MgA}^* = [K_{HA}^*(NaMgCl) \ / \ K_{HA}^*(NaCl) - 1]/[Mg]_F \tag{14}$$

This equation is not exact since part of the changes in K_{HA}^* are related to changes in the activity coefficient of H^+ and HA (Millero and Thurmond, 1983).

The K_{HA}^* is related to the thermodynamic value, K_{HA}, by

$$K_{HA}^* = K_{HA} \ (\gamma_{HA}/\gamma_H\gamma_A) \tag{15}$$

where γ_i are the total or stoichiometric activity coefficients. Values of γ_A can be determined from a rearrangement of equation (15)

$$\ln\gamma_A(exp) = \ln K_{HA} - \ln K_{HA}^* + \ln\gamma_{HA} - \ln\gamma_H \tag{16}$$

The experimental values of γ_A in a NaCl and NaMgCl media are related to the ion pairing constant by

$$\gamma_A(\text{NaMgCl}) = \gamma_A(\text{NaCl})[1 + K^*_{\text{MgA}}[\text{Mg}]_F]^{-1} \qquad (17)$$

It is also possible to use Pitzer's equations to represent the activity coefficients of γ_A, γ_H and γ_{HA} for the components of equation (15). If all the interaction terms are available, it is possible to estimate K^*_{HA} in a given ionic media. The comparisons of the pK^*_1 and pK^*_2 for the ionization of carbonic acid measured and estimated using Pitzer's equations are shown in Figures 1. a) and b) (Thurmond and Millero, 1982). As is quite apparent from these figures, the estimates are greatly improved when higher order terms are used (Θ, Ψ, $^E\Theta$ and $^E\Theta^1$).

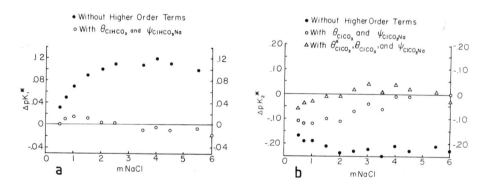

FIGURE 1. a) Comparison of the measured and calculated values of pK^*_1 for H_2CO_3 in NaCl solutions at 25°C. b) Comparison of the measured and calculated values of pK^*_2 for H_2CO_3 in NaCl solutions at 25°C.

From measurements of pK^*_{HA} in various ionic media over a wide concentration range, it is possible to estimate the interaction parameters β^0, β^1, C^ϕ and Ψ for A^- with various cations and anions (Millero, 1983a,b; Peiper and Pitzer, 1982; Thurmond and Millero, 1982; Millero and Thurmond, 1983).
Recently, we have determined values of $B_{MgA} + EC_{MgA}$ from the experimental measurements of pK^*_{HA} in NaMgCl and NaCaCl solutions (Millero,

1983a, b). Since the measurements were made only at I = 0.7, it was not possible to separate B_{MgA} + EC_{MgA} into its components β^0, β^1 and C^ϕ. Using the equations of Pitzer, the differences in γ_A in the two media are related to $\beta^0_{MgA} \overset{\sim}{=} B_{MgA} + EC_{MgA}$ by

$$\beta^0_{MgA} = \Delta \ln \gamma_A / 2m_{Mg} \qquad (18)$$

The reliability of these Pitzer parameters (at I = 0.7) can be demostrated by comparing the measured and calculated activity coefficients of ions in seawater (Millero, 1983b). This comparison is given in Table 1. The agreement is quite good except for the F$^-$ ion whose experimental value may be in error.

Table 1. Comparisons of the measured and calculated values of γ_i of ions in seawater (S = 35 and t = 25°C.

Ion	Meas.	Calc.	Δ
H$^+$	0.590	0.586	0.004
Na$^+$	0.668	0.666	0.002
K$^+$	0.625	0.619	0.006
NH$^+$	0.619	0.624	−0.005
Ca^{2+}	0.191	0.213	−0.022
Sr^{2+}	0.190	0.212	−0.022
F$^-$	0.348	0.238	0.110
Cl$^-$	0.666	0.666[a]	0
OH$^-$	0.242	0.224	0.018
HS$^-$	0.673	0.663	0.010
B(OH)$_4^-$	0.419	0.427	−0.008
HCO3$^-$	0.579	0.569	0.007
H$_2$PO$_4^-$	0.492	0.511	−0.019
SO$_4^{2-}$	0.103	0.109	−0.006
CO$_3^{2-}$	0.040	0.038	0.002
HPO$_4^{2-}$	0.046	0.045	0.001
PO$_4^{3-}$	1.4×10^{-5}	0.9×10^{-5}	0.5×10^{-5}

a) Assigned Value.

4. MEASUREMENTS OF pK_1^* AND pK_2^* FOR CARBONIC ACID

To determine reliable Pitzer parameters valid over a wide range of ionic strength, we have initiated a program to measure the pK_{HA} of acids in various ionic media. In our recent measurements, we have concentrated on the carbonate system due to its importance in interacting with metals in natural waters. We have completed measurements of pK_1^* and pK_2^* in NaCl and NaMgCl solutions from 0.5 to 6.0 m (Thurmond and Millero, 1982; Millero and Thurmond, 1983). Our results in NaCl were shown in Figures 1a and 1b. If our results in NaMgCl solutions are treated in terms of ion pairing using eq. (14), we obtain the values of $K_{MgHCO_3}^*$ shown in Figure 2a. If corrections are made for the differences in γ_H and eq. (17) is used to determine the β's, the results for pK_1 cannot be interpreted in terms of ion pair formation (γ_{HCO_3} in MgCl is greater than in NaCl). It is possible, however, to determine the Pitzer parameters $\beta^0_{MgHCO_3}$ and $\beta^1_{MgHCO_3}$ from the data. This is shown in Figure 2b.

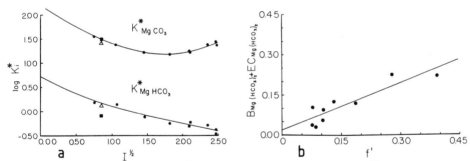

FIGURE 2. a) Values of log $\hat{K}_i$ for the formation of $MgHCO_3^+$ and $MgCO_3$ ion pairs versus the square root of ionic strength.   b) Values of $[B_{Mg(HCO_3)_2} + EC_{Mg(H_2CO_3)}]$ versus f' = (1/2 I) [1 - (1 + $2I^{1/2}$) exp $(-2I^{1/2})$] where I is the ionic strength.

The values of $K_{MgCO_3}^*$ determined from eq. (17) can be used to determine K_{MgCO_3} and the γ_{MgCO_3}, using

$$\log[\ K_{MgCO_3}^* / \gamma_{Mg}\gamma_{CO_3}\ ] = \log K_{MgCO_3} - kI \qquad (19)$$

where k is the salting coefficient for $MgCO_3$. A plot of the left hand side of eq. (19) versus I is shown in Figure 3. We obtain pK_{MgCO_3} = 3.0 and k = 0.056. Our value of pK_{MgCO_3} is in good agreement with the value of 2.9 ± 0.1 selected by Millero and Schreiber (1982). The salting coefficient is the same as found for other neutral ion pairs such as $MgSO_4$ ° (Millero and Schreiber, 1982) and unionized acids such as H_3PO_4 (k = 0.052, Pitzer and Silvester, 1976).

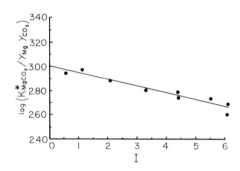

FIGURE 3. Values of log ($K^*_{MgCO_3}/\gamma_{Mg}\gamma_{CO_3}$) versus ionic strength.

If Pitzer parameters are generated from the experimental values of γ_{CO_3} in NaMgCl solutions, four parameters are needed (β^0, β^1, β^2 and C^ϕ) for the $MgCO_3$ interactions. Thus, the ion pairing model is preferable to the Pitzer formalism for the interactions of Mg^{2+} and CO_3^{2-} (it requires only two parameters).

The reliability of the Pitzer and ion pairing parameters derived from our measurements is demonstrated in Figures 4 a) and b). The Pitzer parameters for Mg-HCO_3 interactions yield values of pK_1^* that agree on the average to ± 0.004 with the measured values. A similar comparison with the measurements of Dyrssen and Hansson (1973) and Pytkowicz and Hawley (1974) is shown in Table 2.

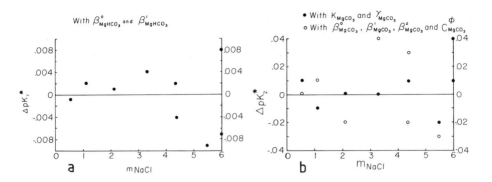

FIGURE 4. a) Comparison of the measured and calculated values of pK_1^* for H_2CO_3 in NaMgCl solutions at 25°C. b) Comparison of the measured and calculated values of pK_2^* for H_2CO_3 in NaMgCl solutions at 25°C.

Table 2. Comparisons of measured and calculated values of pK_1^* and pK_2^* in NaMgCl solutions.

I	[Mg]	ΔpK_1^* a	ΔpK_2^* b	ΔpK_2^* c	Ref.
0.72	0.05	0.001	0.00	0.00	d
	0.04	–	0.16	0.14	
	0.08	0.01	0.32	0.13	
	0.16	0.01	0.73	0.04	
	0.24	0.02	1.22	0.03	
0.73	0.03	0.01	0.01	0.01	e
	0.05	0.02	0.05	0.01	

a) ΔpK_1^* = meas – calc using $\beta^0_{Mg(HCO_3)_2}$ and $\beta^1_{Mg(HCO_3)_2}$ (Millero and Thurmond, 1983).
b) ΔpK_2^* = meas – calc using $\beta^0_{MgCO_3}$, $\beta^1_{MgCO_3}$, $\beta^2_{MgCO_3}$ and $C^\phi_{MgCO_3}$ (Millero and Thurmond, 1983).
c) ΔpK_2^* = meas – calc using the ion pairing constant $pK_{MgCO_3}o = 3.00$ and $\log \gamma_{MgCO_3}o = 0.0560\ I$.
d) Pytkowicz and Hawley (1974).
e) Dyrssen and Hansson (1973).

    The predicted values are in good agreement with the measured values. The agreement at m = 0.24 demonstrates the reliability of the parameters for pure $MgCl_2$ solutions.

    The Pitzer parameters for Mg-CO_3 interactions yield values of pK_2^*

that agree on the average to ± 0.02 with the measured values. A similar comparison in Table 2 shows large differences at high Mg^{2+} concentrations. If the ion pairing parameters are used, the calculated values of pK_2^* agree on the average to ± 0.01 with the measured values. The comparisons in Table 2 support the use of the ion pairing parameters even for pure $MgCl_2$ solutions. In future work, we plan to continue these measurements for other metals and other acids.

5. MEASUREMENTS OF pK_{sp}^* FOR CARBONATE MINERALS

Recently, we have made measurements on the solubility of $CaCO_3$ (calcite), $SrCO_3$ (strontianite) and $BaCO_3$ (witherite) in NaCl solutions from 0.1 to 6.0 m at 25°C (Millero et al., 1983). The directly measured values of pK_{sp}^* have been extrapolated to infinite dilution using activity coefficients estimated from Pitzer's equations with higher order interaction terms (Θ and Ψ). Values of $pK_{sp} = pK_{sp}^* - \ln (\gamma_M \gamma_{CO_3})$ calculated at various ionic strengths are shown in Figure 5. Thermodynamic values of $pK_{sp} = 9.46 ± 0.03$, $9.13 ± 0.03$ and $8.56 ± 0.04$ were found respectively, for $CaCO_3$, $SrCO_3$ and $BaCO_3$. These results are in good agreement with literature data (Millero et al., 1983). Since parameters for the interactions of CO_3^{2-} with Ca^{2+}, Sr^{2+} and Ba^{2+} were not used, our results indicate that they are not necessary at low values of P_{CO_2}.

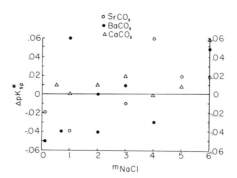

FIGURE 5. Comparisons of the measured and calculated solubility product for $SrCO_3$, $BaCO_3$ and $CaCO_3$ in NaCl solutions at 25°C.

6. EFFECT OF MEDIA ON THE FORMATION OF METAL COMPLEXES

To determine the activity of trace metals in natural waters, it is necessary to have stoichiometric association constants (K_A^*) in the natural water of interest. Values of K_A^* can be estimated by using values measured in a given media (e.g. $NaClO_4$) at the same ionic strength (Dyrssen and Wedborg, 1974), or by using thermodynamic constants and making estimates for the activity coefficients (Zirino and Yammamoto, 1972). Both methods assume that K_A^* is only a function of ionic strength, which is not the case (Sillen and Martell, 1964). This has clearly been demonstrated for the formation of lead chloro complexes by Byrne and Millero (1983) in HCl, $HClO_4$, NaCl, $NaClO_4$, $MgCl_2$ and $CaCl_2$ media.

Recently, we have demonstrated (Millero and Byrne, 1983) how Pitzer's equations can be used to make extrapolations to infinite dilution and represent the differences in the measured values of K_A^* in various ionic media for the formation of lead chloro complexes. This is demonstrated in Figures 6 a), b) and 7 for the stepwise formation of $PbCl^+$, $PbCl_2^{\,o}$ and $PbCl_3^{\,-}$ complexes. The different slopes in these figures are related to differences in the activity coefficients of $PbCl^+$, $PbCl_2^{\,o}$ and $PbCl_3^{\,-}$ in

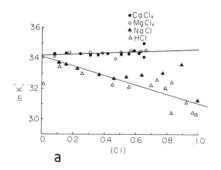
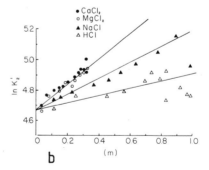

FIGURE 6. a) Values of the corrected stoichiometric association constant for the formation of $PbCl^+$ in various ionic media at 25°C. b) Values of the corrected stoichiometric association constant for the formation of $PbCl_2^{\,o}$ in various ionic media at 25°C.

198

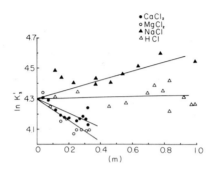

FIGURE 7.   Values of the corrected stoichiometric association constant
for the formation of $PbCl_3^-$ in various ionic media at 25°C.

the various ionic media. For seawater (S = 35), the Pitzer equations
give $\gamma_{PbCl} = 0.600$, $\gamma_{PbCl_2} = 0.751$ and $\gamma_{PbCl_3} = 0.456$. The values of
PbCl and $PbCl_3$ can be compared to $\gamma_{NH_4} = 0.601$ and $\gamma_{B(OH)_4} = 0.442$
(Millero, 1983a). The value for the neutral ion pair $PbCl_2$ is a lot
lower than values for other neutral pairs (e.g. CaCO3). This is probably
related to the structure of the ion pair (Cl-Pb-Cl). In our future work,
we plan to extend these calculations to the formation of other trace
metal complexes in ionic media.

In this paper, we have briefly reviewed our work on the use of
Pitzer's equations to estimate the activity of metals at high ionic
strengths. Further measurements are needed on other systems to extend
these calculations to other trace metal systems.

ACKNOWLEDGEMENTS

This work was supported by the Oceanographic section of the National
Science Foundation (OCE-8120659) and the Office of Naval Research
(N00014-80-C-0042).

REFERENCES

- Byrne, R. H. and W. L. Miller, 1983. Medium composition dependence of lead (II) complexation by chloride ion. Am. J. Sci. - in press.
- Dickson, A. G. and M. Whitfield, 1981. An ion-association model for estimating acidity constants (at 25°C. and 1 atm total pressure) in electrolyte mixtures related to seawater (ionic strength < 1 mol $kg^{-1}H_2O$). Mar. Chem. 10: 315-333.
- Dyrssén, D. and I. Hansson, 1973. Ionic medium effects in seawater - a comparison of acidity constants of carbonic acid and boric acid in sodium chloride and synthetic seawater. Mar. Chem. 1: 137-149.
- Harvie, C. E. and J. H. Weare, 1980. The prediction of mineral solubilities in natural waters: the $Na-K-Mg-Ca-Cl-SO_4-H_2O$ system from zero to high concentration at 25ºC. Geochim. Cosmochim. Acta 44: 981-997.
- Krumgalz, B. S. and F. J. Millero, 1982. Physico-chemical study of the Dead Sea waters. I. Activity coefficients of major ions in Dead Sea water. Mar. Chem. 11: 209-222.
- Mayer, J. E., 1950. The theory of ionic solutions. J. Chem. Phys. 18: 1426-1436.
- Millero, F. J., 1974. Seawater as a multicomponent electrolyte solution. In: E. D. Goldberg (Ed), The Sea, Ideas and Observations. John Wiley & Sons, Inc., New York, New York, pp. 3-80.
- Millero, F. J., 1983a. Use of models to determine ionic interactions in natural waters. Thallassia Jugoslavica, in press.
- Millero, F. J., 1983b. The estimation of the pK_{HA}^{*} of acids in seawater using Pitzer equations. Geochim. Cosmochim. 47: 2121-2129.
- Millero, F. J. and R. H. Byrne, 1983. Use of Pitzer's equations to determine the media effect on the formation of lead chloro complexes. Geochim. Cosmochim. Acta, in press.
- Millero, F. J. and D. R. Schreiber, 1982. Use of the ion pairing model to estimate activity coefficients of the ionic components of natural waters. Amer. J. Sci. 282: 1508-1540.
- Millero, F. J. and V. Thurmond, 1983. The ionization of carbonic acid in Na-Mg-Cl solutions at 25°C. J. Solution Chem., 12 :401-412.
- Millero, F. J., P. Milne and V. Thurmond, 1983. The solubility of calcite, strontianite and witherite in NaCl solutions at 25°C. Geochim. et Cosmochim. Acta, in press.
- Peiper, J. C. and K. S. Pitzer, 1982. Thermodynamics of aqueous carbonate solutions including mixtures of carbonate, bicarbonate and chloride. J. Chem. Thermodyn. 14: 613-638.
- Pitzer, K. S., 1973. Thermodynamics of electrolytes. I. Theoretical basis and general equations. J. Phys. Chem. 77: 268-277.
- Pitzer, K. S., 1975. Thermodynamics of electrolytes. V. Effects of higher order electrostatic terms. J. Solution Chem. 4: 249-265.
- Pitzer, K. S., 1979. Ion interaction approach. In: R. M. Pytkowicz (Ed), Activity coefficients in electrolyte solutions. Vol. I, CRC Press, Inc., Boca Raton, Fla., pp. 157-208.
- Pitzer, K. S., and J. J. Kim, 1974. Thermodynamics of electrolytes. IV. Activity and osmotic coefficients for mixed electrolytes. J. Am. Chem. Soc. 96: 5701-5707.

- Pitzer, K. S. and G. Mayorga, 1973. Thermodynamics of electrolytes, II. Activity and osmotic coefficients for strong electrolytes with one or both ions univalent. J. Phys. Chem. 77: 2300-2308.
- Pitzer, K. S. and G. Mayorga, 1974. Thermodynamics of electrolytes. III. Activity and osmotic coefficients for 2:2 electrolytes. J. Solution Chem. 3: 539-546.
- Pitzer, K. S. and L. F. Silvester, 1976. Thermodynamics of electrolytes. VI. Weak electrolytes including H_3PO_4. J. Solution Chem. 5: 269-278.
- Pytkowicz, R. M. and J. E. Hawley, 1974. Bicarbonate and carbonate ion-pairs and a model of seawater at 25°C. Limnol. Oceanogr. 19: 223-234.
- Sillen, L. G. and A. E. Martell, 1964. Stability constants of metal-ion complexes. The Chemical Society.
- Thurmond, V. and F. J. Millero, 1982. Ionization of carbonic acid in sodium chloride solutions at 25°C. J. Solution Chem. 11: 447-456.
- Whitfield, M., 1975. The extension of chemical models for seawater to include trace components at 25°C and 1 atm. pressure. Geochim. Cosmochim. Acta, 39: 1545-1557.
- Whitfield, M., 1979. Activity coefficients in natural waters. In: R. M. Pytkowicz (Ed), Activity coefficients in electrolyte solutions. Vol. II, CRC Press, Inc., Boca Raton, Fla.
- Zirino, A. and S. Yamamoto. 1972. A pH-dependent model for the chemical speciation of copper, zinc, cadmium, and lead in seawater. Limnology Oceanography 17, 661-671.

PART III APPLICATION TO NATURAL WATERS

MEASUREMENT OF THE TRACE METAL COMPLEXING CAPACITY OF MAGELA CREEK WATERS

B.T. HART and M.J. JONES

1. INTRODUCTION

Trace metal speciation in natural waters is an area of increasing interest in environmental chemistry and water quality management. When a trace metal is added to a natural water it is distributed between several physico-chemical forms, which may have quite different environmental effects (Florence & Batley 1980; Hart 1982a). For example it is now generally accepted that the free uncomplexed forms of trace metals such as copper, cadmium and zinc, are the forms most acutely toxic to a range of aquatic animals and plants (Davies et al. 1976; Lake et al. 1979; Sunda & Ferguson 1983; Petersen 1982).

The added trace metals may be complexed by a range of dissolved inorganic or organic ligands, or by colloidal and particulate matter. The ability of a natural water to complex added trace metals is called the complexing capacity. Measurement of this complexing capacity is obviously important if the impact of added trace metals is to be assessed.

This paper reports work undertaken in which a complexometric titration technique, employing two methods (anodic stripping voltammetry (ASV) and an ion selective electrode (ISE)) to determine the uncomplexed metal, was used to determine the copper complexing capacity in natural waters from the Magela Creek system in northern Australia.

2. THEORY

The complexometric method involves titration of the water sample with ionic copper. After each addition of the metal, a short equilibration time is allowed and then the "unreacted" copper (i.e. copper not involved in forming a complex) is determined using either ASV or ISE. A plot of "free" metal concentration versus the total copper concentration for such a titration is shown in Figure 1. In the initial stages very little free copper is determined (i.e. the slope is very small), because most of the added metal is being complexed. However, once the complexing capacity is exceeded (i.e. when the copper concentration is

Kramer, C.J.M. and Duinker, J.C. (eds.), Complexation of Trace Metals in Natural Waters. ISBN 90-247-2973-4
© 1984 Martinus Nijhoff/Dr W. Junk Publishers, The Hague/Boston/Lancaster.
Printed in the Netherlands.

greater than the ligand concentration), the slope increases since all the added metal remains in an uncomplexed form.

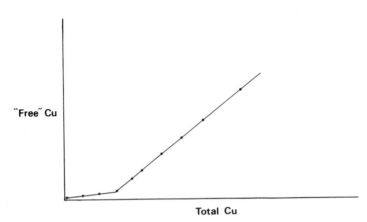

Figure 1:   Free copper versus total copper for a complexing capacity titration.

In those waters where only one ligand (L) exists, and only 1:1 complexes (ML) are formed, the reaction between the metal ion (M) and the ligand is given by equation 1, together with the conditional stability constant (*K) (charges have been omitted for clarity).

$$M + L \rightleftharpoons ML \qquad \text{where} \quad {}^*K \quad = \frac{[ML]}{[M][L]} \quad \dots\dots\dots\dots (1)$$

The total filterable metal concentration is given by:

$$[M_f] = [M] + [MI] + [ML]$$

where

   [M]   = concentration of free (hydrated) metal ions

   [MI]   = concentration of metal-inorganic complexes (e.g. MOH, MCO_3)

   [ML]   = concentration of metal-organic complexes (and metal-colloid if present)

At equilibrium these species will be related to each other as shown equation 2

$$[ML] = \frac{*K [M] [L^t]}{1 + *K [M]} \quad \dots\dots\dots\dots\dots(2)$$

where $[L^t]$ is the total ligand concentration or complexing capacity.

Rearrangement of equation 2 leads to equation 3.

$$\frac{[M]}{[ML]} = \frac{[M]}{[L^t]} + \frac{1}{*K [L^t]} \quad \dots\dots\dots\dots\dots(3)$$

A plot of $[M]/[ML]$ versus $[M]$ results in a straight line with slope $1/[L^t]$ and an intercept $= 1/*K [L^t]$.

It must be emphasised that $*K$ is a conditional stability constant and, although closely related to the thermodynamic stability constant, it is really only applicable under those conditions of pH, ionic strength, etc, at which it was determined. It is also worth noting that complexing capacity is generally determined on filtered samples that may still contain some colloidal surfaces. In the analysis given above, such colloidal surfaces are assumed to make up part of the single general ligand (L) which dominates the complexation. Also in filtered samples no account can be taken of the capacity of the removed particulate matter to adsorb or complex added metals. This can be a serious omission since particulate matter can represent an important pathway in trace metal cycling in natural waters (Hart 1982b).

As discussed more fully below we have interpreted the ASV and ISE data on the basis of a two ligand system. Theoretical relationships governing the titration of a two ligand system (L_1 and L_2) with one metal (M) to form two 1:1 complexes (ML_1 and ML_2) have been developed by Ruzic (1982).

In this case:

$$[M_f] = [M] + [MI] + [ML_1] + [ML_2] \quad \dots\dots\dots\dots\dots(4)$$

At equilibrium:

$$\frac{[M]}{[M_f] - [M]} = \frac{1}{L} \left\{ [M] + \left(\frac{[M]}{A} + \frac{1}{*K_1 *K_2} \right) \Big/ \left((M) + \frac{1}{B} \right) \right\} \quad \dots\dots(5)$$

where:

$$L = [L_1^t] + [L_2^t]$$

$$A = \frac{L}{\dfrac{[L_1^t]}{{}^*K_1} + \dfrac{[L_2^t]}{{}^*K_2}}$$

$$B = \frac{L}{\dfrac{[L_1^t]}{{}^*K_2} + \dfrac{[L_2^t]}{{}^*K_1}}$$

For relatively large amounts of added M (i.e. in the range of the I experiment), equation 5 simplifies to:

$$\frac{[M]}{[ML]} = \frac{[M]}{L} + \frac{1}{AL} \quad \dots\dots\dots\dots\dots\dots\dots\dots\dots\dots\dots (6)$$

Thus in this case a plot of $[M]/[ML]$ versus $[M]$ should give a straight with gradient $1/L$ and intercept $1/A\,L$.

3. EXPERIMENTAL

3.1 Sampling

The samples for complexing capacity determination were taken from Isla billabong (waterhole) during two Dry seasons, in September, October an November, 1981 (5 samples) and in September, 1982 (1 sample). This billabong waterhole is situated in the Magela Creek system in northern, tropic Australia. Flow only occurs in the Wet season (December-May) and progressive over the Dry season the stream dries up to leave a series of isolated billabong of which Island billabong is one.

At the commencement of the Dry season, Island billabong water is general of low conductivity (15-20 uS/cm), pH 6.0 to 7.0, with cationic dominance

Na > Mg > Ca > K and anionic dominance HCO_3 > Cl > SO_4 (Hart & McGregor 1980; Walker & Tyler 1983). Over the Dry season evaporation and groundwater ingress results in conductivity increasing up to 40-60 uS/cm, pH decreasing to around 5.0 and the anionic dominance changing to Cl = SO_4 >> HCO_3; cationic dominance changes little.

3.2 Complexometric titration

Anodic stripping voltammetry: A PAR 303 polarographic analyser coupled with a PAR 303 static mercury drop electrode in the hanging mercury drop electrode mode was employed. Deposition was carried out for 5 minutes at -0.9 V vs Ag/AgCl with the solution stirred for all but the last 15 seconds of the deposition time. Ten mL of the sample was placed in the cell and buffered to pH 6.0 by the addition of 0.1 mL of 1.0 M sodium acetate buffer. The ionic strength of the solution was then adjusted to 0.1 M with KNO_3 to make the experimental conditions as comparable as possible with the ion selective method. Merck Suprapur reagents were used. High purity nitrogen was bubbled through the sample for 8 minutes to remove oxygen and, during the analysis, was directed over the top of the sample. After completing the voltammogram the solution was spiked with copper solution and allowed to equilibrate for 10 minutes; nitrogen was bubbled through the solution during this time. The deposition and stripping steps were repeated. This procedure was continued until sufficient data were obtained to draw a titration curve of peak current (i_p) vs added copper concentration.

A computer program COMPCAP was used to analyse the experimental data and produce the total copper-binding ligand concentration [L_t] and the conditional stability constant $^*K$ (Hart & Davies 1981).

Ion selective electrode: The apparatus used consists of a Radiometer pH-stat, a Radiometer Research pH meter, an Orion copper ISE (plus reference electrode) and a constant temperature water bath (25 $\pm$ 0.1°C). Sample (47.5 mL, four fold concentrated using rotary evaporation at 25°C under vacuum) plus 2.5 mL of 2 M KNO_3 was placed in the cell and maintained at pH 6.00 (+ 0.01) with a pH-stat using 0.1 M NaOH as titrant. High purity nitrogen was bubbled continuously through the solution. Mixing was achieved by the bubbling and by a Teflon-coated magnetic stirring bar. The free copper concentration was monitored with the ISE.

The methodology used in the titration was analogous to the ASV titration outlined above. Free copper was added, the sample equilibrated for a certain period, and the unreacted copper measured. The equilibration time was nominally

10 minutes, however at low $[Cu_t]$ it was often necessary to wait longer than th
for a stable EMF value. The EMF reading was recorded when the rate of change w
< 0.1 mV in 2 minutes. The range of copper additions was from 1.0 x 10^{-6} M
7.0 x 10^{-4} M.

For each copper addition, the total copper concentration ($[Cu_t]$) v
calculated. A correction was made for the increase in volume due to the copp
and NaOH additions. The last five additions were used to calibrate the ISE i
regressing the EMF against log $[Cu_t]$ (r generally > 0.9998). In the region
high $[Cu_t]$ it was assumed that the proportion of complexed copper was very sm
and $[Cu_t]$ approximated $[Cu^{2+}]$. The calibration curve was then used to calcul
the $[Cu^{2+}]$ corresponding to each $[Cu_t]$. $[CuL]$ was obtained by mass balance,
i.e. $[CuL] = [Cu_t] - [Cu^{2+}]$. The COMPCAP program was then used to obtain
values of L and K (see equation 6) from a plot of $[Cu^{2+}]/[CuL]$ vs $[Cu^{2+}]$.

4. RESULTS & DISCUSSION

A comparison between the ASV and ISE methods for determining complexi
capacity was undertaken using filtered (< 0.4 um) water taken from Isla
billabong during the 1981 dry season. The values obtained for complexi
capacity (L_t) and stability constant ($^*K$) by each technique are given in Tab
1.

Since the ISE method requires that the samples have a high ionic streng
(0.1 M KNO_3), all samples analysed by ASV also had their ionic strength adjuste
to the same value to allow valid comparisons to be made.

The ISE-determined complexing capacities were considerably higher (5 to
times) than those determined by the ASV method (Table 1). Also, the stabili
constants of the copper complexes formed by these ligands were quite differen
The ASV method gave log $^*K$ values in the range of 7.5 to 8.1 while the IS
method gave values of 5.7 or 5.8; all experiments were done at pH 6.0.

The obvious explanation for these observations is that the two method
determine different copper binding ligands. This is shown diagramatically i
Figure 2 where two different copper binding ligands (L_1 and L_2) are assumed
Since the copper ISE reacts only to free copper ions, it gives a measure of th
total complexing capacity which consists of both L_1 and L_2 (c.f. equation 6)
The ASV method however determines all 'labile' copper species, which wil
include free copper ions, most inorganic complexes (MI) and the weaker organi
complexes (ML_2); because of this, the ASV determined complexing capacity wil
consist only of the stronger binding ligands (L_1). There may also be coppe

TABLE 1:   COMPLEXING CAPACITY OF ISLAND BILLABONG WATER USING A
COMPLEXOMETRIC TITRATION METHOD WITH ASV AND ISE DETECTION.

Method	Date	$[L^t]$		$\log {}^*K$	$[Cu_i]^{(a)}$
		(uM)	(ug/L)		(ug/L)
ASV[(b)]	22.09.81	0.18	11.5	8.0	0.94
	06.10.81	0.09	5.5	8.1	0.75
	20.10.81	0.14	9.2	8.0	1.25
	03.11.81	0.25	15.6	7.9	0.59
	17.11.81	0.16	10.3	7.5	1.16
ISE[(c)]	22.09.81	1.47	93.4	5.7	
	06.10.81	0.95	60.4	5.8	
	20.10.81	1.10	69.9	5.7	
	03.11.81	1.19	75.6	5.7	
	17.11.81	0.92	58.5	5.8	

(a) $[Cu_i]$ = initial filterable copper concentration
(b)   Deposition potential −0.9 V; pH 6.0;
electrolyte 0.1 M KNO_3; 0.01 M CH_3COONa
(c)   pH 6.0; electrolyte 0.1 M KNO_3

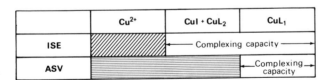

Figure 2:   Diagrammatic representation of the differences in copper
complexing capacity measurements by ASV and ISE.

CuI = Copper-inorganic complexes
CuL_2 = ASV-labile copper-organic complexes
CuL_1 = Copper-organic complexes not detected by either ISE or ASV

species which are kinetically labile, that is copper-organic complexes th<
dissociate in the diffusion layer surrounding the mercury drop.

Data from the ASV and ISE experiments has been combined to describe a t
ligand system which explains the complexation of added copper and can be used
predict the free (toxic) copper concentration for any give added amount. T
procedure is as follows:

(i)   use the ASV data to obtain $[L_1^t]$ and $^*K_1$ (equation 3);

(ii)   use the ISE data to obtain $L = ([L_1^t] + [L_2^t])$ and A (equation 6);

(iii) combine (i) and (ii) to obtain $[L_2^t]$ and $^*K_2$.

The converted data for the Island billabong samples analysed are given
Table 2.

These data may then be used to calculate the speciation of copper in ea<
water. This is illustrated in Figure 3, which shows the relationship between t
percentage free copper (Cu^{2+}) and total copper concentration over the range 1C
to 10^{-5} M. The values of L_1, L_2, $^*K_1$ and $^*K_2$ used in the calculations were t
mean values derived from the data for Island billabong. This diagram shc
clearly that for copper concentrations in the range found naturally (i.e. up
10^{-7}M, 6.4 ug/L), only 7-13% of the copper is in the free ionic form and th
almost all the binding is due to ligand L_1. In the region representative o<
polluted system (i.e. 10^{-7} - 10^{-6} M, 6.4 - 64 ug/L), there is a significa
increase in the proportion of the total copper that is in ionic form and alsc
marked difference in the calculated speciation depending upon whether L_1 on
or L_1 and L_2 are used. For example, at a total copper concentration of 10^{-6}M
calculate 82% to be in the free ionic form if only L_1 is used, while if b<
ligands are used more copper is bound leaving some 55% in the free ionic f<
(Figure 3).

The conclusion to be drawn from these results is that in the region
copper concentrations found naturally in the Magela Creek system, the
determined complexing capacity (i.e. L_1 and $^*K_1$) can be satisfactorily used
explain the binding of added copper. However, at higher copper concentratio
such as those that may be used in toxicity experiments, it is necessary t
both ligands be used to adequately explain the copper speciation.

Some preliminary work was undertaken to provide information on the nat
of the ligands responsible for the complexation. A sample taken from Isl
billabong in September 1982 was filtered (0.4 um Nuclepore) and the cop<
complexing capacity determined. Part of this filtered solution was also pas<
through a YM10 Amicon ultrafilter (nominal cutoff 10 000 dalton) and

TABLE 2:   TRANSFORMED ASV AND ISE DATA TO PRODUCE THE
            CONSTANTS CONSISTENT WITH A TWO LIGAND SYSTEM
            (SEE TEXT FOR METHOD).

Date	$[L_1^t]$ (uM)	$\log {}^*K_1$	$[L_2^t]$ (uM)	$\log {}^*K_2$
22.09.81	0.18	8.0	1.29	5.6
06.10.81	0.09	8.1	0.86	5.8
20.10.81	0.14	8.0	0.96	5.6
03.11.81	0.25	7.9	0.94	5.6
17.11.81	0.16	7.5	0.76	5.7
Mean	0.16	7.9	0.96	5.7

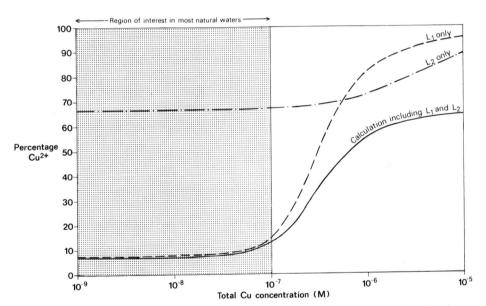

Figure 3:   Calculation of the proportion of ionic copper assuming complexation by
             (a) L_1 only, (b) L_2 only and (c) L_1 and L_2.
             $[L_1] = 0.16\,\mu M$, $[L_2] = 0.96\,\mu M$, $\log {}^*K_1 = 7.9$, $\log {}^*K_2 = 5.7$, $[Ca] = 10\,\mu M$, $[Mg] = 10\,\mu M$, $[CO_3]_{total} = 10\,\mu M$, $pH = 6.0$).

complexing capacity of this ultrafiltered solution also determined. A 50
portion of the filtered solution was u.v. irradiated for 12 h (550 W lamp, 25
H_2O_2 added) and the complexing capacity determined. The results are given
Table 3.

The most obvious conclusion to be drawn from these data is that the copp
complexing capacity in filtered water from Island billabong is entirely due
organic matter, since after u.v. irradiation no complexing capacity
measurable. Also, the greater part (62 %) of the complexation in the 0.4
filtered solution was by organic matter less than 10 000 dalton in size. T
latter result is by no means universal since results for other billabongs fr
the Magela Creek system showed that, although there was a significant degree
similarity in the complexing capacities for 0.4 um filtered solutions,
distribution of this complexing capacity between organic matter less than
greater than 10 000 dalton molecular weight varied considerably (Hart & Jon
unpublished data).

TABLE 3:   COMPLEXING CAPACITY OF A FRACTIONATED ISLAND BILLABONG
            SAMPLE USING COMPLEXOMETRIC TITRATION WITH ASV DETECTION.

Fraction[a]	(L^t)		$\log {}^*K$
	(uM)	(ug/L)	
0.4	0.21	13.1	7.9
YM10	0.13	8.1	8.2
u.v.	N.D.	N.D.	N.D.

(a)   0.4 - filtered through 0.4 um Nuclepore
       filter; YM10 - 0.4 filtrate through a
       YM10 ultrafilter; u.v. - 0.4 um filtrate
       u.v. irradiated for 12 h
(b)   N.D. - not detectable

ACKNOWLEDGEMENT

This work was funded by the Office of the Supervising Scientist.

REFERENCES

- Davies, P.H., Goettl, J.P., Sinley, J.R. and Smith, N.F., 1976. Acute and chronic toxicity of lead to Rainbow trout *Salmo gairdneri* in hard and soft water. Water Res. 10: 199-206.
- Florence, T.M. and Batley, G.E., 1980. Chemical speciation in natural waters. CRC Crit. Rev. Anal. Chem. 9: 219-296.
- Hart, B.T., 1982a. Trace metals in natural waters. I. Speciation. Chem. Aust. 49: 260-265.
- Hart, B.T., 1982b. Uptake of trace metals by sediments and suspended particulates: A review. Hydrobiologia 91: 299-313.
- Hart, B.T. & McGregor, R.J., 1980. Limnological survey of eight billabongs in the Magela Creek catchment, northern Australia, Aust. J. Mar. Freshwater Res. 31: 611-626.
- Hart, B.T. & Davies, S.H.R., 1981. Copper complexing capacity of waters in the Magela Creek system, northern Australia, Environ. Technol. Letts. 2: 205-214.
- Lake, P.S., Swain, R. and Mills, B., 1979. Lethal and sublethal effects of cadmium on freshwater crustaceans. AWRC Tech. Paper No. 37, Aust. Govt. Printing Service, Canberra.
- Petersen, R., 1982. Influence of Cu and Zn on the growth of a freshwater alga, *Scenedesmus quadricauda*: the significance of chemical speciation. Environ. Sci. Tech. 16: 443-447.
- Ruzic, I., 1982. Theoretical aspects of the direct titration of natural waters and its information yield for trace metal speciation. Anal. Chim. Acta 140: 99-113.
- Sunda, W.G. & Ferguson, R. L., 1983. Sensitivity of natural bacterial communities to additions of copper and to cupric ion activity: a bioassay of copper complexation in seawater. In: Proc. NATO Adv. Res. Symp. on Trace Metals in the Oceans. Erice, Sicily, March-April 1981.
- Walker, T.D. & Tyler, P.A., 1983. Chemical characteristics and nutrient status of billabongs of the Alligator rivers region, Northern Territory, Research Report, Office of the Supervising Scientist, Sydney (in press).

DETERMINATION OF LIGAND CONCENTRATIONS AND CONDITIONAL STABILITY CONSTANTS
IN SEAWATER. COMPARISON OF THE DPASV AND THE MnO_2 ADSORPTION TECHNIQUES.

C.M.G. VAN DEN BERG, P.M. BUCKLEY AND S. DHARMVANIJ.

1. INTRODUCTION

Various methods have been developed which intend to investigate
organic metal speciation by extraction of the inorganic metal fraction or
of the combined organic fraction. Extraction techniques, using either
liquid/liquid or liquid/solid separating techniques, never obtain 100%
recovery of the isolated material, and comparison of such methods can give
rather contradictory results. For instance, both Hirose et al. (1982) and
Kremling et al. (1981) make use of XAD-2 resin to adsorb the dissolved
organic fraction. Kremling et al. (1981) studied the effect of pH on the
adsorption efficiency and found that at pH 2 more organic material was
collected than at neutral pH. Also the collected organic copper fraction
was found to be greater at low pH (3%) than at neutral pH (7%) even though
one might expect a considerable degree of complex dissociation at low pH.
Hirose et al. (1982) on the other hand found a much greater organic copper
fraction at the natural pH, of around 98%. It is evidently advisable to
determine metal speciation in situ. i.e. without any preceding concentration
procedures.

We decided to compare the technique based on adsorption of inorganic
metal on MnO_2 (Van den Berg, 1982) with the voltammetric determination of
the labile metal ion concentration (Shuman and Cromer, 1979). Both techniques
aim to determine metal speciation in situ as a function of the total metal
concentration. Interestingly the two methods differ discretely in their
approach : the voltammetric measurement of labile metal concentrations is
based on the kinetic separation of apparently electrochemically irreversible
and inert organic metal complexes from electrochemically labile inorganic
complex ions. The MnO_2 adsorption technique on the other hand allows
equilibrium to be obtained between adsorbing metal ions and competing
organic complexing ligands, before the dissolved metal fraction is separated
from the adsorbed fraction by filtration.

Kramer, C.J.M. and Duinker, J.C. (eds.), Complexation of Trace Metals in Natural Waters. ISBN 90-247-2973-4
©1984 Martinus Nijhoff/Dr W. Junk Publishers, The Hague/Boston/Lancaster.
Printed in the Netherlands.

2. EXPERIMENTAL

MnO$_2$ adsorption technique (Van den Berg, 1982, b)

Sample aliquots were equilibrated overnight with 5 x 10^{-5}M MnO$_2$ and with added amounts of copper or zinc (Zn-65). Then the samples were filtered and acidified. Dissolved copper was determined by DPASV while the dissolved zinc concentration was determined by scintillation counting of Zn-65.

DPASV technique (similar to Shuman and Cromer (1979)

Added quantities of copper or lead were equilibrated overnight in 40 ml sample aliquots. Labile metal concentrations were determined by rotating glassy carbon disk electrode with a pre-plated film of mercury. The plating potential was -900mV for both metal ions. This plating potential was selected because it was found that in these conditions the sensitivity in acidified seawater and in UV-irradiated seawater of natural pH were identical. It was thus easy to calibrate the sensitivity in terms of free, inorganic metal. Values for ligand concentrations and conditional stability constants were evaluated following the theory of Van den Berg (1982, a) and Ruzic (1982). Samples: were obtained from the estuaries of the rivers Dee and Ribble, and a surface seawater sample was collected from the South Atlantic during a cruise of the R.R.S. Discovery in May, 1982. Samples were kept frozen until analysis. The salinities of the samples were respectively 23, 25 and 36‰.

TABLE 1 Comparison of ligand concentrations (C_L) and conditional stability constants obtained by the DPASV and the MnO$_2$ techniques.

Sample	MnO$_2$ C_L/M	log K_{ML}	DPASV C_L/M	log K_{ML}
Ribble				
Cu	3.1 x 10^{-7}	10.9	2.5 x 10^{-7}	7.1
Pb, L1	–	–	6.2 x 10^{-9}	8.4
L2			4.9 x 10^{-7}	6.4
Zn-65	1.8 x 10^{-8}	8.7	–	
Dee				
Cu	1.9 x 10^{-7}	10.6	2.2 x 10^{-7}	7.5
Pb	–	–	3.1 x 10^{-8}	7.1
Zn-65	6 x 10^{-9}	8.5	–	
Atlantic				
Cu, L1	3.1 x 10^{-8}	9.9	6.0 x 10^{-8}	9.7
L2	8.7 x 10^{-8}	9.0	1.2 x 10^{-7}	8.6
Pb	–		4.9 x 10^{-8}	8.4
Zn-65	3.0 x 10^{-8}	7.4	–	

3. RESULTS AND DISCUSSION

The detected ligand concentrations, C_L, and conditional stability constants, K_{ML}, are given in Table 1 .

The complexing capacity for copper could be determined successfully by both methods. Ligand concentrations were similar, although generally higher concentrations were obtained by DPASV. Both techniques found greater ligand concentrations in the estuarine samples than in the sample from the Atlantic. Values for the conditional stability constants were generally lower by DPASV than by MnO_2. These differences (lower values for K_{ML} obtained by DPASV) might indicate some degree of complex dissociation during collection at the rotating disk electrode, because complex dissociation would cause the apparent concentration of complexed metal to decrease. Comparison of metals : The apparent ligand concentrations were not found to be the same for each metal. The highest ligand concentrations were found with copper, the lowest with zinc. In part this result may be due to the different sensitivity of the methods for each metal. Use of the radio-tracer Zn-65 enabled a very high sensitivity to be obtained for that metal, while simultaneously only those ligands which form relatively strong complexes can bind a significant fraction of zinc. However, to a large extent these different complexing capacities may have been caused by metal competition for a limited number of ligands. Therefore the ligand concnetration determined with zinc in the estuary of the river Dee was only 6×10^{-9} M, while with copper and lead much greater ligand concentrations were found. The ligand concentration for zinc was greatest in the sample from the Atlantic (though the total ligand concentration was smallest as determined by copper) because the concentrations of competing trace metals were smallest.

REFERENCES

- Hirose, K., Y. Dokiya and Y. Sugimura, 1982, Determination of conditional stability constants of organic copper and zinc complexes dissolved in seawater using ligand exchange method with EDTA. Mar. Chem. 11 : 343-354.
- Kremling, K., A. Wenck and C. Osterroht, 1981. Investigations on dissolved copper-organic substances in Baltic waters. Mar. Chem. 10 : 209-219.
- Ruzic, I., 1982. Theoretical aspects of the direct titration of natural waters and its information yield for trace metal speciation. Anal. Chem. Acta. 140 : 99-113.

- Shuman, M.S. and J.L. Cromer, 1979. Copper association with aquatic fulvic and humic acids. Estimation of conditional stability constants with a titrimetric anodic stripping voltammetric procedure. Env. Sci. Techn. 13 : 543-545.
- Stone, A.T., 1983. The reduction and dissolution of Mn(III) and Mn(IV) oxides by organics. Thesis, Cal. Inst. Technol., Calif., Report No. AC-1-83.
- Sunda, W.G., S.A. Huntsman and G.R. Harvey, 1983. Photoreduction of manganese oxides in seawater and its geochemical and biological implications. Nature, 301 : 234-236.
- van den Berg, C.M.G., 1982, a. Determination of copper complexation with natural organic ligands in seawater by equilibration with MnO_2.1. Theory. Mar. Chem. 11 : 307-322.
- van den Berg, C.M.G., 1982, b. Determination of copper complexation with natural organic ligands in seawater by equilibration with MnO_2.II. Experimental procedures and application to surface seawater. Mar. Chem. 11 : 323-342.

COMPLEXATION CAPACITY AND CONDITIONAL STABILITY CONSTANTS FOR COPPER OF
SEA- AND ESTUARINE WATERS, SEDIMENT EXTRACTS AND COLLOIDS

CEES J.M. KRAMER and JAN C. DUINKER

1. INTRODUCTION

Trace metals in the environment occur in a variety of forms in solution, in colloids and particulates. Speciation studies try to distinguish between different dissolved forms (valence state, ionic, (in)organic complexed). Speciation has become important in studies concerning transport processes and the fate of trace elements. The toxic effects of trace elements are very much related to their form in the environment (see eg. Langston & Bryan, this volume).

An often used technique to study these processes is Differential Pulse Anodic Stripping Voltammetry (DPASV). It has sufficient sensitivity for natural water analysis and it also allows a distinction between electro-chemically labile and non-labile forms of trace metals like Cu, Pb, Zn and Cd. The technique has been used to determine the Complexation Capacity of a sample by titration with ionic (labile) metal, usually copper (CC_{Cu}) (Chau, 1973); resulting in the determination of the amount of organic ligands, able to complex labile metal into non-labile forms. The CC_{Cu} is usually expressed as the concentration of metal. By theoretical treatment of the data, conditional stability constants ($*K$) can be calculated (eg. Shuman & Woodward 1973, Ruzic 1982, this volume). Reviews on the Complexation Capacity of natural waters for trace metals have been published by Hart (1981) and Neubecker & Allen (1983).

Only few CC_{Cu} determinations of samples from the marine environment have been published (summarized by Neubecker & Allen, 1983). In this paper examples of the CC_{Cu} determined for different marine compartments are given: North Sea water, estuarine water samples of the river Scheldt and Humic Substance obtained by extraction of marine sediments. Conditional stability constants ($*K$) are presented, calculated using the method of Ruzic (1982).

The 'dissolved' fraction contains all material passing filters with a

Kramer, C.J.M. and Duinker, J.C. (eds.), Complexation of Trace Metals in Natural Waters. ISBN 90-247-2973-4
©*1984 Martinus Nijhoff/Dr W. Junk Publishers, The Hague/Boston/Lancaster.*
Printed in the Netherlands.

a pore size of 0.45 μm. It is well established that this includes colloids, ranging from 0.01-0.1 μm (Steinnes, 1983). It is difficult however to distinguish experimentally between truely dissolved and colloidal material (Stumm & Morgan 1981, p 384). (In)organic colloids might be expected to be important in the transport of trace metals because of possible complexation or adsorption reactions. Studies of colloids or large organic molecules (Humics) in the marine environment have involved laboratory mixing experiments of fresh- with saline waters; only few experiments have been carried out in the natural environment (e.g. Sholkovitz, 1976; Sholkovitz et al., 1978). We have studied complexation reactions of colloidal material, originating from an estuarine environment (Scheldt). Filtered samples obtained at various salinities were concentrated using Ultra Filtration techniques. The CC_{Cu} of the concentrate was determined several times, during concentrati

In this paper it will be shown that the choice of electrode (Hanging Mercury Drop Electrode, Rotating and Jet-stream Mercury Film Electrodes) affects the resulting CC_{Cu}. Differences in the two existing CC determination methods using DPASV, the direct titration method (Duinker & Kramer, 1977) and the equilibration method (e.g. Plavsic et al, 1982) will be demonstrated for different samples from the marine environment.

2. MATERIAL & METHODS

2.1. Experimental

Voltammetric measurements (DPASV) were made with a PAR 174 Polargraphic Analyser with a PAR 315 Electro Analysis Controller (E.G.&G.). Three types of mercury working electrodes were used: a Hanging Mercury Drop Electrode (HMDE Metrohm E 410), a Rotating Mercury Film Electrode (RMFE) (Sipos et al, 1974) and a Jet-stream Mercury Film Electrode (JMFE) according to Magjer and Branica (1977). The film electrodes were used in the 'preformed' mode. The mercury film was applied each day to the Glassy Carbon (GC) surface (Tokay, 6 mm ∅) using t_{dep} = 20 min, 8×10^{-5} mol Hg/dm^3, 1500 RPM for the RMFE and t_{dep} = 10 min, 8×10^{-5} mol Hg/dm^3, 50 Hz for the JMFE. These settings represent optimum conditions determined for these electrode systems (Kramer et al, 1984). An Ag/AgCl (sat. KCl) reference electrode and a platinum wire counter electrode were used in a Metrohm silanized 25ml glass cell (HMDE) or 100 ml FEP teflon cell (RMFE, JMFE). Typical parameter settings were t_{cond} = 90 s, t_{dep} = 180 s, E_{cond}=0 mV, E_{dep} = -900 mV and scan speed 5 mV/s (10 mV/s for the Scheldt estuary). For the analytical procedure see Duinker & Kramer (1977).

Samples were analysed at natural pH, controlled with a pH-stat consisting of a Metrohm pH meter (E632) and Inpulsomat (E473). This controlled a solenoid valve, spiking the sample with CO_2 (Kramer and Manshanden, 1982). Sediment extracts were analysed at fixed pH = 8.0 ± 0.03.

Water samples were taken with a model 1080-30 Go-Flo sampler (General Oceanics), filtered over 0.45 membrane filters (Sartorius) stored at 4°C in polythene bottles and analysed on board. Sediment samples (10 g ww subsamples of a core taken with a PVC tube lined with polythylene foil), were extracted with 25 cm³ 0.1 N NaOH for 2 hours (Kramer & Duinker, 1980). Spikes of the extracts were added to off-shore seawater prior to the CC_{Cu} determination.

Colloids were concentrated using Ultra Filtration techniques. The initially high concentration of particulate matter in 10 dm³ (estuarine) samples was reduced by flow-through centrifugation (Hereaus Labofuge 1500, 7000 RPM; 330 ml/min). Subsequently membrane filtration (0.45 µm, Sartorius) was carried out. The resulting 10 dm³ 'dissolved' sample was concentrated over a Amicon Hollow Fiber membrane filter (H1 P5-20) with a theoretical cut-off of M.W. 5000 (∿ 1.3 nm). The system is presented schematically in fig. 1.

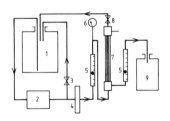

FIGURE 1. Set up for concentration of colloids.
1) concentrate, 2) pump, 3) control valve by pass, 4) pre-filter, 5) flowmeter, 6) pressure gauge, 7) U.F. hollow fiber, 8) back-pressure valve 9) filtrate.

The system was cleaned with 0.1 N HCl and bidest, using a back pressure of 70 kPa. The pressure over the hollow fiber membrane, by a peristaltic pump (Watson/Marlow ltd.) is controlled with a bypass. As the system consists of a cross flow filter, clogging of the UF membrane is minimal. A cross flow of 0.8-0.9 l/min was used. The applied pressure (175 kPa) resulted in a filtration rate of 20-30 ml/min. Subsamples were taken from the concentrate (1 in fig. 1) during the concentration process.

2.2. Methods of Complexation Capacity determination (DPASV)

The determination of the Complexation Capacity is based on the titration of the sample with ionic metal, usually Cu(II). The electrochemically labile fraction was measured at a mercury electrode and i_p was plotted against the M_{added}. An example is given in Figure 2. Part of the ionic copper added to the sample is complexed into non-labile complexes, resulting in a lower lope than in a calibration plot at pH 2 (A). When no free copper complexing ligands are left in the sample, all added labile copper will be measured, and the slope will be identical to that of the calibration plot (B).

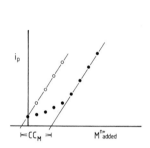

FIGURE 2. Schematic diagram for the calculation of the CC_M, the calibration curve (o) and sample (•).

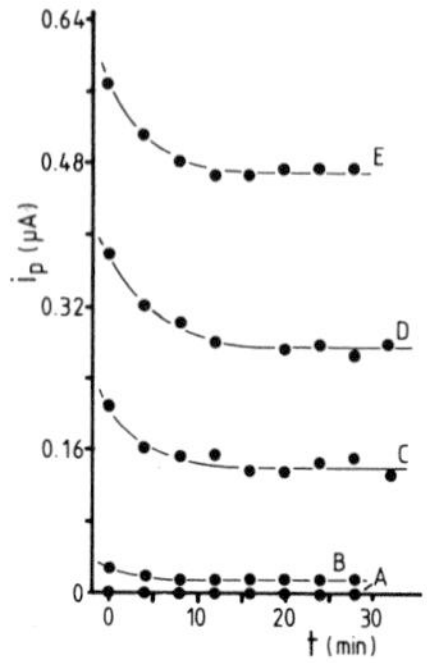

FIGURE 3. Different spikes of copper added to subsamples from the river Rhine estuary, i_p (HMDE) followed in time. Initial spikes of: A) 25 and 50, B) 75, C) 100, D) 125 and E) 150.10^{-8} mol Cu^{2+}/dm^3.

Extrapolation of B and a line through i_p of the sample drawn parallel to B result in the CC_{Cu} value (fig. 2). If M_{found} is plotted instead of i_p, the angle between B and the axes should be 45°; any deviation would indicate incorrect measurements. If slow complexation kinetics result in a nonlinear behaviour of B, a theoretical treatment of the data can result in an estimation of CC (Ruzic 1982, this volume).

From stability constant data (K) it can be calculated that especially copper will be able to form complexes in the marine environment. For this reason one usually determines the CC for copper. As the ligands do not necessarily form complexes with each trace metal, the CC will be specific for the metal considered.

Two experimental methods can be used using DPASV for the determination of CC_{Cu}.

2.2.1. <u>Direct titration method (CC_{Cu}(direct))</u>. The labile copper concentration in a sample, after a single spike with ionic copper, may decrease towards a constant value. This is demonstrated in fig. 3. Six aliquots of a sample from the Rhine estuary were spiked with different amounts of Cu^{2+}. It appears that within the time required for the first measurement (4 min) all labile copper was complexed in the samples spiked with 25 and 50×10^{-8} mol/dm³. With higher spikes, the decrease in labile Cu concentration with time can be followed. In all cases the system reached equilibrium within 20 min.

In the direct titration method a voltammetric measurement is made immediately after spiking several times in succession (Duinker & Kramer 1977). As the entire titration takes 1-1½ h, the first few additions will have reached equilibrium and the important slope of labile copper (B in fig. 2) can be constructed accurately.

Advantages of the method are the short time required for the CC analysis and the possibility of analysing a small sample ($\sim$ 5 cm³). However, slow complexation reactions may not be included in the CC determination.

2.2.2. Equilibration method (CC_{Cu} (eq)). To overcome the problem of slow complexation kinetics, the spiked sample is equilibrated for several hours or days. In this study we used equilibration times >18 h. Samples in cleaned volumetric (polypropylene) flasks were spiked with different amounts of ionic trace metal and analysed after equilibration. Long analysis time and a large sample ($\sim$ 2dm³) for cleaning and analysis are required. The use of many flasks might involve increased risk of contamination. As the complexation reactions will have reached equilibrium the resulting CC will represent a value that is related more to natural conditions than the previous one.

In this paper the CC_{Cu}-direct method is used, unless stated otherwise. The CC_{Cu} and *K are calculated from the plot M_{found} vs $M_{found}/M_{total}-M_{found}$, according to Ruzic (1982). The plots indicated in all cases the typical shape of the case for 1:1 complex formation (Ruzic, this volume).

3. RESULTS AND DISCUSSION

3.1. Complexation Capacity of North Sea samples

During two cruises in Oct. 1981 and Oct. 1982 surface water samples were taken in the southern part of the North Sea (fig. 4) and analysed for the CC_{Cu} immediately on board. Different electrode types were used (HMDE and RMFE in 1981, JMFE in 1982). In both years the two titration methods were compared. The results are given in table 1.

TABLE 1. Complexation Capacity for copper and conditional stability constants of North Sea samples, with different electrodes and titration (direct and after equilibration).

Sample	CC_{Cu}	($\times 10^{-8}$ mol/dm³)					$S (\times 10^{-3})$	pH
	HMDE		RMFE		log *K (RMFE)			
	direct	eq.	direct	eq.	direct	eq.		
81/005	2.0	3.8	4.5	–	7.3	7.8	34.50	8.23
009	0.7	3.4	3.5	–	8.0	8.0	34.95	8.11
018	2.1	4.1	3.1	4.8	7.9	8.3	34.85	8.01
028	–	3.6	3.7	8.7	8.0	7.8	33.30	8.10
030	1.8	4.9	–	–	–	–	31.10	8.08
036	–	5.4	3.8	7.1	8.4	8.5	34.65	8.14
039	2.0	2.2	2.6	3.9	7.5	8.5	35.30	8.10

Sample	JMFE		log *K (JMFE)		$S (\times 10^{-3})$	pH
	direct	eq.	direct	eq.		
82/001	8.4	10.3	7.6	7.7	31.61	7.98
/002	8.0	9.8	7.6	7.7	32.54	8.21
/003	4.4	8.2	7.8	7.9	34.71	8.05
/004	4.4	–	7.8	–	34.66	7.75
/005	6.4	8.0	7.7	7.9	33.81	7.93

For all three electrode types the CC_{Cu}(eq) > CC_{Cu}(direct). This fits well with the expectations on the basis of the two experimental approaches. There are however insufficient data for an evaluation of the fractions fast and slow complexation reactions.

The results obtained in 1981 allow a direct comparison between HMDE and RMFE. CC_{Cu} for the RMFE is larger than for the HMDE. These electrodes cannot be compared directly with the JMFE, which was used in 1982. Data for off-shore stations sampled both in 1981 and 1982 show higher values for the JMFE than for the RMFE. These observations may be directly related to the trend in effective diffusion layer thickness (δ) HMDE > RMFE > JMFE (Plavsic et al, 1982). Any regional variations should preferably be studied with the same electrode.

Most stations represent offshore water type. Only few stations are directly influenced by fresh water sources (e.g. stations 81/028, 82/001 and 002). There seems to be a tendency that coastal waters have higher CC_{Cu} values than open sea samples (eg. stations 81/009, 018 and 039, 82/003 and 004). These CC_{Cu} results compare well with data obtained by Plavsic et al, (1982). They found for Adriatic Seawater a CC_{Cu} of 13 and 15.10^{-8} mol/dm³, using the same technique and electrode. Other results for the CC in the marine environment include Kerr and Quinn (1980), solubilisatic technique, Atlantic ocean water: $27 - 50.10^{-8}$ mol/dm³ ; Gillespie and Vaccaro (1978), biological technique, Sargasso Sea 5.10^{-8} mol/dm³ and Saanich Inlet estuary 19.10^{-8} mol/dm³.

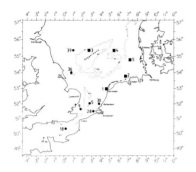

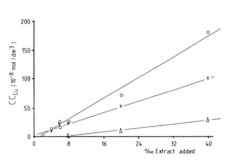

FIGURE 4. Sampling stations at the North Sea during October 1981 (●) and October 1982 (■).

FIGURE 5. The Complexation Capacity (CC_{Cu}) in relation to different amounts of alkaline extracts of sediments (Humic Substance) added to seawater, for three different samples.

Srna et al (1980) found using ASV 18.10^{-8} mol/dm³ for coastal Californian water and $< 3 \times 10^{-8}$ mol/dm³ for off-shore waters. Van den Berg (1982) found 17×10^{-8} mol/dm³ (Irish sea) and 11×10^{-8} mol/dm³ (Atlantic) using the MnO_2 adsorption technique.

For the conditional stability constants (*K) no large differences were observed for the North Sea samples (table 1). At lower salinities a lower *K is usually observed. These results are slightly higher than the log *K = 7.50 reported by Plavsic et al. (1982) using Ruzic (1982) method and the CC_{Cu} (eq) technique. From the data for the HMDE no *K could be calculated, the *K obtained with the RMFE show higher values than found for comparable water samples analysed with the JMFE. The use of the MnO_2 adsorption technique resulted in marked higher log *K values for Irish Sea (log *K = 9.8) and Atlantic Ocean (log *K = 9.9) samples (van den Berg, 1982).

3.2. Complexation Capacity of marine sediment extracts

Humic Substance in marine sediments are important, potential ligands for organometallic complexes. The high organic matter content in alkaline extracts (0.1 N NaOH) caused problems in the CC_{Cu} determination in the extracts themselves. Instead, determinations were made in seawater to which extracts had been added. Fig. 5 shows CC_{Cu} in relation to the (small) amounts of three different extracts added to seawater. In each case, a linear relation was measured.

The vertical distribution of fluorescence methods and CC_{Cu} in extracts of a core, taken in the western Wadden Sea are given in Table 2.

TABLE 2 Vertical distribution of fluorescence in mFl, Complexation Capacity (CC_{Cu}) their ratios and conditional stability constants in sea-water, to which 2% extracts of sediments had been added (pH 8.0).

depth (cm)	fluorescence (mFl)	CC_{Cu} (10^{-8} mol/dm³)	mFl:CC_{Cu}	log*K
0-5	295	15	20	8.2
5-10	360	17	21	7.9
10-15	80	8	10	8.3
15-20	70	12	6	7.7
20-25	40	10	4	7.3
25-30	30	6	5	7.5
30-35	35	6	6	7.6
35-40	25	3	8	7.5

Of this core the top 10 cm obtained a higher concentration of fluorescing compounds (∿ Humic Substance) than the deeper layers. The CC_{Cu} of the top layers also show a higher value, but the decrease with depth is much less than for fluorescence. Obviously, the amount of fluorescing material, present in marine sediments, is only partly available for formation of copper complexes. The possible reservoir of organic ligands present in marine sediment is illustrated by the considerable CC_{Cu} resulting from the addition of only 2% extract.

From the log*K data it can be seen that a maximum in the conditional stability constants is obtained at 10-15 cm depth. Apperently the nature of the Humic Substance with respect to complexation varies within the sediment column.

3.3. Complexation Capacity in the Scheldt estuary.

In the Scheldt estuary surface samples were taken in January at high water slack at 10 stations (fig. 6). Salinities covered the entire estuarine range. The distribution of the suspended matter, fluorescent matter content, oxygen content and pH is geven in fig, 7. The CC_{Cu} obtained a JMFE is illustrated in fig. 8. The CC_{Cu} (direct) data are based on 3 measurements, the CC_{Cu} (eq) on one observation. CC_{Cu} (direct) shows no conservative behaviour, (in constrast to the e.g. dissolved fluorescing organic matter (fig. 7)). A rapid decrease is observed between salinities 0.7 and 10×10^{-3}, it is not clear from the data whether, from then on, the behaviour is conservative. The CC_{Cu} (eq) shows about the same pattern

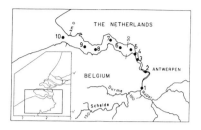

FIGURE 6. Sampling stations at the
river Scheldt estuary.

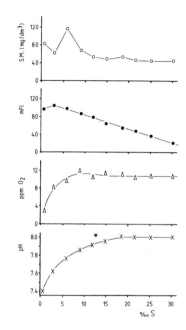

FIGURE 7. Suspended matter, fluorescence, oxygen content and pH at the
different stations in the river
Scheldt estuary.

(fig. 8b). The higher values are to be expected, as discussed before.
From low to high salinity, a relative decrease (87 to 64%) is observed
for the amounts of complexes (rapidly) formed during the CC_{Cu} (direct)
determination. The conditional stability constants calculated at various
salinities are given in Table 3.

TABLE 3. Complexation Capacity (CC_{Cu}JMFE) and conditional stability
constants (*K) in the river Scheldt estuary

$S(\times10^{-3})$	CC_{Cu} (10^{-8}mol/dm³)		log*K	
	direct	eq.	direct	eq.
0.7	29.9	–	7.4	–
3.0	24.5	27.6	7.7	7.3
5.9	22.1	–	7.5	–
9.2	17.3	–	7.3	–
14.9	15.9	–	7.4	–
18.7	14.6	18.8	7.4	7.5
21.7	15.2	22.6	7.3	7.5
25.2	13.0	18.1	7.2	7.6
30.3	10.8	17.3	7.4	8.2

The two determination methods used for the calculation of the conditional
stability constants give slightly different results. With the CC_{Cu} (direct)
method a gradual decrease in log*K is observed (7.5 to 7.2). Although few

data are available for the CC_{Cu} (eq) determination to reverse seems to occur. The CC_{Cu}(eq) results in little higher log*K data, no decrease is observed. The complexes with slow formation rates, only included in the measurements with prolonged equilibration time, might caunt for this increase in conditional stability constants.

3.4. Complexation to colloidal material

Using ultrafiltration (with e.g. a theoretical cut-off of 5000 MW = ∿ 1,3 nm) colloids and large organic molecules are retained by the filter. The experiments were designed to concentrate the colloids in the bulk solution for detailed study. It is assumed here that their properties do not change during the concentration process, involving low concentration factors. The complexation capacity of some samples and concentrates were determined. CC_{Cu} shows a linear relation with the concentration factor for samples of the upper part of the estuary (fig. 9). This contrasts with

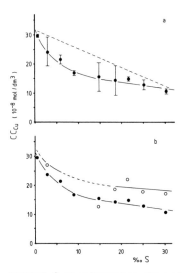

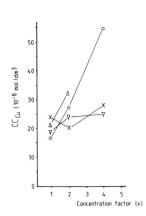

FIGURE 8 Complexation Capacity (CC_{Cu}) for the different stations in the river Scheldt estuary.
a) CC_{Cu} (direct) as mean values for 3 measurements.
b) Direct (●) and equilibration (o) measurements.

FIGURE 9 Complexation Capacity (CC_{Cu}) of different UF concentration steps of river Scheldt estuarine samples at salinities of 0,7 (o), 5,9 (Δ), 12 (∇) and 30,3 $x10^{-3}$(x).

samples at S= 12 and 30 x 10^{-3} , where no increase in the CC_{Cu} is observed with increasing concentration factor. It is well known that colloids, present in the riverwater, tend to flocculate in the estuary. Non conservative behaviour of dissolved (colloidal) Humic Acids was demonstrated in laboratory mixing experiments (Sholkovitz 1976) and in the field (Sholkovitz 1978). It is usually assumed that these flocculation processes have been completed around $S = 10 \times 10^{-3}$. As the flocs are retained by the 0.45 μm filter, they do not play a role in the CC_{Cu} determination of the samples with a higher salinity.

From fig. 8 and fig. 9 it can be concluded that colloids (or large > 5000 MW organic molecules) play an important role in the binding of copper in the upper part of the estuary. Unfortunately the ASV technique can not discriminate between complexation or adsorption with colloids.

4. CONCLUSIONS

The possibilities of the determination of the complexation capacity for copper of several compartments of the marine environment is shown. It appears that little variation occurs for the North Seawater samples except for areas under anthropogenic influence. Sediments may contain a substantial amount of organic matter, able to complex trace metals. Colloidal material in the fresh water part of the estuary shows important complexation behaviour.

It is demonstrated that different mercury working electrodes (HMDE, RMFE or JMFE) show different characteristics. Comparison of Complexation Capacity data obtained with different electrodes should be carried out with care.

ACKNOWLEDGEMENTS

Thanks are due to A. Baks, B.T.W. Holtkamp and G. de Niet for their assistance with the different analyses. The crews of the RV Aurelia and RV Navicula are acknowledged for their help at sea.

REFERENCES

-Berg, C.M.G. van den, 1982. Determination of copper complexation with natural organic ligands in seawater by equilibration with MnO_2. II experimental procedures and application to surface seawater. Mar. Chem. 11: 323-342.
-Chau, Y.K. 1973. Complexing capacity of natural water - its significance and measurement. J. Chromatog. Sci 11: 579.
-Duinker, J.C. and C.J.M. Kramer, 1977. An experimental study on the speciation of dissolved Zn Cd Pb and Cu in river Rhine and North Seawater, by DPASV. Mar. Chem. 5: 207-228.
-Gillespie, P.A. and R.F. Vaccaro, 1978. A bacterial bioassay for measuring the copper chelation capacity of seawater. Limnol.Oceanogr. 23: 543-548.

228

-Hart, B.T, 1981. Trace metal complexing capacity of natural waters: a review. Environ. Technol. Lett. 3 :95-110.

-Kerr, R.A. and J.G. Quinn, 1980. Chemical comparison of dissolved organic matter isolated from different oceanic environments. Mar. Chem. 8 : 217-229.

-Kramer, C.J.M. and J.C. Duinker, 1980. Complexing of copper by sediment extracted humic and fulvic material. Thalassia Jugusl. 16 :251-258.

-Kramer, C.J.M. and G.M. Manshanden, 1982.Improvements in automatic electrochemical analyses, using the PAR 374. NIOZ int. report 1982-4.

-Kramer, C.J.M., Yu Guo-hui and J.C. Duinker. Optimalisation and comparison of four mercury working electrodes in DPASV speciation studies. in prep.

-Magjer, T. and M. Branica, 1977. A new electrode system with efficient mixing of electrolyte. Croatica Chem. Acta 49 : L1-L5.

-Neubecker, T.A. and H.E. Allen, 1983. The measurement of complexation capacity and conditional stability constants for ligands in natural waters. Water Res. 17: 1-14.

-Plavsic, M., D. Krznaric and M. Branica, 1982. Determination of the apparent copper complexing capacity of seawater by ASV. Mar. Chem. 11: 17-31.

Ruzic, I, 1982. Theoretical aspects of the direct titration of natural waters and its information yield for trace metal speciation. Anal. Chim. Acta 140 : 99-113.

-Sholkovitz, E.R, 1976. Flocculation of dissolved organic and inorganic matter during the mixing of river water and seawater. Geochim. Cosmochim. Acta, 40 : 834-845.

-Sholkovitz, E.R., E.A. Boyle and N.B. Price, 1978. The removal of dissolved humic acids and iron during estuarine mixing. Earth Planet. Sci. letts. 40 : 130-136.

-Shuman, M.S. and G.P. Woodward jr, 1977. Stability constants of copper-organic chelates in aquatic samples. Env. Sci.   Techn. 11: 809-813.

-Srna, R.F., K.S. Garrett, S.M. Miller and A.B. Thum, 1980. Copper complexation capacity of marine water samples from southern California. Environ. Sci. Technol. 14 : 1482-1486.

-Steinnes, E. Phisical separation techniques in trace element speciation studies 1983. In: Trace element speciation in surface waters and its ecological implications. GC Leppard Ed. Plenum, NY.

-W. Stumm amd J.J. Morgan. Aquatic Chemistry 1981 2nd ed. Wiley Interscience New York.

INVESTIGATIONS ON THE COMPLEXATION OF HEAVY METALS WITH HUMIC
SUBSTANCES IN ESTUARIES

W. HAEKEL

1. INTRODUCTION

Humic substances are believed to play an important role in
the speciation of heavy metals in fresh water systems.

Investigations presented in this report concern interaction
of trace metals and dissolved humic material during the estuarine mixing.

Normally concentrations of humic substances in natural waters
are too low to perform detailed laboratory experiments.

Methods like adsorption on resins (Mantoura and Riley, 1975)
or extraction with organic solvents (Eberle, 1973) normally
used for enrichment of humics from waters are very drastic.
Changes in properties of humic material brought about by changes
in salinity could not be detected in the following experiments.

The presented scheme (see Fig. 1) describes a moderate procedure for enrichment and investigation of humic material. All
steps within this scheme allow control of pH and salinity.

2. PROCEDURE

2.1. Sample pretreatment.

10 l water samples were prefiltered through 0.7 µm Whatman
GF/F glass-fibre filters. In order to avoid separation of colloidal humic material 0.7 µm filters were used instead of conven-
tionally applied 0.45 µm filters.

Biological growth in the filtrate was checked by addition
of 20 mg/l sodiumazide. Waters treated this way are stable with
respect to spectral characteristics, molecular-size distribution,
and complexing properties for more than 9 months.

Kramer, C.J.M. and Duinker, J.C. (eds.), Complexation of Trace Metals in Natural Waters. ISBN 90-247-2973-4
© 1984 Martinus Nijhoff/Dr W. Junk Publishers, The Hague/Boston/Lancaster.
Printed in the Netherlands.

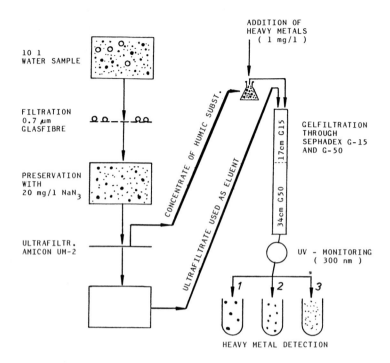

FIGURE 1. Scheme for enrichment and investigation of dissolved humic substances

2.2. Ultrafiltration.

Enrichment of macromolecular substances - i.e. mainly humic material - was performed by means of ultra-filtration with an Amicon UM-2 membrane. This membrane has a reported retention limit of 1000 molecular-weigth-units.

A 10 l stock container and a 400 ml stirred cell with a fil-tration area of 4.5 cm were used. Filtration was performed und pressure of 3 bar nitrogen. Enrichment was continued up to a concentration of 200 mg/l humics (i.e. 100 mg/l DOC) in the retentate.

2.3. Gelchromatography.

10 ml aliquots of a humic concentrate were separated into different molecular-size fractions by means of gelfiltration.

A column of 20 mm internal diameter and 600 mm length filled with 17 cm of Sephadex G-15 gel and 34 cm of G-50 gel was used.

Filtration through G-15 gel allows the separation of humic substances from ions like free or inorganically complexed trace-metals. In a second step molecular-size separation of humics was performed by means of the G-50 gel layer.

24 h prior to gelfiltration aliquots to be analysed were spiked with heavy metals (1 mg/l Cd, Pb, Cu and Hg).

The humic free ultrafiltrate was used as eluent and the con-centration of humic molecules in the eluate was monitored by measuring the UV-absorption at 300 nm. Solutions were collec-ted in a fraction sampler.

An exact calibration with respect to molecular-size of the humics is not possible because of chemical interactions with the gel matrix (Gjessing, 1971).

Prior to each chromatographic run a preconditioning of the column is recommended. For general cleaning 20 ml of a solution of 5g Triton X-100 and 0.5g EDTA in 1 l of NaOH (pH 11) were passed through the column followed by 20 ml of 0.2 m $CaCl_2$ solution. The column was equilibrated with the eluent before a new analysis was started.

2.4. Detection of heavy metals.

The analysis for heavy metals in different fractions of the gelfiltrate was done by means of DPASV for Cd, Pb and Cu and by cold vapor AAS for Hg.

3. RESULTS AND DISCUSSION
3.1. Simulation of a model estuary.

Conditions of a natural estuary were simulated in laboratory experiments by adding different amounts of sea-salt (ASTM-Note 1141-52) to a 1 to 5 concentrate of bog-water humics. Bog-water was collected out of a drainage ditch in Holmer Moor - northern Germany.

In solutions with higher concentration of salt a precipitation of up to 20% of the humic material was observed. For the sample with an addition of 36g/l of salt refering to seawater condi-

232

tions the decrease of UV-absorbance versus time is shown in
Fig. 2. These results show that the flocculation of humic material by salt seems to be a relatively slow process.

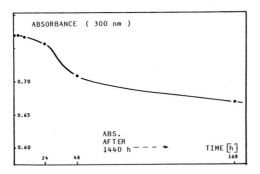

FIGURE 2. Decrease
of UV-absorbance
versus time brought
about by floccula-
tion with 36g/l salt

Changes in the molecular-size distribution brought about by
different amounts of sea salts are shown in Fig. 3.

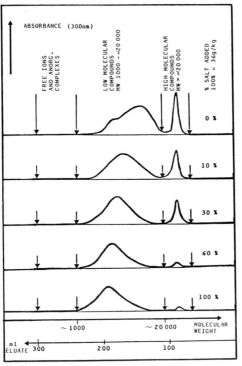

FIGURE 3. Sephadex
G-15/50 gelchromato-
graphic spectra at
different salt con-
centrations.
Shifts of low-molecu
peaks are due to
changes in chemical
interactions betweer
gel and humic sub-
stances caused by
different concentra-
tions of salts
(Sada, 1979).

As can be related from these chromatograms, the diminishing
the high-molecular fraction is the reason for the decrease in tl
total UV-absorbance. Data listed in table 1 support this assump
tion.

TABLE 1. Decrease of UV-absorbance by flocculation with 36g/l salt

	E_{360} initial	E_{360} after 2 months
high molecular humic fraction	0.220·	0.036
low molecular humic fraction	0.260	0.252

Complexing properties of the different molecular-size fractions were determined by spiking the sample solutions prior to
the addition of salt with 1 mg/l of Cd, Pb, Cu and Hg.

During the experimental procedure these heavy metals can be:
1. coprecipitated with flocculated humic material
2. complexed by different molecular-size fractions of dissolved
   humic substance - If this complexation is sufficiently strong
   and complexes are kinetically inert, these metals should be
   found in corresponding humic fractions of the gelfiltrate -
3. uncomplexed or complexed inorganically - In this case heavy
   metals are found in fractions after the humic material -
4. adsorbed on the gelmatrix.

Results of the heavy metal determinations are listed in table 2.
An extended discussion is found in Haekel (1982).

TABLE 2. Percentages of spiked heavy metals found in different
fractions of the model experiment

	Salt added g/l	% uncomplexed or adsorbed on gel	% bound to low-molecular humics	% bound to high-molecular humics	coprecipitaded with humics
Cd	0	100	–	–	–
	36	75	24	–	1
Pb	0	42	10	48	–
	36	12	–	–	88
Cu	0	51	33	16	–
	36	73	–	–	27
Hg	0	66	20	14	–
	36	60	5	3	32

Cadmium. Nearly no complexation with humic substances was found in freshwater and only a small complexing capacity of the low-molecular fraction could be observed in artificial seawater. Nearly no Cd was bound to the precipitated high-molecular fraction.

Lead. A selective bonding to high-molecular humics was found in fresh water. In sea water Pb was nearly completely coprecipitat with this fraction.

Copper. Contrary to Pb, Cu was preferentially complexed by low-molecular humics. Besides a coprecipitation with high-molecular substances in seawater, Cu is released from the low-molecular complexes by an excess of major ions.

Mercury. Hg was linked to low- and high-molecular humics. In seawater a coprecipitation of 32 percent could be observed.

Different molecular-size fractions of humic substances do not only differ in complexing properties but in spectral characteristics. A comparison of UV and fluorescence spectra is shown in Fig. 4 and 5.

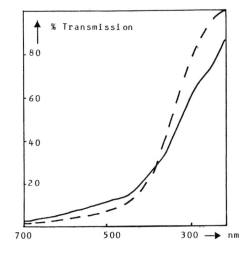 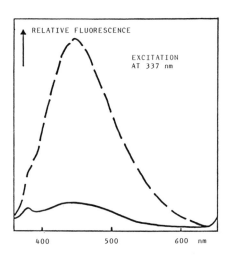

FIGURE 4. UV-spectra of low — — and high —— molecular humics

FIGURE 5. Fluorescence spectra of low — ∙— and high —— molecular humics

These spectral differences might be an indication of variations in chemical structures causing different complexing properties.

3.2. Elbe estuary

To test wether results from the model experiments can be transfered to conditions in a real estuary, samples out of the Elbe
estuary were investigated.

Sampling areas and corresponding gelchromatograms of humic
concentrates are shown in figure 6.

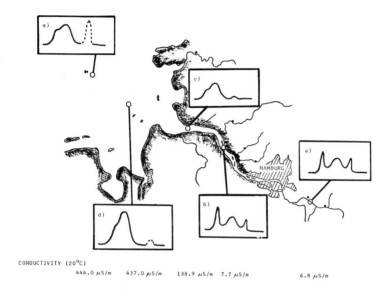

CONDUCTIVITY (20°C)

444.0 μS/m   437.0 μS/m   138.9 μS/m   7.7 μS/m   6.8 μS/m

FIGURE 6. Sample sites and corresponding gelchromatograms in
the Elbe estuary
Samples from position a) Geesthacht and b) Glückstadt are
freshwater while sample c) Cuxhaven is a typical sample out
of a zone of mixing of fresh- and sea-water.
Next two samples d) and e) are marine waters out of the German
Bight.

Magnified Chromatograms are shown in figure 7.

Chromatograms are more structured than the ones found for
bogwater humics. Peaks on the left-hand side can be attributed
to high concentrations of low-molecular substances. Dotted
peaks of the marine samples do not correspond to humic material but to silicic acid.

Most remarkable effect again is the vanishing of the high-
molecular peak. This effect is similar to the results found
with bog-water humics.

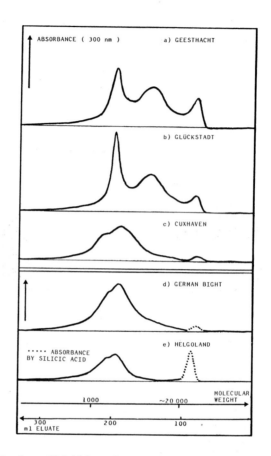

FIGURE 7. Sephadex Gl5/50 Gelchromatograms of Elbe-water humic
Sample a), b) and c) concentration factor 1 : 25
Sample d) and e)   concentration factor 1 : 200

Complexing properties were tested for the organic material
of samples at Glückstadt and Cuxhaven.
Results are shown in figure 8.
Data shown in these figures can be interpreted as follows.
Copper. Similar to bog-water experiments, Cu was found to be
preferentially complexed by low molecular substances. The
amount of complexation is reduced by an excess of major ions
in the brackish water. Uncomplexed copper is adsorbed on the
gel matrix.
Cadmium. In both samples Cd was found in fractions after the

humic material. Therefore kinetically stable and strong complexes of Cd with humics did not exist.

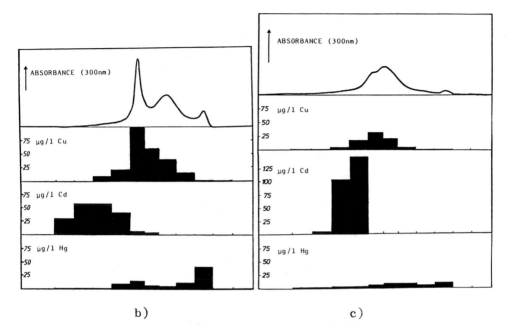

b)   c)

FIGURE 8. Gelspectra (compare Fig. 7) and heavy metal concentrations in different fractions
b) sample at Glückstadt (salinity 0.4 $^o/oo$)
c) sample at Cuxhaven   (salinity 9.9 $^o/oo$)
further information see text

Mercury. In sample b) a selective bonding of Hg to high-molecular substances could be observed. The vanishing of this
humic fraction in brackish water c) caused a reduction of
complexation capacity for this element.
Lead. Difficulties arose because of complete adsorption of
lead on the gel column. Therefore no significant results were
obtained. Differences to the model experiments still have to
be investigated.

4. CONCLUSION

Interactions of various heavy metals with humic substances
do not only differ in the amount and stability of complexation
but in a selective affinity to different molecular-size frac-

tions.

Therefore attention should be drawn to the molecular-size distribution of humic material when these substances are used as complexing agents.

A selective bonding of Hg to high molecular humics which are precipitated during the mixing of freshwater and sea-water could perhaps be one explanation for an enrichment of this heavy metal in the sediments of an estuary.

Nevertheless concentration of heavy metals used for spiking in the described experiments are very high compared to natural concentrations. Therefore data can only be regarded as maximum complexation capacities. Complexing properties at the natural concentration level might be different.

REFERENCES

- Eberle, S.H. and K.H. Sweer, 1973. Bestimmung von Huminsäure und Ligninsulfonsäure im Wasser durch Flüssig-Flüssigextrakt Vom Wasser 41: 27-44.
- Haekel, W., 1982. Untersuchungen zur Schwermetallbindung dur₵ Huminstoffe in Ästuarien. Thesis , University of Kiel, FRG. Gjessing, E.T., 1971. Effect of pH on the filtration of aquat humus using gels and membranes, Schweiz.Z.f.Hydrol. 33: 592-₵
- Mantoura, R.F.C. and J.P. Riley, 1975. The analytical concen- tration of humic substances from natural waters. Anal.Chim.Acta: 76 97-106.
- Sada, A., 1979. Salt effects on adsorption of aromatic compoι in sephadex G-25 chromatography. J.of Chromatography 177: 35ℑ

TRACE METAL CONCENTRATIONS IN THE ANOXIC BOTTOM WATER OF
FRAMVAREN

DAVID DYRSSEN, PER HALL, CONNY HARALDSSON, ÅKE IVERFELDT AND
STIG WESTERLUND

1. INTRODUCTION

The land-locked Norwegian fjord Framvaren is NE of
Farsund. The depest part, our sampling station at 180 m, is
situated at $58^O09.27'$N and $6^O45.0'$E. Dr. Jens Skei at the
Norwegian Institute of Water Research in Oslo started in 1979
an international co-operation on a major investigation of
Framvaren of our times. Our poster at the Texel meeting showed
the depth profiles of in situ temperature, density, oxygen-
-total sulphide, alkalinity, pH, loss of sulphate, phosphate,
ammonia, and the trace metals (except mercury) in unfiltered
water. In addition the values of δ^{34}S for sulphate and sulphide were shown.

In this article we shall give the results of manganese,
iron, cobalt, nickel, copper, zinc, cadmium, mercury and lead
from our sampling on June 20-21, 1982 and available sulphide
data for 1979-1983.

Sulphide complexation of trace metals in anoxic waters
has been discussed in recent years by Dyrssen and Wedborg
(1980), Jacobs and Emerson (1982), Boulegue et al., (1982),
Kremling (1983) and Dyrssen (1983). Transport mechanisms in
fjords such as vertical mixing and particle sedimentation have
been treated by Dyrssen and Svensson (1982), Wassman (1983) and
Skei (1983).

2. RESULTS AND DISCUSSION

The trace metal data are presented in figure 2 at the end
of the article as depth profiles of Fe, Mn, Ni, Zn, Cu, Co,
Pb and Hg. The redoxcline at 14-18 m is denoted with a hatched
bar.

Kramer, C.J.M. and Duinker, J.C. (eds.), Complexation of Trace Metals in Natural Waters. ISBN 90-247-2973-4
©*1984 Martinus Nijhoff/Dr W. Junk Publishers, The Hague/Boston/Lancaster.*
Printed in the Netherlands.

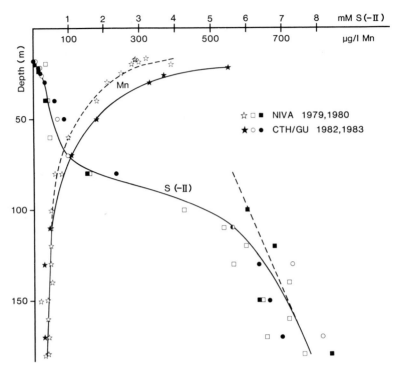

FIGURE 1. Sulphide and manganese profiles in Framvaren. Data by the Norwegian Institute for Water Research (NIVA) and by the Chalmers University of Technology and the University of Göteborg (CTH/GU).

2.1. Sulphide and manganese concentrations

The depth profiles of total sulphide and manganese are shown in figure 1. The dashed manganese curve is a result by NIVA (Knutzen et al., 1981). Using the NIVA results for 1979 and 1980 and our results for 1982 and 1983 the scatter in the sulphide data does not allow the conclusion that the sulphide concentrations in the bottom water below 90 m have increased over that period (the mean value is 0.1 mM/y with one sigma error of ± 1.1). On the contrary, it looks as if the bottom water has reached a steady state. The limiting slope of the depth profile (dashed line in Fig. 2) gives dC/dz = 21.75 mmol m^{-4}. Using a sedimentation rate of 2 mol C m^{-2}y^{-1}, found for the stagnant water mass in Lindåspollene (Wassmann, 1983) the reaction

$$SO_4{}^{2-} + 2CH_2O \rightarrow HS^- + HCO_3{}^- + CO_2 + H_2O$$

gives a release flux of 2.74 mmol sulphide per m^2 per day. Using the simple diffusion equation

release flux $= -K_z dC/dz$

according to Almgren et al. (1975), cf. Dyrssen and Svensson (1982), one obtains

$K_z = 2.74/21.75 = 0.126 \ m^2/d$

For a stagnant period for Byfjorden, Svensson (1980) found $K_z = 0.04 - 0.52 \ m^2/d$. For Byfjorden the release rate was $4.9 - 5.8$ mmol sulphide per m^2 per day (Dyrssen and Svensson, 1982). Thus it seems that vertical mixing also occurs in the bottom water of Framvaren. Above 90 m K_z is most likely considerably larger.

In the case of manganese it seems that manganese(IV)-hydroxide at the interface between oxic and anoxic water at 18-19 m is reduced to dissolved Mn^{2+}. Manganese(II)carbonate is most likely not formed since at 22 m

$$[Mn^{2+}][HCO_3^-]/[H^+] = 0.3$$

which is considerably lower than the solubility product of 10 calculated by Dyrssen and Wedborg (1980). Thus vertical mixing brings manganese(II) up to the outflowing oxic water where it is oxidized back to manganese(IV)hydroxide, but vertical mixing also mixes some Mn^{2+} down into the anoxic water. The depth profile is determined by this mixing and the release from the particulate material, the net flux of which decreases towards the bottom (cf. Wassmann, 1983).

2.2. Sulphide complexation

The most reliable work on metal sulphide complexation has probably been carried out for mercury and silver by Schwarzenbach and Widmer (1963 and 1966). Using the stability constants for silver one may calculate the equilibrium concentrations for Ag_2S in the bottom water where pH = 7, $p[H_2S] \approx p[HS^-] \approx 2.5$. Such a calculation shows that the dominating species $[Ag(HS)_2^-]$ $\approx 19nM$ and that $[Ag(HS)_2^-]/[AgHS] \approx 25$. Thus the total soluble silver would be close to 20 nM. Since Cu(I) and Ag(I) have a very similar chemistry it may be safe to state that the copper concentration in the bottom water of Framvaren (11 ± 1.8 ng/l or 0.17 ± 0.03 nM) is well below the solubility of Cu_2S. A recalculation by Dyrssen (1983) of the cadmium sulphide solubility data of Ste-Marie et al. (1964) seems to give accetable equilibrium constants for the following equilibria:

$$CdS(s) + 2H^+ \rightleftharpoons Cd^{2+} + H_2S \qquad (\log K = -4.64)$$
$$CdS(s) + H_2S \rightleftharpoons Cd(HS)_2 \qquad (\log K = -4.57)$$
$$CdS(s) + HS^- \rightleftharpoons CdHS_2^- \qquad (\log K = -3.93)$$

Using these equilibria the cadmium concentration in the Framvaren bottom water should be $[Cd(HS)_2] + [CdHS_2^-] = 0.91 \cdot 10^{-7} + 3.98 \cdot 10^{-7} = 489$ nM, which is considerably larger than the measured concentration 1.7 ± 0.2 ng/l or 0.015 ± 0.002 nM). Likewise for mercury the three equilibria

$$HgS(s) + H_2S \rightleftharpoons Hg(HS)_2 \qquad (\log K = -5.97)$$
$$HgS(s) + HS^- \rightleftharpoons HgHS_2^- \qquad (\log K = -5.28)$$
$$HgS(s) + HS^- \rightleftharpoons HgS_2^{2-} + H^+ \qquad (\log K = -13.58)$$

give the following equilibrium concentrations: $[Hg(HS)_2] = 3.63$ nM, $[HgHS_2^-] = 17.8$ nM and $[HgS_2^{2-}] = 0.89$ nM or a total concentration of 22.3 nM. Again this concentration is considerably higher than the highest concentration found (0.46 nM).

Dyrssen and Wedborg (1980) used the tendency for sulphide forming metal ions to react with dithizone (HDz) in order to inter(extra)polate the stability constants for the complexes $M(HS)_2$ and MHS_2^- from existing data for cadmium and mercury. The following data are taken from Dyrssen (1983):

Metal	$\log K_{ex}$	$\log K_{22}$	$\log K_{12}$	$\log K_s$
Fe	-3.4	-4.63	-10.91	2.8
Co	1.5	0.08	-6.18	-0.4
Ni	-1.19	-2.54	-8.82	1.5 (-2.2)
Cu	10.53	8.53	2.29	-15.2
Zn	2.26	0.72	-5.55	-3.8 (-3.2)
Pb	0.38	-1.06	-7.33	-6.6

The values of $\log K_s$ in parentheses are taken from Kremling (1983).

The constants represent the following equilibria:

$$M^{2+} + 2HDz\,(CCl_4) \rightleftharpoons MDz_2\,(CCl_4) + 2H^+ \qquad (K_{ex})$$
$$M^{2+} + 2H_2S \rightleftharpoons M(HS)_2 + 2H^+ \qquad (K_{22})$$
$$M^{2+} + 2H_2S \rightleftharpoons MHS_2^- + 3H \qquad (K_{12})$$
$$MS\,(s) + 2H^+ \rightleftharpoons M^{2+} + H_2S \qquad (K_s)$$

The conclusion is the same as for Cu(I), Cd and Hg that the trace metal concentration in the bottom water of Framvaren are 0.8 to 6.9 log units below the solubility curves (cf. Dyrssen, 1983).

This can be explained in at least three ways.

1. The metal concentration is controlled by the particle flux (sedimentation rate) and the vertical mixing.

2. We find 0.04 to 0.30 mg/l of extractable sulphur. Thus the metal sulphides could be stabilized by a pyrite reaction $FeS\,(s) + S\,(s) \rightleftharpoons FeS_2\,(s)$
    NIVA (Knutzen et al., 1981 and Skei, 1983) found framboidal pyrite in the anoxic water.

3. The activity of the metal sulphide MS is lower when it is scavenged by FeS or FeS_2. The activity of MS should be equal to its mol fraction.

244

FIGURE 2. Trace metal depth profiles in Framvaren. The hatched area at 14-18 m marks layer between oxic and sulphidic waters.

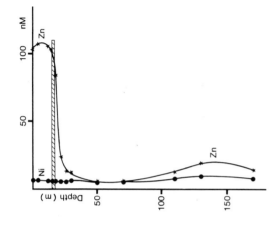

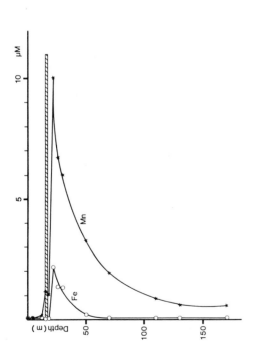

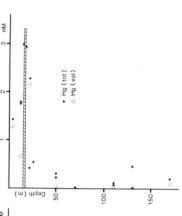

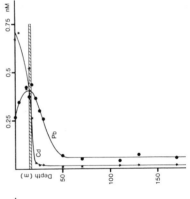

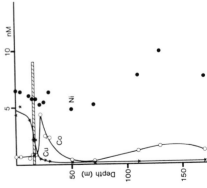

REFERENCES

-Almgren, T., Danielsson, L.-G., Dyrssen, D., Johansson, T. and Nyquist, G., 1975. Release of inorganic matter from sediments in a stagnant basin. Thalassia Yugoslav.11:19-29.
-Boulegue, J., Lord III, C.J. and Church, T.M., 1982. Sulfur speciation and associated trace metals (Fe,Cu) in the pore waters of Great Marsh, Delaware. Geochim.Cosmochim.Acta 46:453-464.
-Dyrssen, D. and Wedborg, M., 1980. Major and minor elements, chemical speciation in estuarine waters. Ch. 3 in "Chemistry and Biochemistry of Estuaries. John Wiley, pp. 452.
-Dyrssen, D. and Svensson, T., a982. On the calculation of release rates from stagnant basin sediments. Chem.Geol. 36:349-351.
-Dyrssen, D., 1983. Metal complexation in sulphidic seawater. Submitted to Mar.Chem.
-Jacobs, L. and Emerson, S., 1982. Trace metal solubility in an anoxic fjord. Earth and Planetary Science Letters 60:237-252.
-Knutzen, J., Ormerod, K., Rygg, B., Skei, J., Sørensen, K., 1981. Et biogeokjemisk studium av en permanent anoksisk fjord - -Framvaren ver Framsund. Norsk institutt for vannforskning (NIVA), Oslo 3, Norway.
-Kremling, K., 1983. The behavior of Zn,Cd,Cu,Ni,Co,Fe, and Mn in anoxic Baltic waters. Mar.Chem.13:87-108.
-Schwarzenbach, G. und Widmer, M., 1963. Die Lösligkeit von Metallsulfiden. I. Schwarzes Quecksilbersulfid. Helv.Chim. Acta 46:2613-2628.
-Schwarzenbach, G. und Widmer, M., 1966. Die Lösligkeit von Metallsulfiden. II. Silbersulfid [1]. Helv-Chim.Acta 49:111-123.
-Skei, J., 1983. Geochemical and sedimentological considerations of a permanent anoxic fjord - Framvaren, south Norway. J-Sedim.Geol.36:131-145.
-Ste-Marie, J., Torma, A.E. and Gübeli, A.O., 1964. The stability of thiocomplexes and solubility products of metal sulphides. I. Cadmium sulphide. Can.J.Chem.42:662-668.
-Svensson, T., 1980. Water exchange and mixing in fjords. Matematical models and field studies in the Byfjord. Thesis, Department of Hydraulics, Chalmers University of Technology, Göteborg, Rep.Ser. A:7.
-Wassmann, P., 1983. Sedimentation of organic and inorganic particulate material in Lindåspollen, a stratified, land-locked fjord in Western Norway. Mar.Ecology.Prog.Ser. 13:237-248.

THE CHEMISTRY OF ALUMINUM IN AN ACIDIC LAKE IN THE ADIRONDACK REGION OF
NEW YORK STATE, USA

C.T. DRISCOLL and G.C. SCHAFRAN

1. INTRODUCTION

Elevated levels of aluminum in acidic surface waters have been
attributed to the dissolution of soil minerals by acidic atmospheric
deposition (Dickson, 1978; Johnson et al., 1981). There is currently
considerable interest in aqueous chemistry of aluminum because of its
role, a) as a toxicant to aquatic organisms (Baker and Schofield 1982),
b) in the cycling of orthophosphate and organic carbon (Dickson, 1978),
and c) as a pH buffer (Dickson, 1978; Driscoll and Bisogni, 1984). In
this study we evaluated the chemistry of aluminum in an acidic drainage
lake.

2. STUDY SITE AND METHODS

The study site, Dart Lake, is located in the Adirondack region of
New York State, U.S.A. (74°52'W, 43°48'N). Samples were collected for
water quality analysis at the inlet, outlet and at seven depths from
a pelagic sampling station approximately every two weeks over an annual
cycle (10/25/1981-11/21/82).

Aqueous aluminum was fractionated into labile monomeric aluminum (IMAl),
non-labile monomeric aluminum (OMAl) and acid soluble aluminum (ASAl)
(Driscoll, 1984) and detected using the procedure of Barnes (1975).
Water samples were analyzed for all major solutes using standard methods
(Driscoll and Schafran, 1984). Thermodynamic calculations used in this
study were made with a modified version of the chemical equilibrium model
MINEQL (Westall et al., 1976). The thermochemical data used in our cal-
culations are summarized elsewhere (Driscoll, 1984). To evaluate the
distribution of IMAL we calculated aquo aluminum (Al^{3+}), hydroxide
(Al-OH), fluoride (Al-F), sulfate ($Al-SO_4$) complexes of aluminum and

Kramer, C.J.M. and Duinker, J.C. (eds.), Complexation of Trace Metals in Natural Waters. ISBN 90-247-2973-4
©1984 Martinus Nijhoff/Dr W. Junk Publishers, The Hague/Boston/Lancaster.
Printed in the Netherlands.

aluminum base neutralizing capacity (Al-BNC; Table 1). We define hydrogen
ion and aluminum base neutralizing capacity (H-Al-BNC) as the amount of
strong base required to increase the pH of a liter of aluminum solution
to 8.3. Further details on the study site, research program and analytical
methods are available elsewhere (Driscoll and Schafran, 1984).

3. RESULTS AND DISCUSSION

Like many other investigators (Dickson, 1978; Johnson et al., 1981;
Driscoll et al., 1984), we observed an exponential increase in aluminum,
and particularly IMAl, with decreases in solution pH. Variations in hydrogen
ion and aluminum BNC were strongly correlated to changes in NO_3^- concentrations (H-Al-BNC=0.94(NO_3^-) + 2.4; in meq m^{-3}, r^2 = 0.54, p < 0.0001). Note
that this correlation is linear with a slope close to one and an intercept
near the origin. Although SO_4^{2-} was the dominant anion in Dart Lake, no
statistically significant relationship with H-Al-BNC was observed. H-Al-BNC
was positively correlated with organic anion concentration (H-Al-BNC = 33.1
($RCOO^-$) -478; where $RCOO^-$ represents organic anion concentration in meq·m^{-3},
r^2 = 0.13, p < 0.0001) and no statistically significant relationship was
observed with Cl^-.

Table 1   FORMS OF INORGANIC ALUMINUM AND BASE NEUTRALIZING CAPACITY

Aquo aluminum;	Al^{3+} = $[Al^{3+}]$
Hydroxide bound aluminum;	Al-OH = $[Al(OH)^{2+}]$ + $[Al(OH)_2^+]$ + $[Al(OH)_4^-]$
Fluoride bound aluminum;	Al-F = $[AlF^{2+}]$ + $[AlF_2^+]$ + $[AlF_3]$ + $[AlF_4^-]$ + $[AlF_5^{2-}]$ + $[AlF_6^{3-}]$
Sulfate bound aluminum;	Al-SO$_4$ = $[AlSO_4^+]$ + $[Al(SO_4)_2^-]$
Aluminum base neutralizing capacity;	Al-BNC = $3[Al^{3+}]$ + $2[Al(OH)^{2+}]$ + $[Al(OH)_2^+]$ + $3[Al-F]$ + $3[Al-SO_4]$ - $[Al(OH)_4^-]$
Hydrogen ion and aluminum base neutralizing capacity;	H-Al-BNC = Al-BNC + $[H^+]$ - $[OH^-]$

During high flow conditions associated with snowmelt, elevated levels
of NO_3^-, hydrogen ion and aluminum were introduced to Dart Lake through
drainage water. Under these acidic conditions, aqueous aluminum was extremel
conservative and transported through the lake with minimal retention. During
periods of stratification, microbially mediated assimilation of nitrate serve
to neutralize H-Al-BNC, resulting in the formation of particulate aluminum

which was deposited to lake sediments. Thus, transformations of nitrate were extremely important in regulating short-term changes in the H-Al-BNC of Dart Lake.

In Dart Lake, fluoride and organic complexes were the predominant forms of aqueous aluminum (Table 2). Al^{3+}, Al-OH and ASAl were present at lower but significant levels, while $Al-SO_4$ was insignificant. The concentrations and relative distribution in Dart Lake were similar to values reported by Johnson et al. (1981) for a stream in the Hubbard Brook Experimental Forest (HBEF) in New Hampshire and by Driscoll et al. (1984) for Adirondack surface waters.

In Dart Lake a substantial portion of mononuclear aluminum appeared to be complexed with organic ligands. We observed a weak but statistically significant empirical relationship between OMAl and dissolved organic carbon concentration (OMAl = 0.018 DOC + 1.84, in $mmol \cdot m^{-3}$, $r^2 = 0.14$, p < 0.0001). Similar empirical relationships have been reported by Driscoll et al. (1984) for Adirondack lakes and streams and by Driscoll (1984) for HBEF streams.

Table 2   MEAN CONCENTRATION AND STANDARD DEVIATION OF ALUMINUM
(in $mmol \cdot m^{-3}$) AND RELATIVE DISTRIBUTION AS TOTAL AND
MONOMERIC ALUMINUM

Aluminum Form	Concentration	Relative Distribution as Total Aluminum	Relative Distribution as Monomeric Aluminum
Total Aluminum (TAl)	14.8 ± 4.4	—	—
Monomeric Aluminum (MAl)	11.5 ± 3.7	0.78	—
Labile Monomeric Aluminum (lMAl)	8.1 ± 3.2	0.54	0.70
Al^{3+}	1.9 ± 1.7	0.13	0.17
Al-OH	2.5 ± 1.3	0.17	0.22
Al-F	3.6 ± 0.7	0.24	0.31
$Al-SO_4$	0.1 ± 0.1	0.01	0.01
Non-labile Monomeric Aluminum (OMAl)	3.5 ± 0.1	0.24	0.30
Acid Soluble Aluminum (ASAl)	3.1 ± 2.1	0.22	—

The correlation between OMAl and DOC was considerably weaker for Dart Lake than these other systems, which may be attributed in part to the limited

range of DOC (220 - 420 mmol·m^{-3}) in our study. Also DOC is a relatively
coarse parameter because it encompasses a variety of organic solutes which
vary greatly in capacity to complex aluminum.

The toxicity of aluminum to fish is very dependent on the speciation
of aluminum (Baker and Schofield, 1982). Moreover the non-conservative
nature of aluminum, which was particularly pronounced in Dart Lake during
stratification, may influence the cycling of critical nutrients like
phosphorus and organic carbon in acidic lake systems. Therefore any assessment of the effects of aluminum in acidic lakes should be made with an under
standing of the speciation and transport of aluminum.

4. ACKNOWLEDGEMENTS

This project was funded in part by the U.S. Environmental Protection
Agency/North Carolina State University Acid Precipitation Program (APP0094-
1981). This paper has not been subjected to the EPA's required peer review
policy and therefore does not necessarily reflect the views of the Agency,
no official endorsement should be inferred. Contribution No. 27 of the
Upstate Freshwater Institute.

REFERENCES

- Baker, J.P. and Schofield, C.L. 1982. Aluminum toxicity to fish in
  acidic waters. Wat. Air Soil Poll. 18:289-309.
- Barnes, R.B. 1976. The determination of specific forms of aluminum
  in natural water. Chem. Geol. 15:177-191.
- Dickson, W. 1978. Some effects of the acidification of Swedish lakes.
  Verh. Internat. Verein. Limnol. 20:851-856.
  Driscoll, C.T. 1984. A procedure for the fractionation of aqueous
  aluminum in dilute acidic waters. Int. J. Environ. Anal. Chem (in pres
- Driscoll, C.T. and Bisogni, J.J. 1984. Weak acid/base systems in
  dilute acidified lakes and streams in the Adirondack region of New
  York State. In: Modeling of Total Acid Precipitation Impacts.
  J.L. Schnoor, ed. Ann Arbor Science, Ann Arbor MI, 53-72.
- Driscoll, C.T. and Schafran, G.C. 1984. The chemistry, transport
  and fate of aluminum in a dilute acidic lake. Final report for project
  APP0094-1981) USEPA/NCSU Acid Precipitation Program, Raleigh, NC.
- Driscoll, C.T., Baker, J.P., Bisogni, J.J. adn Schofield, C.L. 1984.
  Aluminum speciation in dilute acidified surface waters of the Adirondac
  region of New York State. In: Acid Precipitation: Geological
  Aspects, O.R. Bricker, ed. Ann Arbor Science, Ann Arbor, MI, 55-75.
- Johnson, N.M., Driscoll, C.T., Eaton, J.S., Likens, G.E. and McDowell,
  W.H. 1981. Acid rain, dissolved aluminum and chemical weathering
  at Hubbard Brook Experimental Forest, New Hampshire. Geochim.
  Cosmochim. Acta. 45:1421-1437.
- Westall, J.C., Zachary, J.L. and Morel, F.M.M. 1976. MINEQL: A
  computer program for the calculation of chemical equilibrium of
  aqueous systems. Civil Engineering Department, Massachusetts
  Institute of Technology, Technical Note No. 18.

CHEMICAL PROCESSES AT THE PARTICLE-WATER INTERFACE;
IMPLICATIONS CONCERNING THE FORM OF OCCURRENCE OF SOLUTE
AND ADSORBED SPECIES

LAURA SIGG, WERNER STUMM and BETTINA ZINDER

1. INTRODUCTION

In recent years it has become more obvious that the reactions
of solutes with particles are of importance in regulating the
solute concentrations of many reactive elements and controlling
the geochemistry of most trace elements.

In this chapter we will

(1) emphasize that we need to know the form of occurrence of
metal ions in natural water systems in order to understand the
factors that control the concentration of metal ions, their
chemical and biological reactivity and toxicity and their ultimate fate;

(2) review the processes occurring at the interface between the
solution and the solid surface and show that aquatic particles contain functional groups whose acid-base and other coordinative
properties are similar to those of their counterparts in soluble
compounds; more specifically, that the reactions between the metal
ions in solution and the functional groups at the solid surface
are coordination reactions that can be treated by mass law
equations;

(3) document that particle surfaces - because of their ability
to rival with solute complex formers in tieing up metal ions - are
important scavengers and sinks for heavy metal ions in natural
water systems;

(4) illustrate with the help of simplified equilibrium models,
why it is very difficult to determine with analytical leaching
methods the site of "adsorbed" trace elements; and finally

(5) to propose an electrochemical method to differentiate
analytically between dissolved and particulate concentrations.

Kramer, C.J.M. and Duinker, J.C. (eds.), Complexation of Trace Metals in Natural Waters. ISBN 90-247-2973-4
© *1984 Martinus Nijhoff/Dr W. Junk Publishers, The Hague/Boston/Lancaster.*
Printed in the Netherlands.

1.1. Chemical speciation.

Fig. 1 illustrates the various forms of occurrence of Cu(II)
in natural water systems. One would wish to have analytical
possibilities to distinguish between the various solute and "adsorbed" species or to identify the solid or surface sites (organic surface, iron(III)oxide, aluminum silicates) in which the
metal ion is present or bound to. Usually, the evidence for a
particular form of occurrence is circumstantial and is based on
complementary evaluations together with kinetic and thermodynami
considerations. The ion selective electrode (ISE), if it were
sufficiently sensitive, would permit the measurement of the free
metal ion activity; otherwise no single simple method permits
unequivocal identification of a species. Bioassays with algae
have been used to determine free Cu^{2+}. (Sunda and Guillard 1976)
Furthermore, it is operationally very difficult to distinguish
between dissolved and particular concentrations. (Colloids are
often sufficiently small to pass through the pores of membrane
filters.) Although ASV is an extremely sensitive technique, one
must be aware that it measures the electrochemically labile, i.e
electrochemically (within the diffusion layer) available metal
ion concentration (Davison and Whitfield 1977, Sipos et al. 1980
Turner and Whitfield 1979, Buffle 1981, Buffle et al. 1984)
which in the case of Cu(II) includes for example the carbonato
and glycine complexes. (At low pH, the ASV measurement gives
values that are usually representative of total soluble metal
ion concentrations.) Furthermore, no simple analytical recipes
are available to determine with sufficient selectivity the site
of the metal ion in the adsorbed or solid phase. It is extreme
difficult to distinguish analytically between the various specie
or to identify unequivocally the site of "adsorbed" species.

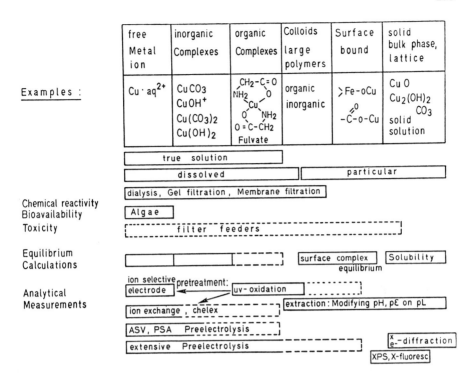

FIGURE 1. The Forms of Occurrence of Cu(II) in
Natural Water Systems.

2. THE ROLE OF PARTICLES IN REGULATING THE CONCENTRATIONS OF METAL IONS

Particles play a dominating role in regulating the concentration of most reactive elements in natural waters. This is especially important in fresh waters; more than 95 % of the heavy metals that are transported from land to the sea occur in the form of particulate matter (Martin and Whitfield (1983)).

The concentration of metals is typically much larger in the solid phase than in the solution phase. Thus, the buffering of metals is much higher in the presence of particles than in their absence. Particles as scrubbing agents are the major sinks for metals and other reactive elements.

2.1. The Oxygen-Metal Ion Bond.

Hydrous oxide surfaces as well as organically coated and organic surfaces contain functional surface groups ($\equiv$Me-OH, $\equiv$R-OH, $\equiv$R-COOH) that act as coordinating sites on the surface. Surface complex formation constants have been determined. These constants can be used to estimate the extent of surface binding (adsorption) or the distribution coefficient between particles and the solution as a function of pH and solution variables (Fig. 2) (Stumm et al. 1976, 1980; Schindler 1981; Stumm and Morgan 1981; Whitfield 1981; Li 1981).

Oxide
$$>M-OH + Me^{2+} \rightleftharpoons >M-OMe + H^+ \qquad ; \quad {}^*K_1{}^s$$

organic surface
$$>R-OH + Me^{2+} \rightleftharpoons >R-OMe + H^+ \qquad ; \quad {}^*K_1{}^s$$

Oxide
$$>M-OH + A_2{}^- \rightleftharpoons >M-A^- + OH^- \qquad ; \quad K_1{}^s$$
$$>M-OH + HB \rightleftharpoons >M-B + H_2O \qquad ; \quad {}^*K_1{}^s$$

$$>Fe-OH + HPO_4{}^{2-} \rightleftharpoons >Fe-OPO_3H + OH^-$$

FIGURE 2. <u>Typical Surface Coordination Equilibria.</u> With the help of equilibrium constants, the extent of surface binding (adsorption) as a function of pH and solution variables can be estimated.

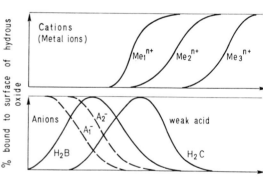

These equilibrium constants can be related to the oxygen-metal i bond strength (as indicated for example by an electronegativity function) or to corresponding equilibrium constants in solution (for example, the tendency to form a MeOSi< bond is related to tendency to form a solute MeOH complex) (Balistieri et al. 1981 Schindler 1981). Recent spectrometric evidence using EPR and EN measurements confirm that such metal ion adsorption is typicall due to "inner sphere" surface complexes (Rudin and Motschi 1983

2.2. Sequestering Surfaces vs. Complex Formation in Solution.

With the help of equilibrium constants, simple models can be established to evaluate the competition between the coordination sites of soluble ligands and those of surfaces for metal ions. An example is illustrated in Fig. 3 where we estimate the proportion of free, carbonato complex and organically complex-bound metal ions as well as surface-bound ("adsorbed") metal ions.

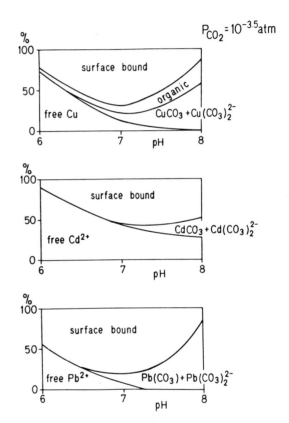

FIGURE 3. Competition between Surface Sequestration and Complex Formation in Solution.
Fresh water: pCa = 4.0; pMg = 4.3; pSO_4 = 4.3 and pCl = 3.0 plus 10^{-5} M salicylate (as organic complex-forming model substance) and silica (with 10^{-5} M > SiOH groups).

This example illustrates that surfaces can tie up significant proportions of trace metals even in the presence of an organic chelate former (in our example salicylate). This calculation also shows that organic complex formers change above all the speciation of Cu(II). In a real lake or ocean, the effects of particles on

the regulation of metal ions are enhanced because the continuo[us]
settling particles (phytoplankton and particles introduced by
rivers) act like a conveyor belt in transporting (Fig. 4) reac[tive]
elements.

PARTITION COEFFICIENTS (sedimenting particles/wate[r)
AND RESIDENCE TIME

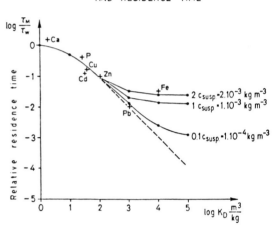

FIGURE 4. The Distri-
bution Coefficients
between particles and
water or between rock
and water determine
the geochemical fate
of reactive elements.

a) Lake Constance,
from Sigg et al.(1983).

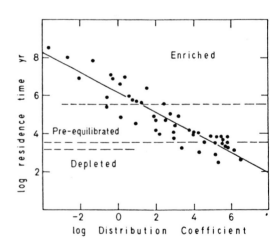

b) Ocean,
from Whitfield (1981).

2.3. The Scavenging of Metals by Particles in Lakes.

Indeed, the partitioning of metals and other reactive elements between particles and water is the key parameter in establishing the residence time and thus in turn the residual concentrations of these elements in the ocean and in lakes (Fig. 4). Thus, the geochemical fate of metals is controlled by the chemical process occurring between the solid surfaces and the water. The more reactive an element is in a lake or ocean, the more will it be bound to particles and the more rapidly will it be removed and the shorter will be its residence time (Schindler 1981, Li 1981, Whitfield 1981).

Lakes, despite being polluted with metal ions (from riverine and atmospheric inputs) ten to hundred times as much as oceans, are nearly as much depleted in these trace metal concentrations as are the oceans (example Table 1).

TABLE 1   ATMOSPHERIC DEPOSITION AND SEDIMENTATION RATE OF HEAVY METALS IN THE NORTH ATLANTIC AND IN LAKE CONSTANCE.

PARTICULATE FLUX NORTH ATLANTIC		Cu	Pb	Cd	Zn
ATMOSPHERIC DEPOSITION	ng cm^{-2} yr^{-1}	25	310	-	130
SINKING SUSPENDED PARTICLES	ng cm^{-2} yr^{-1}	234	330		1040
CONCENTRATIONS SEA	ng l^{-1}	100 (30-300)	3 (1-15)	10 (1-120)	100 (10-600)
FLUX LAKE CONSTANCE					
ATMOSPHERIC DEPOSITION	ng cm^{-2} yr^{-1}	714	11'000	20	8'400
SINKING SUSPENDED PARTICLES	ng cm^{-2} yr^{-1}	6500	9500	(100)	36'000
CONCENTRATION LAKES	ng l^{-1}	300-800	50-100	6-20	1000-4000
CONCENTRATION SWISS RIVERS	ng l^{-1}	1000-3000	300-1000	20-100	10'000

Data for North Atlantic from P. Buat-Menard and R. Chesselet (1979)
Data for Lake Constance from L. Sigg et al. (1983)

Thus, elimination mechanisms in lakes are more dynamic than in oceans; larger productivities and higher particle sedimentation rates are primarily responsible for the more efficient scavenging

(Fig. 4). The uptake of metals occurs by phytoplankton, and to smaller extent by other particles. The input of phosphorus into lakes (influencing the production of biogenic particles) is a major factor in controlling the sedimentation rate of biogenic particles and, in turn, in removing "biophile" heavy metals. (Sigg et al. 1983). Other potential scavenging and metal regeneration cycles operate near the sediment-water interface. Subsequent to early epidiagenesis in the partially anoxic sediments, Fe(II) and Mn(II) and, depending on redox conditions, other elements are released by diffusion from the sediments to the overlying water where they become oxidized to Fe(III) and Mn(III,IV). These higher valent Fe and Mn oxides are important conveyors of heavy metals.

3. SPECIATION OF SOLID AND SURFACE PHASES

In order to understand the processes at the particle-water interface, the buffering of metal ions, the mechanisms and kinetics of release or of binding of metal ions we need to know the speciation of the solid or surface phases, i.e., the identifications of the sites of surfaces and solids that contain these metals or to which these metals are attached to.

Unfortunately, the difficulties encountered in attempting the speciation of solid phase are even more difficult than those experienced in the speciation of the metals in solution. Analytical methods, especially those used in soil chemistry, have been adapted and further developped by various researchers to permit sequential selective extractions for partitioning particulate trace metals into the various chemical forms of occurrence. But it is very difficult to test and evaluate the reliability of the sequential extraction procedures used for this purpose. The speciation task is perhaps more amenable to interpretation in situations where bulk solid phases, e.g., in sediments after partial diagenesis, prevail. Important regulations of the concentration of reactive elements occur, however, at particle-water interface in non-sedimentary environments, i.e., under conditions where

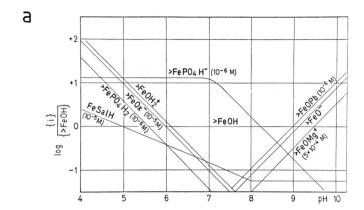

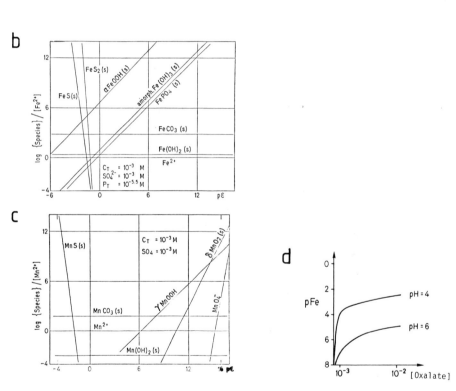

FIGURE 5. "Titration" of Model Surface (patterned after goethite, FeOOH) with acid, reductant and complex-forming ligand, respectively. The activity ratios or pFe ($-\log[Fe(III)]$) are plotted as a function of pH (Fig. 5a), pε (Fig. 5b, c) and [oxalate] (Fig. 5d), respectively; they are based on equilibrium model calculations.

suspended solid phase is characterized by a diversified mixture
of solid components many of which are amorphous or biological a
by surfaces often covered with hydrous oxides and organic
coatings. Surface complex formation with various oxygen donor-
groups are of particular importance in binding heavy metals.

As in the case for the speciation in solutions, no single
method permits unequivocal identification of the site of "adsor
bed" species. Analytical investigations together with thermodynamic and kinetic insights can provide circumstantial evidence
for a particular form of occurrence.

3.1. <u>"Titrating" Hydrous Oxides with H^+, e^- and Ligands.</u>

In order to illustrate some of the difficulties involved
the speciation of solids, we will consider a simplified model
surface (patterned after FeOOH(s)) which is brought into equili
brium with some representative constituents of a fresh water.
Since extraction methods depend on leaching the solids with aci
(or bases), reductants or complex formers, we will try to asses
from a theoretical point of view, how the surface changes its c
position if it is "titrated" with H^+, e^- and ligands. We first
consider the change in composition of a FeOOH surface as a func
tion of pH (Fig. 5a). Above pH = 7, heavy metal ions are bound
the dissociated surface hydroxo groups. At lower pH, they becc
displaced by H^+ and surface complex-forming ligands. Hydrous c
surfaces (and organically coated >R-OH surfaces) loose their te
dency to bind metal ions at low pH values. In this range, howev
they tie up phosphates, silicates and hydroxy carboxylates, e.c
salicylates, fulvates.

That an oxide surface "coated" with organic ligands is not
oblivious to other ligands is exemplified by Fig. 6a) where sa
cylate surface-bound to a goethite surface is replaced by H_2PO
Fulvates and humates may similarily be replaced by $H_2PO_4^-$ or
other ligands; but one needs to make allowance for the fact tha
these - usually polymeric - substances may become bound to
the surface at least partially also by "hydrophobic bonding".

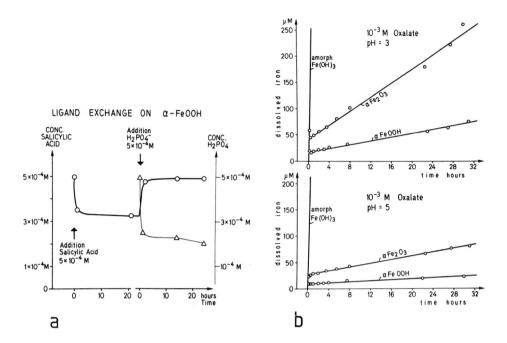

FIGURE 6. Kinetic Aspects of Surface Speciation.
Fig. 6a.Ligand exchange at surface: Salicylate (as a
simplified model for fulvate) incipiently bound to a
FeOOH surface is rapidly replaced at pH ≳ 4 by $H_2PO_4^-$.
Fig. 6b. The rate of dissolution of Fe(III) oxide or
hydroxide and the concommittant release of metal ions
or ligands (originally bound to the surface) depends
critically on the structure of the particular oxide.

Heavy metal ions are bound chemically to surface of hydrous
oxides and organic matter. As shown in Figs. 5a, 2 and 4 the extent of this binding critically depends on pH and on the presence
of other ligands available. Reduction of pH to about 5 will liber-
ate a large fraction of heavy metal ions; it will also dissolve
calcium carbonates but carbonates are remarkably - in comparison
to iron(III) and Mn(III, IV) oxides - inefficient in binding heavy
metals.

Iron- and manganese oxides become reduced and cations and
anions bound to them dissolved at appropriate pε (Fig. 5b, c).
Depending on the oxide and its structure a different pε level has
to be chosen for selective oxide dissolution.

The dissolution of the oxides by complex forming ligands is another possible way to dissolve an iron-oxide coating. The equ librium dissolution depends on pH and oxalate concentration. (Fig. 5d).

The rate of dissolution however, is very much affected by th structure of the Fe(III) oxides (Fig. 6b), while amorphous iron(oxide or iron(III) hydroxide becomes dissolved rapidly, goethit and hematite resist rapid dissolution under these conditions.

The few simple model cases given here illustrate how compli- cated the surface speciation can be. Analytical information nee to be completed by thermodynamic and kinetic considerations.

4. DISTINGUISHING ANALYTICALLY BETWEEN SOLUBLE AND PARTICULATE HEAVY METAL CONCENTRATIONS

As pointed out, (Fig. 1), distinguishing chemical species be tween being "dissolved" and "particulate" is no trivial matter, especially for heavy metal ions occurring typically in concentr tions much smaller than $10^{-8}M$ and in presence of excesses of other ions. The pore sizes of membrane filters provide an opera tional rather than a conceptual distinction. Metal ions, especi ly in fresh waters, are often bound to colloids which may have particle size smaller than $100Å$, sufficiently small to pass thr a membrane filter. Furthermore, any filtration operation is fra with possibilities for contamination and adsorption loss. The k procedure will involve minimum sample manipulation. Changes in concentration of particles and colloids, almost unavoidable dur the filtration steps or the centrifugation, introduce changes ir the partition of the chemical species between the liquid and th often electrically charged surface phases, and this in turn may cause additional errors in the determination of solute species.

4.1. Electrochemical Measurements Permit the Differentiation between Soluble and Particulate Metal Ions.

Voltammetric measurements allow basically to distinguish between electrochemically labile (species electrochemically available within the diffusion layer) and non-labile species (Turner and Whitfield 1979, Davison 1978). In experiments with model colloids, voltammetric measurements are able to differentiate between dissolved species and those bound to particles or colloids (Gonçalves et al, submitted) without prefiltrating the water samples. Potentiometric stripping analysis, PSA (Jagner 1978), a method related to ASV that involves no current measurements, can also be used for metal determinations in the presence of particles. In solutions containing silica suspensions, concentrations of Pb^{2+} measured by PSA were compared with those measured by ion selective electrode (Sigg and Jagner, unpublished results); it could be shown that PSA did not measure the particle-bound fraction and that the colloidal particles did not interfere with the measurements (Fig. 7).

FIGURE 7. pH-Dependent Adsorption of Pb^{2+} on Silica Particles, the residual concentration in solution was measured by PSA in the presence of the silica particles. The drawn-out curves have been calculated with the help of surface complex formation equilibrium constants (Sigg and Jagner, 1980, unpublished results).

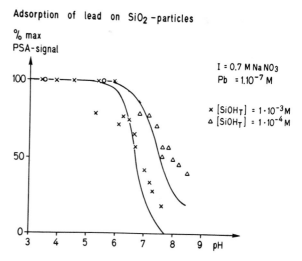

Adsorption of lead on SiO_2-particles

4.2. Particles are Part of "Complexing Capacity".

If a water sample is titrated with a metal ion, the plot of the peak current (representative of the labile metal concentration) vs. [Me added] gives information of the metal binding by non-labile complex formers; that includes metal binding by particles. The shift in the titration curve has often been called the "complexing capacity". Obviously, particles are an important part of "complexing capacity" in natural waters (Fig. 8).

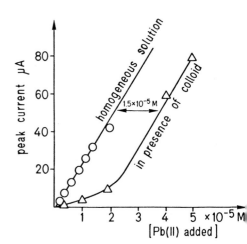

FIGURE 8. "Complexing Capacity by MnO₂ Colloids (4.2 mg/l); pH = 4.8 (from Gonçalves, Sigg and Stumm, submitted). Particles, similar as solute non-labile complex formers, cause a shift in the titration curve. The shift is related to the capacity for metal ion bonding.

5. CONCLUDING REMARKS

5.1. The concentration of most trace elements in natural waters is controlled by the reaction of solutes with solid surfaces.

Surface coordination reactions with organic functional groups and with hydrous oxides at the particle-water interface are particularly important.

5.2. Available analytical extraction and other partitioning methods are not sufficiently selective to determine unequivocally the site of "adsorbed" trace elements.

5.3. Particles, especially colloidal ones, typically encountered in larger concentrations in freshwaters than in seawater, contribute to the complexing capacity.

Electrochemical methods (ASV and PSA) may be superior to membrane filtration in analytically differentiating between dissolved and particulate concentrations.

REFERENCES

- Balistieri, L., P.G. Brewer and J.W. Murray, 1981. Scavenging residence times of trace metals and surface chemistry of sinking particles in the deep ocean. Deep Sea Res., 28 A: 101-121.
- Buat-Menard, P., R. Chesselet, 1979. Variable influence of the atmospheric flux on the trace metal chemistry of oceanic suspended matter. Earth Planet, Sci. Lett. 42: 399.
- Buffle, J., 1981. Speciation of trace elements in natural waters. Trends Anal. Chem., 1: 90.
- Buffle, J., A. Tessier and W. Haerdi, 1984. Interpretation of trace metal complexation by aquatic organic matter, this volume.
- Davison, W., 1978. Defining the electroanalytically measured species in a natural water sample. J. Electroanal. Chem., 87: 395.
- Davison, W. and M. Whitfield, 1977. Modulated polarographic and voltammetric techniques in the study of natural water chemistry. J. Electroanal. Chem., 75: 763-789.
- Gonçalves, S. M-de L., L. Sigg and W. Stumm (submitted). Voltammetric differentiation between solute and particular concentrations of heavy metal ions.
- Jagner, D., 1978. Instrumental approach to potentiometric analysis of some heavy metals. Anal. Chem., 50: 1924.
- Li, Y.H., 1981. Ultimate removal mechanisms of elements from the ocean. Geochim. Cosmochim. Acta, 45: 1659.

266

- Martin, J.M. and M. Whitfield, 1983. The significance of the river input of chemical elements to the ocean. In: Trace Metals in Sea Water, C.S. Wong, ed., Plenum Press, N.Y.
- Rudin, M. and H. Motschi,in press. A molecular model for the structure of copper complexes on hydrous oxide surfaces. J. Colloid and Interf. Sci.
- Schindler, P.W., 1981. Surface complexes at oxide water interfaces. In: Adsorption of inorganics at solid liquid interfaces, M.A. Anderson and A.J. Rubin, eds., Ann Arbor Science, Ann Arbor.
- Sigg, L., M. Sturm, J. Davis and W. Stumm, 1983. Metal transfer mechanisms in lakes. Thalassia Jugoslavica, 18: 293-311.
- Sigg, L., M. Sturm, L. Mart and H.W. Nürnberg, 1982. Schwermetalle im Bodensee. Naturw. 69: 546.
- Sipos, L., P. Valenta, H.W. Nürnberg, M. Branica, 1980. Voltammetric determination of the stability constants of the predominant labile lead complexes in seawater. In: Proc. Int. Expert Discussion Lead Occurrence, Fate and Pollution in the Marine Environment, Rovinj 1977, Pergamon Press, Oxford: 61-76.
- Stumm, W. and J.J. Morgan, 1981. Aquatic Chemistry, Wiley Interscience, N.Y., 780 pp.
- Stumm, W., H. Hohl and F. Dalang, 1976. Interaction of metal ions with hydrous oxide surfaces. Croat. Chem. Acta, 48: 491.
- Stumm, W., R. Kummert and L. Sigg, 1980. A ligand exchange model for the adsorption of inorganic and organic ligands. Croat. Chem. Acta, 52: 291.
- Sunda, W. and R.L.L. Guillard, 1976. The relationship between cupric ion activity and the toxicity of copper to phytoplankton. J. Mar. Research, 34: 511.
- Turner, D.R., M. Whitfield, 1979. The reversible electrodeposition of trace metal ions from multi ligand systems. J. Electroanal. Chem., 103: 43, 61.
- Whitfield, M., 1981. The world ocean; mechanism or machination? Interdiscipl. Sci. Rev., 6: 20.

A HETEROGENEOUS COMPLEXATION MODEL OF THE ADSORPTION OF TRACE METALS ON NATURAL PARTICULATE MATTER

Alain C.M. BOURG and Christophe MOUVET

1. INTRODUCTION

Trace metals are present in, and transported through, aquatic systems both in dissolved and particulate forms. This duality of phase contributes to the complexity in the understanding of their environmental geochemistry. Turekian (1977) stressed the significance of solid particles which, from soils to marine sediments, via rivers and winds, are the vectors of trace metals during the course of their geological migration. The relative mobility of a trace metal depends on its physical form (dissolved versus particulate). Similarily, the physical form, more than the distribution of the dissolved species might be responsible for its toxic or beneficial effect on biota (e.g., filter feeders and seizers will respond predominantly to particulate species).

Exchanges across the water-solid matter interface occur in all aquatic systems (groundwater, rivers, lakes, estuaries and oceans). The mechanisms responsible for these phase changes are: (1) precipitation/dissolution, (2) adsorption/desorption, (3) biological uptake and excretion, and (4) scavenging/remobilization (during flocculation-coagulation/stabilization of colloidal and particulate suspensions, adsorption/desorption of organic macromolecules and precipitation/dissolution) (Figure 1).

This paper presents a description of mechanism (2) in terms of a heterogeneous complexation/dissociation of trace metals with the surfaces of natural particulate matter. Introduction of this concept, together with precipitation processes, into speciation calculations for natural riverine and estuarine aquatic systems makes it possible to identify the potential importance of adsorption/desorption mechanisms as geochemical controls of dissolved trace metal concentrations.

Kramer, C.J.M. and Duinker, J.C. (eds.), Complexation of Trace Metals in Natural Waters. ISBN 90-247-2973-4
© *1984 Martinus Nijhoff/Dr W. Junk Publishers, The Hague/Boston/Lancaster.*
Printed in the Netherlands.

268

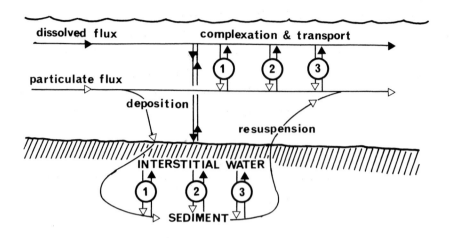

FIGURE 1. Processes responsible for phases changes in aquatic systems;
①precipitation/dissolution, ②adsorption/desorption, ③biological
uptake and release.

2. ADSORPTION AS HETEROGENEOUS COMPLEXATION

2.1. Surface Sites

The components of natural particulate matter have two kinds of electrica
charge. The first is intrinsic and originates from lattice defects due
to isomorphous substitution of minerals such as clays (e.g., Al replacing
Si, or Mg replacing Al). The second comes from amphoteric surface sites
such as the hydroxyls of silica and alumina (located at the edges of clay
layers) and of Fe and Mn hydroxides and oxides and the carboxyls or ammonia
of particulate organic matter. The charges of this latter category are ver
sensitive to the pH of the aqueous environment of the particles. They vary
according to the following acid-base reactions (S = surface site):

$$SH + H^+ \rightleftharpoons SH_2^+ \quad \text{and} \quad SH \rightleftharpoons S^- + H^+ \tag{1}$$

2.2. Adsorption of trace metals

The adsorption of trace metals on simple solids (e.g., alumina, silica
or amorphous iron oxyhydroxide) has been interpreted in terms of the format
of surface complexes with protonated surface hydroxyl groups (Hohl and Stum

1976; Schindler et al., 1976; Davis and Leckie, 1978).

Laboratory experiments have shown that the adsorption behavior of divalent metals on natural aquatic suspended matter and bottom sediments is similar to that on simple solid oxides and hydroxides (Bourg, 1982a,b; Lion et al., 1982; Mouvet and Bourg, 1983). The adsorption at low pH by montmorillonite (Figure 2) is due to its intrinsic charge (mostly localized on the surface

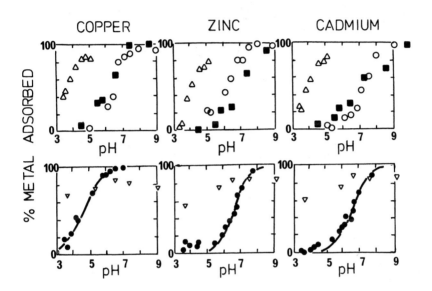

FIGURE 2. Adsorption of trace metals on various solids; total metal = 5 x 10^{-6} M, solid = 1 g l^{-1}, o: amorphous silica, Δ: Gironde Estuary suspended matter (fluid mud), ■: kaolinite, ▼: montmorillonite, ●: Meuse River bottom sediments.

of basal layers). This charge seems to be accessible to the metal in the case of montmorillonite but not of kaolinite (Figure 2). Even though montmorillonite is an important component of natural solids (in quantity as well as in exchange capacity), its adsorption behavior at low pH was not identified in aquatic suspensions and sediments. This phenomenon may result from the presence of organic matter in all natural waters. These macromolecules can easily adsorb on the basal layers (Grim, 1968), the effective surface charge thus becoming amphoteric (Bourg, 1982a).

The adsorption of Cd, Cu and Zn on bottom sediments of the Meuse River, Belgium, and on suspended matter (*crème de vase*) of the Gironde Estuary,

France, was explained by a classical complexation but with surface ligands (Mouvet and Bourg, 1983; Bourg, 1984b). The reaction can be written under the general form:

$$(SH)_n + M^{2+} \overset{*\beta_n^{surface}}{\rightleftharpoons} (S)_n M^{(2-n)+} + nH^+ \tag{2}$$

For most solid hydrous oxides and metals, n can be equal to 1 or 2. For a natural sediment it is very likely that several kinds of surface sites occur on a given particle. The surface site can be Al and Si from clays, Fe and Mn from oxide coatings or C from particulate organic matter. However a reasonable fit (lines in Figure 2) of the experimental data was obtained with only one constant (n=1) for copper, zinc and cadmium (Mouvet and Bourg, 1983)

Table 1. Surface constants of natural particulate matter
(as $\log_{10} *\beta_1^{surface}$ in $I = 0.01$ M $NaNO_3$)

Metal[a]	Meuse River bottom sediments	Gironde Estuary fluid mud (*crème de vase*)
Mg^{2+}	-5.2	-2.3
Ca^{2+}	-6.5	-2.8
Cd^{2+}	-3.7	-1.5
Cu^{2+}	-1.8	-0.9
Ni^{2+}	-3.8	-1.6
Pb^{2+}	-1.7	-0.85
Zn^{2+}	-3.6	-1.3

[a] Mg, Ca, Ni and Pb constants were estimated using a correlation of the form described in equation (5).

2.3. Discussion

The advantages of the model developed above compared to the traditional use of K_D are two-fold (Bourg, 1982a):

(1) The K_D parameters describe the partition of metals into solid and liquid phases without distinguishing between the various processes possibly involved (e.g., precipitation and adsorption). They are not constants. They only quantify the physical partition (particulate versus dissolved) in a given system (for which many parameters such as pH, turbidity and dissolved complexing agents should be specified).

$$K_D = \frac{\text{metal per unit weight solid phase}}{\text{metal per unit volume liquid phase}} \tag{3}$$

The knowledge of the surface constants $*\beta^{surface}$ together with that of the solubility products K_{sp}, on the other hand, permits a better identification of the mechanisms controlling the dissolved metals. The surface constants $*\beta^{surface}$ make it possible to relate the quantity of metal adsorbed to the pH of the medium, the turbidity (surface site concentration) and the dissolved free metal concentration.

$$*\beta_1^{surface} = \frac{[\text{adsorbed metal}][H^+]}{[\text{surface sites}][M^{z+}]} \tag{4}$$

(2) The utilisation of a classical ligand concept to account for adsorption processes makes it possible to offer a coherent (in the mathematical sense) treatment of dissolved, adsorbed and precipitated species in aquatic systems. It is however quite clear that a better knowledge of the reversibility (i.e., desorption) and kinetics involved is still necessary for complete reliance upon model calculations.

The surface constants must be determined in the laboratory for the type(s) of particulate matter present in the system under investigation (see Mouvet -and Bourg, 1983, for an example). It is fortunately not necessary to measure the constants for all trace and major metals since one can use the following correlation, where A and B are characteristic of one type of solid (Schindler et al., 1976; Balistrieri et al., 1981; Mouvet and Bourg, 1983).

$$\log *\beta^{surface} = A \log *\beta^{hydrolysis} + B \tag{5}$$

As for hydrous oxides, the total concentration of surface sites and the pH dependence of the concentration of protonated surface sites can be obtained by acid base titration of the particulate matter (see Mouvet and Bourg, 1983 and Schindler and Kamber, 1968 for more details).

3. MODELING A RIVER: THE MEUSE (HOLLAND)

In Eysden, the Royal Institute for the Purification of Wastewaters routinely surveys the quality of the Meuse River water upon entering Holland. The Meuse is an alkaline river with an average pH value of 7.8. It contains relatively high levels of trace elements. The actual data used in the model calculations is presented in detail elsewhere (Mouvet and Bourg, 1983; Bourg and Mouvet, 1984).

The dissolved speciation of Cd, Cu, Ni, Pb and Zn can be calculated with a thermodynamic model computer program. The possible saturation of important solid minerals can then be quantified by means of the saturation index S.I.:

$$S.I. = \log \frac{\text{Ionic Activity Product}}{\text{Solubility Product}} \qquad (6)$$

The dissolved concentration of metal is controlled by the precipitation/ dissolution of a given mineral if the S.I. of this mineral is equal to zero. Positive S.I. values indicate supersaturation. The S.I. values of the minerals closest to saturation for each metal in the 13 samples studied are given in Figure 3. With the possible exception of willemite (Zn_2SiO_4) for half of the samples, they all seem to be undersaturated with respect to the solid minerals investigated: $CdCO_3$, $Cd(OH)_2$, $Cu_2CO_3(OH)_2$, $Cu(OH)_2$, $CuCO_3$, $Ni(OH)_2$, $NiCO_3$, $PbCO_3$, $Pb_3(OH)_2(CO_3)_2$, $Pb(OH)_2$, $ZnCO_3$, $Zn(OH)_2$ and $Zn_5(OH)_6(CO_3)_2$. The precipitation/dissolution of none of these solids can be the factor controlling the dissolved content in Cd, Cu, Ni, Pb and Zn in the samples studied. Adsorption as a possible geochemical control mechanism was thus investigated.

Laboratory experiments performed on Meuse River sediments (Mouvet and Bourg, 1983) provided estimated values of surface constants (Table 1) by direct determination or by extrapolation using equation (7), as well as of the maximum surface site concentrations (C_0 = 1.3 mol kg^{-1}).

$$\log *\beta_1^{surface} = 0.945 \log *\beta_1^{hydrolysis} + 5.6 \qquad (7)$$

The adsorbed trace metals in the Meuse River samples were estimated by measuring the particulate metal mobilized at low pH and by heating at 45°C. Using the computer model ADSORP (Bourg, 1982b), the calculated adsorbed fractions were much smaller than the observed values. The best agreement was obtained for C_0 = 8.0 mol kg^{-1}. In Figure 4 the sum of the calculated concentration (C) of surface species is compared to the observed value (D) of adsorbed Cd and Cu. The better the model fits, the closer to 1 the ratio C/D is. Compared to the other metals (Mouvet and Bourg, 1983), cadmium is in poorest agreement. However, the generally low values of the calculated

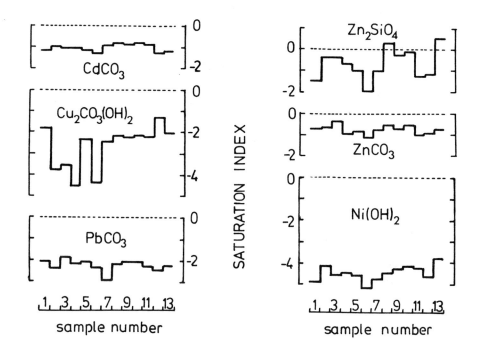

FIGURE 3.   Saturation indexes of minerals closest to saturation

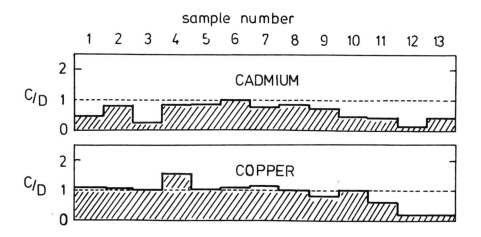

FIGURE 4.   Ratio of calculated (C) and observed (D) values of adsorbed metal concentrations.   (Copper data from Mouvet and Bourg, 1983.)

adsorbed fraction could come from the fact that even though the existence of the adsorption of cadmium chloride and sulfate has been recently qualitatively demonstrated (Benjamin and Leckie, 1982), it has not been included in the model calculations.

The adjustment of the value of C_o from 1.3 to 8.0 mol kg^{-1} can be justified as an empirical calibration performed on 5 metals in 13 samples in order to obtain the best agreement between observed and calculated adsorbed concentrations. It should however be noted that 1.5 and 4 mol kg^{-1} are considered high exchange capacity values for clays and soil organic matter, respectively. Limiting values of about 15 to 25 mol kg^{-1} can be calculated for carboxyl or phenolic groups in filamentous monolayer particulate organic matter (Bourg and Mouvet, 1984). It is unlikely, however, that this applies to the bulk of the suspended particulate matter in the Meuse River.

The high experimental C_o value can be explained by the dissolution of some of the constituents of the sediments during the acid base titration. This means in turn that the calculated values of the surface constants have been underestimated. However, C_o is always associated with the $*\beta_1^{surface}$ in the computation of the concentration of adsorbed metal;

$$[\text{adsorbed metal}] = (*\beta_1^{surface} [\text{surface sites}]) [M^{z+}]/[H^+] \qquad (8)$$

One needs thus only to evaluate the product $*\beta_1^{surface} [\text{surface sites}]$ by empirical fitting using several samples and metals (for a more elaborate discussion see Bourg and Mouvet, 1984).

4. MODELING AN ESTUARY

Estuaries present strong gradients in chemical parameters as well as electrical and hydrodynamic conditions, both capable of modifying the nature of particulate matter. For example, it was observed in the Gironde Estuary, France, that:

(1) particles contained 5 to 15% particulate organic matter and 5 to 6% Fe and Mn oxy-hydroxides in the very low salinity zone and lower than 5% particulate organic matter and 6 to 7.5% Fe and Mn oxy-hydroxides in the middle of the estuary (Etcheber et al., 1983), and

(2) particles from the zone of the turbidity maximum (middle of the estuary on a log scale of salinity) presented specific surfaces twice as

large as particles from either end of the estuary (Bourg, 1983b).

This variability in the nature of suspended matter makes modeling more hazardous in estuaries than in rivers. A first attempt on the Gironde Estuary will be presented in detail elsewhere (Bourg, 1984a). The surface constants experimentally determined on fluid mud (Table 1) are used here to compare laboratory experiments and model calculations.

The effect of increasing the marine nature of water upon the adsorption of Cd was studied (Bourg,1983). The adsorption edge is shifted towards higher pH, with a smaller slope, when Cd is complexed with chloride ions and displaced a step further in the presence of Ca, a competitor for adsorption sites (Figure 5a). These experimental observations are very similar to those obtained by model calculations (Figure 5b). Cadmium, which is one of the most weakly adsorbed divalent trace elements investigated

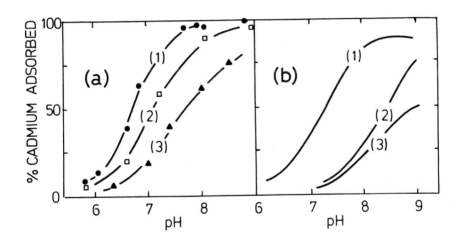

FIGURE 5. Experimental (a) and calculated (b) adsorption of cadmium in simulated estuarine conditions; (1): in 0.001 M $NaNO_3$, (2): in 0.1 M NaCl, (3): in 0.1 M NaCl + 0.01 M $CaCl_2$.

(Table 1) and one of the most strongly complexed with chloride ions, is, therefore, the trace metal whose adsorption/desorption behavior should be most affected by increasing salinity. Therefore, the identification of possible desorption of trace metals in estuaries should start with cadmium.

Salomons (1980) performed an experimental investigation of the adsorption of Cd and Zn on Rhine River sediments. For example, the fraction of adsorbed Zn as a function of the chlorinity is given in Figure 6b for various

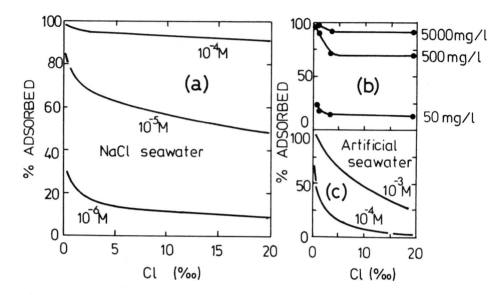

FIGURE 6. Zinc adsorbed and chlorinity; (b): experimental values for Rhine River sediments, from Salomons, 1980, (a) and (c): calculated values using surface constants of Gironde Estuary suspended matter.

turbidities (50 mg/1 to 5,000 mg/1). If the total number of surface sites is estimated at 1 mole kg^{-1}, giving thus 5×10^{-3} M to 3×10^{-3} M, model calculations can be carried out using the surface constants of the fluid mud from the Gironde (Figure 6c). The calculated adsorbed fraction decreases more rapidly than is observed. A new calculation for artificial NaCl sea water however, results in a much better agreement between the model and the laboratory experiment (Figure 6a). The discrepancy between values of surface site concentrations estimated from Salomons' experiments and those necessary to obtain a good agreement with the model calculations can be treated as described at the end of the preceding section (3).

5. CONCLUSION

The inclusion of adsorbed species in mathematical models makes it possible to compare calculated and observed physical distributions of metals. The dissolved/adsorbed distribution of real river samples and of artificial estuarine samples can be modeled fairly well when adsorbed species are taken into consideration as (heterogeneous) surface complexes.

With the exception of willemite (Zn_2SiO_4) in a few samples, no mineral investigated seems to be able to control the dissolved concentrations of Cd, Cu, Ni, Pb and Zn in the Meuse River at Eysden (Holland). Adsorption processes can therefore be proposed as a mechanism of geochemical control.

The changing nature and properties of particulate matter in a salinity gradient make it more difficult to predict the dissolved/adsorbed distribution of trace metals in estuaries. Laboratory experiments can however ascertain, in a qualitative manner, the potential importance of desorption processes in such systems.

ACKNOWLEDGEMENTS

This work was started in Switzerland (Univ. of Bern) and completed in France (National Geological Survey, Orléans for A.C.M.B. and Univ. of Metz for C.M.). Many thanks to Anna Kay Bourg for editing the English of the text and for typing the manuscript.

REFERENCES

-Balistrieri, L., P.G. Brewer and J.W. Murray, 1981. Scavenging residence times and surface chemistry. Deep Sea Res. 28A: 101-121.
-Benjamin, M.M. and J.O. Leckie, 1982. Effects of complexation by Cl, SO_4 and S_2O_3 on adsorption behavior of Cd on oxide surfaces. Environ. Sci. Technol. 16: 162-170.
-Bourg, A.C.M., 1982a. Un modèle d'adsorption des métaux traces par les matières en suspension et les sédiments des systèmes aquatiques naturels. C.R. Acad. Sci. Paris 294: 1091-1094.
-Bourg, A.C.M., 1982b. ADSORP, a chemical equilibria computer program accounting for adsorption processes in aquatic systems. Environ. Technol. Lett. 3: 305-310.
-Bourg, A.C.M., 1983. Role of fresh water/sea water mixing on trace metal adsorption phenomena. In: C.S. Wong, E. Boyle, K.W. Bruland, J.D. Burton and E.D. Goldberg (Eds), Trace Metals in Sea Water, Plenum Press, New York, pp. 195-208.
-Bourg, A.C.M., 1984a. Specific surface and trace elements of the turbidity maximum in macrotidal estuaries. In preparation.
-Bourg, A.C.M., 1984b. Behavior of dissolved Cd, Cu and Zn in estuarine water. In preparation.
-Bourg, A.C.M. and C. Mouvet, 1984. Geochemical control of dissolved trace elements in the Meuse River. In preparation.
-Davis, J.A. and J.O. Leckie, 1978. Surface ionization and complexation at the oxide-water interface. 2. Surface properties of amorphous iron oxyhydroxide and adsorption of metal ions. J. Colloid Interface Sci. 67: 90-107.
-Etcheber, H., A.C.M. Bourg and O. Donard, 1983. Critical aspects of selective extractions of trace metals from estuarine suspended matter. Fe and Mn hydroxides and organic matter interactions. Proc. Internat. Conf. Heavy Metals in the Environment, Heidelberg. Vol. 2: 1200-1203.
-Grim, R.E., 1968. Clay Mineralogy, 2nd ed., Mc Graw-Hill.

-noni, H. and W. Stumm, 1976. Interaction of Pb^{2+} with hydrous $\gamma-Al_2O_3$. J. Colloid Interface Sci. 55: 281-288.

-Lion, L.W., R.S. Altmann and J.O. Leckie, 1982. Trace-metal adsorption characteristics of estuarine particulate matter: evaluation of contributions of Fe/Mn oxide and organic surface coatings. Environ. Sci. Technol. 16: 660-666.

-Mouvet, C. and A.C.M. Bourg, 1983. Speciation (including adsorbed species) of copper, lead, nickel and zinc in the Meuse River. Observed results compared to values calculated with a chemical equilibrium program. Water Res. 17: 641-649.

-Salomons, W., 1980. Adsorption processes and hydrodynamic conditions in estuaries. Environ. Technol. Lett. 1: 356-365.

-Schindler, P.W., B. Fuerst, R. Dick and P.U. Wolf, 1976. Ligand properties of surface silanol groups. Surface complex formation with Fe^{3+}, Cu^{2+}, Cd^{2+} and Pb^{2+}. J. Colloid Interface Sci. 55: 469-475.

-Turekian, K.K., 1977. The fate of metals in the oceans. Geochim. Cosmochim. Acta 41: 1139-1144.

KINETIC SPECTROSCOPY OF METAL FULVIC ACID COMPLEXES

COOPER H. LANGFORD AND DONALD S. GAMBLE

1. INTRODUCTION

Kinetics characteristics of the simple labile complexes of metal ions (e.g. Mg(II), Zn(II), Cd(II), or Ca(II)) which are formed with simple inorganic ligands (carbonate, sulfate, hydroxide) have been fairly well known for many years (Eigen and Wilkins, 1965; Langford and Gray, 1965). However, a large part of the problems of natural water metal ion speciation involves binding of the ions to colloidal ligands such as hydrous oxide colloids of Al(III) or Fe(III) and humic colloids, e.g. humic and fulvic acids. Traditional inorganic kinetic studies of complex formation have not dealt with such species.

Studies have recently been initiated of the reactions of colloidal complexes with standard colourimetric metal ion reagents (Langford et al, 1977). In ordinary analysis the colourimetric reagent comes to equilibrium with the metal species and indicates the total metal. In our studies, kinetics of formation of the coloured species are recorded and the various rates of formation observed indicates the speciation of the original sample. Although the specific rates are those for analytical conditions after the reagent solution is added, they give an indication of the labilities of the original sample species; monomeric aquo ions react rapidly, collidal ligands yield up their metal ions more slowly.

To make the kinetic analysis reproducible, with a simple rate law, the colourimetric reagents must be added in large excess and must control kinetically important parameters such as pH and ionic strength. Consequently, the kinetic solution

Kramer, C.J.M. and Duinker, J.C. (eds.), Complexation of Trace Metals in Natural Waters. ISBN 90-247-2973-4
© 1984 Martinus Nijhoff/Dr W. Junk Publishers, The Hague/Boston/Lancaster.
Printed in the Netherlands.

is no longer a natural water after addition of the reagent and the observed kinetics which characterize species which were present in the original water do not correspond to natural water kinetics. The connection to the interesting kinetics is indicated by the theory below.

2. THEORY

The problem is exhibited by the example of Fe(III) bound to a fulvic acid (FA) which is reacted with the colourimetric reagent sulfosalicyclic acid (SSA).

$$FeFA \underset{k(2)}{\overset{k(-2)}{\rightleftharpoons}} Fe(III) + FA \tag{1}$$

$$Fe(III) + SSA \xrightarrow{k(1)} FeSSA \text{ (red)} \tag{2}$$

The reaction of equation (2) is well understood and has a known rate law and rate. With excess SSA, (2) becomes pseudo first order. Since $k(1)$ is commonly larger than $k(-2)$, the rate of formation of FeSSA from FeFA reduces to equation (3).

$$d[FeSSA]/dt = k(-2)[FeFA] \tag{3}$$

The constants $k(-2)$ are those which may often be connected by theoretical arguments from the analytical solution conditions to allow conclusions on the lability of the species under the original sample conditions.

The only problem with equations (1) to (3) is that FeFA does not designate a single complex. FA is a mixture. Thus we should write equations for the sum over components, i. These are shown in equations (4) and (5) where (5) is the integrated form.

$$d[FeSSA]/dt = \Sigma k(-2i)[FeFA(i)] \tag{5}$$

$$P(inf)-P(t) = \Sigma A(0,i)[\exp(-k(i)t)] + X \tag{6}$$

If the experimental signal, P, is the optical absorbance then A(0,i) is the initial value of the concentration of the i th component in units consistent with P. X designates a time independent term which is either a component fast on the time scale of the kinetic expression or a spectrophotometric blank.

The problem with the application of (6) is that the parameters k(i) and A(0,i) must be extracted from the P(t) function. If only two or three discrete componenets are expected, this is not too difficult using non-linear regression (Mak and Langford, 1983). However, it is very difficult to establish the number of components that should be assumed. An alternative to non-linear regression which identifies components as "peaks" in a kinetic spectrum obtained by an approximate Laplace transform of the time function has been introduced by Olson and Shuman (1983). Although it is very attractive in principle, it leaves similar problems because the approximations make a mixture with discrete components appear to have a continuous distribution of components when the calculations are carried out at a practical level.

In what follows, data will be fit to two or three component fits to equation (6) and the suitability of this number of components judged on the basis of the stability of the constants to moderate variations of conditions. If constants are stable they will be judged to define fairly well defined components (see Fe examples). If derived constants vary with small changes of composition, the components will be judged to represent a distribution and the constants obtained as averages over a part of the distribution.

3. KINETIC STUDIES OF IRON SPECIATION

The species expected to account for Fe in oxygenated waters include principally hydrous oxide colloids and organic complexes because of the limited solubility of monomeric Fe(III). Some exemplary results for the extraction of Fe(III) from hydrous oxide colloids first alone, and then in

combination with fulvic acids is shown in Table I. In the
experiments reported, SSA was not used. Instead, a reagent
system of ferrozine (a very sensitive Fe(II) reagent) combined
with the reducing agent hydroxylamine hydrochloride was
employed for high sensitivity.

TABLE I.   Iron speciation in solutions at pH = 6 as indicated
            kinetically. (Samples prepared from acid Fe(III)
            stocks, slowly neutralized and aged for 24 hrs.)
            See Wong and Langford (1981).

	"Fe(OH)$_3$" Colloids	Fe(III)/FA*
larger k	$1.5 \times 10^{-1}s^{-1}$	$1.49 \ s^{-1}$
conc.	$2 \times 10^{-6}M$	$4.27 \times 10^{-5}M$
lesser k	$4.9 \times 10^{-3}s^{-1}$	fast $(>5s^{-1})$
conc.	$4.8 \times 10^{-5}M$	$0.5 \times 10^{-5}M$

*A "1:1" molar ratio based on phenol-carboxylate sites of FA.

    We see in the table that there are well defined rates
for release of Fe(III) from the hydrous oxides which are
slower than the release from organic complexes. This probably
correlates well with an increase of lability of Fe at the
sample natural water pH's. It is interesting to note that the
"availability" of Fe is purely kinetic. Light scattering and
filtrating results demonstrated that the organic colloids had
larger particle sizes. Tipping et al (1982) demonstrated
similar distinctions of Fe isolated from lake water.

4. SPECIATION OF Al(III)
    Table II collects a few of the results on Al(III)
samples prepared in a way very similar to the Fe(III) samples
of Table I. Again, we see that the fulvic acid labilizes the
metal in comparison to the hydrous oxide colloids. However,
this case also illustrates another point. We find here that
small changes in the ratios of FA to Al(III) leads to a change
in the value of the rate coefficient fitted. This is the

situation which clearly indicates that the components
identified are not well defined but that the kinetic
parameters represent averages over a distribution of related
species. In this case the minimum value of "i" in equation
(5) is clearly more than the number of components required to
give a good parameter fit to the data of one run. However, it
is important to emphasize that no kinetic run gave data that
would allow objective identification of more than two rate
coefficients from one run.

TABLE II. Kinetic analysis of Al(III) species* prepared at
pH = 5 or 6 and analyzed by calcien blue (CB)
spectrofluorimetry at pH = 5. See Mak and Langford
(1982).

Al(III) Total M	FA Total M	pH	$k(i)$ s^{-1}	$k(2)$ s^{-1}
2.0×10^{-5}	0	5.0	1.4×10^{-3}	-
2.0×10^{-6}	0	5.0	3.1×10^{-4}	-
2.0×10^{-5}	2.0×10^{-5}	5.0	2.7×10^{-2}	3.4×10^{-3}
2.0×10^{-5}	2.0×10^{-5}	6.0	2.4×10^{-2}	2.0×10^{-3}
2.0×10^{-6}	2.0×10^{-6}	6.0	1.9×10^{-2}	1.9×10^{-3}
2.0×10^{-6}	5.0×10^{-5}	5.0	-	7.7×10^{-4}

*Note that in no case is Al(III) recovery equal to 100%.

REFERENCES

-Eigen, M., and R.G. Wilkins, 1965. Advances in Chemistry
Series, 49, 55.
-Langford, C.H. and H.B. Gray "Ligand Substitution
Processes", W.A. Benjamin, New York, 1965. Chap. 3.
-Langford, C.H., R. Kay, G.W. Quance and T.R. Khan, 1977.
Analytical Letters, 10, 1247.
-Mak, M.K.S., and C.H. Langford, 1982. Canadian Journal of
Chemistry, 60, 2023.
-Mak, M.K.S., and C.H. Langford, 1983. Inorganica Chimica
Acta, 70, 237.

-Olson, D.S., and M.S. Shuman, 1983. Analytical Chemistry, 55, 1103.
-Tipping, E., C. Woolf and M. Ohnstad, 1982. Hydrobiologia, 92, 383.
-Wong, S.M., and C.H. Langford, 1981. Canadian Journal of Chemistry, 59, 181.

THE INFLUENCE OF ANTIBIOTICS ON THE ADSORPTION KINETICS OF
54-MN AND 59-FE ON SUSPENDED PARTICLES IN RIVER ELBE WATER

R.-D. WILKEN

1. INTRODUCTION

Many pollutants in river water are mainly transported by
suspended particles. It depends on many physical and chemical
parameters how these pollutants are distributed between water
and suspended particles. Much effort has been made for measuring this distribution.

Our examination has dealt with the determination of influences on the adsorption kinetics of some metal ions from water
to suspended particles. We used $54\text{-}MnCl_2$ and $59\text{-}FeCl_3$ as tracers

2. PROCEDURE

For this purpose we took samples of Elbe river water from
the estuary in August 1981 and then in 1982 from the upper
Elbe and treated them with a mixture of both isotopes about
15 minutes after sampling. The amount of the radioactive Manganese was about 250 times lower than the inactive Manganese
in solution. The amount of 59-Fe added was about a third of
the inactive Fe in solution. 10 µl of the tracer solution were
added to 500 ml of the sample.

The samples were shaken in darkness. 5 minutes after adding
the tracers the first aliquot was taken, the phases separated
by centrifugation and the radioisotopes measured by γ-spectros-
copy in the water. In this way the adsorption pattern could be
followed by taking about 30 subsamples over a period of 300
hours.

The typical run of the decrease of the tracer concentrations
in the water phase are shown in the fig. 1 and 2. A quick adsorption in the first 20 hours and than only a slow adsorption
can be seen.

Kramer, C.J.M. and Duinker, J.C. (eds.), Complexation of Trace Metals in Natural Waters. ISBN 90-247-2973-4
©*1984 Martinus Nijhoff/Dr W. Junk Publishers, The Hague/Boston/Lancaster.*
Printed in the Netherlands.

Under this curve we can analyze three different reactions, which have three different reaction speeds. The first reaction has a half life of few minutes and is as quick as an adsorption of ions on an ionexchanger. The second reaction speed has an half life of about 5 hours, whereas the third reaction has a much longer half life of about 150 hours (fig. 3).

This slow reaction can be explained by the slow precipitation of active and inactive ions. The reaction with a half life of some hours could not so easily explained.

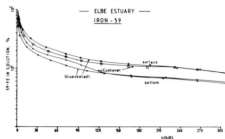

FIGURE 1. 59-Fe uptake by Elbe estuary SPM

The results of the same experiments with water from the upper Elbe showed a first half life of about 60 hours after the quick adsorption due to ion exchange, and a second reaction with a half life of about 300 hours, which is twice as long as for the sample from the estuarine water.

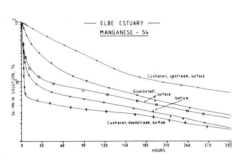

FIGURE 2. 54-Mn uptake by Elbe estuary SPM

The final value after 300 hours for the samples from the estuary shows, that 95% of the Mn and Fe tracers are adsorbed on the suspended particles. Looking at the distribution of the inactive Mn and Fe, it should be about 99%. That means, that about 4% of Fe and Mn ions are situated in places not reached by the solution.

In the samples from the upper Elbe river we can find only 70% of exchangeable Mn and Fe. By this way we can estimate the load

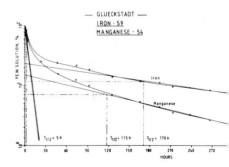

FIGURE 3. Kinetic behaviour of 59-Fe and 54-Mn uptake by SPM

of exchangeable ions on suspended particles.

Looking at the turnovers, we see in the first reactions 75% in estuarine water and 60% in upper Elbe river water. In the second, slower reaction, we have a turnover of about 25% in estuarine to 40% in upper Elbe water.

With water from the upper Elbe we made some experiments to test the influence of bacteria inhibiting chemicals on the reaction kinetics. The results are shown in fig. 4 and 5. Considering the 59-Fe adsorption, formaldehyde is acting in such a way, that it is only impairing the exchange but is not able to stop it. With NaN_3 there is a quicker exchange than in the untreated sample. By adding penicillin, which deactivates the Gram-positive bacteria for about 200 hours, the turnover in the first reactions is reduced from 65% in the untreated sample to only 10%.

The effect of streptomycin on the Fe adsorption is smaller: it reduces the value from 65 to 40%.

Regarding the 54-Mn tracer, the turnover of 60% for the first reactions are reduced slightly to 50% by penicillin; streptomycin has a much bigger effect: it reduces the turnover from 60 down to only 10%.

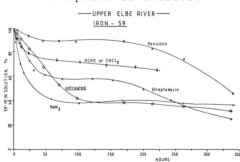

FIGURE 4. 59-Fe uptake by upper Elbe SPM with and without bacteriocids.

3. DISCUSSION

The influence of antibiotics on the adsorption kinetics of Mn and Fe on suspended particles is evident. The question is if these effects are due to complexation of the ions by the antibiotics, or the antibiotics are occupying the adsorption sides on the suspended particles so that the ions cannot reach them or the bacteria inhibiting qualities are

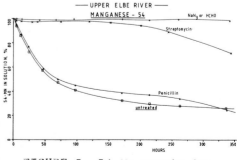

FIGURE 5. 54-Mn uptake by upper Elbe SPM with and without bacteriocids.

responsible for the effects. Bacteria effects are discussed contrarily by Forster et al (1971), Wolfe et al (1975) and Eaton (1979). Looking at the complexation constants of penicillin and streptomycin, both from literature data (Sillén and Martell (1964)) and own experiments with an ion selective electrode the effects probably are small and cannot explain the decrease in the turnovers.

It cannot be excluded that the antibiotics are occupying the binding sides of the suspended particles, but this would be astonishing, because two very different chemical species, (streptomycin is a sugar derivate, whereas penicillin is a derivate of a thiazol) are then acting in the same manner.

TABLE 1. The influence of antibiotics on samples from the upper Elbe

MANGANESE :

UNTREATED: 60 % WITH $T_{1/2}$ OF 50 H
PENICILLIN : 50 % WITH $T_{1/2}$ OF 55 H — 10 % IS CAUSED BY GRAM POSITIVE BACTERIA
STREPTOMYCIN : 10 % WITH $T_{1/2}$ OF 10 H — 50 % IS CAUSED BY GRAM NEGATIVE BACTERIA

IRON :

UNTREATED: 65 % WITH $T_{1/2}$ OF 70 H
PENICILLIN : 10 % WITH $T_{1/2}$ OF 60 H — 55 % IS CAUSED BY GRAM POSITIVE BACTERIA
STREPTOMYCIN : 40 % WITH $T_{1/2}$ OF 30 H — 25 % IS CAUSED BY GRAM NEGATIVE BACTERIA

We can therefore postulate, as it is summarized in table 1, that Gram-positive bacteria influenced about 55% of the Fe adsorption and only 10% of the Mn adsorption. 25% of the Fe adsorption and 50% of the Mn adsorption is caused by the Gram-negative bacteria. Generally speaking, the influence of bacteria on the adsorption kinetics of metal ions cannot be neglected in the first 20 hours of reaction.

REFERENCES

-Eaton, A., 1979. Removal of 'soluble' iron in the Potomac river estuary. Estuarine and Coastal Marine Science 9:41-49
-Forster, W.O., D.A. Wolfe, F.G. Lowman and R. McClin, 1971. In: Proc. 3rd Nat. Symp. on Radioecologa. Oak Ridge, Tenn., May 10.-12.
-Sillén, L.G. and A.E. Martell,1964.
Stability constants of metal-ion complexes. The Chemical Society, London. Section I, 754 pp.
-Wolfe, D.A., W.O. Forster, R. McClin and F.G. Lowman,1975. Trace element interactions in the estuarine zone of the Ana river, Puerto Rico.In: IAEA, Combined effects of radioactive chemical and thermal releases to the environment.IAEA Proc. STI/PUB/404, Vienna, 357 pp.

EFFECTS OF DISSOLVED ORGANIC COMPOUNDS ON THE ADSORPTION OF TRANSURANIC
ELEMENTS

T.H. SIBLEY, J.R. CLAYTON, JR., E.A. WURTZ, A.L. SANCHEZ AND J.J. ALBERTS

1. INTRODUCTION

During the past ten years a great deal of research has been conducted to
determine whether dissolved organic material affects the biogeochemical
cycling of metals. It is now well established for some metals that organic
ligands can alter both adsorption to suspended particulates (Davis and
Leckie 1978) and the toxicity and bioavailability (Davey et al. 1973; Giesy
et al. 1977) of the metals by forming soluble complexes with the metal. Much
of this research has concentrated on divalent or trivalent cations that are
toxic at trace concentrations (e.g. Cu, Cd) or metals that are essential
micronutrients and may limit phytoplankton growth (e.g., Mn, Fe). Additional
elements that are neither micronutrients nor toxicants should also be
considered in order to better understand their biogeochemical cycling.
Long-lived radionuclides that are produced in nuclear power reactors such as
the transuranic elements are a particular concern. These anthropogenic
elements have been introduced into the environment recently (since the
1940's), are characterized by long-lived isotopes, and are reported to be
highly carcinogenic and mutagenic.

Because of their concentration in nuclear wastes, the most important
transuranic elements in the environment are plutonium (Pu), americium (Am),
curium (Cm) and neptunium (Np). All of these are produced by neutron capture
reactions in power reactors and may accumulate in waste depositories. Except
for Np, which occurs primarily as the monovalent ion NpO_2^+, these elements
bind strongly to suspended particulates with partition coefficients that
typically exceed 10^4 (Sanchez et al. 1982; Watters et al. 1980). Because
fallout from atmospheric testing of nuclear weapons released significant
quantities of plutonium into the environment, it is the best studied of the
transuranic elements (Hanson 1980). Nevertheless, there is relatively little
information on the chemical interactions between Pu and organic ligands

Kramer, C.J.M. and Duinker, J.C. (eds.), Complexation of Trace Metals in Natural Waters. ISBN 90-247-2973-4
© *1984 Martinus Nijhoff/Dr W. Junk Publishers, The Hague/Boston/Lancaster.*
Printed in the Netherlands.

(Watters et al. 1980; I.A.E.A. 1981). Even less information is available for the other transuranic elements. Complexation should affect the migration of transuranics from nuclear waste facilities through groundwater as well as the adsorption and consequent transport and bioavailability in aquatic systems.

In this paper we present some results of laboratory experiments that were designed to obtain initial information on the effect of organic ligands on the adsorption of Np, Pu, Am, and Cm to natural sediments. We also review the limited information that has been published previously describing transuranic-organic interactions.

2. METHODS AND MATERIALS

We conducted laboratory adsorption experiments by adding radioisotopes to filtered water that was collected from a variety of freshwater environments in the United States and measuring adsorption to sediments from the same environments. Some water quality characteristics of these environments are given in Table 1; additional information is available

TABLE 1. Some water quality characteristics of water and sediments.

Collection Site	pH	Salinity o/oo	Hardness mg L^{-1}	Sediment Surface Area m^2 g^{-1}
Hudson River				
mp 0.1	8.2	21.0	-------	12.34
mp 18.6	8.3	10.7	-------	10.95
mp 43.3	8.5	3.2	-------	9.24
mp 59.8	8.2	<1	-------	12.24
Cattaraugus Creek	7.8	fresh	202	10.75
Lake Washington	7.8	fresh	42	31.4
PAR Pond	6.6	fresh	<10	50.1
Pond B	6.2	fresh	<10	29.8
Soap Lake				
surface	9.85	------	4,920	
deep	9.50	------	<50,000	

elsewhere (Schell et al. 1981; Sibley et al. 1984; Sanchez 1983). They include alkaline $HCO_3^-/CO_3^=$ buffered water; organic rich, acidic systems; and an hypersaline, high alkalinity lake.

Adsorption experiments were conducted using the constant shaking method (Duursma and Bosch 1970) at a pH that corresponded to that of the natural system at the collection time and a sediment concentration of either 40 or 200 mg l^{-1}. Water was filtered (0.45 um Millipore) at the time of collection

and sediments were sieved to obtain silt and clay size particles (<63 um).
After sediments, radionuclides and ligands were added to filtered water and
shaken for designated times, samples were collected and filtered to separate
soluble (<0.45 um) and particulate phases for radiological counting. Liquid
scintillation spectrometry was used to measure alpha emissions of Pu and Cm
and gamma spectroscopy was used for Np and Am (Wang et al. 1975).

Repetitive sampling during the experiments allowed us to determine when
the system approached a steady state. Some of our results are presented as
distribution coefficients (K_d), the ratio of radioactivity in one gram of
particulates to the radioactivity in one milliliter of solution.
Distribution coefficients are assumed to represent equilibrium conditions
although it is usually doubtful that they do. In our experiments they
indicate that the ratio does not change significantly between consecutive
sampling times, although it might change over longer time periods. It is
important to remember that as the K_d value increases the soluble
concentration of radionuclide decreases.

3. RESULTS AND DISCUSSION

3.1. Addition of organic ligands

The effect of specific organic ligands can be determined by comparing
adsorption of radionuclides in experimental systems to which the ligand has
been added to systems without the ligand. Clayton et al. (1982) investigated
the effect of several ligands on the adsorption of [241]Am to natural
sediments in freshwater systems. Initial experiments were conducted with
different concentrations of EDTA, a ligand that forms strong chelates with
Am (Schultz 1976), to evaluate what concentration of ligand was necessary to
observe a significant effect. Figure 1 shows the results obtained for
sediments and water from Cattaraugus Creek. At high concentrations of added
ligand (10^{-4}M) there is a significant decrease in the amount of radionuclide
adsorbed and corresponding increase in the soluble phase. This affect might
also have occurred at the 10^{-6}M concentration although the magnitude of the
effect is considerably less.

Adsorption of [241]Am appears to be enhanced at the concentration of
10^{-8}M EDTA. We would not expect this result if the only important
interaction was between EDTA and Am. However, EDTA forms complexes with most
other metals and interactions between EDTA and those metals may produce
secondary interactions, such as increasing the number of available

292

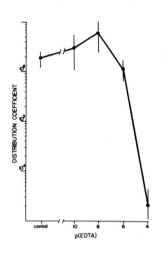

FIGURE 1. Distribution coefficient of ^{241}Am for different concentrations of added EDTA. p(EDTA)= -log (Molar Conc.) Error bars are one standard deviation propagated counting error.

adsorption sites, that could affect the sediment:water partitioning of Am. In our experimental systems we cannot identify these interactions although at low concentrations of ligands they appear to be more important than the direct inter- action with Am.

The concentration of transuranic elements in our experiments is significantly higher than typical environmental concentrations. How- ever, these experimental concentra- tions of transuranics will generally be much lower than the naturally occurring concentrations of trace elements that are present in our water samples. Since the organics we add will react both with the trans- uranics and with other metals that may be present at higher but unknown concentrations, we cannot define the transuranic-organic reaction in quantitative chemical terms. We can, however, describe the effects of ligands qualitatively and identify some environmental characteristics that will alter these effects.

The effects of several different ligands were determined in sediment:water systems from Lake Washington using ligand concentrations of 10^{-4}M and sediment concentrations of 200 mg l^{-1} (Table 2). Again, EDTA increased the concentration of soluble Am and reduced the distribution coefficient nearly two orders of magnitude relative to the controls. K_d values in systems containing acetate and salicylate did not differ dramatically from the controls. Addition of 1-nitroso-2-napthol or 1,10 phenanthroline enhanced adsorption, suggesting that different sites of the ligand simultaneously bind to the sediments and complex with Am, effectively attaching Am to the particulates and increasing K_d value (Davis and Leckie 1978).

TABLE 2. Distribution coefficients for ^{241}Am in sediment-water systems with added organic ligands. All values must be multiplied by 10^5. Humic acid concentration was 10.8 mg L^{-1} and other ligand concentrations were 10^{-4}M for Lake Washington and 10^{-5}M for the Hudson River estuary.

Added Ligand	Hudson River Estuary mp 0.1	mp 18.6	mp 43.3	Lake Washington
None	>2.3	2.6	>2.1	0.23
Humic acids	>1.7	1.3	0.4	
Glycolic acid	>2.6	4.2	2.6	
EDTA	1.2	0.1	0.1	0.01
1,10-phenanthroline	>2.0	3.8	2.7	2.53
1-nitroso-2-napthol	>2.5	>1.9	>2.1	0.78
Acetic acid				0.56
Salicylic acid				0.29

A similar set of experiments was conducted for sediment-water systems from three locations in the Hudson River estuary. In these experiments the sediments and radionuclide were allowed to equilibrate for 144 hours before the organic ligand was added. The concentrations of soluble Am following the addition of EDTA and humic acid are shown in Figure 2. The most interesting feature on these graphs is that the effect of adding these ligands is much greater at the stations with lower salinity because of the non-specificity of the ligand. That is, at the more saline stations the natural concentrations of other elements (e.g. Ca, Mg) that can compete with Am for the ligands are greater than at the upstream stations, and much of the ligand is presumably complexed with those elements. Therefore, there is a smaller effective

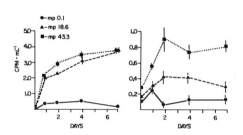

FIGURE 2. Soluble activity of ^{241}Am after adding 10^{-5}M EDTA (left) or 10.8 mg L^{-1} Aldrich Humic Acids (right) to experimental systems from the Hudson River estuary. Error bars are two standard deviations propagated counting errors.

concentration of ligand to complex the Am and the effect on Am adsorption is correspondingly reduced. The magnitude of the ligand effect differed among experimental systems for all ligands. Glycolic acid and 1,10-phenanthroline produced increased K_d values at mp 18.6 and mp 43.3 (Table 2). For 1-nitroso-2-napthol the concentration of Am in the soluble phase was always below detection limits so we could not identify differences between stations.

Another point to notice in Figure 2 is that the maximum effect of the ligand does not occur until 48 to 96 hours after the addition of the ligand. Sanchez (1983) showed that the percent adsorption of Pu also can increase for several days. However, many laboratory experiments to evaluate adsorption are conducted over shorter time periods and ignore these reaction kinetics. Furthermore, in natural waters we know neither the chemical characteristics of the available organic ligands nor the concentrations of competing metals. Therefore, experiments conducted at high concentrations of sediments or transuranics to better define the adsorption reaction may not provide reliable information about the behavior of transuranics in natural systems.

3.2. Effect of natural organics

One way to evaluate the role of natural organics is to compare adsorption in systems that have different concentrations of dissolved organics. Wahlgren and Orlandini (1982) reported that distribution coefficients of Pu(IV) were linearly and inversely correlated with dissolved organic carbon content of the water and not significantly correlated with any other limnological parameter that they measured in freshwater environments. The percent of dissolved Pu in reduced oxidation states was also dependent upon the total concentration of dissolved organic matter. Simpson et al. (1980, 1982) and Anderson et al. (1982) reported exceptionally high concentrations of soluble Pu in Mono Lake, California which they attributed to complexation with high concentrations of carbonates (0.3 uM) in the water column. Complexation of Pu with both organic and inorganic ligands is poorly understood and requires considerably more experimental research (Schwab and Felmy 1983). However, it appears that carbonates are more important for solubilizing Pu in hypersaline, alkaline lakes (Sanchez 1983) while organic ligands are more significant in freshwater environments (Wahlgren and Orlandini 1982).

Sibley et al. (1983) compared adsorption of Cm in sediment/water systems from blackwater (acidic, organic rich, freshwater) environments to adsorption in typical $HCO_3^-/CO_3^=$ environments. The blackwater systems consistently had lower K_d values than the carbonate buffered systems, suggesting that dissolved organics complex with Cm and reduce adsorption. However, the differences were less than an order of magnitude and could have been attributed to slight differences in pH among the different experimental systems. Table 3 provides a similar comparison for the K_d values of ^{237}Np in blackwater and carbonate buffered systems. As with Cm there are no major differences in the K_d values between carbonate buffered and blackwater systems. It appears from comparing these different sediment/water systems that natural concentrations of organic compounds may be less significant for determining the K_d values of Cm or Np than for Pu.

TABLE 3. Distribution coefficients for ^{237}Np in sediment-water systems from blackwater (PAR Pond, Pond B) and bicarbonate buffered systems. Values must be multiplied by 10^2, errors are one standard deviation around mean of triplicate experiments.

	24 Hours	96 Hours
Lake Michigan	2.51 + 0.21	3.57 + 0.40
Cattaraugus Creek	1.77 + 0.50	2.17 + 0.55
PAR Pond	2.08 + 0.40	3.05 + 0.67
Pond B	1.40 + 0.04	1.89 + 0.87

Another way to evaluate the role of dissolved organics is to remove the organics and compare adsorption with and without the organics. We did this by ultrafiltering water through Amicon H5P10 Hollow Fiber Bundles to remove any materials larger than 10 nm. For sediments that were treated with a variety of extraction procedures to change their surface properties, and for untreated sediments adsorption of Cm always increased following ultrafiltration of the water (Sibley and Alberts 1983). This suggests that dissolved organic compounds increase the solubility of Cm, and decrease adsorption, in some organic rich systems.

Adsorption of Cm and Np to sediments in filtered and ultrafiltered water from blackwater ponds is shown in Figure 3. For Cm, ultrafiltration of the water increases adsorption. However, for Np, which does not adsorb nearly as much as Am, Pu or Cm the amount adsorbed was greater in filtered water than in ultrafiltered at all sampling times during the 160 hours of this experiment. Although additional experiments need to be performed in this system, we believe the higher adsorption of Np in filtered water is caused by reduction of Np(V) to Np(IV) rather than by complexation

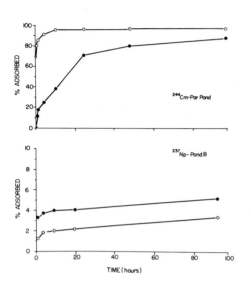

% ADSORBED

^{244}Cm-Par Pond

% ADSORBED

^{237}Np- Pond B

TIME (hours)

FIGURE 3. Adsorption of ^{244}Cm and ^{237}Np to sediments in filtered (closed circles) and ultrafiltered (open circles) water from blackwater environments.

reactions. Nash et al. (1981) have reported that humic acids will reduce Np(V) to Np(IV) which adsorbs more readily. These results agree with previous reports on the reduction of Pu(V) to Pu(IV) by natural organic compounds (Bondietti et al. 1976; Sanchez 1983). Thus, organic ligands can affect the biogeochemical cycling of transuranics both by complexation and by reducing some elements to lower oxidation states.

Another indication that complexation with organics decreases adsorption was obtained by allowing ^{244}Cm to equilibrate with filtered water prior to introducing sediments.

Figure 4 shows the percent soluble ^{244}Cm during 96 hours of adsorption to natural sediments and kaolinite for systems in which Cm was allowed to equilibrate with the water for 48 hours and for systems that had no prior equilibration. If equilibration in solution occurs prior to adsorption the soluble concentration remains higher throughout the experiment. This suggests that complexes are formed during the equilibration period which do not adsorb readily and persist for several days of adsorption. Without equilibration the added Cm rapidly adsorbs to the sediments and the soluble percentage remains lower throughout the experiment.

3.3. Thermodynamic models

The development of thermodynamic equilibrium models has been very useful for understanding and interpreting the environmental chemistry of some trace metals. These models depend upon the availability of reliable stability constants for the important complexes, a consistent set of constants for the metals and ligands being modeled and a realistic interpretation of the model

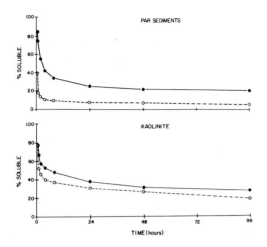

FIGURE 4. Soluble ^{244}Cm during 96 hours of adsorption to sediment or kaolinite in a PAR Pond water. Closed circles are for system in which radioisotope equilibrated with solution for 48 hours prior to adsorption.

results. Although considerable research has been conducted on the chemistry of transuranic elements in simple solutions and at high concentrations (Keller 1971) relatively little information is available at environmental concenrations. Only recently have there been attempts to assemble environmentally significant thermodynamic information for Pu (Cleveland 1979; Rai et al. 1980). Schwab and Felmy (1983) have recently prepared a critical reevaluation of the published constants for soluble Pu species. We are unaware of a similar compilation for any other transuranic element. Even for Pu there is uncertainty in many of the values of the relevant equilibrium constants (Watters et al. 1980) and there are few constants for organic ligands. It has been argued (Simpson et al. 1982; Anderson et al. 1982) that there are insufficient data to reasonably model these elements. However, information on Pu complexation with humic acids (Alberts et al. 1980) and Pu adsorption (Sanchez 1982) to selected particulate surfaces indicate that modeling may be feasible for some systems.

Alberts et al. (1980) reported conditional stability constants for Pu(IV) with natural stream organic matter. Using Scatchard plots they obtained two binding sites which they assigned to carboxylic acid and phenolic groups, and calculated stability constants for both. For site 1 they reported a site concentration of 0.2 mM per gram of organic matter with a stability constant $K_1 = 3.7 \times 10^9$; $K_2 = 5.9 \times 10^8$ with 1.1 mM sites per gram of organic matter. Sanchez (1983) studied the adsorption of Pu(IV) on

298

natural sediments, and goethite that had been well characterized by Balistrieri and Murray (1981, 1982). He determined adsorption constants for Pu(IV) on goethite and adapted the stability constants of Alberts et al. (1980) to model the effects of dissolved organic carbon on P··(IV) adsorption to goethite in Soap Lake. Figure 5 compares model predictions with his experimental results. Although a reasonable fit to the data is obtained, we cannot be certain of the stability constants because the percent

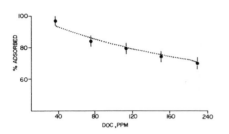

FIGURE 5. Adsorption of Pu(IV) to goethite as a function of dissolved organic carbon. Line is MINEQL prediction. Error bars are one standard deviation for triplicate experimental results after 240 hours of adsorption.

adsorbed is so high at all concentrations of dissolved organic carbon. Nevertheless, these results suggest that we can progress toward modeling transuranic speciation in natural environments.

REFERENCES

-Alberts, J.J., D. Lutkenhoff, R.A. Geiger and D. Gant, 1980. Stability constants determined for complexes of ^{233}U, ^{237}Pu, ^{237}Np and ^{99}Tc with natural stream organic material. In: Savannah River Ecology Laboratory, University of Georgia, Annual Report 1980.
-Anderson, R.F., M.P. Bacon and P.G. Brewer, 1982. Elevated concentrations of actinides in Mono Lake. Science 216: 514-516.
-Balistrieri, L.S. and J.W. Murray, 1981. The surface chemistry of goethite (α-FeOOH) in major ion seawater. Amer. J. Sci. 281: 788-806.
-Balistrieri, L.S. and J.W. Murray, 1982. The adsorption of Cu, Pb, Zn, and Cd on geothite from major ion seawater. Geochim. Cosmochim. Acta 46: 1253-1265.
-Bondietti, E.A., S.A. Reynolds and M.H. Shanks, 1976. Interaction of plutonium with complexing organics in soils and natural waters. pp. 273-287. In: Transuranium Nuclides in the Environment. IAEA, Vienna.
-Clayton, J.R., T.H. Sibley and W.R. Schell, 1982. Effects of selected organic compounds on radionuclide adsorption to sediments in freshwater systems. Bull. Environm. Contam. Toxicol. 28: 409-415.
-Cleveland, J.M., 1979. Critical review of plutonium equilibria of environmental concern. pp. 321-338. In: E.A. Jenne (Ed), Chemical Modelling in Aqueous Systems. American Chemical Society, Washington, D.C.

-Davey, E.W., M.J. Morgan and S.J. Ericksen, 1973. A biological measurement of the copper complexation capacity of seawater. Limnol. Oceanogr. 18: 993-997.

-Davis, J.A. and J.O. Leckie, 1978. Effects of adsorbed complexing ligands on trace metal uptake by hydrous oxides. Environ. Sci. Technol. 12: 1309-1315.

-Duursma, E.K. and C.J. Bosch, 1970. Theoretical, experimental and field studies concerning diffusion of radioisotopes in sediments and suspended particles of the sea. Part B. Methods and experiments. Neth. J. of Sea Res. 4: 395-469.

-Giesy, J.P., G.J. Leversee and D.R. Williams, 1977. Effects of naturally occuring aquatic organic fractions on cadmium toxicity to Simocephalus serrulatus (Daphnidae) and Gambusia affinis (Poeciliidae). Water Research 11: 1013-1020.

-Hanson, W.C. (Ed), 1980. Transuranic Elements in the Environment. Technical Information Center, U.S. Dept. of Energy, DOE/TIC-22800.

-I.A.E.A., 1981. Techniques for identifying transuranic speciation in aquatic environments. Proc., Tech. Committee Meeting, Ispra, 24-28 March 1980. IAEA, Vienna.

-Keller, C., 1971. The chemistry of the transuranium elements. Verlag-Chemie GmbH, Germany, 675 pp.

-Nash, K., F. Sherman, A.M. Fiedman and J.C.Sullivan, 1981. Redox behavior, complexing, and adsorption of hexavalent actinides by humic acid and selected clays. Environ. Sci. Technol. 15: 834-837.

-Rai, D., R.J. Serne and J.L. Swanson, 1980. Solution species of plutonium in the environment. J. Environ. Qual. 9: 417-420.

-Sanchez, A.L., 1983. Chemical speciation and adsorption behavior of plutonium in natural waters. Ph.D. Dissertation, University of Washington, Seattle, Washington. 191 pp.

-Sanchez, A.L., W.R. Schell and T.H. Sibley, 1982. Distribution coefficients for plutonium and americium on particulates in aquatic environments. pp. 188-203. In: Migration in the Terrestrial Environment of Long-Lived Radionuclides from the Nuclear Fuel Cycle. IAEA, Vienna.

-Schell, W.R., T.H. Sibley, A.E. Nevissi, A.L. Sanchez, J.R. Clayton, and E.A. Wurtz, 1981. Distribution coefficients for radionuclides in aquatic environments. U.S.Nuclear Regulatory Commission. Vol. 1. NUREG/CR-1852. 21 pp.

-Schultz, W.W., 1976. The chemistry of americium. NTIS. U.S. Department of Commerce, Springfield, Virginia. 291 pp.

-Schwab, A.P. and A.R. Felmy, 1983. Review and reevaluation of Pu thermodynamic data. Presented at the American Chemical Society, National Meeting, Seattle, Washington. April 1983.

-Sibley, T.H. and J.J. Alberts, 1983. Adsorption of ^{244}Cm to extracted sediments. Submitted to J. Environ. Qual.

-Sibley, T.H., E.A. Wurtz and J.J. Alberts, 1984. Partition coefficients for ^{244}Cm on freshwater and esturarine sediments. Submitted to Environ. Science Technol.

-Simpson, H.J., R.M. Trier, G.R. Olsen, D.E. Hammond, A. Ege, L. Miller and J.M. Melack, 1980. Fallout plutonium in an alkaline, saline lake. Science 207: 1071-1073.

-Simpson, H.J., R.M. Trier, J.R. Toggweiler, G. Mathieu, B.L. Deck, C.R. Olsen, D.E. Hammond, C. Fuller and T.L. Ku, 1982. Radionuclides in Mono Lake, California. Science 216: 512-514.

-Wahlgren, M.A. and K.A. Orlandini, 1982. Comparison of the geochemical
behavior of plutonium, thorium and uranium in selected North American lakes
pp. 757-774. In: Migration in the Terrestrial Environment of Long-lived
Radionuclides from the Nuclear Fuel Cycle. IAEA, Vienna.

-Wang, C.H., D.L. Willis and W.D. Loveland, 1975. Radiotraces methodology i
the biological, environmental and physical sciences. Prentice-Hall,
Englewood Cliffs, N.J., 480 pp.

-Watters, R.L., D.N. Edgington, T.E. Hakonson, W.C. Hanson, M.H. Smith, F.W
Whicker and R.E. Wildung, 1980. pp. 1-44 In: W.C. Hanson (Ed), Transuranic
Elements in the Environment. U.S. Dept. of Energy, DOE/TIC-22800.

PART V INTERACTION WITH ORGANICS

INTERPRETATION OF TRACE METAL COMPLEXATION BY AQUATIC ORGANIC MATTER

J. BUFFLE, A. TESSIER, W. HAERDI

1. INTRODUCTION

Although a classification of aquatic organic matter is difficult, an examination of recent literature reviews (Williams (1975), Reuter and Perdue (1977), Buffle (1984)) indicates that the major groups of organic ligands are polysaccharides, proteins and peptides, "pedogenic" (soil derived) refractory organic matter (PROM) and "aquogenic" (formed in situ in the water body) refractory organic matter (AROM). Altogether, the last two groups form the so-called water fulvic and humic acids which represent a large proportion (~ 70-80%) of the organic matter in natural waters. The characteristics of PROM resemble those of soil fulvic acids (SFA: extracted from soil; Schnitzer (1978)), but differ markedly from those of AROM and of humic and fulvic fractions of sediments (Buffle (1984)). The complexation data reported in the literature concern mostly SFA and the fulvic fractions of PROM and AROM.

Several authors, particularly Mantoura (1981), have recently reviewed the complexation of trace metals in natural waters. Figures 1 and 2 show values of complexation capacity ($\widetilde{CC}$) and equilibrium "constant" (called hereafter equilibrium "quotient", $\widetilde{K}$ (see section 3)), reported in the literature for the complexation of Cu(II). Those figures indicate that:

- most methods give $\widetilde{CC}$ values between 0.6 and 6.0 mM per gram of organic carbon (mM/gC), which is roughly 5 times lower than the total acidity of organic matter;
- bioassays lead to $\widetilde{CC}$ values that are systematically lower than those of most other methods;
- anodic stripping voltammetry (ASV or DPASV) generally gives low, but above all, dispersed values of $\widetilde{CC}$;
- the dispersion of log $\widetilde{K}$ values is very large (up to 6-7 log units at a given pH).

The aim of this paper is to discuss these last two aspects and to explain tentatively the dispersion of the data reported in the literature. As most

Kramer, C.J.M. and Duinker, J.C. (eds.), Complexation of Trace Metals in Natural Waters. ISBN 90-247-2973-4
© *1984 Martinus Nijhoff/Dr W. Junk Publishers, The Hague/Boston/Lancaster.*
Printed in the Netherlands.

302

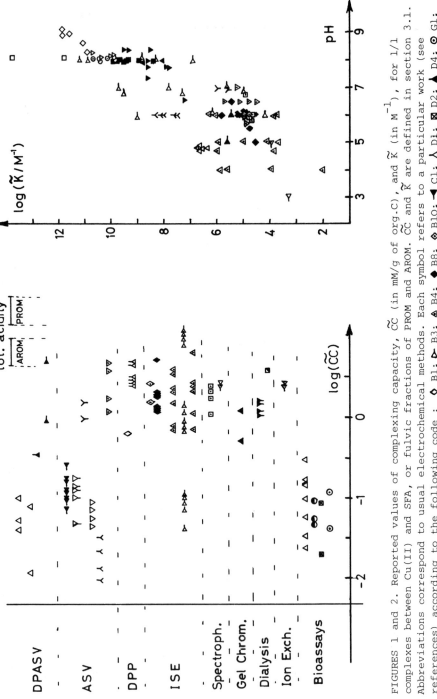

FIGURES 1 and 2. Reported values of complexing capacity, $\tilde{C}C$ (in mM/g of org.C), and $\tilde{K}$ (in M^{-1}), for 1/1 complexes between Cu(II) and SFA, or fulvic fractions of PROM and AROM. $\tilde{C}C$ and $\tilde{K}$ are defined in section 3.1. Abbreviations correspond to usual electrochemical methods. Each symbol refers to a particular work (see references) according to the following code : ◇ B1; ◺ B3; ◺ B4; ◆ B8; ◈ B10; ▼ C1; ◺ D1; ⊠ D2; ▲ D4; ⊙ G1; ◑ G4; ⅄ H1; ▽ H4; □ H5; ▲ M2; □ R2; ⅄ R3; ▽ S3; ◺ S4; ⅄ S5; ◺ S6; ⅄ S7; ▷ S8; ◪ T1; ▽ T2; ⊙ V1; ▼ V2, Temperature and ionic strength are generally 25°C and 0.1M. For all $\tilde{C}C$ values : 4 < pH < 8.8 (usually 6 < pH < 8).

studies do not discriminate clearly between PROM and AROM, and because PROM
is quite similar to SFA, complexation by these three groups of organic matter
are discussed together.

2. DETERMINATION OF $\widetilde{CC}$ BY ASV OR DPASV

The ASV determination comprises two steps: the deposition, during which
the metal ion, M, is reduced and deposited for a time t_d in (or on) the
electrode, at a constant potential, and the stripping during which the metal
is reoxidized as a result of a positive scanning of the potential, giving rise
to a current vs potential peak (e.g. Buffle (1981), Hoffman et al. (1981),
Shuman and Woodward (1977)). When the corresponding peak current, i_p, is
plotted against the total concentration of M, $|M|_t$, a straight line with a
slope of s_M^{nc} is obtained in a non complexing medium (calibration curve), but a
broken curve (figure 3a) is generally observed in the presence of complexing
organic matter, L (e.g. Chau and Wong (1976), Hart and Davies (1981)), due to
the formation of ML complexes.

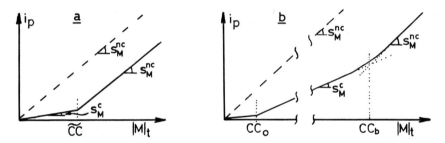

FIGURE 3. Schematic representation of the titration curves of dissolved
organic matter (L) with a metal ion (M) (see text): a) formation of non (or
partially) reducible complexes (chemically or physically slow proces). b)
formation of labile, slowly diffusing, complexes with $|L|_t / |M|_t < 1000$.

The observed decrease in the current (compared to the calibration curve)
may be due to one of the three basic following processes or to their
combination (typical cases are given in Table 1):
a) ML dissociates slowly and hence it is only partly (or not at all) reduced
   during the deposition step (e.g. Davison (1978), Shuman and Woodward
   (1977), Chau and Wong (1976));
b) ML is labile but diffuses slowly, thus contributing less than M (or not at
   all) to the overall reduction current, during the deposition step (Saha et
   al. (1979), Buffle and Greter (1979), Greter et al. (1979));

c) ML is labile, but the total concentration of L in the bulk solution, $|L|_t$, is not much larger than $|M|_t$, so that L is saturated with M at the electrode surface, during the stripping step.

2.1. Surface saturation

The last effect (c) results from the much larger concentration of M at the electrode surface, $|M|_t^\circ$, during the stripping step than in the bulk solution, $|M|_t$. For a hanging mercury drop electrode $|M|_t^\circ$ is given by (Buffle 1981):

$$|M|_t^\circ = 3 \cdot t_d \cdot |M|_t \cdot \sqrt{D_{ox} \cdot D_R} / \delta \cdot r \qquad (1)$$

where D_{ox} and D_R are the diffusion coefficients of the oxidized (in the aqueous solution) and reduced (in the mercury drop) forms of M, δ is the thickness of the diffusion layer, and r is the radius of the mercury drop. Using $\delta = 20\mu m$, $r = 0.03$ cm and $D_{ox} = D_R = 10^{-5}$ cm^{2}s^{-1}, equation (1) becomes:

$$|M|_t^\circ = 0.5\ t_d \cdot |M|_t \qquad [t_d : s] \qquad (1a)$$

With equations (1) or (1a), one estimates that $|M|_t^\circ / |M|_t \sim 50 - 200$ for $t_d \sim$ few minutes. This ratio is even larger for mercury film electrodes. Hence L may be easily saturated with M at the electrode surface (during the stripping step), even if this is far from being the case in the bulk solution, causing a broadening of the peak and consequently a decrease in i_p (Buffle (1981)). This "surface effect" may finally produce i_p vs $|M|_t$ titration curves with similar shapes as those obtained in cases a) and b) above, but with significant mechanistic differences; all the cases considered correspond to saturation of L, but i) in different parts of the system (electrode surface for c) and bulk solution for a) and b) respectively) and ii) during different steps (stripping for c) and deposition for a) and b)). It is worth noting that the "surface effect" may be significant when $|L|_t / |M|_t < 1000$ (equation (1)) as often observed in natural waters (Buffle (1984)).

2.2. Discrimination between surface and bulk saturation of L

Discrimination between saturation of L with M in the bulk solution and at the electrode surface is easily obtained by studying the influence of t_d and

$|M|_t$ on the peak current, i_p, and on the potential, E_p, as it is shown for typical examples in Table 1. For instance, changing t_d does not modify the fraction of reducible metal ion so that, for cases I and L2 (Table 1), linear i_p vs t_d curves must be obtained, contrasting with the broken curves (like in figure 3a) obtained for i_p vs $|M|_t$. In those two cases, the slope of the i_p vs t_d lines depends either on the dissociation rate of ML (case I) or on its diffusion rate (case L2). Non linear curves must however be obtained for both i_p vs t_d and i_p vs $|M|_t$ in the case where L is saturated with M at the surface (case L3). Indeed, equation (1) shows that t_d and $|M|_t$ have an equal influence on $|M|_t^\circ$. For labile, fast diffusing, synthetic complexes, S-shape curves were indeed obtained for i_p vs t_d (Buffle (1981)). Interestingly, the same type of curves were reported by Bhat et al. (1981) for the i_p vs $|Cu|_t$ curves with SFA.

The importance of a "surface effect" with natural organic matter is clearly shown in figure 4 for the complexation of Pb(II) in a representative sample of PROM. Detailed polarographic studies (by direct reduction methods: Buffle and Greter (1979), Greter et al. (1979)) showed that the complexes formed were labile and slowly diffusing, and that both complexes and free ligand adsorbed onto the Hg electrode (case L4 in Table 1). A similar behaviour was obtained for Cu (Dauthuille (1982)) and Cd (van Leeuwen (1978)) complexes with Fluka humic acids. Based on the detailed ASV studies, results of figure 4 are interpreted as follows:

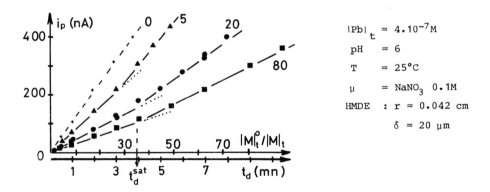

FIGURE 4. i_p vs t_d curves for the complexation of Pb(II) by a representative PROM sample (n° 50 in Buffle et al. (1980)). Concentration of PROM is given on each curve, in mg·dm^{-3} of organic matter. $|Pb|_t^\circ$ computed from equation (1) with $D_{PbL} = 1.3 \ 10^{-6}$ cm^2·s^{-1} and $D_{Pb^\circ} = 1.4 \ 10^{-5}$ cm^2·s^{-1}.

Table 1. Discrimination between various typical electrochemical behaviours, from the change in i_p and E_p with t_d and $|M|_t$.*

(1a): Characteristics of the various cases.

Case	Determining step	Dissociation rate of ML	Diffusion rate of ML	$|L|_t/|M|_t$	Adsorption
L1	---	Fast	Fast	> 1000	No
I	Dissociation of ML (chemically inert)	→ 0	Fast	> 1000	No
L2	Diffusion of ML (physically inert)	Fast	→ 0	> 1000	No
L3	Surface saturation	Fast	Fast	< 1000	No
L4	Slow diffusion + surface saturation	Fast	Slow	< 1000	Yes

(1b): Changes in i_p^c and E_p^c with t_d and $|M|_t$.

Case	Value of ΔE_p (mV)	Change in ΔE_p when $t_d \uparrow$ or $|M|_t \uparrow$	$i_p^c = f(t_d)$	$i_p^c = f(|M|_t)$
L1	ΔE_p^m	Cte	Linear (no break)	Linear (no break)
I, L2	0	Cte	Linear (no break)	b-c $\begin{cases} \text{1 break at } \quad |M|_t = |L|_t \\ i_p^c = \text{cte} \cdot |M| \end{cases}$
L3	$0 \lessapprox \Delta E_p \lessapprox \Delta E_p^m$	↘0	S-c	S-c ($i_p^c \neq$ cte $\cdot |M|$)
L4	$\Delta E_p > 0$	↘0	b-c (1 break)	b-c (2 breaks)

*: L1-L4: systems chemically labile; I: system chemically inert;
$\Delta E_p = E_p^{nc} - E_p^c$; E_p^{nc}, E_p^c = peak potentials of M in non complexing and complexing media respectively; i_p^c = peak current in complexing medium.
$\Delta E_p^m = \frac{RT}{2F} \ln (|M|_t/|M|)$; ↑ = increase; ↘X = tends to X by decreasing values; S-c = S-shaped curve; bc = broken curve. M_t, M = total and free metal ions concentrations respectively.

- for $t_d < t_d^{sat}$, i_p is lower in the presence than in the absence of PROM due to: i) a slow diffusion of the complexes during_the_deposition_step and ii) a partial saturation of L with M at the electrode surface during_the stripping_step;

- t_d^{sat} corresponds to a complete saturation with M of both adsorbed ($|L|_{ads}$) and dissolved ($|L|°$) L at the surface. Because $|L|_{ads} >> |L|°$ and because the Hg surface is fully covered by L_{ads}, even for the lowest value of $|L|_t$ reported in figure 4 (Buffle and Cominoli (1980)), t_d^{sat} is approximately independent of $|L|_t$;

- for $t_d > t_d^{sat}$, the decrease of i_p in the presence of PROM is essentially due to the slow diffusion of PbL during_the_deposition_step. Indeed figure 4 shows that, with increasing $|L|_t$, the slope of the i_p vs t_d line decreases. As discussed above, this slope depends on the average diffusion coefficient, $\bar{D}$, of the diffusing species (Pb^{2+} and PbL). Increasing $|L|_t$ leads to a decrease in the proportion of the fast diffusing species (Pb^{2+}) and hence in $\bar{D}$ (figure 4). This change in $\bar{D}$ allowed the calculation of the diffusion coefficient of PbL ($D_{PbL} = 1.3 \ 10^{-6} \ cm^2.s^{-1}$) and of the equilibrium quotient (log $\tilde{K}_{PbL} = 5.6$) (according to the theory given in Heyrowsky and Kuta (1966)); those values agree with previously determined values (Buffle and Greter (1979), Greter et al. (1979)).

2.3. Role of the "surface effect" in the interpretation of complexing capacity determined by ASV

According to the above discussion, two types of break in the i_p vs $|M|_t$ curves are expected for labile, slowly diffusing complexes of M with PROM, AROM and SFA (figure 3b); a first break, CC_o, should correspond to the saturation of L (adsorbed or non adsorbed) with M at the electrode surface, and a second, CC_b, (with $CC_b >> CC_o$) should correspond to the saturation of L with M in the bulk solution. This phenomenon produces a change in the slope due to the increase in $\bar{D}$, as $|M|$ increases with $|M|_t$; in practice, this change in the slope is gradual, making difficult an accurate determination of CC_b (e.g. Bhat and Weber (1982), Bhat et al. (1981), Shuman and Woodward (1977)).

Obviously CC_o should decrease when the deposition time t_d is increased (since $|M|_t°$ increases with t_d (equation (1)), whereas CC_b should be independent of t_d. With this in mind, the possible role of "surface effect" on the reported $\widetilde{CC}$ values can be examined by plotting $\widetilde{CC}$ values as a function of t_d (figure 5). Figure 5 shows that most of the values of $\widetilde{CC}$ determined by ASV are lower than those determined by other methods, and that they decrease with an increase in t_d. Although this observation might be fortuitous and due to other reasons than electrode "surface effect" (e.g. differences in the nature of the organic matter), the general trend shown in figure 5 suggests that "surface effect" should be considered in interpreting ASV measurements of bulk complexation by aquatic organic matter.

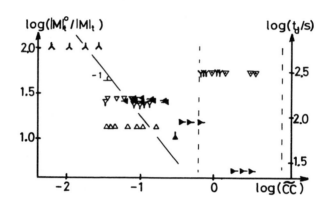

FIGURE 5. Dependence of $\widetilde{CC}$ on t_d for complexation of Cu(II) by PROM, AROM and SFA. For symbols: see figure 1. Dashed lines: range of $\widetilde{CC}$ values obtained by methods other than ASV. Full line with slope -1 given for comparison. Log ($|M|_t^o$ / $|M|_t$) estimated from equation (1) with: δ = 20 μm, r = 0.04 cm, D_{ox} = 2 10^{-6} cm$^2 \cdot$ s^{-1}, D_R = 1.06 10^{-5} cm$^2 \cdot$ s^{-1}.

3. INTERPRETATION OF EQUILIBRIUM QUOTIENTS, $\widetilde{K}$; ROLE OF THE METAL TO LIGAND RATIO

3.1. Relationship between $\widetilde{CC}$ and $\widetilde{K}$

It is often difficult to compare log $\widetilde{K}$ values reported in the literature in particular because of the various experimental conditions used, such as nature of organic matter, pH, temperature, ionic strength, ligand concentration, reaction time, or metal to ligand ratio. The important role of the last factor was underlined by some workers (Gamble and Schnitzer (1973), Sunda and Hanson (1979), Gamble et al. (1980)). The latter authors developped a theoretical model for SFA, assuming an infinity of different complexing sites, each of them being present at an infinitely small concentration, $\delta |L|$, and forming 1/1 complexes with M. For such polyfunctional ligands, only the so-called differential equilibrium function K has a thermodynamic meaning:

$$K = \delta |ML| / \delta |M| \cdot \delta |L| \qquad (2)$$

In practice one determines an average equilibrium function:

$$\bar{K} = |ML| / |M| \cdot |L| = |ML| / |M| \cdot \left[\{L\}_t \cdot \widetilde{CC} - |ML| \right] \quad (3)$$

where $|ML|$ is the overall concentration of bound M ($|ML| = |M|_t - |M|$), and $\{L\}_t$ is the total concentration of ligand in gC/l. The relationship between K and $\bar{K}$ was derived by Gamble et al. (1980) for SFA, assuming the existence of only salicylic and phtalic types of sites. This theory was later extended to AROM and PROM (Buffle (1984)) in order to take into account their larger number of possible types of sites, and the fact that the minor sites, being often the strongest, are the least negligible. In particular, it results from those studies that K may be easily computed, for any $|ML|/\{L\}_t$ ratio, from a single titration curve at a constant pH, without the need to specify any value of $\widetilde{CC}$. On the opposite, $\bar{K}$ is strongly dependent on the value of $\widetilde{CC}$ used for its computation (figure 6).

Figure 6 shows typical K and $\bar{K}$ functions obtained from the ISE titration curve of a PROM sample. Interestingly, for all the tested samples, straight lines with slopes close to -2 were obtained for log K vs log ($|ML|/\{L\}_t$) curves. K and $\bar{K}$ are always found to be decreasing functions of $|ML|/\{L\}_t$ as a result of the polyelectrolytic properties and of the polyfunctional character of this type of organic matter (for more details see Gamble et al. (1980), Buffle (1984)). However, it can be seen from figure 6b that, for $(\log(\widetilde{CC})-0.7) < (\log(|ML|/\{L\}_t)) < \log \widetilde{CC}$, the variation of the log $\bar{K}$ function is only within the experimental error. In practice, its average value is generally considered as the logarithm of the equilibrium "constant" (preferably called here equilibrium quotient), log $\widetilde{K}$. Obviously, $\widetilde{K}$ is valid only in the $|ML|/\{L\}_t$ range defined by the chosen value of $\widetilde{CC}$ and the tolerated experimental error; in general, the corresponding range of $|ML|$ is considered as the concentration of the particular "sites" corresponding to log $\widetilde{K}$. From figure 6 and the above discussion, the following points can be stressed:

- $\widetilde{K}$ or $\widetilde{CC}$ alone have no physical meaning, as any arbitrary change in the chosen value of $\widetilde{CC}$ (e.g. $\widetilde{CC}_1$, $\widetilde{CC}_2$, $\widetilde{CC}_3$, etc.) produces a parallel shift of the $\bar{K}$ function (and of $\widetilde{K}$) along the thermodynamic (K) curve;

310

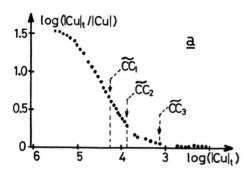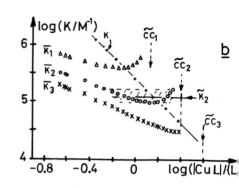

FIGURE 6. Variation of K, K, and $\widetilde{K}$ (b), computed from the titration curve (a of a PROM sample (n° 10 in Buffle et al. (1980b) with Cu(II) at pH = 6, T = 25°C, μ = 0.1M, $\{L\}_t$ = 30 mg·dm^{-3}. In figure 6b: ordinate = K, $\overline{K}$ or $\widetilde{K}$; $\overline{K}_1$, $\overline{K}2$, $\overline{K}3$ functions are computed with arbitrarily choosen values of $\widetilde{CC}_1$, $\widetilde{CC}_2$ $\widetilde{CC}_3$; |ML| /$\{L\}_t$ in mM/gC; ⧸⧸ ⧸⧸⧸⧸ ⧸⧸ = experimental errors.

- the so-called "number_of_sites" (e.g. as defined in the Scatchard method) represents mostly the number of couples ($\widetilde{K}$, $\widetilde{CC}$) necessary to fit the whole range of the titration curve. Since the range of |ML| /$\{L\}_t$), in which K is considered as constant, directly depends on the accepted experimental error, this also influences the experimentally found "number of sites";

- only a couple_($\widetilde{K}$,_$\widetilde{CC}$) (and not one of the parameters alone) can be related to the thermodynamic K function: it just locates a point close to K in figure 6b. From figure 6 it can be seen that the log $\widetilde{K}$ vs log $\widetilde{CC}$ function is a curve parallel to log K vs log (|ML| /$\{L\}_t$) but vertically shifted by about + 0.5 log units. It must be emphasized, however, that the determination of the K function is always preferable to the measurement of $\widetilde{K}$ and $\widetilde{CC}$: it is not more difficult to obtain, but much more reliable for comparison purposes, and theoretically sound.

3.2. Comparison of literature data

Figure 7 compares experimental values of log $\widetilde{K}$ obtained for Cu(II), at a given pH. It clearly shows that log $\widetilde{K}$ increases gradually when log $\widetilde{CC}$ decreases. Dispersion of the points can be considered relatively small when the various origins of the water samples, as well as the various methods and conditions used for measuring $\widetilde{CC}$ and $\widetilde{K}$ are taken into account. Interestingly, the average slopes of the two groups of points, for fresh waters and SFA at pH 8.0 - 8.3 on the one hand, and at pH 5.7 - 6.0 on the other hand, are close

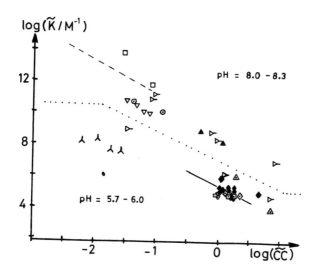

FIGURE 7. Comparison of reported data for the complexation of Cu(II) by SFA, PROM, and AROM. □ and ----: sea water organic matter; other data points and ———: SFA and fresh water organic matter; see figure 1 for symbols and other details; lines are the K functions at pH 6 from figure 6b (———) and at pH 8.1 from the data of Hirose et al. (1982) (----).

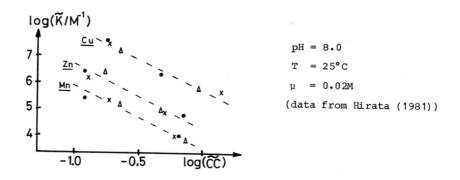

FIGURE 8. Log $\tilde{K}$ vs log $\tilde{CC}$ for complexation by sedimentary organic matter. ●, Δ and x refer to three different extracts of organic matter.

(about -2) to those of the thermodynamic K functions computed from single titration curves at the same pH (pH 6: full line = K function from figure 6; pH 8.1: dashed line computed from data of Hirose et al. (1982) and Sugimura (personal communication, 1982)). Similar results were obtained at other pH values, but less experimental data are available. Figure 8 shows that the same trend (and the same slope) is observed with organic matter from other origin (sedimentary) and with various metals (Cu, Zn, Mn).

It is clear from figures 7 and 8 that a comparison of $\widetilde{K}$ (or K) values is meaningful only at constant $|ML|/\{L\}_t$. Variation of $\widetilde{K}$ with pH and with the nature of metal ions is discussed in more details elsewhere (Buffle (1984)). In any case, from the overall compilation of reported complexation data, it appears that a large proportion of the dispersion of log $\widetilde{K}$ in figure 2 is explained by variations in the $|ML|/\{L\}_t$ ratios. More generally pH, $|ML|/\{L\}$ and the nature of M seem to be the key factors influencing the degree of complexation. The source of organic matter (within the groups including SFA and fulvic fractions of PROM and AROM), the ionic strength and the temperature seem to have much less influence.

4. CONCLUSION

The results of section 3 suggest that complexation by natural organic matter is mainly governed by two important factors: the polyfunctional character, initially mentioned and studied by Gamble (e.g. Gamble et al. (1973, 1980)) and the polyelectrolytic properties (e.g. Wilson and Kinney (1977), Gamble et al. (1980)) of the organic matter. As discussed in more details elsewhere (Buffle (1984)), changes in the value of K are primarily due to the former property at low $|ML|/\{L\}_t$ ratios, whereas the latter is also important for $|ML|/\{L\}_t > 1$ mM/gC. Nevertheless, other factors, such as aggregation phenomenon or mixed complexes formation, which are not considered here, could also play a significant role.

Computation of the differential equilibrium function K seems to be presently the best available mean to interpret and compare complexation data on a rigorous basis. Indeed, organic matter interactions with trace metals are better described by continuous functions rather than by single equilibrium constant values. Such continuous functions may seem, at first sight, difficult to handle, particularly for modeling. However, this apparent difficulty would drop if further studies confirmed, as suggested by figures 6 and 7, that the log K vs log ($|ML|/\{L\}_t$) functions are well approximated by straight lines. In any case, when complexation properties of organic matter are characterised by $\widetilde{K}$ and $\widetilde{CC}$ parameters, it is most important to report, besides the detailed experimental conditions: i) both values $(\widetilde{K}, \widetilde{CC})$ determined by the same method in the same conditions, and ii) the concentration of organic matter (e.g. DOC), since valid comparison are based on $\widetilde{CC}$ values expressed as concentration of bound metal per weight of organic matter.

ACKNOWLEDGEMENTS

       J. Buffle wishes to thank D. Gamble and C.H. Langford for very stimulating discussions and W. Stumm for his continuous and helpful encouragements and suggestions. This work was partly realized during a sabbatical leave of J. Buffle at INRS-Eau (Université du Québec). It was supported by the Swiss National Foundation (Project No 2413-0.82).

REFERENCES (For the meaning of symbols in parentheses, after the year of
           publication, see figure 1).

-Baccini, P., U. Suter (1979) (B1). Melimex, an experimental heavy metal pollution study. Chemical speciation and biological availability of copper in lake water. Schw. Z. Hydrol. 41: 291-314.

-Bhat, G.A., J.H. Weber (1982) (B2). Cadmium binding by soil derived fulvic acid measured by anodic stripping voltammetry. Anal. Chim. Acta 141: 95-103.

-Bhat, G.A., R.A. Saar, R.B. Smart, J.H. Weber (1981) (B3). Titration of soil derived fulvic acid by copper (II) and measurement of free copper (II) by anodic stripping voltammetry and copper (II) selective electrode. Anal. Chem. 53: 2275-2280.

-Bresnahan, W.T., C.L. Grant, J.H. Weber (1978) (B4). Stability constants for the complexation of copper (II) ions with water and soil fulvic acids measured by an ion selective electrode. Anal. Chem. 50: 1675-1679.

-Buffle, J. (1981) (B5). Calculation of the surface concentration of the oxidized metal during the stripping step in the anodic stripping techniques and its influence on speciation measurements in natural waters. J. Electroanal. Chem. 125: 273-294.

-Buffle, J. (1984) (B6). The natural organic substances and their metal complexes in aquatic systems. In H. Sigel (Ed.) : Circulation of metals in the environment. Vol. 18 of the series: Metal ions in biological systems. M. Dekker, N.Y., Basel.

-Buffle, J., A. Cominoli (1980) (B7). Voltammetric study of humic and fulvic substances. Part IV: Behaviour of fulvic substances at the mercury water interface. J. Electroanal. Chem. 121: 273-299.

-Buffle, J., P. Deladoey, F.L. Greter, W. Haerdi (1980) (B8). Study of the complex formation of copper (II) by humic and fulvic substances. Anal. Chim. Acta 116: 255-274.

-Buffle, J., F.L. Greter (1979) (B9). Voltammetric study of humic and fulvic substances. Part II. J. Electroanal. Chem. 101: 231-251.

-Buffle, J., F.L. Greter, W. Haerdi (1977) (B10). Measurement of complexation properties of humic and fulvic acids in natural waters with lead and copper ion-selective electrodes. Anal. Chem. 49: 216-222.

-Chau, Y.K., P.T.S. Wong (1976) (C1). Complexation of metals in natural waters. In R.W. Andrew, P.V. Hodson, D.E. Konasewitch (Ed.). Toxicity to biota of metal forms in natural waters. Proceedings Workshop , held in Minnesota (7-8/10/1975). International Joint Commission's Research Advisory Board.

-Dauthuille, P. (1982) (D1). Etude électrochimique des complexes Cu(II)-acides fulviques en solution aqueuse. Thèse. Ecole nationale supérieure de chimie de Paris.

-Davey, E.W., M.J. Morgan, S.J. Erickson (1973) (D2). A biological measurement of the copper complexation capacity of sea water. Limnol. Oceanogr. 18: 993-997.

-Davidson, W. (1978) (D3). Defining the electroanalytically measured species in a natural water sample. J. Electroanal. Chem. 87: 395-404.

-Duinker, J.C., C.J.M. Kramer (1977) (D4). An experimental study on the speciation of dissolved zinc, cadmium, lead and copper in river Rhine and North Sea water, by differential pulsed anodic stripping voltammetry. Mar. Chem. 5: 207-228.

-Gächter, R., J.S. Davis, A. Mares (1978) (G1). Regulation of copper availability to phytoplancton by macromolecules in lake water. Environ. Sci. Technol. 12: 1416-1421.

-Gamble, D,S., M. Schnitzer (1973) (G2). The chemistry of fulvic acid and its reactions with metal ions. In P.C. Singer (Ed.). Trace metals and metalorganic interactions in natural waters. Ann Arbor Science Pub. Inc., Ann Arbor, Mich.

-Gamble, D.S., A.W. Underdown, C.H. Langford (1980) (G3). Copper(II) titration of fulvic acid ligand sites with theoretical, potentiometric and spectrophotometric analysis. Anal. Chem. 52: 1901-1908.

-Gillespie, P.A., R.F. Vaccaro (1978) (G4). A bacterial bioassay for measuring the copper chelation capacity of sea water. Limnol. Oceanogr. 23: 543-548.

-Greter, F.L., J. Buffle, W. Haerdi (1979) (G5). Voltammetric study of humic and fulvic substances. Part. I. J. Electroanal. Chem. 101: 211-229.

-Hart, B.T., S.H.R. Davies (1981) (H1). Copper complexing capacity of waters in the Magela creek system, Northern Australia. Environ. Technol. Letters 2: 205-214.

-Heyrovsky, J., J. Kuta (1966) (H2). Principles of polarography, Academic Press, N.Y., London.

-Hirata, S. (1981) (H3). Stability constants for the complexes of transition metal ions with fulvic and humic acids in sediments measured by gel filtration. Talenta 28: 809-815.

-Hirose, K., Y. Dokiya, Y. Sugimura (1982) (H4). Determination of conditional stability constants of organic copper and zinc complexes dissolved in sea water using ligand exchange method with EDTA. Mar. Chem. 11: 343-354.

-Hoffman, M.R., E.C. Yost, S.J. Eisenreich, W.J. Maier (1981) (H5). Characterization of soluble and colloidal phase metal complexes in river water by ultrafiltration. A mass-balance approach. Environ. Sci. Technol. 15: 655-661.

-Mantoura, R.F.C. (1981) (M1). Organo-metallic interactions in natural
waters. In E.K. Duursma, R. Dawson (Ed.). Marine organic chemistry,
Elsevier Oceanography Series 31. Elsevier Sci. Publ. Cy., Amsterdam. N.Y.

-Mantoura, R.F.C., J.P. Riley (1975) (M2). The use of gel filtration in
the study of metal binding hy humic acids and related compounds. Anal.
Chim. Acta 78: 193-200.

-Reuter, J.H., E.M. Perdue (1977) (R1). Importance of heavy metal-organic
matter interaction in natural water. Geochim. Cosmochim. Acta 41: 326-334.

-Ryan, D.K., J.H. Weber (1982) (R2). Copper (II) complexing capacities of
natural waters by fluorescence quenching. Environ. Sci. Technol. 16: 866-
872.

-Saha, S.K., S.L. Dutta, S.K. Chakravarti (1979) (S1). Polarographic study
of metal-humic acid interaction. Determination of stability constants of
cadmium and zinc-humic acids at different pH. J. Indian Chem. Soc. 56:
1129-1134.

-Schnitzer, M. (1978) (S2). Humic substances: chemistry and reactions. In
M. Schnitzer, S.U. Khan, Soil organic matter. Developments in soil science
8. Elsevier Sci. Publ. Co., Amsterdam, N.Y.

-Schnitzer, M., E.H. Hansen (1970) (S3). Organo-metallic interactions in
soils: an evaluation of methods for the determination of stability
constants of metal-fulvic acids complexes. Soil Sci. 109: 333-340.

-Shuman, M.S., G.P. Woodward (1977) (S4). Stability constants of copper-
organic chelates in aquatic samples. Environm. Sci.Technol. 11: 809-813.

-Shuman, M.S., J.L. Cromer (1979) (S5). Copper association with aquatic
fulvic and humic acids. Estimation of conditional formation constants
with a titrimetric anodic stripping voltammetry procedure. Environ. Sci.
Technol. 13: 543-545.

-Srna, R.F., K.S. Garrett, S.M. Miller, A.B. Thum (1980) (S6). Copper
complexation capacity of marine water samples from southern California.
Environ. Sci. Technol. 14: 1482-1486.

-Stokes, P., T.C. Hutchinson (1976) (S7). Copper toxicity to phytoplancton,
as affected by organic ligands, other cations and inherent tolerance of
algae to copper. In R.W. Andrew, P.V. Hodson, D.E. Ko nasewitch, Toxicity
to Biota of Metal Forms in Natural Waters. Proc. Workshop held in Minnesota
(7-8/10/1976). International Joint Commission's Research Advisory Board.

-Sunda, W.C., P.J. Hanson (1979) (S8). chemical speciation of copper in
river water. Effect of total copper, pH, carbonate, and dissolved organic
matter. In E.A. Jenne (Ed.). Chemical modeling in aqueous systems.
American Chemical Society, Washington.

-Truitt, R.E., J.H. Weber (1981) (T1). Determination of complexing capacity
of fulvic acid for copper (II) and cadmium (II) by dialysis titration.
Anal. Chem. 53: 337-342.

-Truitt, R.E., J.H. Weber (1981) (T2). Copper(II) and cadmium(II) binding
abilities of some New Hampshire freshwaters determined by dialysis
titration. Environ. Sci. Technol. 15: 1204-1208.

-Van den Berg, C.M.G. (1982). (V1). Determination of copper complexation
with natural organic ligands in sea water, by equilibration with MnO_2.
Part II. Mar. Chem. 11: 323-342.

-Van den Berg, C.M.G., J.R. Kramer (1979) (V2). Determination of complexing
capacities of ligands in natural waters and conditional stability constants
of the copper complexes by means of manganese dioxide. Anal. Chim. Acta
106: 113-120.

-Van Leeuwen, H.P. (1978) (V3). Pulse polarography of heavy metal ions in
the presence of natural complexing agents. Lecture at 29th ISE Meeting,
Budapest.

316

- Williams, P.J. le B. (1975) (W1). Biological and chemical aspects of dissolved organic material in sea water. In J.P. Riley, G. Skirrow (Ed.). Chemical Oceanography, vol. 2, Academic Press, London.
-Wilson, D.E., P. Kinney (1977) (W2). Effects of polymeric charge variations on the proton metal ion equilibria of humic materials. Limnol. Oceanogr. 22: 281-289.

SIGNIFICANCE OF DISSOLVED HUMIC SUBSTANCES FOR HEAVY METAL
SPECIATION IN NATURAL WATERS

B.RASPOR , H.W.NÜRNBERG , P.VALENTA and M.BRANICA

1. INTRODUCTION

In natural waters trace heavy metals are distributed between the
dissolved and solid phase. In the dissolved phase heavy metals
can exist as hydrated ions, labile complexes predominatly with
the respective anionic inorganic constituents of water and
nonlabile complexes with components of the dissolved organic
matter (DOM). Among these complexing DOM components humic and
fulvic acids are considered to be of major significance in speciation of heavy metal traces by DOM in natural waters.

The aim of this study was to provide the experimental data
from which the significance of humic substances for complexation of trace heavy metals in certain water types could be deduced.
For this purpose the distribution of trace amounts of Zn, Cd and
Pb between various chemical forms in the presence of inorganic
components of water and dissolved organic ligands of synthetic
origin have been studied previously (Raspor et al., 1977; 1978;
1980 a; 1980 b; 1981 and Nürnberg and Raspor, 1981). Based on the
complexation of heavy metal traces (Zn, Cd, Pb) by ligands of
synthetic origin, NTA and EDTA, which have a well defined composition and a chelate structure, the distribution of the existing metal and ligand species, the side reactions with the macrocomponents of water, the rate of complex formation and in general the
mechanism of trace metal complex formation in water media was
established. Knowing the factors which influence the extent and
the rate of trace metal complex formation in saline and fresh water
types it was possible to measure the interaction of the same trace
metals with the humic substances. These are polymeric ligands of
unknown structure, molecular weight, kind and number of complexing sites per molecule.

Kramer, C.J.M. and Duinker, J.C. (eds.), Complexation of Trace Metals in Natural Waters. ISBN 90-247-2973-4
©1984 Martinus Nijhoff/Dr W. Junk Publishers, The Hague/Boston/Lancaster.
Printed in the Netherlands.

318

The used humic substances were two humic acids of marine sedi
ment origin and a humic and fulvic acid from the sediment of a
tropical estuary. Humic substances from different locations were
investigated, to see if there is any correlation between the ori
and therefore composition of the humic substance and its complex
properties in sea and lake water.

Although the humic substances were of marine and estuarine
origin, their interaction with heavy metals was also studied
in lake water in order to determine the influence of lower
ionic strength and lower ionic macrocomponent levels existing
in the fresh water medium on the same interaction previously
measured in sea water.

2. EXPERIMENTAL

2.1. Instrumentation

Voltammetric measurements were carried out with the PAR polar
graph, model 174 A, in conjunction with the automatic unit, PAR
315 A. All measurements were performed in the thermostated quart
cell, 5o ml volume at $25^\circ \pm 1^\circ C$. A three electrode system was
applied, consisting of the HMDE as working electrode (Metrohm
29o E), a Pt-wire as counter electrode and a Ag/AgCl reference
electrode connected to the solution via a salt bridge filled
with o.7 M NaCl.

Before measurement the solution was deaerated with 99.999 %
argon. A small partial pressure of CO_2 was adjusted in the argon
before it entered the cell, passing it through a suspension of
magnesium hydroxycarbonate in borate buffer of pH 8.

For the voltammetric determination of Zn, Cd and Pb different
deposition potentials were adjusted, Zn -1.2 V, Cd -o.9 V and Pb
-o.8 V. During the plating of the trace metal the solution was
stirred with a teflon-coated magnetic bar at 8oo rpm. For a concentration of 10^{-7}M deposition in stirred solution lasted 6o s,
10^{-8}M 12o s. Subsequently, after 15 s quiescent time, the amalga
formed during the deposition step was oxidized in the differenti
pulse anodic stripping mode (DPASV) employing a train of rectan-
gular voltage pulses of 5o mV height, 57 ms duration, o.5 s clocl
time and 2 mV/s scan rate.

2.2. Natural water types

Sea water was sampled in the coastal area of the Adriatic and Ligurian sea, fresh water in Lake Ontario. At first the water samples were filtered through a Sartorius membrane filter of o.45 μm pore size. The ionic strength of sea water is o.7 M and that of the lake water o.oo7 M. The concentrations of the ionic macrocomponents Na, Ca, Mg and Cl which influence specifically the extent and the rate of heavy metal complex formation were determined for each water type.

2.3. Origin and treatment of humic substances

Humic acid isolated from the sediment in the shallow water of the Lim Fjord (Adriatic sea) will be abbreviated as HAL. It was extracted according to the procedure of Desai and Ganguly (197o) and Huljev (197o). Humic acid isolated from the bottom sediment of the deep Norwegian sea (1 m below the sediment surface) will be abbreviated as HAN. The humic and fulvic acid from the same sediment of the tropical estuary at Mahakam (Borneo) will be assigned as HAM and FAM, respectively. HAN, HAM and FAM were extracted according to the procedure of Kononova and Belachikova (196o).

2.3.1. Characterization of humic substances. The humic substances were characterized by elemental analysis (C, H, N, S, Cl), molecular size distribution at pH 7 in $10^{-2}M$ phosphate buffer, visible and UV spectra of their NaOH solutions, infrared spectrum in KBr as matrix material, the trace metal contents (Al, Cu, Zn, Pb and Cd) and the adsorbability at the mercury drop electrode from the sea water (Raspor et al., in press a).

HAN and HAL are characterized by higher nitrogen content compared to the humic acid HAM of estuarine origin. HAN is in comparison with the other two humic acids the most condensed i.e. humified material, having the highest content of carbon and aliphatic groups. It is a high molecular substance (> 2o ooo daltons) and due to the highest hydrophobic character it adsorbs the most.

The characteristics of HAM are opposite to those of HAN, while HAL has the characteristics in between of the two. A high Al-content was determined in HAN and HAM.

The fulvic acid FAM has a lower carbon content and stronger
hydrophylic character. Aliphatic groups are hardly observed and
in comparison with the humic acids its adsorption is the least
pronounced. The predominant fraction in the FAM sample has the
apparent molecular weight of 2o ooo daltons. The content of trace
metals Cu, Cd, Pb and Zn is higher in it than in the corresponding
humic acid (HAM) from the same sampling site.

3. VOLTAMMETRIC METHODOLOGY

The voltammetric approach for the investigation of stable complexes with organic ligands MeL_m differs from that for the study
of labile trace metal complexes. The latter are usually formed
with inorganic ligands, as Cl^-, OH^-, CO_3^{2-}, SO_4^{2-}, and produce a
common reversible response. The stability and composition of
such labile complexes are determined via the half wave potential
shift (Branica et al., 1977). The stabile metal-organic complexes
are investigated in a different way. The complex formed with
organic ligands of synthetic (NTA, EDTA) or natural origin (humic
substances) yields an irreversible response at a substantially
more negative potential than the labile complexed species ΣMeX_j
of the same trace metal, where several labile chemical forms under
go a reversible electrode reaction at practically the same potent:
(see scheme 1). At this stage it should be emphasized that voltam
try is a species sensitive method and therefore suited for invest
gating the existing chemical forms of heavy metals in natural wat

$$MeX_j \rightleftharpoons Me^{2+} \quad + \quad H_nL^{m+n} \rightleftharpoons MeL^{m+2} \quad + nH^+ \qquad \text{homogenenous}$$

$$\Big\Updownarrow 2n \qquad\qquad\qquad \Big\Downarrow \alpha n \qquad\qquad\qquad \text{reaction}$$

$$Me(Hg) \qquad\qquad\qquad Me(Hg) \qquad\qquad\qquad \text{heterogeneous}$$

The potentials at which the reversible electrode reaction of
the labile complexed metal and the irreversible electrode reactio
of the organically bound metal takes place are well separated.
Therefore, it is possible to observe the formation of inert type
of complexes, following the decrease of the concentration of the
labile forms of the metal. Sensitivity and accuracy are higher fo
monitored reversible response than for the irreversible process.

Due to the fact, that the inert type of complex is formed, the concentration of the reactants is low, the plating time is short (6o or 12o s), the dissociation of the chelate in the electrode vicinity remains negligible.

4. RESULTS AND DISCUSSION

The extraordinarily high sensitivity of differential pulse anodic stripping voltammetry enables to study the complexation of heavy metals at or close to their natural trace levels (Nürnberg, 1982).

The applied voltammetric measuring procedure is illustrated in Fig.1, where the formation of the Zn(II) complex with humic acid (HAL) is followed by the decrease of the response corresponding to the labile complexed Zn(II), remaining uncomplexed by HAL. It is also assigned as Zn_{ionic}. The reversible response of Zn_{ionic} is the

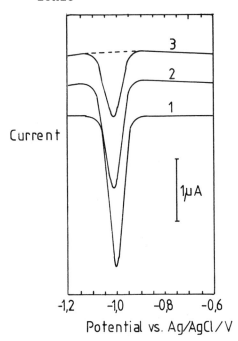

FIGURE 1. Differential pulse anodic stripping voltammograms of $4.7 \times 10^{-7} M$ Zn(II) in Ligurian sea water, pH 8.1, after addition of: 1) 0.0, 2) 3.o and 3) 6.o mg HAL/dm^3.

cumulative one of the hydrated metal ion and labile complexes ZnX_j, where X represents inorganic ligands Cl^-, OH^-, CO_3^{2-} and SO_4^{2-}. The height of the Zn_{ionic} peak before any ligand addition corres-

ponds to 1oo % of Zn_{ionic} (viz. Fig.1, curve 1). Zn(II) complexed
with HAL is not electroactive in the potential range up to -1.o V
but only at more negative potentials. Therefore, after each HAL
addition, the concentration of Zn_{ionic} will correspondingly decrease. It was experimentally determined that after each ligand
addition up to 3o min are necessary until the complexation equilibrium is attained. The decrease of the Zn_{ionic} peak is expresse
in percent with respect to the initial peak height before any lig
addition. A series of such measurements can be graphically presen
as the dependence of the amount of Zn_{ionic} on the totally added l
gand concentration. This dependence has the form of a "titration
curve", as it is presented in Fig.2.

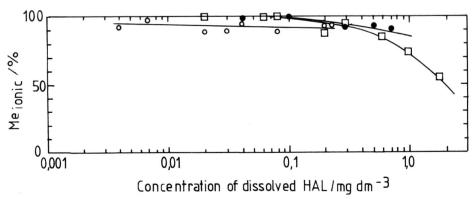

FIGURE 2. Voltammetric "titration curve" presenting the percentag
of umcomplexed Cd (●), Pb (o) and Zn (□) as function of increasir
concentration of the dissolved humic acid (HAL) in Ligurian sea
water. At pH ≈8 a total concentration of $3.3 \times 10^{-8}M$ Zn(II) was
originally present in the water, while $3 \times 10^{-9}M$ Cd(II) and $1.9 \times$
$10^{-8}M$ Pb(II) were adjusted.

It can be noticed from Fig.2 that the voltammetric "titration
curve" for $3.3 \times 10^{-8}M$ Zn(II) could be recorded up to 5o % comple
ation of Zn(II) by HAL, while for Pb(II) and Cd(II) only the init
part of the curve, i.e. up to 1o % complexation could be determir
Further evaluation for Pb(II) and Cd(II) at higher HAL concentrat
was prohibited, due to adsorption of HAL on the electrode surface
at bulk concentrations above o.3 and o.7 mg dm^{-3}, respectively. 1
amount of the adsorbed humic acid depends on the electrode potent

and its bulk concentration. It was determined by out-of-phase
ac-voltammetry, that the measurements of Zn are not influenced by
the adsorption of humic acid, because the preconcentration potential of -1.2 V adjusted for Zn_{ionic} is close to the desorption potential of humics and only a low degree of adsorbed humic acid
remains on the surface at this potential.

It can be concluded that in sea water at the usual low Cd, Pb
and Zn concentration and up to o.3 mg HAL dm^{-3} only maximal 1o %
of the total dissolved trace metal concentrations will be complexed.
Due to the low concentration of dissolved humic acid in the open
sea, which hardly exceeds o.2 mg dm^{-3} it follows that the speciation
of Cd, Pb and Zn will remain usually unaffected by dissolved humic
material and will be predominantly influenced by the formation of
labile complexes ΣMeX_j with the prevailing inorganic salinity components (Nürnberg, 1983; Sipos et al., 198o).

From a general point of view these results provide an example
of the potentialities of advanced voltammetric modes in studying
and revealing the fundamental problems of heavy metal speciation
at or close to their actual trace level in natural waters
(Nürnberg and Raspor, 1981; Nürnberg and Valenta, 1983;
Valenta, 1983; Nürnberg, 198o, 1983).

The speciation of heavy metal in the presence of dissolved organic matter in two water types was limited to the measurements
of Zn only, because for this metal the plating step was not
interfered by the adsorption of humic acid. Comparative investigations were performed in sea water and lake water with the
well defined synthetic ligands as NTA and EDTA. The results
of the evaluated "titration curves" are summarized in Table 1.
The "titration curves" of Zn(II) recorded with different ligands can be seen in the paper by Raspor et al. (in press b)
The data correspond to 5o % complexation degree and are the
required ligand concentrations (in mg dm^{-3}) to bind 5o % of the
totally dissolved Zn(II) in the studied natural water type.

The efficiency for binding Zn(II) decreases along the sequence
EDTA>NTA>Humic acid>Fulvic acid (viz. Table 1). Ligands like EDTA
and NTA are more effective complexing agents than the ligands of
natural origin. The change in the salinity components of water

TABLE I.

Concentrations on the mg dm^{-3} scale of synthetic (NTA, EDTA) and of natural (FAM, HAM, HAL, HAN) ligands L required in saline and fresh waters to compl 5o % of the total dissolved Zn(II) concentration

Water type	pH	Ligand	$[L_{total}]$ mg dm^{-3}	$[ZnL] = [Zn_{ionic}] = 5o\%[Zn(II)]$ moles dm^{-3}
Ligurian sea	7.8	NTA	o.52	$1.o \times 10^{-7}$
	7.8	EDTA	o.1o	$1.o \times 10^{-7}$
Adriatic sea	7.6	FAM	54.o	2.1×10^{-7}
Ligurian sea	8.2	HAM	6.3	1.2×10^{-7}
	8.1	HAL	4.9	1.4×10^{-7}
	8.2	HAN	4.7	1.6×10^{-7}
Lake Ontario	7.5	NTA	o.13	2.1×10^{-7}
	7.8	EDTA	o.o8	$2.o \times 10^{-7}$
	8.2	FAM	13.o	$o.4 \times 10^{-7}$
	8.2	HAM	5.1	1.1×10^{-7}
	8.3	HAL	2.8	1.3×10^{-7}
	8.5	HAN	4.2	$1.o \times 10^{-7}$

exerts a higher influence on the complexation of Zn(II) with N EDTA or fulvic acid than with humic acids.

The amount of humic acids of quite different origin, needed complex 5o % of the present total Zn(II), falls into a narrow range on the mg dm^{-3} scale. It should be noted, that this be-haviour is observed inspite the fact, that different isolation procedures were applied for HAL and for HAN and HAM samples. Buffle and Greter (1979) have reported as well a rather similar behaviour of fulvic acids independent of their origin. There is a correlation between the binding of Zn(II) by humic acids and their nitrogen content, which is 4.2 % for HAL and HAN and 1.7 % for HAM. From Table 1 can be seen, that HAL and HAN are better complexing agents for Zn(II) than HAM. This might be re: to the electrophilic character of Zn(II) having thus a preferer for coordination with N-atoms which carry a free electron pair

A concentration of 1 mg dm^{-3} or more of dissolved humic
acid will be required in sea water to achieve a complexation
degree above   1o % of the usual trace levels of Zn(II), Cd(II)
and Pb(II). In most parts of the sea the dissolved humic acid
concentration will hardly exceed o.2 mg dm^{-3}. Therefore, dissolved
humic acids will not significantly contribute to the speciation
of these heavy metals in the open sea. The situation can be
different in certain coastal waters, a number of estuaries, rivers
and lakes if concentrations higher than 1 mg dm^{-3} of dissolved
humic acids will be present.

Fulvic acid has a lower complexing capacity for Zn(II), than
the humic acids, both in sea water and lake water, which is probably a consequence of the relatively high number of acidic
groups in fulvic acid.

At a constant concentration of the reactants, the extent of
Zn(II) complexation with the dissolved humic acid was experimentally observed altering the pH. The concentration of the
formed Zn-humate increases about 3o % when the pH is altered
from 7.2 to 8.4. It was proved, that over the same pH range
the amount of the adsorbed HAL does not change. The with rising
pH increased concentration of Zn-humate can therefore be explained
by additional proton dissociation, most probably from amino groups
present in the humic acids. In this manner new complexing sites
are formed. It is obvious, that Zn-humate formation is very
pH-sensitive as the amount of the complexed Zn(II) increases
in the stated rather narrow pH range around 8. Therefore measurements of different authors can only be compared at the same pH
value.

ACKNOWLEDGEMENTS

    This work is part of the joint research project "Environmental
Research in the Aquatic Systems" of the Center for Marine Research
Zagreb, Rudjer Bošković Institute, Zagreb and the Institute of
Applied Physical Chemistry, Nuclear Research Center (KFA), Juelich,
in the bilateral German-Yugoslav Agreement. Financial support by
the International Bureau of KFA, Juelich, and the Self-managed
Authority for Scientific Research, SR Croatia, Yugoslavia, is
gratefully acknowledged.

326

REFERENCES

- Branica, M., Novak, D.M. and Bubić, S. 1977. Application of Anodic Stripping Voltammetry to Determination of the State of Complexation of Traces of Metal Ions at Low Concentration Levels. Croat.Chem.Acta 49, 539-547
- Buffle, J. and Greter, F.-L. 1979. Voltammetric Study of Humic and Fulvic Substances, Part II. Mechanism of Reaction of the Pb-Fulvic Complexes on the Mercury Electrode. J. Electroanal. Chem. 1o1, 231-251
- Desai, M.V.M. and Ganguly, A.K. 197o. Interaction of Trace Elements with the Organic Constituents in the Marine Environment, Bhabha Atomic Research Center, Report Nr. 488, Bombay, India
- Huljev, D. 197o. Interaction of Metal Ions with Humic Acid in Aqueous Solutions. M.Sc.Thesis, Univ. Zagreb.
- Kononova, M.M. and Belachikova, N.P. 196o. A Study of Soil Humic Substances by Fractionation, Soviet Soil Sci. 4, 1149-1155
- Nürnberg, H.W. 198o. Features of Voltammetric Investigations of Trace Metal Speciation in Sea Water and Inland Waters. Thalass. Jugosl. 16, 297-315
- Nürnberg, H.W. 1982. Voltammetric Trace Analysis in Ecological Chemistry of Toxic Metals. Pure Appl. Chem. 54, 853-878
- Nürnberg, H.W. 1983. Voltammetric Studies on Trace Metal Speciation in Natural Waters. Part II. Applications and Conclusions for Chemical Oceanography and Chemical Limnology. In: Leppard, G.G. (Ed.), Trace Element Speciation in Surface Waters and its Ecological Implications. Plenum Press, New York-London, pp.211-23o
- Nürnberg, H.W. and Raspor, B. 1981. Applications of Voltammetry in Studies of the Speciation of Heavy Metals by Organic Chelators in Sea Water. Environm.Technol. Letters 2, 457-483
- Nürnberg, H.W. and Valenta, P. 1983. Potentialities and Applications of Voltammetry in Chemical Speciation of Trace Metals in the Sea. In: Wong, C.S., Boyle, E., Bruland, K.W., Burton, J.S. and Goldberg, E.D. (Eds.), Trace Metals in Sea Water, Plenum Press, New York-London, pp.671-697
- Raspor, B., Valenta, P., Nürnberg, H.W. and Branica, M. 1977. Polarographic Studies on the Kinetics and Mechanism of Cd(II)-Chelate Formation with EDTA in Sea Water. Thalassia Jugosl. 13, 79-91
- Raspor, B., Valenta, P., Nürnberg, H.W. and Branica, M. 1978. The Chelation of Cadmium with NTA in Sea Water as a Model for Typical Behaviour of Trace Metal Chelates in Natural Waters. Sci. Tot. Environm. 9, 87-1o9
- Raspor, B., Nürnberg, H.W., Valenta, P. and Branica, M. 198o a. The Chelation of Pb by Organic Ligands in Sea Water. In: Branica, M. and Konrad, Z. (Eds.), Lead in the Marine Environment, Pergamon Oxford, pp. 181-195
- Raspor, B., Nürnberg, H.W., Valenta, P. and Branica, M. 198o b. Kinetics and Mechanisms of Trace Metal Chelation in Sea Water. J. Electroanal. Chem. 115, 293-3o8
- Raspor, B., Nürnberg, H.W., Valenta, P. and Branica, M. 1981. Voltammetric Studies on the Stability of Zn(II)-Chelates with NTA and EDTA and the Kinetics of their Formation in Lake Ontario Water. Limn. Oceanogr. 26, 54-66

- Raspor, B., Nürnberg, H.W., Valenta, P. and Branica, M. (in press a). Voltammetric Study of Trace Metal Interaction with Humic Substances of Different Origin in Sea Water and Lake Water, Part I. Characterisation of Humic Substances.
- Raspor, B., Nürnberg, H.W., Valenta, P. and Branica, M. (in press b). Voltammetric Study of Trace Metal Interaction with Humic Substances of Different Origin in Sea Water and Lake Water, Part II. Voltammetric Investigations on the Trace Metal Complex Formation in the Dissolved Phase.
- Sipos, L., Raspor, B., Nürnberg, H.W. and Pytkowicz, R.M. 1980. Interaction of Metal Complexes with Coulombic Ion Pairs in Aqueous Media of High Salinity. Mar. Chem. 9, 37-47
- Valenta, P. 1983. Voltammetric Studies on Trace Metal Speciation in Natural Waters. Part I. Methods. In: Leppard, G.G. (Ed.), Trace Element Speciation in Surface Waters and its Ecological Implications. Plenum Press, New York-London, pp. 49-69

COMPLEXATION CAPACITIES OF HUMIC SUBSTANCES ISOLATED FROM FRESHWATER
WITH RESPECT TO COPPER(II), MERCURY(II), AND IRON(II,III)

F.H. FRIMMEL, A. IMMERZ AND H. NIEDERMANN

1. INTRODUCTION

Humic substances (HUS) are the predominant part of organic carbon (DOC)
dissolved in unpolluted water (Sontheimer and Gimbel, 1977). From their
general structures their abilities for heavy metal complexation can be derived.
This property is of considerable practical interest, for complexation may
affect the transport of metals in aquatic systems and through treatment plants.
It may also affect biological uptake and the toxicity of metals, especially in
freshwater systems (Stevens, 1982; Christman and Gjessing, 1983). Therefore
quantitative methods for the characterization of heavy metal complexation in
aquatic systems are needed.

2. GENERAL EXPERIMENTAL APPROACH

Besides the calculation of stability constants (Bresnahan, et al., 1978)
and rate constants (Shuman and Michael, 1978) for metal association and dissociation at different sites of the humic material, the overall binding
capacities with respect to different metal ions are of general interest.
Methods for their determination have been reviewed recently (Florence and
Batley, 1980; Hart, 1981; Neubecker and Allen, 1983).

Complexation capacity (CC) is a function of several parameters including
pH, metal and ligand concentrations, ionic strength, and temperature.

These parameters are also used for the general description of the quality
of aquatic systems. Since most aquatic systems are much too complicated for a
straightforward correlation of these parameters to an individual complexation
reaction, an approach for a better understanding of natural complexation
reactions should combine the characterization of the original water with the
information obtainable from taking these systems apart and determination of
the separate units. The resulting model should fit reality as well as possible
(Fig. 1). Model reactions can also help to confirm these results.

Although looking at the original system includes the advantage of an un-

Kramer, C.J.M. and Duinker, J.C. (eds.), Complexation of Trace Metals in Natural Waters. ISBN 90-247-2973-4
©1984 Martinus Nijhoff/Dr W. Junk Publishers, The Hague/Boston/Lancaster.
Printed in the Netherlands.

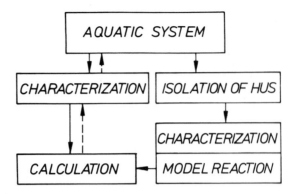

FIGURE 1. General scheme for the determination of the complexation ability of humic substances and its influence on aquatic systems.

disturbed realistic sample, conclusions have to be drawn carefully since the correlation of data might be complicated by a number of uncontrolled inter-actions. Looking at the isolated parts of a system can minimize these inter-actions, and leads in general to more reliable data. However, this approach does often imply a drastic change of the original reaction system according to the method applied, and by this limits the meaning of the operationally defined values.

2.1. Isolation procedure

The method for the isolation of humic material (mainly fulvic acids) from freshwater was based on the procedure of Mantoura and Riley (1975). It uses adsorption/desorption on a column packed with polystyrene resin, and yields nearly metal-free concentrated solutions of humic substances. The complexatic capacities of HUS can be determined by defined titrations or reactions. This experimental approach includes some advantages and drawbacks concerning the quality of the humic material and its characterization reactions.

Advantages: Minimized disturbance from nonhumic ligands.
Minimized disturbance from metal ions originally present.
Comparable results for products from different origin.
Wide choice of suitable analytical methods, since the concen-trated sample covers most detection limits.
Minimized disturbance from electrode reactions (adsorption) and kinetic effects by applying differential pulse polaro-graphy with the dropping mercury electrode (DPP/DME).
Minimized condensation reactions by maintaining HUS in the dissolved phase.

Disadvantages: Operational definition of HUS according to the method.
Undefined denaturation of HUS at different pH values during isolation.

Possible influence of the concentration step on the molecular
structure of the HUS.
Changed matrix and therewith changed reaction conditions compared with aquatic systems.

2.2. Characterization of humic substances

For a general characterization of isolated HUS, the dissolved organic carbon
(DOC) and spectroscopic absorption (A) at 254 and 436 nm were measured (Wölfel
and Sontheimer, 1974; Sontheimer, 1978).Proton capacity between the pH values
of 4.3 and 8.2 (H^+-Cap.), complexation capacities with respect to Cu^{2+} (Cu(II)-
CC), Hg^{2+} (Hg(II)-CC), and the strongly bound iron which even remains at pH 2.2
in the core of the HUS (Fe(str.)) were used for the characterization of the
complexation abilities (Frimmel, 1978; Frimmel, et al., 1980, 1982; Frimmel and
Niedermann, 1980).

2.3. Stability of solutions containing isolated humic substances

Khairy and Ziechmann (1981) have shown that high pH values can change the
properties of dissolved humic material. Therefore it is advisable to neutralize
isolated HUS after their elution as soon as possible.

The influence of light and temperature on the optical properties of HUS
extracts are shown in figure 2 and 3. Different dilutions (1:10 to 1:250) of a
HUS stock solution (containing 536 mg/l DOC) were kept at 4°C in the dark ($\textcircled{n}$)
and at room temperature in daylight ($\textcircled{\textcircled{n}}$) in colorless stoppered glass bottles
at pH 10. The change of the optical absorption (A(254nm) and A(436nm)) of the
solutions was measured with time, and refers to the original concentration of
the stock solution assuming the validity of Beer's law. It can be seen that the
samples kept cool and dark are fairly stable compared with the ones kept in
daylight and at room temperature. The biggest changes are observed in the concentrated samples stored in daylight and in the diluted samples stored in the
dark. Daylight does affect the yellow color (436nm) significantly more than the
UV absorption (254nm). The Cu(II)-CC measured for the 1:10 dilution decreases
within 500 h for about 20%, and seems to be slightly more affected in the sample
kept in daylight. The cool sample showed an increased value of 1.3 μmol/mgC after
the first 18 h. This coincides with a temporary maximum of the UV absorption.

3. USE OF DIFFERENT REFERENCE METALS

For determining CCs, Cu^{2+} is most widely used as reference metal, mainly
for the following reasons:

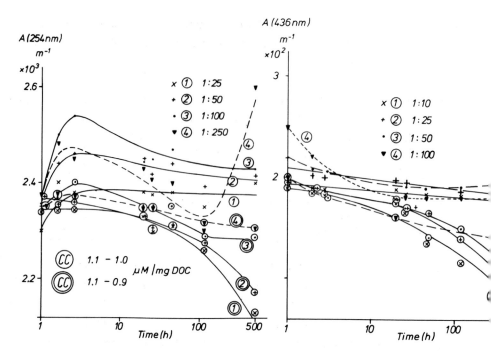

FIGURE 2 and 3. Influence of storage on the UV absorptivity at 254 nm, and
the yellow color (absorptivity at 436 nm) of aquatic humic material.

a) Cu(II) can easily be detected.
b) Cu plays a significant role in aquatic systems.
c) Cu(II) forms fairly stable complexes with many O- and N-containing
   chelating ligands.

Most of the functional groups in HUS suited for complexation therefore
should react with Cu^{2+} and could be determined by that means. However, some
groups can be expected to be more specific for complex formation with other
metal ions. E.g. Fe and Hg are not only of general importance in aquatic
systems but also show some significance in coordination chemistry.

3.1. Copper(II) complexation capacity

A wide variety of methods for the determination of Cu(II)-CC has been reported, ranging from spectroscopic and electrochemical methods to physical
separation techniques (Hart, 1981; Neubecker and Allen, 1983). In this work
we used a titration technique with polarographic Cu^{2+} detection.

3.1.1. Titration with polarographic detection. As pointed out earlier the
isolation step for HUS from aquatic systems supplies fairly concentrated
solutions. This allows the application of DPP/DME as analytical method. A

known amount of Cu(II)$_i$ is titrated with increasing amounts of a HUS stock
solution, for which the DOC-concentration is known. Typical reaction conditions are given in Table 1.

TABLE 1. Reaction conditions for the determination of the Cu(II)-CC by means
of DPP/DME titration.

TITRIMETRIC CONDITIONS		POLAROGRAPHIC SETTINGS	
Total volume:	10 ml	Initial potential:	-0.15 V
Cu(II) total:	10 µmol/l	Final potential:	+0.20 V
HUS added:	variable	Modulation ampl.:	50 mV
	(5 to 50 µl)	Scan rate:	5 mV/s
Buffer: 0.01 M acetate (pH 6.8)		Drop size:	medium
Temperature:	25°C	Drop time:	0.5 or 1 s
Detection:	DPP/DME	Reference electrode:	Ag/AgCl/KCl(3M)

The application of DPP/DME - though not as sensitive as voltammetric methods
(Branica et al., 1977; Nürnberg, 1978) - show some advantages for the determination of free and labile Cu(II) in HUS-solutions. The relatively small life-
time of the mercury drop ($\geqslant 0.5$s) and the even smaller measurement time
minimize disturbances which come from uncontrolled electrode reactions (e.g.
adsorption of HUS onto the Hg surface) and kinetics. However, fast reactions
as described by Olson and Shuman (1983) might still have an effect.

3.1.2. <u>Detection of disturbances</u>. An experimental method to control the
validity of the titration procedure uses the peak potential, which has to be
constant for the duration of the experiment. In addition the peak shape has to
be symmetrical (Figure 4). Further information on the reliability of the
results can be gained from the influence on the signal of drop size and drop
time in connection with the scan speed. In case of strong electrochemical
influences of the HUS on the current-potential peak, the determination of
fractions separated by liquid chromatography may lead to reliable results for
parts of the gross sample (Frimmel and Sattler, 1982).

Best quantitative data can be achieved by the standard addition method.
Figure 5 shows independent calibration curves for Cu(II) in the presence of
different amounts of HUS (A to D). To minimize the influence of the complexation reaction between HUS and Cu(II) the standard additions were made on top
of an initial Cu(II) concentration of 10 µmol/l.

Titration of a given amount of HUS with increasing amounts of Cu(II) show,
after the equivalent point of complexation, linear branches having slopes
identical with the ones in Figure 5 for the same DOC-concentration.

334

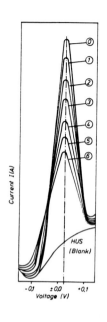

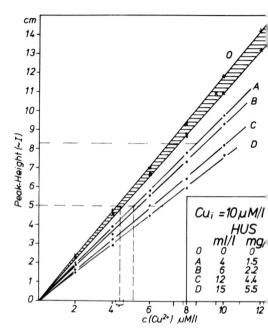

FIGURE 4 (left). Polarographic peaks for the titration of Cu(II) (⓪) with increasing amounts of humic substances (① to ⑥).

FIGURE 5 (right). DPP/DME calibration curves for Cu(II) in the presence of different amounts of humic substances.

The segment O reflects the calibration without any HUS in different electrolytes (10^{-2}M KClO$_4$, pH 6.8, line near to O; 10^{-2}M acetate, pH 6.8, line near to A). The experiment suggests that an exact titration curve should not only rely on the calibration curve for the solution without HUS (a in Figure 6) The standard addition procedure at each step of the titration (shown by b and c in Figure 6) gives a correct titration curve from which the Cu(II)-CC (100 – 87.5 = 12.5 = 1.25 μmol/mgC) can be derived. It can be up to 50% smal than the uncorrected values.

3.1.3. <u>Influence of the pH-value</u>. A buffer system (10^{-2}M acetate) was normally used for pH control during the titration. It is important not to exceed the buffer capacity by the addition of the increments, since a change of the pH of the complexation reaction would include a change of the determi CC. Figure 7 shows a typical curve for the pH dependence of the Cu(II)-CC. Th inflections are due to the superimposed increase of the solubility of Cu(II) and the decreased conditional stability of the formed Cu(II)-HUS complexes as

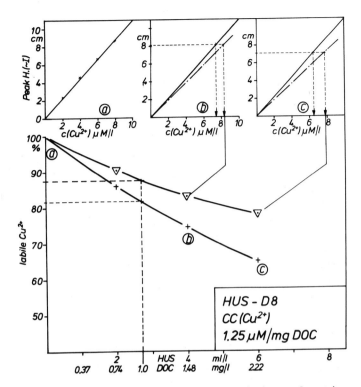

FIGURE 6. Polarographic determination of the Cu(II) complexation capacity of isolated humic substances (corrected values).

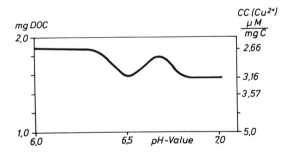

FIGURE 7. Typical influence of the pH-value on the determination of the Cu(II) complexation capacity of isolated humic substances.

the pH decreases. In the range of pH 7.0 to 6.0 the Cu(II)-CC changes for about 20%.

3.1.4. Influence of earth alkaline ions. The meaning of the Cu(II)-CC for the original aquatic system is strongly dependent on the earth alkaline ions and their influence on the complexation reactions. Cu(II)-CCs have been determined at different earth alkaline ion concentrations. The hardness of th solutions, including the ratio of Ca^{2+} and Mg^{2+}, were adjusted according to the hardness of the original aquatic system (Figure 8). It is obvious that th Cu(II)-CC decreases as the concentration of Ca^{2+} and Mg^{2+} increases, showing difference of about 25% for a solution free of alkaline earth ions compared with a solution which is as hard as the original surface water (point 1 at th abscissa). The inflections of the curve might be due to different binding sit and hence different stable complexes. More than 5 times the original hardness seems to have no additional influence on the results of the method.

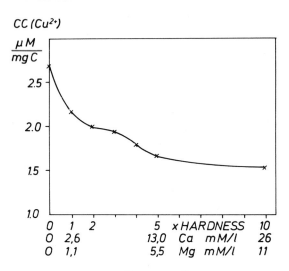

FIGURE 8. Typical influence of Ca^{2+} and Mg^{2+} on the determination of the Cu(II) complexation capacity of isolated humic substances.

3.2. Mercury(II) complexation capacity

Mercury concentrations in aquatic systems and its distribution in dissolve and undissolved species cannot be explained by purely inorganic reaction mechanisms. The interaction of Hg(II) with organic matter will be dominated by the reactions with mercapto groups, Hg(II) and the ligand both being "soft in terms of Pearson's acid and base theory (SHAB).

3.2.1. Model reactions with thio-group ligands. The reaction of Hg(II) wit cystein (Cys) in the range of mmol/l followed by UV irradiation (λ_{max} = 254 yields black HgS. This is not unexpected, for the Hg-Cys complexes are known to be fairly stable (K_1 = 10^{14}; β_2 = $10^{20.5}$; Sillen and Martell, 1971) and

HgS has one of the lowest solubility products known ($L_{SO} = 3 \cdot 10^{-54}$).

If the UV-digestion is performed in oxygen-containing solution, some HgS will redissolve. The reactions can be written in the general form:

Complexation: $\quad Hg(II)aq + HS-R \rightleftharpoons (Hg-S-R)OH + HOH \qquad (1)$

UV-reaction: $\quad (Hg-S-R)OH \xrightarrow[(254 \text{ nm})]{UV} HgS + R' \qquad (2)$

Oxidation: $\quad HgS + O_2 + 2 \ OH^- \longrightarrow Hg(II)aq + SO_4^{2-} + 2 \ H^+ \qquad (3)$

Excluding oxygen, the complexation reaction (1) can be characterized by the amount of photochemically formed HgS (2), which can be removed from solution by membrane filtration. After the phase separation mercury concentrations can be reliably determined by atomic absorption spectrometry (cold vapour method). HgS was digested by using acid at high temperature. Typical reaction conditions are given in Table 2.

Table 2. Reaction conditions for the determination of Hg(II)-CC by means of photochemical decomposition reactions.

COMPLEX FORMATION		PHOTO REACTION	
Reaction volume:	1000 ml	Irradiation (λ max):	254 nm
Hg(II) inital:	1 μmol/l	Irradiation time:	5 to 120 min
Ligand concentration:	variable	Temperature:	12°C
pH-value:	3 to 6.2		
Temperature:	12°C		
Equilibration time:	10 min		

The kinetics of the reaction (2) is shown for the Hg(II)-Cys complex in Figure 9. The reaction time for 50% yield ($t_{1/2}$) and the final yield (c_f/c_o), expressed as ratio of the final concentration of dissolved mercury (c_f) and the initial mercury concentration (c_o), are characteristic data. It is obvious that the presence of additional heavy metals had no significant influence on the photochemical reaction. The behaviour of the metals themselves is also given in Figure 9. Whereas the concentrations of dissolved Cu(II), Zn(II), Cd(II) do not change for the whole reaction time, there is a decrease in the concentrations of dissolved Pb(II) and Fe(II/III). In these cases, the final yield of the Hg(II)-Cys reaction is slightly smaller (-7%). This might be due to some adsorption of the ligand on the undissolved metal species.

Characteristic reaction data for some model ligands (L) are given in Table 3. and allow some interpretation of the coordination sphere of the Hg(II). The more stable the Hg(II)-L complexes are, the longer is $t_{1/2}$. It is also apparent that the photochemical HgS formation is hindered as the ligand gets more bulky.

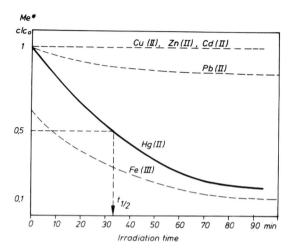

FIGURE 9. Kinetics of the photochemical decomposition of the Cu(II)-Cys comp:

Table 3. Reaction data for the photochemical decomposition of Hg(II) complexe
with different thio-ligands.

Ligand (L)	L:Hg (initial)	Hg reacted c_f/c_o	$t_{1/2}$ min	Hg-coordination (suggestion)
2-mercapto-propionic acid	1:1	0.94	17.5	a
Thioglycolic acid	1:1	0.85	18.8	a
Thiomalic acid	1:1	0.88	25.0	b (a)
3-mercapto-propionic acid	1:1	0.84	29.5	b
Cystein	1:1	0.82	33.5	c
	2:1	0.81	50.0	d
Cystin	1:1	0.86	53.5	d[*]
Cystamine	1:1	0.58	73.5	d
Gluthatione	1:1	0.66	86.0	e (c,b)

[*] after symmetrical ligand splitting by UV irradiation, which is included in $t_{1/2}$.

3.2.2. <u>Reactions with isolated humic substances</u>. Use of HUS as ligands for the photochemical HgS formation leads to a significant increase in $t_{1/2}$, which might be due to shielding effects. The yields after irradiation of 120 min are strongly dependent on the concentration of the HUS applied (Figure 10). At low concentrations there is a linear correlation of the yield and the DOC. At higher concentrations the yield levels off. For the determination of a Hg(II)-CC only the low concentration range is suitable. According to that, Hg(II)-CC has to be defined as the molar concentration of mercapto groups per mg C of the HUS which lead to HgS in the low concentration range and under the photochemical reaction conditions given above.

System control reactions which contain only Hg(II) resp. HUS with and without UV irradiation are neccessary to detect and correct disturbances from adsorption etc.

Table 4 shows some of the results for HUS isolated from rivers and lakes. It is obvious that the Hg(II)-CC values do not reflect the whole sulfur content of the HUS, determined by elemental analysis (S(HUS)). However, there seems to be a correlation between the surprisingly high concentrations of dissolved Hg (Hg*) in the brown water (BAN) of an unpolluted prealpine surrounding, and the present amount of HUS capable for Hg complexation.

Table 4. Hg(II) complexation capacities of isolated aquatic humic substances and related parameters

		MA	DO	STA	BAN
Hg*	nmol/l	1 to 9	1	1	1 to 3
DOC(HUS)	mg/l	1 to 2	0.6 to 1	1 to 2	3 to 5
Hg(II)-CC μmol/mgC		0.15 to 0.4	0.2 to 0.35	0.2 to 0.4	0.1 to 0.3
S(HUS)	μmol/mgC	0.4	0.6	0.5	0.35

3.3. Strongly bound iron

Isolation of HUS according to the method described above includes acidification of the water sample to pH 2.2. At this pH-value some percentage of the originally dissolved and complexed iron remains bound to the core of the HUS, which owing to its properties must be complexed in a very stable form. There have been suggestions that the possible coordination sphere resembles hydroxamate complexes. However, the exact structure is yet unknown.

Whether there are still other metals bound strongly to the HUS at low pH values cannot be answered at the moment. Analytical techniques for these low concentrations are under investigation.

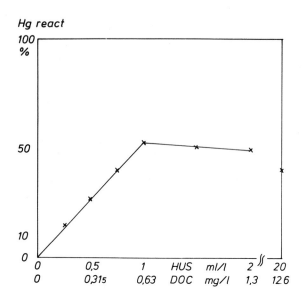

FIGURE 10. Influence of the HUS-concentration on the reaction yield of the photochemical HgS formation.

4. RELATION TO AQUATIC SYSTEMS

Isolated HUS have to be discussed in close connection with the original aquatic system owing to the effect their complexation abilities will have on the actual dissolved cations. Figure 11 shows some of the characteristic data of aquatic systems (left side), and their isolated HUS (right side). The DOC, Fe^* and Me^* concentrations refer to 0.45 μm membrane filtration. Me^* includes the metals Cu, Zn, Cd, Pb, and Hg. The river Main (MA) is known to be fairly polluted. The Danube (DO), the river Isar (IS), lake Starnberg (STA) and lake Kochel (KOC) are moderately polluted. The brown-water (BAN) was sampled from a small prealpine lake which was for the first period of the investigation unpolluted, and has changed in the last two years (Frimmel and Niedermann, 1982). Samples were taken over a period of three years at different seasons. Sampling and water analysis were performed according to standard methods (Fachgruppe Wasserchemie, 1982). The columns in Figure 11 represent the minimum and the maximum values. The abreviations on the abscissa reflect the seasons for which they were determined (Sp = spring; s = summer; F = fall; W = winter).

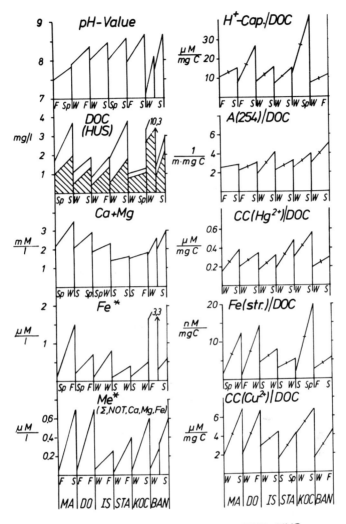

FIGURE 11. Comparison of the complexation capacities of isolated humic substances with data of the original aquatic systems.

5. CONCLUSIONS

- Humic substances (HUS) are present in all natural aquatic systems. They deliver a significant contribution to the complexation reactions.
- The isolation of HUS, necessary for their characterization, can include some major denaturation of the material.

- Dissolved samples kept under N_2 at $4^{\circ}C$ in the dark at neutral pH are fairly stable for at least three weeks.
- For the characterization of the complexation abilities of HUS the determination of complexation capacities (CC) is useful.
- CC never should be reported or interpreted without including the reference metal used and the method applied.
- As operationally defined parameters different CC cannot be directly compared with one another.
- Differential pulse polarography with a dropping mercury electrode is a suitable method for the determination of Cu(II)-CC for isolated HUS. However, possible disturbances have to be examined carefully.
- pH and earth alkaline ions show significant influences on the values of the so determined Cu(II)-CC.
- Complexation between Hg(II) and thio-groups can be characterized by a photochemical reaction with HgS as reaction product.
- The application of this reaction to HUS results in a Hg(II)-CC.
- There is evidence for stable iron-HUS complexes even at pH 2.2.
- CC for isolated HUS could be determined in the range of 1.6 to 6.5 $\mu mol/mgC$ for Cu(II), 0.1 to 0.4 $\mu mol/mgC$ for Hg(II). The strongly bound iron was about 0,005 to 0.015 $\mu mol/mgC$.
- The situation in some fresh waters show that the content of dissolved heavy metals in the aquatic system is less than the maximum concentrations which can be complexed by their isolated HUS.

ACKNOWLEDGEMENTS. The authors would like to acknowledge financial support for this work by the Deutsche Forschungsgemeinschaft, Bonn - Bad Godesberg (Grant Fr 536/1-4), and the assistance of Dr. D.S. Millington (Department of Environmental Sciences and Engineering, University of North Carolina at Chapel Hill) in the preparation of the manuscript.

REFERENCES

- Branica, M., Novac, D.M. and Bubič, S., 1977. Application of Anodic Stripping Voltammetry to Determination of the State of Complexation of Traces of Metal Ions at Low Concentration Levels. Croat. Chem. Acta 49: 539-547.
- Bresnahan, W.T., Grant, C.L. and Weber, J.H., 1978. Stability Constants for Complexation of Copper(II) Ions with Water and Soil Fulvic Acids Measured by an Ion Selective Electrode. Anal. Chem. 50: 1675-1679.
- Christman,R.F. and Gjessing, E.T. (Ed.), 1983. Aquatic and Terrestrial Humic Materials. Ann Arbor Science Publ., Ann Arbor, Michigan.

- Fachgruppe Wasserchemie in der Gesellschaft Deutscher Chemiker (Ed.), 1982. Deutsche Einheitsverfahren zur Wasser-, Abwasser- und Schlammuntersuchung. Verlag Chemie, Weinheim.
- Florence, T.M. and Batley, G.E., 1980. Chemical speciation in natural waters. CRC Crit. Rev. Anal. Chem. 9: 219-296.
- Frimmel, F.H., 1978. Die Aussagekraft der Pufferkapazität. Hydrochem. hydrogeol. Mitt. 3: 57-73.
- Frimmel, F.H. and Niedermann, H., 1980. Komplexierung von Metallionen durch Gewässerhuminstoffe. I. Ein Braunwassersee als Huminstofflieferant. Z. Wasser Abwasser Forsch. 13: 119-124.
- Frimmel, F.H. and Sattler, D., 1982. Komplexchemische Charakterisierung isolierter Gewässerhuminstoffe nach gelchromatographischer Fraktionierung. Vom Wasser 59: 335-350.
- Frimmel, F.H., Sattler, D. and Quentin, K.-E., 1980. Photochemischer Abbau von Quecksilber-Thioverbindungen in sauerstoffhaltigem und sauerstofffreiem Wasser. Vom Wasser 55: 111-120.
- Frimmel, F.H., Immerz, A. and Niedermann, H., 1982. Heavy Metal Interaction with Aquatic Humus. Intern. J. Environ. Anal. Chem. 14: 105-115.
- Hart, B.T., 1981. Trace metal complexing capacity of natural waters. Environ. Technol. Lett. 2: 95-110.
- Khairy, A.H. and Ziechmann, W., 1981. Die Veränderung von Huminsäuren in alkalischer Lösung. Z. Pflanzenernaehr. Bodenk. 144: 407-422.
- Mantoura, R.F.C. and Riley, J.P., 1975. The Analytical Concentration of Humic Substances from Natural Waters. Anal. Chim. Acta 76: 97-106.
- Neubecker, T.A. and Allen, H.E., 1983. The measurement of complexation capacity and conditional stability constants for ligands in natural waters. Water Res. 17: 1-14.
- Nürnberg, H.W., 1978. Potentialities and Applications of Advanced Polarographic and Voltammetric Methods in Aqueous and Marine Trace Metal Chemistry. In: Ahlberg, P. and Sundelof, L.-O. (Ed.), Structure and Dynamics in Chemistry. Almqvist & Wiksell, Stockholm.
- Olson, D.L. and Shuman, M.S., 1983. Kinetic Spectrum Method for Analysis of Simultaneous, First-Order Reactions and Application to Copper(II) Dissociation from Aquatic Macromolecules. Anal. Chem. 55: 1103-1107.
- Shuman, M.S. and Michael, L.C., 1978. Application of Rotating Disk Electrode Techniques to the Measurement of Copper-Organic Complex Dissociation Rate Constants in Marine Coastal Samples. Environ. Sci. Technol. 12: 1069-1072.
- Sillen, L.G. and Martell, A.E., 1971. Stability Constants of Metal-Ion Complexes. The Chemical Society, London.
- Sontheimer, H., 1978. Summarische Parameter bei der Beurteilung der Eigenschaften von Oberflächenwasser. Hydrochem. hydrogeol. Mitt. 3: 15-29.
- Sontheimer, H. and Gimbel, R., 1977. Untersuchungen zur Veränderung der Fracht an organischen Wasserinhaltsstoffen mit der Wasserführung am Beispiel des Rheins. GWF, Wasser, Abwasser 118: 165-173.
- Stevens, F.J., 1982. Humus Chemistry. John Wiley & Sons, New York.
- Wölfel, P. and Sontheimer, H., 1974. Ein neues Verfahren zur Bestimmung von organisch gebundenem Kohlenstoff im Wasser durch photochemische Oxidation. Vom Wasser 43: 315-325.

COMPLEXATION OF Cu^{2+} WITH HUMIC SUBSTANCES IN RELATION TO DIFFERENT EXTRACTION
PROCEDURES OF SANDY AND SILTY MARINE SEDIMENTS

REMI W.P.M. LAANE and CEES J.M. KRAMER

1. INTRODUCTION

The interest in humic substances in sediments has increased considerably in
recent years. They are important for fixation and mobilization of inorganic
material, especially (heavy) metal ions (Reuter & Perdue, 1977; Boniforti, 1978;
Longacre Phelps 1979; Srna et al, 1980, Saar & Weber, 1982; Tada & Suzuki, 1982).
The study of 'easily exchangeable' forms is essential when considering transport
of humic metal complexes between sediment and water. Many extraction methods
have been published involving various combinations of extraction solvents, pH,
temperature etc. (Rashid & King, 1969; Felbeck, 1971; Schnitzer & Kahn, 1972;
Saito & Hayano 1979; Hatcher et al, 1980; Poutanen & Morris, 1983). All methods
are operationally defined and it is not clear whether we can prefer any specific
one. However, it is expected that nearly all the humic substances can be extracted with alkaline extractants.

Laane & Koole (1982) found different humic substances when extracting silty
marine sediments with NaOH and seawater respectively. Processes in the environment involving humic matter at the water-sediment interface take place at a
considerably lower pH than that of NaOH solutions.

In this paper, the copper complexation capacity and the fluorescence spectra
of the humic substances from Wadden Sea sediments extracted with NaOH and seawater are compared.

2. MATERIAL & METHODS

Sandy and silty Wadden Sea sediments (10 gram wet weight) were extracted as
follows: with 25 cm^3 seawater (1), 25 cm^3 0.1 NaOH (2), both during 20 h, 11
consecutive extractions with 25 cm^3 fresh water (3) and with 25 cm^3 fresh
0.1 NaOH (4). Different extraction periods were applied in 3 and 4.

The fluorescence spectra were recorded at pH 8.4 at roughly equal concentrations of the fluorescent material in the different extracts. For description of
the methods see Laane (1982) and Kramer (this volume).

Kramer, C.J.M. and Duinker, J.C. (eds.), Complexation of Trace Metals in Natural Waters. ISBN 90-247-2973-4
© 1984 Martinus Nijhoff/Dr W. Junk Publishers, The Hague/Boston/Lancaster.
Printed in the Netherlands.

3. RESULTS

Figure 1 shows the fluorescence of the material extracted from sandy and silty sediments with seawater and NaOH, followed in time. The pH during extraction was constant, and no shift was observed in the wavelength of the emission maximum. A rapid increase in fluorescence was observed in all four cases (within 30 min) with no further change for the seawater extracts over the remaining period. In contrast, a continuous increase was found for the NaOH extracts. Most fluorescent material was extracted from silty than from sandy sediments.

Figure 2A shows the fluorescence of the material extracted from the 2 sediments with seawater or 0.1 N NaOH during 11 consecutive extracts involving fresh extractants. Three consecutive extractions appeared to be sufficient to extract most of the 'easily exchangeable' fluorescing material. However, especially for the alkaline extractant, even after 7 changes, a considerable amount of fluorescing matter could still be extracted. At extended extraction times fluorescence increased only for the NaOH extracts. Consecutive extractions during such extended periods (24 - 72 h) finally resulted in decreased fluorescence. The absorbance of the NaOH extract of silty sediment generally followed the fluorescence pattern with an important exception after the last change of solution. Also here no shift in the emission maximum wavelength was observed.

An example of the fluorescence spectra of different extraction procedures is given in Fig. 3. The emission maximum of extracted material from a silty sediment occured at 465 nm when using seawater, at 476 nm for one extraction with 0.1 N NaOH after 1 h and at 520 nm for an extract after 11 consecutive extractions with 0.1 N NaOH.

TABLE 1. Dissolved Organic Carbon and Complexation Capacity of different extracts

Extract	DOC $_3$ mg C/dm	CC(x 10^{-8} MCu) 1% extract	CC:DOC
Seawater (1h)	19	38	2
0.1N NaOH (1h)	34	25	0.7
0.1N NaOH (11 changes)	775	110	0.14

Concentrations of dissolved organic carbon and copper complexation capacity of different extracts are given in Table 1. Eleven medium changes result in a high DOC concentration, and a low CC:DOC ratio, contrasting with seawater and 1 hour NaOH extracts.

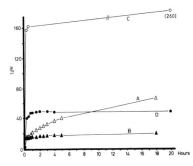

FIGURE 1. Amounts of fluorescence (mFl) of the organic matter extracted from sandy (A and B) and silty (C and D) sediments with seawater (closed symbols) or with 0.1 N NaOH (open symbols).

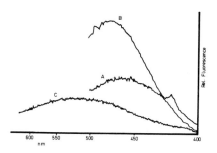

FIGURE 3. Fluorescence spectra (exc. 365 nm) of organic matter extracted from a silty sediment with A) seawater, B) 0.1 N NaOH after 1 hour, C) after 11 changes of 0.1 N NaOH.

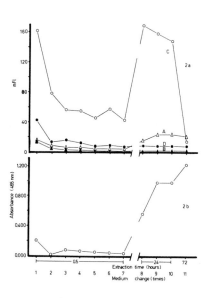

FIGURE 2. Consecutive medium changes and different extraction times. a) Fluorescence of organic matter extracted from sandy (A and B) and silty (C and D) sediments with seawater (closed symbols) or 0.1 N NaOH (open symbols). b) Absorbance (465 nm) of the 0.1 N NaOH extract of silty sediment.

4. DISCUSSION

Based on the results described above it is concluded that:

- 0.1 N NaOH extracts more humic matter from sediments than seawater.
- Fluorescence maximum and copper complexation capacity characteristics differ between seawater and 0.1 N NaOH extracts.
- The larger amount of DOC extracted with NaOH at extended contact periods has relatively low copper complexation capacity.

The fluorescence spectra of the organic matter extracted with seawater agree with those found for seawater (Laane, 1982). However, those obtained after prolonged extraction and consecutive medium changes with 0.1 N NaOH is probably due to production of new humic matter, from carbohydrates and proteins (Kalle, 1963). However, this new organic matter has no relation to the 'easily exchangeable' organic matter in the environment.

CONCLUSION

In order to extract material that may represent the easily exchangeable
organic matter from sediments under natural conditions in the marine environmen
(e.g. for the study of the interaction between heavy metals and humic material)
it is suggested that use is made of seawater rather than of NaOH solutions.

REFERENCES

- Boniforti , R., 1978. Metal analysis in aquatic sediments. A review
  Thalassia Jugosl., 14: 281-301.
- Felbeck, G.T. jr., 1971. In: Soil biochemistry, vol, 2. Eds. A.D. McLaren &
  J. Skujins, Marcel Dekker Inc. New York.
- Hatcher, P.G., D.L. van der Hart and W.L. Earl, 1980 . Use of solid-state [13]C
  NMR in structural studies of humic acids and humin from Holocene sediments.
  Org. Geochem., 2: 87-92.
- Kalle, K., 1963. Über das Verhalten und die Herkunft der in den gewässern un
  in die Atmosphäre vorhandenen himmelblauen Fluoreszenz. Dt. Hydrogr. Z., 16:
  154-166.
- Laane, R.W.P.M., 1982. Chemical characteristics of organic matter in the
  waterphase of the Ems-Dollart estuary. Thesis. R.U. Groningen. 134 p.
- Laane, R.W.P.M. and L. Koole., 1982. The relation between fluorescence and
  dissolved organic carbon in the Ems-Dollard estuary and the western Wadden
  Sea. Neth. J. Sea Research, 15: 217-227.
- Longacre Phelps., 1979. Cadmium sorption in estuarine mud-type sediment and
  the accumulation of cadmium in the soft-shell clam, mya arenaria. Estuaries
  2: 40-44.
- Poutanen, E,-L. and R.J. Morris., 1983. The occurrence of high molecular
  weight humic compounds in the organic-rich sediments of the Peru continenta
  shelf. Ocean. Acta : 6, 21-28.
- Rashid, M.A. and L.H. King., 1969. Molecular weight distribution measurement
  on humic and fulvic acid fractions from marine clays on the Scotian shelf.
  Geochim. et Cosmochim. Acta., 33: 147-151.
- Reuter, J.H. and E.M. Perdue., 1977. Importance of heavy metal-organic matte
  interactions in natural waters. Geochim. et Cosmochim. Acta: 41, 325-334.
- Saar, R.A. and J.H. Weber., 1982. Fulvic acid: modifier of metal-ion
  chemistry. Env. Sci. Technol., 16:510A-517A.
- Saito, Y. and S. Hayano., 1979. Application of high-performance aqueous gel
  permeation chromatography to humic substances from marine sediments.
  J. of Chrom., 177: 390-392.
- Schnitzer, M. and S.U. Khan., 1972. In: Humic substances in the environment
  Ed: A. Douglas McLaren, Marcel Dekker. Inc. New York, pp. 1-7.
- Srna, R.F., K.S. Garrett, S.M. Miller. and A.B. Thum., 1980. Copper complexa
  tion capacity of marine water samples from southern California. Env. Sc.
  & Technol., 14: 1482-1486.
- Tada, F. and S. Suzuki., 1982. Adsorption and desorption of heavy metals in
  bottom mud of urban rivers. Water Res., 16: 1489-1494.

THE INTERRELATIONSHIP OF AGGREGATION AND CATION BINDING OF FULVIC ACID

Donald S. GAMBLE, Cooper H. LANGFORD and Alan W. UNDERDOWN

1. INTRODUCTION

Practical problems associated with metal ions in soil solutions and surface waters have led to the investigation of two related phenomena. They are the effects of dissolved humic materials on metal ion speciation and on the metal ion complexing capacity of the water. Several years of work by Wershaw and coworkers have demonstrated that dissolved humic materials can aggregate (Wershaw et al., 1967, 1970, 1973, 1973). In addition research by Schnitzer (Schnitzer et al,, 1965; Stevenson, 1976) and earlier workers (Broadbent et al., 1952; Coleman et al., 1956; Beckwith, 1959) proved that humic materials make the most important contribution to complexing capacity. A number of authors have suggested that aggregation and cation binding are related (Stevenson, 1976; Senesi et al., 1977; Ghosh et al., 1981). A preliminary report has outlined the application of Rayleigh light scattering to the investigation of this relationship (Underdown et al., 1981). The objective of this review is to determine the extent to which stoichiometricly exact chemistry can be deduced from the available experimental evidence. The Armadale Bh horizon fulvic acid is the only humic sample available to date with which this may be done. It is unique because of the functional group research of Schnitzer and coworkers (Schnitzer et al., 1962, 1965), and because of the subsequent measurements of the numbers of acidic functional groups and their spectrum of weak acid K_A values (Gamble, 1970, 1972).

2. KNOWN STOICHIOMETRIES

Figure 1 postulates a generic type of structure that illustrates those features related to cation binding and aggregation. The salicylic acid type of bidentate chelation sites and the extensive hydrogen bonding should especially be noted. Table 1 and Figure 2 show the exact numbers of these functional groups and the distribution of differential equilibrium constants for the carboxyl groups (Gamble, 1972).

Kramer, C.J.M. and Duinker, J.C. (eds.), Complexation of Trace Metals in Natural Waters. ISBN 90-247-2973-4
© *1984 Martinus Nijhoff/Dr W. Junk Publishers, The Hague/Boston/Lancaster.*
Printed in the Netherlands.

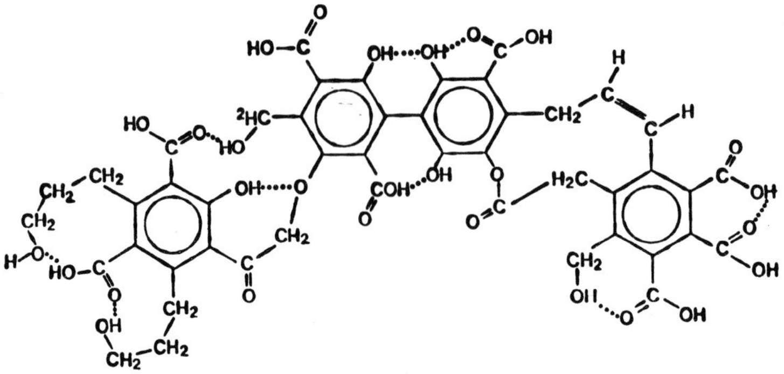

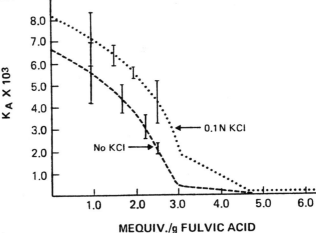

FIGURE 1. A potulated generic type of structure in Armadale fulvic acid.

TABLE 1. Armadale Bh Fulvic Acid: Stoichiometric Properties (meq/g)

Type A Carboxyl Groups	Type B Acidic Functional Groups	Salicylic Type Chelation Sites (maximum possible)
5.0	2.7	3.

The stoichiometric information implies that the effects of Cu(II) complexing on aggregation should be determined by the types of complexing sites in Figure 3. Because the salicylic acid types can only be intramolecular, they cannot contribute to aggregation. On the other hand, the dicarboxylic sites might perhaps be either intermolecular or intramolecular. The relative amounts of the two will determine exactly the effect of Cu(II) complexing on aggregation.

FIGURE 2. Type A carboxyl groups, Armadale fulvic acid: differential equilibrium constant spectrum.

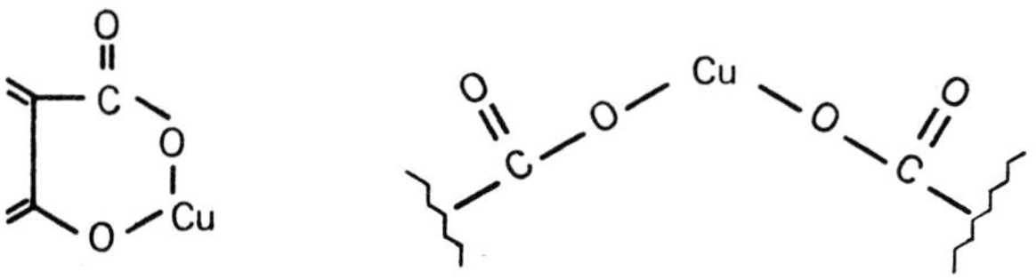

Salicylic
Type

Dicarboxylic Type

FIGURE 3. Categories of chelation sites in Armadale fulvic acid.

3. AGGREGATION EXPERIMENTS, MONITORED BY RAYLEIGH LIGHT SCATTERING

Aggregation was monitored by means of two light scattering functions
(Underdown et al., 1981). These are the Rayleigh ratio at 90°, R_{90}, and
the dyssymmetry ratio $D_{45/135}$, defined by the equations below (Huaglin,
1972).

EQUATIONS

RAYLEIGH RATIO

$$R_\theta = \frac{2\pi^2 \, \tilde{n}_0^2 \, [\frac{d\tilde{n}}{dc}]^2}{N_A \, \lambda^4} \, P(\theta) \, M_w C$$

DYSSYMETRY RATIO

$$D_{45/135} = \frac{\langle R_{45} \rangle}{\langle R_{135} \rangle}$$

R_{90} detects changes in weight average particle weights, while $D_{45/135}$
provides information about the polydispersity of the sample. Three
categories of experiments were monitored with R_{90} and $D_{45/135}$.

3.1. Protonation experiments

Figure 4 shows the aggregation effects of acid and inorganic salt.
An increase in $(1-\alpha_A)$, the mole fraction of the carboxyl groups that are
protonated, favours aggregation. The effect is small in the absence of
inorganic salt. In 0.1 M KCl, the unfiltered fulvic acid aggregates strongly
above $(1-\alpha_A) = 0.6$. The microfilter fraction reflects the behaviour of the
whole sample. The carboxyl groups evidently play a key role, even though a
sharply defined stoichiometry does not emerge from the experiment.

3.2. Cu(II) chelation experiments

The number of the salicylic type sites in Figure 3 available for chelating Cu(II) can be exactly predicted at any pH, by using the functional group analyses of Schnitzer (1962, 1965), and the published acid dissociation data (Gamble, 1970, 1972). The predictions for pH 3.6 and 6.0 have been checked experimentally with the compleximetric titration curves in Figure 5. Very close agreement is seen between the predicted and experimental values listed in Table 2. Evidently a sharp stoichiometric distinction can be observed, between the intramolecular and intermolecular chelation sites of Figure 3.

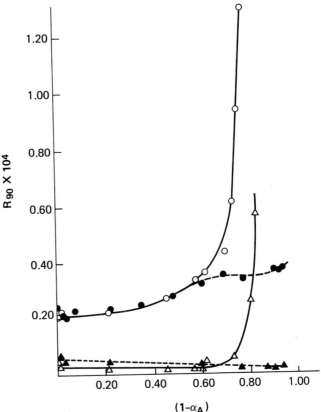

FIGURE 4. The effects of acid and inorganic salt on aggregation. Unfiltered sample: ●, no KCl; O, 0.1 M KCl. 0.2 μm filter: ▲ , no KCl, △ , 0.1 M KCl.

The dyssymmetry ratio experiment in Figure 6 leads to the qualitative observation that the range of particle sizes decreases, as the fulvic acid is aggregated by Cu(II) chelation. This implies that smaller polymer

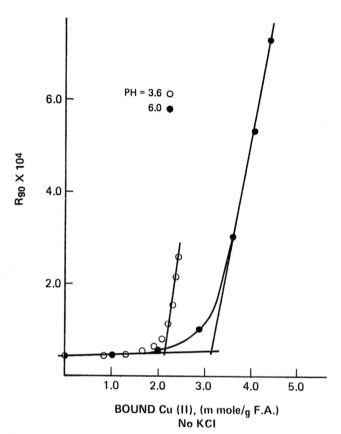

FIGURE 5. Compleximetric Cu(II) titration of Armadale fulvic acid.

TABLE 2. Compleximetric End Points: Cu(II) titration of Armadale fulvic acid, monitored by R_{90} measurements. See Fig. 4.

pH = 3.6; (m moles/g)		pH = 6.0; (m moles/g)	
Prediction	Experiment	Prediction	Experiment
2.24	2.16	3.0	3.1
±10.%	±5.%	±10.%	±5.0%

molecules are selectively aggregated. The content of oxygen bearing functional groups generally increases with the decrease of the molecular weight of humic materials.

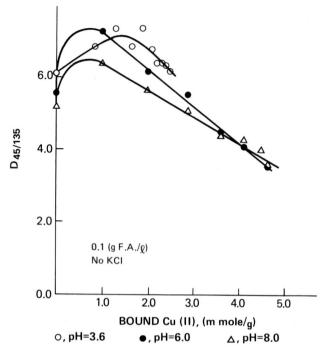

FIGURE 6. The decrease of polydispersity during Cu(II) chelation

3.3. Microfiltration experiments

Filtration experiments were used as an independent check on the evidence that the fulvic acid is polydisperse. The photometric absorbance of fulvic acid at 465 nm is an approximately linear function of concentration (Underdown et al., 1981). The A_{465} and R_{90} curves for the filtration experiment in Figure 7 monitor different properties of the same filtre fraction. The small change in A_{465} indicates a small decrease in mass of the sample. At the same time, there is a large loss of scattering particles. The experiment implies that large particles account for only a small portion of the total mass.

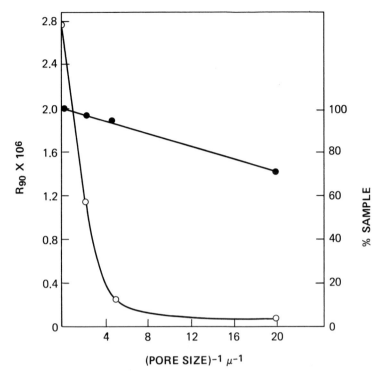

FIGURE 7. Filtration experiment. ●, small loss of mass monitored by A_{465}. O, large loss of big particles, monitored by R_{90}.

4. CONCLUSIONS

The acidic functional groups are the key to the interrelationship between cation binding and the aggregation of dissolved fulvic acid. These functional groups are unevenly distributed among the molecular weight fractions of the polydisperse mixture. From their known stoichiometries for the Armadale Bh horizon fulvic acid, aggregation measurements can be directly correlated with cation binding. Three important observations are obtained from the experiments.

–In 0.1 M KCl, aggregation increases steeply when the carboxyl groups are more than 60 mole % protonated.

–Compleximetric Cu(II) titrations monitored by Rayleigh light scattering can distinguish between intramolecular chelation sites, and aggregating intermolecular "pseudo chelation" sites.

–Large particles that cause most of the Rayleigh light scattering account for only a small portion of the mass of the sample.

REFERENCES

-Beckwith, R.S. 1959. Titration curves of soil organic matter. Nature,
v. 184, p 745-746.

-Broadbent, F.E. and Bradford, G.K. 1952. Cation-exchange groupings in the
soil organic fraction. Soil Sci., v. 74, p 447-457.

-Coleman, N.T., McClung, A.C. and Moore, D.P. 1956. Formation constants for
Cu (II)-peat complexes. Science, v. 123, p 330-331.

-Gamble, D.S. 1970. Titration curves of fulvic acid: the analytical chemist
of a weak acid polyelectrolyte. Can. J. Chem., v. 48, 2662-2669.

-Gamble, D.S. 1972. Potentiometric titration of fulvic acid: equivalence
point calculations and acidic functional groups. Can. J. Chem., v. 50, p
2680-2690.

-Ghosh, K. and Schnitzer, M. 1981. Fluorescence exicitation spectra and vis
behaviour of a fulvic acid and its copper and iron complexes. Soil Sci. So
Amer. Proc., v. 21, p 368-373.

-Huaglin, M.B., 1972. Light scattering from polymer solution. Academic
Press: New York.

-Schnitzer, M. and Desjardin, J.S. 1962. Molecular and equivalent weights
of the organic matter of a podzol. Soil Sci. Soc. Amer. Proc., v. 26,
p 362-365.

-Schnitzer, M. and Skinner, S.I.M. 1965. Organo-metallic interactions in
soils: 4, carboxyl and hydroxyl groups in organic matter and metal retenti
Soil Sci., v. 99, p 278-284.

-Senesi, N., Chen, Y. and Schnitzer, M. 1977. Aggregation dispersion phenom
in humic substances: in Soil organic matter studies, International Atomic
Energy Agency, Vienna, v. 2, p 143-155.

-Stevenson, F.J. 1976. Binding of metal ions by humic acids: in Nriagu, J.O
ed., Environmental biogeochemistry, v. 2, Ann Arbor Science Pub. Inc., Ann
Arbor, Mich., p 519-540.

-Underdown, A.W., Langford, C.H. and Gamble, D.S. 1981. Light scattering of
a polydisperse fulvic acid. Anal. Chem., v. 53 p 2139-2140.

-Wershaw, R.L., Burcar, P.J., Sutula, C.L. and Wiginton, B.J. 1967. Sodium
humate solution studied with small-angle X-ray scattering. Science, v.
157, p 1429-1431.

-Wershaw, R.L., Heller, S.J. and Pinckney, D.J. 1970. Measurement of the
molecular size of a sodium humate fraction. J. Advan. X-ray anal., v. 13,
p 609-617.

-Wershaw, R.L. and Pinckney, D.J. 1973. The fractionation of humic acids
from natural water systems. J. Res. U.S. Geol. Surv., v. 1, p 361-366.

-Wershaw, R.L.and Pinckney, D.J. 1973. Determination of the association and
dissociation of humic acid fractions by small-angle X-ray scattering. J. R
U.S. Geol. Surv. v. 1, p 701-707.

VOLTAMMETRIC STUDIES ON THE SPECIATION OF Cd AND Zn BY AMINO-
ACIDS IN SEA WATER

P.VALENTA, M.L.S. SIMOES GONÇALVES , M. SUGAWARA

1. INTRODUCTION

The speciation of dissolved heavy metals is of great significance for their interactions with suspended matter and sediments and their uptake by aquatic organisms. Although the concentrations of the ecotoxic metals in the dissolved state are
rather low, typically below 1/ug/l, the dissolved state remains
very important. In and from this state occurs to a major extent
the transfer of the toxic metals to and from the interfaces of
suspended particles, sediments and marine organisms usually taking
up these metals but sometimes releasing them also, due to biological regulation effects (Nürnberg 1983 a, b; Valenta, 1983;
Nürnberg and Valenta, 1983).

Among the potentially complexing components of dissolved organic matter (DOM) in sea water, which could influence the uptake
of both metals by marine organisms, are various amino acids. The
speciation of two trace metals, Zn(II) and Cd(II) with two amino
acids, glycine and L-aspartic acid, has been studied by differential pulse polarography (DPP), linear sweep anodic stripping voltammetry (LSASV) and partially also by potentiometry in artificial
and in genuine sea water. The stoichiometric stability constants
$ß_j$ of Zn(II) and Cd(II) complexes with these amino acids have been
determined and the significance of the amino acids among the complexing components of DOM for the speciation of Cd(II) and Zn(II)
in sea water has been tested.

Kramer, C.J.M. and Duinker, J.C. (eds.), Complexation of Trace Metals in Natural Waters. ISBN 90-247-2973-4
© *1984 Martinus Nijhoff/Dr W. Junk Publishers, The Hague/Boston/Lancaster.*
Printed in the Netherlands.

2. PROCEDURE

2.1. Materials and methods

2.1.1. Chemicals and solutions. Almost all chemicals were Merck "Suprapur" grade. It was proved that the content of heavy metals in the chemicals was undetectable at the current sensitivity used. Standard solutions of Zn and Cd were prepared from Merck "Titrisol". The stock solution of $NaClO_4$ was prepared from Merck "Suprapur" NaOH and $HClO_4$. The water supplied by a Milli-Q system (Millipore, Bedford, Mass., USA) contained very small bla: of heavy metals, i.e. o.1 ng Cd l^{-1}, 1 ng Pb l^{-1}, 5 ng Cu l^{-1} an 5o ng Zn l^{-1}. Pure nitrogen (99.999 %) was used to deaerate the studied solutions of o.7 M $NaClO_4$ or the artificial sea water. I the case of real sea water the nitrogen passed first through a washing bottle filled with a suspension of magnesium hydroxide carbonate in a borate buffer (5×10^{-3}M H_3BO_3 + 5×10^{-4}M NaOH) pH 8 to maintain the natural carbonate and bicarbonate concentra tion in sea water.

Artificial sea water was prepared according to Lyman and Fleming (Sverdrup et al. 1963). The sea water samples were filtr ted through a o.45/um pore size membrane filter in a SM 1611 Sar torius device and kept in polyethylene bottles acidified to pH 2 with HCl. Before being analysed they were submitted to UV-radia- tion for 24 h in order to destroy any high-molecular organic matter (Mart et al. 1980).

2.1.2. Instruments. The polarographic measurements were per- formed under potentiostatic control with a PAR polarographic ana lyzer model 174 A, in the differential pulse mode. The pulse hei was 5o mV, the scan rate 2 mV s^{-1} and the drop time 2 s. The mea surements were carried out in a thermostated cell at 2o $\pm$ o.2°C, equipped with a three-electrode system. The working electrode wa a dropping mercury electrode (DME), a platinum wire served as counter electrode and the Ag/AgCl electrode described was the re ference electrode.

2.2. Procedure

Potentiometric measurements were performed in 25 ml of a solution with about 10^{-2}M glycine hydrochloride and 10^{-2}M of th respective heavy metal adjusted to the ionic strength of sea water with o.7 M $NaClO_4$. The titration was performed with 2 M NaOH using the glass electrode.

DPP measurements were performed in 25 ml solution of the heavy metal at the $1o^{-8}$ to $1o^{-6}$M level and of variable concentrations of the ligand adjusted to a given pH with 2 M NaOH. The pH was measured after each addition of the ligand and kept constant adding aliquots of 2 M NaOH.

3. RESULTS AND DISCUSSION

3.1. Evaluation of the stability constants from potentiometric measurements

The potentiometric measurements of the stability constants of Cd(II) and Zn(II) with glycine were performed only in o.7 M $NaClO_4$. The measured potential values were transformed into pH-values according to the Gran method, titrating the titrant NaOH with $HClO_4$ (Gran 1952). The acid dissociation equilibrium constant of glycine has been determined by titration in the same medium when the respective heavy metal was absent. The free ligand concentration [L] was then evaluated from the neutralisation degree and the equations for mass balance in the usual manner. Alternatively, the complex formation curve method was used constructing the estimated theoretical curve for the complex formation function. The evaluation of the constants β_j and ligand numbers j were described in detail previously (Simoes-Goncalves and Valenta, 1982). The stoichiometric β_j-values of Cd(II)-glycine and Zn(II)-glycine complexes are given in Tab.1 and Tab.2.

3.2. Determination of the stability constants from polarographic measurements

Due to the inherent sensitivity of polarography and voltammetry (Nürnberg 1982) the determination could be performed in sea water medium applying DPP at low metal level of $1o^{-8}$ to $1o^{-6}$M corresponding to a moderate to high heavy metal pollution. Thus, the possible competition of other ligands present in sea water had to be taken into account.

Table 1. Stoichiometric stability constans of Zn(II)-glycine and Cd(II)-L-aspartic acid complexes in sea water at 25°C

Medium	Glycine					L-Aspartic Acid				
	Method	pH	Zn-conc. (M)	$\log \beta_{ML}$	$\lg \beta_{ML_2}$	Method	pH	Zn-conc. (M)	$\log \beta_{ML}$	$\log \beta_{ML_2}$
Artificial sea water	DPP	7.7	1.6×10^{-6}	4.1	8.3	DPP	6.7	4×10^{-7}	5.5	1o.9
North Sea water	–	–	–	–	–	DPP	6.7	4×10^{-7}	5.4	1o.7
Pacific sea water	DPP	7.5	1.8×10^{-6}	4.3	7.8	–	–	–	–	–
o.7M NaClO$_4$	DPP	7.8	1.6×10^{-6}	4.1	8.5	–	–	–	–	–
o.7M NaClO$_4$	Potentio-metry	6-7	9.7×10^{-3}	4.38	8.76	–	–	–	–	–

Table 2. Stoichiometric stability constans of Cd(II)-glycine and Cd(II)-L-aspartic acid complexes in sea water at 25°C

Medium	Glycine					L-Aspartic Acid				
	Method	pH	Cd-conc. (M)	$\log \beta_{ML}$	$\lg \beta_{ML_2}$	Method	pH	Cd-conc. (M)	$\log \beta_{ML}$	$\log \beta_{ML_2}$
Artificial sea water	DPP	7.8	7×10^{-7}	4.2	7.3	LSASV	7.5	8.8×10^{-8}	4.7	9.0
Pacific off-shore water	DPP	7.5	7×10^{-7}	4.2	7.3	-	-	-	-	-
Coastal sea water (Ishi-kari,Japan)	-	-	-	-	-	LSASV	7.7	8.8×10^{-8}	4.2	8.7
0.7M NaClO$_4$	DPP	7.9	7×10^{-7}	4.0	7.3	LSASV	8.1	8.8×10^{-8}	4.3	8.1
0.7M NaClO$_4$	Potentio-metry	6.4 -7.8	9.7×10^{-2}	3.88	7.35	-	-	-	-	-

3.2.1. Zn(II) complexes. It was proved that Zn(II) forms with glycine and with L-aspartic acid rather stable complexes. Under the given experimental conditions effects by the dissociation kinetics on the measured voltammetric response remain negligible. Therefore, the equilibrium value of the concentration of Zn(II) complexed by the respective adjusted ligand concentration [L] were determined in a manner analogous to the determination of the stability constant of the Cd(II)-NTA complex (Raspor et al. 1978). However, the calculation procedure was modified, because two complexes of the type ZnL and ZnL_2 are formed. The determination has been performed in the following manner. After addition of certain amounts of the ligand L to the investigated medium that fraction of the overall concentration of Zn(II) which remains complexed by the amino acid was determined by differential pulse polarography at the level of $10^{-7}M$ for L-aspartic acid and $10^{-6}M$ for glycine, respectively (Simoes-Goncalves and Valenta, 1982).

The stoichiometric stability constants β_1 and β_2 were calculated with eq.(1) expressing the dependence of the apparent stability constant K_{ML} on the free ligand concentration [L]:

$$K_{ML} = \frac{[ZnL^+] + [ZnL_2]}{[Zn^{2+}][L^-]} = \beta_1 + \beta_2[L^-] \qquad (1)$$

where ZnL^+ and ZnL_2 are Zn(II) complexes with 1 or 2 ligands, respectively, and Zn^{2+} the uncomplexed Zn. The stability constants β_1 and β_2 can be determined from the intercept and the slope of the best straight line fitting the experimental points according to eq.(1) and the obtained values are listed in Table 1.

3.2.2. Cd(II) complexes. The methodological approach in the polarographic measurement and the evaluation of Cd(II)-glycine and Cd(II)-L-aspartic acid complexes is entirely different from that applied for the Zn-complexes. This is connected with the fact that the Cd(II) complexes behave in a very mobile way in contrast to the Zn(II) species, although their stability constants proved to be comparable with those of the Zn(II) complexes. Thus, after addition of a certain concentration of the ligand L, a commo peak is obtained and not separate peaks for the amount of complexe and uncomplexed Cd(II), respectively. However, with increasing li-

gand concentration the potential of the common unique peak corresponding to the reversible Cd(II) reduction becomes progressively
more negative (Simoes-Goncalves et al. 1983). This behaviour is
similar to that usually observed with mobile inorganic Cd-complexes
CdX_j, where the ligand X corresponds to Cl^-, CO_3^{2-}, HCO_3^- etc.
(Nürnberg, 1983 b, 1984). The electrode process of the Cd-glycine
complexes is still sufficiently reversible, as shown by a peak
width 6o mV at half-peak height. From the dependence of the peak
potential shift ΔE_p on the ligand concentration [L] the stability
constants are evaluated.

For the determination of the stability constants of Cd-L-
aspartic acid complexes the LSASV has been used. Thus, the rather
realistic Cd(II)-concentration of 9×10^{-8}M could be used which
corresponds to a small Cd-pollution (Sugawara et al., 1984; Valenta
and Sugawara, 1981). The method is based on the registration of a
family of LSASV-curves obtained at different cathodic plating potentials adjusted in the potential range of the hypothetical dc-
polarogram of the Cd(II) complex reduction. From the heights of the
LSASV-peaks corresponding to these plating potentials the pseudo-
polarogram (Bubic and Branica, 1973) was constructed and its half
wave potential $E_{1/2}^*$ was measured. The measurements were performed
over an extended range of the ligand concentration [L] and the
evaluation dependence of $E_{1/2}^*$ on the ligand concentration [L] gave
the stability constants of the Cd(II)-L-aspartic acid complexes.

As in both, Cd(II)-glycine and Cd(II)-L-aspartic acid complexes, reversible reduction of Cd(II) takes place, in principle
the DeFord-Hume method (DeFord and Hume, 1951) can be applied
for the evaluation of the stability constants β_1 and ligand
numbers j. For sea water conditions, however, the DeFord-Hume
method cannot be applied directly, because the inorganic species
of Cd formed with salinity components have also to be taken into
account. In practice only the chlorocomplexes are of significance,
as they are responsible for 97 % of the Cd-speciation in sea water
in the absence of organic ligands of significance (Simoes-Goncalves et al., 1981; Nürnberg, 1983 b). The stability constants of
the Cd-chlorocomplexes have to be further corrected for ion pairing
between the cationic and anionic macroconstituents of sea water

(Sipos et al., 198o). Thus, for the to be applied DeFord-Hume treatment the Leden function $F_o(L)$ is given by eq. (2)

$$(2) \quad F_o(L) = \exp\left[(2F/RT)\Delta E\right] = 1 + \beta_1[L] + \beta_2[L]^2 + \beta_1^c[Cl^-] + \beta_2^c[Cl^-]^2$$

where ΔE is the peak potential shift ΔE_p or the pseudo-half wave potential shift $\Delta E_{1/2}^*$, β_1 and β_2 are the stoichiometric stabilit constants of the Cd(II) amino acid complexes, β_1^c and β_2^c the for i pairing corrected stability constants of Cd-chlorocomplexes $CdCl^+$ and $CdCl_2$.

The modified $F_1(L)$ relation $F_1^*(L)$ is defined by eq. (3)

$$(3) \quad F_1^*(L) = \left\{F_o(L) - 1 - \beta_1^c[Cl^-] - \beta_2^c[Cl^-]^2\right\}/[L] = \beta_1 + \beta_2[L]$$

Using a similar treatment as previously, β_1 and β_2 were eva-luated from the best fit of the experimental points of the depen-dence of $F_1^*(L)$ on the free ligand concentration [L] and the obtai values are listed in Table 1.2.

3.3. Conclusions

The results (viz. Table 3) on the complexation of Cd(II) and Zn(II) by the two amino acids, glycine and L-aspartic acid, have shown that in sea water substantially higher [L] values than actu ly present are required for a noticeable complexation. This was t be expected with respect to the rather moderate stability constan of those complexes. Therefore the required amino acid concentrati exceed by orders of magnitude the total amino acid content to be expected for DOM. Therefore amino acids can be usually ruled out as contributors to the speciation of Zn and Cd in the sea. As the other amino acids have comparable stability constants the same conclusion is drawn for them. An exception could be sulphur con-taining amino acids, particularly cysteine.

A special situation might occur in estuarine or coastal waters with particularly high planktonic acitivity. Also, in interstitia water and pore waters of sediments containing substantial DOM levels up to 15o mgC l^{-1}, amino acids could gain importance for the Zn and Cd speciation. The same applies to the surfaces of suspended matter if they are coated with adsorbed amino acids.

Table 3. Required amino acid concentration for a given complexation degree of heavy metal traces in sea water

Complexation degree	Glycine Cd (M)	Zn	L-Aspartic Acid Cd (M)	Zn
5 %	3.7×10^{-3}	5×10^{-4}	9×10^{-4}	2.5×10^{-5}
1o %	7.5×10^{-3}	9×10^{-4}	1.5×10^{-3}	4.5×10^{-5}
2o %	1.4×10^{-2}	1.6×10^{-3}	2.5×10^{-3}	8.5×10^{-5}

Remark: The values refer to sea water having pH 8 and the total chloride concentration of o.55 M.

REFERENCES

- Bubic S, Branica M. 1973. Voltammetric characterization of ionic state of cadmium present in sea water. Thalassia Jugosl., 9: 47-53
- DeFord DD and Hume DN. 1951. The determination of consecutive formation constants of complex ions from polarographic data. J.Am.Chem.Soc. 73: 5321-5323
- Gran R. 1952. Determination of the equivalence point in potentiometric titrations. Part II. Analyst 77: 661-671
- Mart L, Nürnberg HW, Valenta P. 198o. Prevention of contamination and other accuracy risks in voltammetric trace metal analysis with a multicell-system designed for clean bench working. Fresenius Z.Anal.Chem. 3oo: 35o-362
- Nürnberg HW, Valenta P, Mart L, Raspor B and Sipos L. 1976. Applications of polarography and voltammetry to marine and aquatic chemistry. II. The polarographic approach to the determination and speciation of toxic trace metals in the marine environment. Fresenius Z.Anal.Chem. 282: 357-367
- Nürnberg HW. 1982. Voltammetric trace analysis in ecological chemistry of toxic metals. Pure Appl. Chem. 54: 853-878
- Nürnberg HW. 1983 a. Voltammetric studies on trace metal speciation in natural waters. Part II. Application and conclusions for chemical oceanography and chemical limnology. In: GG Leppard (Ed.) Trace element speciation in surface waters and its ecological implications. Plenum Publ. Corp. New York - London, pp. 219-23o
- Nürnberg HW. 1983 b. Investigations on heavy metal speciation in natural waters by voltammetric procedures. Fresenius Z. Anal. Chem. 316: 557-565
- Nürnberg HW, Valenta P. 1983. Potentialities and applications of voltammetry in chemical speciation of trace metals in the sea. In: CS Wong, E Boyle, KW Bruland, D Burton and ED Goldberg (Eds.) Trace metals in sea water. Plenum Press, New York-London, pp. 671-697

366

- Nürnberg HW. 1984. Potentialities of voltammetry for the study of physicochemical aspects of heavy metal complexation in natural waters. this book, pp. 95-115
- Raspor B, Valenta P, Nürnberg, HW and Branica M. 1978. The chelation of cadmium with NTA in sea water as a model for the typical behaviour of trace metal chelates in natural waters. Sci. Tot. Environm. 9: 87-1o9
- Simoes-Goncalves MLS, Vaz, MCTA and Fraústo da Silva JJR. 1981. Stability constants of chloro-complexes of Cd(II) in sea water medium. Talanta 28: 237-24o
- Simoes-Goncalves MLS and Valenta P. 1982. Voltammetric and potentiometric investigations on the complexation of Zn(II) by glycine in sea water. J. Electroanal.Chem. 132: 357-375
- Simoes-Goncalves MLS, Valenta P and Nürnberg HW. 1983. Voltammetric and potentiometric investigations on the complexation of Cd(II) by glycine in sea water. J. Electroanal. Chem. 149: 249-262
- Sipos L, Raspor B, Nürnberg HW and Pytkowicz RM. 198o. Interaction of metal complexes with coulombic ion-pairs in aqueous media of high salinity. Mar.Chem. 9: 37-47
- Sugawara M, Valenta P, Nürnberg HW and Kambara TJ. Voltammetric study on the speciation of Cd(II) with L-aspartic acid in sea water. J. Electroanal. Chem., in press
- Svedrup HU, Johnson MW and Flemming RH. 1963. The oceans. Prentice-Hall. Englewood Cliffs, p. 186
- Valenta P and Sugawara M. 1981. Voltammetric studies on the speciation of trace metals by amino acids in sea water. Rapp. Comm. int. Mer Médit. 27: 165-167
- Valenta P. 1983. Voltammetric studies on trace metal speciation in natural waters. Part I. Methods. In: GG Leppard (Ed.) Trace element speciation in surface waters and its ecological implications. Plenum Publ. Corp. New York - London, pp. 49-69

KINETICS OF THE DISSOCIATION OF CADMIUM–GLUTAMIC ACID COMPLEX

M.L.S. GONÇALVES and   M.M. CORREIA dos SANTOS

1. INTRODUCTION

The complexation of cadmium with aminoacids as glycine, alanine,
L valine, serine and glutamic acid has been studied in seawater conditions
(ionic strength 0.70 M $NaClO_4$, t = 20 + 0.1^{o}C). The stability constants
have been determined by potentiometry and differential pulse polarography
as complementary methods in order to check the accuracy. For all systems,
with the exception of cadmium–glutamic acid complexes, a kinetic contribution was not noticeable in terms of differential pulse polarography,
because the cadmium complexes were sufficiently labile and their equilibria
sufficiently mobile with respect to the time scale of this technique.

The rate constants of the cadmium glutamic acid complex ML has been
determined by cyclic voltammetry in the same experimental conditions.

2. RESULTS AND DISCUSSION

For all the ligands stability constants of ML and ML_2 complexes have
always been determined and the calculated values determined by polarographic
and potentiometric methods agree within the experimental errors[1]. Since
the total cadmium concentration is about 10^{-7}M in the first method and
about 10^{-3}M in the second, the results can therefore be extrapolated to
the seawater levels of heavy metals, as the same type of species should
exist in solution.

The differential polarograms for all the cadmium–aminoacid solutions,
with the exception of the cadmium–glutamic acid complexes, have constant
values for peak currents and peak potentials more negative with increasing
ligand concentration, so the species are labile in polarographic terms.
However, with the cadmium–glutamic acid complexes the differential pulse
and d.c. polarograms show a systematic decrease of i_p (or $\bar{i}_\ell$) and a more
negative E_p (or $E_{1/2}$) with increasing ligand concentration which is typical
of a slow step in homogeneous solution. As it was not possible to determine

Kramer, C.J.M. and Duinker, J.C. (eds.), Complexation of Trace Metals in Natural Waters. ISBN 90-247-2973-4
©1984 Martinus Nijhoff/Dr W. Junk Publishers, The Hague/Boston/Lancaster.
Printed in the Netherlands.

the rate constant of the slow step using the d.c. polarographic results because an adsorption problem was superimposed, we tried the cyclic voltammetric method with hanging mercury electrode[1]. With this technique higher scan rates can be used and so it is possible to separate, at least in principle, the irreversible reduction peaks of the complex and of the adsorbed complex. We have used scan rates between 50 mV/s and 10 000 mV/s, cadmium concentration of 2×10^{-4}M and glutamic acid between 2×10^{-2}M and 10^{-1}M.

The voltagrams showed the reduction peak of the cadmium and of the complex and the oxidation of Cd^0 as can be seen in Fig. 1, peaks I, II and III. For the highest value of scan rate the peak II is being divided in two due to the separation of the peak reduction of the adsorbed complex.

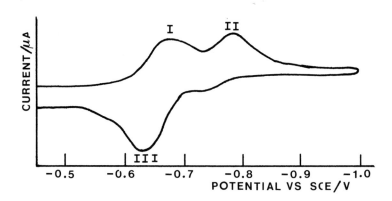

FIGURE 1. Cyclic voltagram of cadmium glutamic acid complexes, $C_M = 2.0 \times 10^{-4}$M and C_L 6.0×10^{-2}M, I = 0.70 M, pH = 7.83; wave I, Cd^{++} reduction, wave II, complex reduction, wave III oxidation of Cd^0.

The normalized peak current of wave I decreases with the scan rate and that of wave II increases with it and tends to a constant value, as can be seen in Fig. 2, which is characteristic of the kinetic behaviour. On the other hand the peak potential of wave I shifts slightly in an anodic direction with increasing scan rate as also predicted by the theory. The peak potential of wave II becomes more negative with increasing scan rate as expected for an irreversible system.

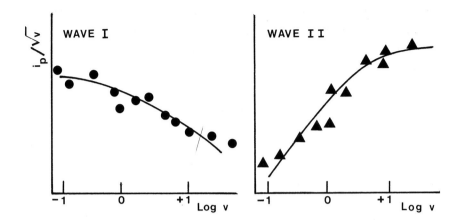

FIGURE 2. Variation of the experimental current functions with scan rate for waves I and II, $C_M = 2.0 \times 10^{-4}$ M, $C_L = 10 \times 10^{-1}$ M, $I = 0.70$ M

As the first peak is reversible and not affected by the reduction of the complex, it has been used for the determination of the rate constant of the slow step according to the semi-empirical expression (2,3) adapted to the existence of two species in solution and assuming the mechanism $CdL_2 \leftrightarrow CdL \underset{k_f}{\overset{k_d}{\rightleftharpoons}} Cd^{++} + 2e \leftrightarrow Cd^{o}$.

$$i_d/i_k = 1.02 + 0.471 \sqrt{\frac{nF}{RT}} \times \frac{\sigma\sqrt{v}}{\sqrt{\ell}} \text{ with } \ell = \rho_d + \rho_f, \; \beta_1 = \frac{k_f}{k_d},$$

$$\rho_d = k_d \frac{\beta_1[L]}{\beta_1[L] + \beta_2[L]^2}, \rho_f = k_d\beta_1[L] \text{ and } \sigma = \beta_1[L] + \beta_2[L]^2$$

$$i_k = (i_p)_I \; (Cd^{++} \text{ reduction})$$

i_d = estimated from the height of cadmium peak without ligand

As a constant value for the rate constant has been determined for several ligand concentrations (Table 1), this means that the assumed hypothesis of the slow step in the dissociation of the ML complex is correct.

TABLE 1. Rate constant of dissociation ($k_d s^{-1}$) of cadmium-glutamic acid complex (ML) at $25 \pm 0.1^\circ C$, $I = 0.70$ M and different ligand concentration.

pH	Cation	Anion	Method	$C_M(M)$	$C_L(M)$	$k_d(sec^{-1})$
7.83	Cd^{++}	Glu	VC	2×10^{-4}	2×10^{-2}	2.1×10^5
"	"	"	"	"	4×10^{-2}	3.2×10^5
"	"	"	"	"	6×10^{-2}	5.2×10^5
"	"	"	"	"	10^{-2}	1.4×10^5

2. CONCLUSIONS

1. The stability constants of Cd + aminoacids are of the same order of magnitude because the chelating groups are the same ($\log \beta_1 \simeq$ $\simeq 3.8$ and $\log \beta_2 \simeq 7.1$)[1].

2. The dominant species in seawater conditions are Cd chlorocomplexes[4]. The interaction with aminoacids is more important in less saline freshwaters (Ca^{++} and Mg^{++} complexes can be neglected), or in estuaries and interstitial waters where the levels of heavy metals and organics are higher[5].

3. In speciation studies the kinetic aspect must also be considered because in open sea the equilibrium is only an approximation.

4. If a complex is less toxic than free metal for an organism the rate of dissociation of the complex has to be considered due to the disturbance near the biological membrane.

ACKNOWKEDGEMENT:This work was partially supported by Junta Nacional de Investigação Cientifica e Tecnológica under research contract no.315.81.57.

REFERENCES

- Simões Gonçalves, M.L.S., Correia dos Santos, M.M., Cadmium Complexes of Aminoacids in Seawater Conditions. Accepted in J.Electroanal. Chem. Nicholson, R.S. and Shain, I., 1964. Theory of Stationary Electrode Polarography. Anal. Chem. 36: 706-723.
- Shuman, M.S. and Shain, I., 1969. Studies of the Chemical Reaction Preceding Reduction of Cadmium Nitrilotriacetic Acid Complexes Using Stationary Electrode Polarography. Anal. Chem. 41: 1818-1825.
- Simões Gonçalves, M.L.S., Abreu Vaz, M.C., and Fraústo da Silva, J.J.R., 1981. Stability Constants of Chlorocomplexes of Cd(II) in Seawater Medium. Talanta 28:237.
- Simões Gonçalves, M.L.S., Valenta, P., and Nurnberg, H.W..Stability Constants in Seawater conditions by polarography and potentiometry. 2-Cadmium-glycine complexes. Accepted in J. Electroanal. Chem.

BINDING OF HEAVY METALS TO POLYMERIC LIGANDS; AN ELECTROCHEMICAL STUDY

ROB F.M.J. CLEVEN and HERMAN P. VAN LEEUWEN[*]

1. INTRODUCTION

A number of naturally occurring metal complexing agents is polyelectrolytic, i.e. they may be considered as macromolecules with many ligand sites
per molecule. It is known that the interaction between charged polymers and
their counterions is largely controlled by non-specific, electrostatic effects
which result from the field around the polyion. We found it worthwhile to
compare the binding of heavy metal ions to natural humic acids with the
characteristic binding of cations to synthetic polyacids. Such a study would
seem to be especially helpful in answering the question whether the frequently
observed dispersion of effective stability constants for metal/humic acid
complexes is (only) caused by a distribution of chemical affinities of the
different binding sites [see e.g. Buffle (1983) and Shuman et al (1983)].

2. MATERIALS AND METHODS

Polyacrylic acid (PAA) and polymethacrylic acid (PMA) were used as
model polyelectrolytes. They were obtained from Polysciences and BDH respectively. Extensive details on the preparation of the solutions and on
the relevant properties are given elsewhere [Cleven (1984)].

The humic acids (HA) solution was prepared from Fluka humic material
(sample code 53680/191699116). The fraction soluble in the pH range 3-9 was
used. It was dialyzed against water using a Spectrapor membrane with a molecular mass cut-off of 3,500. The total number of ligands was determined by
conductometric acid-base titration.

Pulse polarography and conductometry were used simultaneously. The two
techniques yield information on the distribution of metal ions over the free
and the bound state in solutions with complexing agents. Some typical sets
of pulse polarographic data were described earlier by Van Leeuwen et al
(1981, 1983).

Kramer, C.J.M. and Duinker, J.C. (eds.), Complexation of Trace Metals in Natural Waters. ISBN 90-247-2973-4
© 1984 Martinus Nijhoff/Dr W. Junk Publishers, The Hague/Boston/Lancaster.
Printed in the Netherlands.

3. RESULTS AND DISCUSSION

Fig. 1 shows conductometric data for the titration of PAA with different metal salts. The PAA molecules were partly charged: degree of dissociation $\alpha_d = 0.41$. The data are represented as $\Delta\kappa_T$ which expresses the change of the specific conductivity κ with metal addition in the reference case (no polymer present), minus the change of κ with metal addition for the polyelectrolyte solution:

$$\Delta\kappa_T = \kappa(\Delta Me^{2+}) - [\kappa(\Delta Me^{2+}/Pol) - \kappa(Pol)] \tag{1}$$

In Fig. 1 we see that for all sorts of metal ions, both of the earth alkaline type and of the transition metallic type, there is an increase of $\Delta\kappa_T$ with added metal in the first part of the titration curve.

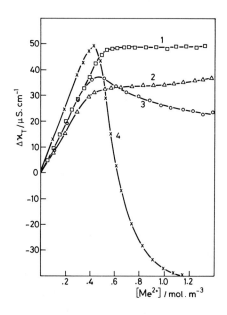

FIGURE 1. Conductometric data for Me^{2+}/PAA; $\overline{M}_{PAA} = 300,000$; $[PAA] = 2.50 \times 10^{-3}$ M; no salt added; $\alpha_d = 0.41$. 1 = Ba^{2+}; 2 = Zn^{2+}; 3 = Cd^{2+}; 4 = Pb^{2+}.

The slopes nearly correspond with the equivalent conductances of the metal ions. This means that here practically all of these ions are immobilized due to interaction with the polyionic species. The fraction f, which denotes the number of bound metal ions per (deprotonated) carboxylate group, can be seen to be between 0.4 and 0.5. Addition of 'inert' electrolyte leads to a decrease of f and this is related to competition from the added cations. Lead(II) ions, and to a very small extent also Cd^{2+} ions, are capable of replacing some

protons from the carboxyle groups. This behavior corroborates with the higher
intrinsic stability constants for simple Pb^{2+}/carboxylate complexes.

The strength of the binding of Zn^{2+} by PAA is indicated in Fig. 2.
On the basis of pulse polarographic data, an operationally defined stability
constant K_1 { = [metal bound]/([metal free] x [carboxylate free]) }
has been evaluated for different polyionic charges. It appears that K_1

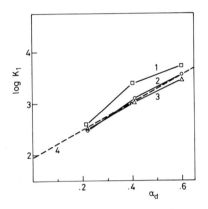

FIGURE 2. Typical plot of the
apparent stability constant K_1
versus α_d for Zn^{2+}/PAA. $\overline{M}_{PAA}$ =
300,000; [PAA] = 2.50 x 10^{-3} M;
[Zn^{2+}]/[PAA] = 0.12 (1), 0.24 (2),
0.32 (3). Dotted line (4) is based
on acid-base titration in the
presence of Zn^{2+} with [Zn^{2+}]/[PAA]
= 0.20.

strongly increases with increasing α_d due to the growing influence of the
electrical field around the polyion. Tentative extrapolation to α_d = 0 seems
to give K_1 values close to the intrinsic stability constant for the monomeric complex.

Results for humic acids solutions show a strong analogy with what was
found for the polyacid model systems. Fig. 3 gives some typical results for
the conductometric titration of HA with Cd^{2+} ions. Binding of Cd^{2+} takes
place till the charge on the HA entities is practically compensated for:
f is close to 0.5 under these conditions. The replacement of protons is
negligibly small and therefore there is no direct relation between the metal
binding capacity and the total number of (protolytic) ligand groups. The
binding of Zn^{2+}, Ca^{2+} and Ba^{2+} is essentially the same as for Cd^{2+}.

To an important extent, the binding of heavy metal ions by (partly
dissociated) HA is governed by cooperative charge effects. The polyelectrolytic nature of HA leads to a substantial electrostatic contribution to the
free energy of binding of cations. For our HA sample we found an increase
of K_1 with increasing charge on the HA entities by more than two decades

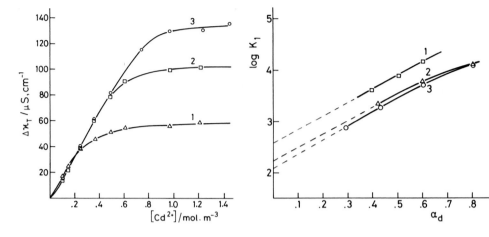

FIGURE 3. Conductometric data for Cd^{2+}/HA; [HA] = 2 x 10^{-3}M; no salt added; α_d = 0.40 (1), 0.60 (2), 0.80 (3).

FIGURE 4. Typical plot of K_1 versus α_d for Me^{2+}/HA. [HA] = 2 x 10^{-3}M; [KNO_3] = 0.05 M; [Me^{2+}] = 1 x 10^{-3}M. 1 = Pb^{2+}; 2 = Zn^{2+}; 3 = Cd^{2+}.

(see Fig. 4). At natural pH values the degree of protolytic dissociation of HA is generally rather high. This means that the effective K_1, for the first metal ions bound, is by a factor of about 100 higher because of the 'poly-electrolyte effect'. With increasing metal-to-ligand ratio, K_1 gradually decreases to the intrinsic value, in a way related to the decrease of the electrical field strength around the HA polyionic species.

REFERENCES

-Buffle, J., 1983. Interpretation of potentiometric and voltammetric data in studies of reactions between trace metals and fulvic compounds. Int. Symp. on Complexation of Trace Metals in Natural Waters, Texel.
-Cleven, R.F.M.J., 1984. Heavy metal/polyacid interaction. Ph.D.thesis, Wageningen.
-Shuman, M.S., B.J. Collins and S.E. Cabaniss, 1983. An affinity spectrum model for metal binding. Int. Symp. on Complexation of Trace Metals in Natural Waters, Texel.
-Van Leeuwen, H.P., W.F. Threels and R.F.M.J. Cleven, 1981. Pulse polarography of heavy metal ions in polyelectrolyte solutions: behavior of Pb^{2+} and Cd^{2+} in PMA solutions. Collection Czech. Chem. Commun. 46: 3027-37.
-Van Leeuwen, H.P., H.L.F.M. Spanjers and R.F.M.J. Cleven, 1983. Polyelectrolytic nature of the binding of heavy metals to humic acids. in Heavy Metals in the Environment, Int. Conf. Heidelberg, CEP Consultants, Edinburgh, p.1219-22.

PART VI BIOLOGICAL RESPONSE

THE RELATIONSHIPS BETWEEN METAL SPECIATION IN THE ENVIRONMENT AND
BIOACCUMULATION IN AQUATIC ORGANISMS

W.J. LANGSTON AND G.W. BRYAN

1. INTRODUCTION

Recent advances in methodology have resulted in a much better
understanding of the speciation of metals in natural waters. However,
despite the realization that the speciation of elements is important in
terms of bioavailability and toxicity, progress in understanding the
relationships between metal levels in organisms and those in the
environment has been slow. Furthermore, emphasis has generally been
placed on the bioaccumulation and toxicity of <u>dissolved</u> metals,
although in many organisms <u>sediment particles</u> or <u>food</u> may be the most
important sources. The ability of aquatic organisms to accumulate
metals from a number of different sources complicates organism-
environment relationships. In addition, these relationships are
markedly influenced by the differing capacities of organisms to
metabolise metals. Thus, variations between the toxicities of metals
to different species depend not only on the bioavailability of the
metal in the environment but also on the physiological state of the
organism. It is also important to evaluate speciation of metals within
the framework of their roles as essential (for example Fe, Cu, Mn, Zn,
Co) or non-essential elements (Hg, Cd, Ag) bearing in mind that some
metals can exert beneficial effects at low concentrations and harmful
ones at higher levels.

Relationships between speciation and the bioavailability and
toxicity of metals to aquatic organisms have been clearly demonstrated
under experimental conditions. In contrast, few of the relationships
between metal species distributions and organisms have been evaluated
in the field, where abiotic factors (salinity, temperature, light, pH,
eH and ligand concentrations) cannot be controlled. Recent evidence
regarding the impact of metal speciation on aquatic organisms under
field and laboratory conditions forms the basis of this review.

Kramer, C.J.M. and Duinker, J.C. (eds.), Complexation of Trace Metals in Natural Waters. ISBN 90-247-2973-4
© *1984 Martinus Nijhoff/Dr W. Junk Publishers, The Hague/Boston/Lancaster.*
Printed in the Netherlands.

2. MECHANISMS OF METAL UPTAKE

Irrespective of the source, transport of metal ions across cell membranes can take place by one of two generalized processes. The first of these, active transport, requires that energy-dependent ion pumps maintain concentration gradients across membranes allowing selective uptake of metals such as Na, K and Ca. It is possible that other metal ions may utilise these pathways if conditions of charge and geometry are suitable. Active uptake may also occur by pinocytosis as observed for iron in mussels by George et al. (1976, 1977).

Results of experiments with a large number of organisms are consistent with a theory of passive uptake for many of the heavy metals about which we are most concerned; that is an initial rapid adsorption onto the surface followed by subsequent diffusion into cells. The lack of temperature dependence or effects of metabolic inhibitors confirm the passive nature of this mechanism. Since metal ions are hydrophilic and plasma membranes hydrophobic, transport across the membrane may be facilitated by ligands, either naturally occurring, or produced by the cell. Once inside the cell, exchange to stronger ligands may take place preventing back diffusion and forming a kinetic trap for the metal. This second type of transport mechanism then, can be envisaged as passage from the aqueous medium to the inside of the cell along a cascade of ligands of increasing binding strengths (Williams, 1981; Simkiss et al., 1982). Evidence for such a mechanism is available. Carpene & George (1981), for example, demonstrated an increase in stability constants from Cd-chlorocomplexes in sea water to carboxyl groups present on the membrane surface to -SH groups associated with metal-binding proteins in mussel gills.

The abundance and strength of metal-ligand binding outside the organism will clearly determine the uptake of metals. Some specific examples can now be considered which illustrate the ways in which speciation determines the toxicity and availability of metals.

3. PROCESSES AFFECTING BIOLOGICAL RESPONSES TO 'DISSOLVED' METALS

3.1. Inorganic speciation

Cadmium complexation, primarily with chloride ions at increasing salinities, reduces the toxicity of dissolved metal to grass shrimp Palaemonetes pugio (Sunda et al., 1978) and eggs of the Atlantic silverside Menidia menidia (Engel et al., 1981) as a result of the decrease in free cadmium ion concentration. Complexation may also explain the lower accumulation rates of cadmium at higher salinities by the brown alga Fucus vesiculosus (Bryan, 1983a) and juvenile oysters, Crassostrea virginica (Engel et al., 1981).

Silver, like Cd, is complexed with Cl^- ions, though more strongly. However, the influences of salinity on availability cannot be explained simply in terms of changes in free ion concentration. Accumulation of silver in P. pugio is more closely related to computed concentrations of the neutral monochlorocomplex (Ag Cl^0) than those of the free ion (Ag^+) or the charged dichlorocomplex ($AgCl_2^-$) (Engel et al., 1981). These results may be explained if the lipid cell membrane is more permeable to neutrally charged species. Furthermore, Gutknecht (1981) has shown that the high rate of transport of Hg(II) through lipid bilayer membranes is due primarily to the permeation of the neutral dichloride complex, $HgCl_2^0$.

Referring to our theoretical uptake model it would seem that the chloride ligand may facilitate the passive diffusion of Ag and Hg through the bilipid layer, whilst at present a free ion model appears more valid for Cd and Cu.

3.2. Competition

Interactions between metal ions are beginning to be evaluated, though much of the data concerns unifactorial experiments rather than examining all possible interacting effects   Cadmium uptake in F. vesiculosus, for example, is significantly reduced in the presence of Zn and may explain unexpectedly low levels of Cd in natural samples collected from areas where Zn contamination is high (Bryan, 1983a). Depletion of essential elements may also occur due to displacement by metals, as with the induction of Mn deficiencies in algal cells by competition from Cu (Huntsman & Sunda, 1981). Also numerous studies have shown that the "hardness" of fresh waters, as dictated by Ca^{2+} concentrations, may influence the toxicity of metals such as Cd^{2+} by competing for uptake sites.

3.3. Biologically mediated transformation of metal species

Some environmentally important elements such as As and Se exist largely in anionic forms in the water column and their oxidation state can significantly influence both bioavailability and toxicity. Arsenic, present in sea water as arsenate, can compete with phosphate for active uptake into algal cells. Reduction to arsenite takes place in the physiological conditions within the cell, probably as protection against further phosphate inhibition, and is accompanied by production of the less toxic methylated species, methylarsonate and dimethyl-arsenate (Sanders & Windom, 1980). These transformations are significant factors in the biogeochemistry of As and in many regions As speciation is different from predictions based on thermodynamic considerations alone.

Bacteria are particularly important in metal species transformations and their control of trace metal equilibria is well documented for Hg. Fresh water bacteria can, for instance, enhance the transport of Hg to the air by the volatilization of Hg^0 (Ramamoorthy et al., 1983), while the most environmentally significant transformation, the conversion of inorganic to methyl mercury by bacteria, has been observed in marine and freshwaters. This conversion is probably the major source of methyl mercury present in aquatic food chains although Rudd et al. (1980) have demonstrated Hg methylation by the intestinal flora of fish.

Microbial methylation of other elements including Pb, Sn and As occurs in cultures but the importance of these events in the field has yet to be established.

Demethylating bacteria have been isolated which balance the cycling of these elements in the environment, whilst the oxidising/reducing activities of bacteria also play significant roles in the cycling of some essential elements. Thus Chapnick et al. (1982) conclude that microbially mediated manganese oxidation is responsible for conversion of Mn^{2+} to insoluble MnO_2, which after burial in reducing sediments can be remobilized once more as soluble Mn^{2+}.

3.4. Organic complexation

Complexation with a variety of organic ligands can significantly affect the availability and toxicity of trace metals. Unspecified, naturally occurring organic matter for example reduces the availability of Cu to shrimp and clams (Crecelius et al., 1982) and reduces the

toxicity of mercury to phytoplankton (Wallace et al., 1982). More
specifically, complexation with polyphenols produced by brown seaweeds
reduces the toxic and growth inhibiting effects of Cu on marine algae
(Ragan et al., 1980). Similarly, humic acids have been shown to
ameliorate the inhibitory effects of Cu and Zn on ^{14}C fixation by
marine algae (Ortner et al., 1983), decrease the rate of Cd uptake in
barnacles (Rainbow et al., 1980), and raise the threshold of the toxic
effects of Cd in algae and fish (Gjessing, 1981).

Complexation with synthetic ligands such as NTA and EDTA can also
reduce the toxicity and availability of metals to many aquatic
organisms, including freshwater algae (Petersen, 1982), mussels
(Watling & Watling, 1982), worms and shrimps (Ray et al., 1979).

Reductions in metal bioavailability can be related to the stability
constant of the metal-ligand association, as shown by Knezovich et al.
(1981) who exposed oyster embryos to Cu in the presence of various
organic materials including humic acids and EDTA. The inhibitory
effects of metals, such as Cu, which have a strong affinity for organic
ligands are, therefore, frequently attributed to the free ionic form
(Cu^{2+}).

Salinity plays an important role in relation to organic as well as
inorganic, complexation. Increasing salinity reduces organic complexation for many metals including Cu, due to increased competition for
available sites from Ca and Mg (Mantoura et al., 1978). In areas of
low organic content, inorganic Cu species, particularly $CuCO_3$, may
become important and biologically significant: carbonate complexation
can, for example, reduce the toxicity of Cu to the crustacean, Daphnia
magna (Andrew et al., 1977).

Algal exudates and break down products exert a significant influence
on trace metal equilibria in natural waters and are responsible for a
variety of metal-associated events, notably the biogenic removal of
dissolved metals such as Cd, Zn and Ni from enriched surface waters of
the oceans and their transport to depths, where subsequent
remobilization may occur (Yeats and Campbell, 1983).

Experimental seawater enclosures have provided useful insights into
the interrelationships between metal speciation and plankton
communities. Wallace et al. (1982) have shown, for example, that 90%
of the total Hg added as $HgCl_2$ to the water column becomes associated

primarily with particulate, colloidal and high molecular weight dissolved organics; the fraction > 10,000 daltons binding the greatest proportion of metal. The particulates become enriched in Hg with increasing depth as the more volatile, non Hg-binding, organics are degraded.

The compounds released by phytoplankton include sulphated poly-saccharides, uronates, polyphenols, amino acids, polypeptides and proteins and may represent a large percentage of the total carbon fixed, thereby significantly influencing the organic complexing capacity of natural waters. It has been suggested by Hatcher et al. (1980) that uronic acid-containing polysaccharides are the most significant components of marine fulvic acids.

The production of exudates in phytoplankton cultures in response to metal exposure is now well documented and can protect the organisms from toxic effects (see Florence et al., this volume). However, these laboratory studies provide only circumstantial evidence of a detoxication mechanism as ligand concentrations may be too low to provide protection in natural waters. It has also been suggested that hydroxamate siderophores, produced predominantly by blue green algae, may be more significant in sequestering essential Fe^{3+} than in their ability to chelate and detoxify Cu (McKnight & Morel, 1980).

Animal products, such as mucus, complex metals and can reduce availability by slowing down diffusion of metals into cells. Chow et al. (1974) demonstrated the metal binding capacity of the mucus secreting epidermis of tuna which, despite representing only 0.25% of the fish's weight, accounts for 52% of the total body Pb burden. Internal production of Ca- and Mg- containing mucus corpuscles may protect fish by limiting the transfer of Cd, Zn and Cu through the intestinal wall (Noel-Lambot, 1981).

4. BIOLOGICAL RESPONSES TO PARTICULATE METALS

Concentrations of trace metals in sediments are so much higher than those in the overlying water that the bioavailability of even a minute fraction may represent an important source for uptake by burrowing organisms. Determining the relative importance of sediments and overlying water as sources of trace metals is sometimes difficult: methods of achieving this have been reviewed by Bryan (1983b). Uptake of metals from sediments may occur following the ingestion of particles in the diet or through uptake of particles over the body surface by pinocytosis, as occurs in some invertebrates. In addition, metals adsorbed on particles may be absorbed following surface contact with the organism, and another potentially important source is provided by metals in the sediment pore waters. The composition of pore water varies with sediment depth and often differs substantially from that of the overlying water. As sediment particles become buried, the reduction of oxides of Fe and Mn leads to their dissolution and to the release of associated trace metals: in particular, concentrations of Mn in pore waters are frequently of the order of 20 μmol.dm^{-3}. Bryan & Hummerstone (1973a) concluded that the two most important factors governing the accumulation of Mn by the estuarine polychaete Nereis diversicolor are the concentration in the pore water and its salinity. At constant salinity the rate of Mn uptake is proportional to the pore-water concentrations. However, with decreasing salinity the rate of uptake is markedly increased and among the factors responsible are: 1) increasing availability of Mn at lower salinities through decreasing inorganic complexation (Mantoura et al., 1978); 2) competition from Ca and Mg for uptake sites decreases at lower salinities; 3) the potential difference across the body wall (negative inside) increases with decreasing salinity (Fletcher, 1970). Furthermore, the worm absorbed Mn similarly from both natural pore water and diluted offshore water spiked with manganese ions, indicating that uptake from the pore water is unaffected by organic complexation. Other metals, including Cu and Zn, are also absorbed from pore water by N. diversicolor, although this may not be so important as it is for Mn (Bryan & Hummerstone, 1971, 1973b; Renfro, 1973). Thus Luoma & Bryan (1982) observed extremely good relationships between levels of Cu in N. diversicolor and the total (HNO$_3$ digest) and 1N HCl extractable concentrations of Cu in surface sediments. Similarly, experimental work with freshwater tubificid worms indicated that sediment-bound copper was a more important source than that in solution (Diks & Allen, 1983).

Uptake of particle-bound metals need not necessarily involve ingestion and digestion in the alimentary tract. It was concluded by Luoma et al. (1982) that in the brown seaweed Fucus vesiculosus direct scavenging of Cu and Pb from particulates provides an important uptake route and may depend on the high selectivity coefficients observed for these metals in Fucus polyphenols (Ragan et al., 1979). In some invertebrates, particles may be taken up over the body surface by pinocytosis. George et al. (1976) showed that particles of Fe oxyhydroxide were absorbed at a significant rate through the gills of Mytilus edulis in this way and then transferred to other tissues by amoebocytes. The uptake of Pb and the metalloprotein ferritin was also observed and this type of uptake process appears to be energy-dependent (George et al., 1977).

Deposit-feeding organisms ingest sediment particles and intuitively this would be expected to be a major route for trace metal uptake, especially in view of the low pH values sometimes encountered in digestive systems. Generally speaking, fine-grained oxidized surface sediments probably provide the most important sources of available metals (Luoma & Davis, 1983). The composition of sediments is so complex and variable that it would appear to be extremely difficult to relate the chemistry of trace metals in sediments to the concentration of metals in the biota. However, Lion et al. (1982) have pointed out that the situation may be simplified if most particles of whatever composition are coated with materials such as organic compounds and oxides of Fe and Mn, thus reducing the heterogeneity of the particle surfaces.

Luoma & Davis (1983) have reviewed the factors which it is necessary to consider when modelling the partitioning of trace metals in oxidized sediments. These parameters include (1) the binding capacity of each sediment component (this depends on the density of binding sites which is, for example, far higher on recently precipitated amorphous Fe oxyhydroxide than on a crystalline form such as goethite): (2) the binding intensity of the metal (Cu, for example, is more strongly bound by organics than Cd): (3) the abundance of each component: (4) parameters such as pH and the concentrations of dissolved inorganic and organic ligands which influence dissolved metal speciation: (5) concentrations of other ions (Ca, for example) which may compete with trace metals for binding sites. Operational determination of the abundances of metal-binding sediment components, including Fe oxyhydroxides and organic matter, have already proved useful in the prediction of trace metal bioavailability in oxidized sediments from British estuaries. Luoma & Bryan (1978) found that concentrations of Pb in the deposit-feeding clam Scrobicularia plana from a range of widely differing estuaries were related not to the sediment-Pb levels but to the ratios of Pb:Fe in 1N HCl extracts of the < 100 μm fraction of the sediments. Similar observations were made in the clam Macoma balthica and it is thought that a higher level of readily extractable amorphous Fe oxyhydroxide in the sediment increases its capacity for binding Pb and reduces the availability of Pb in the digestive tract (Bryan, 1983b). This is consistent with the conclusions of Lion et al. (1982), that operationally defined Fe-Mn oxyhydroxide phases in some sediments are important in the binding of Pb. Observations by Langston (1980) have shown that the availability of sediment As to S. plana is also best related to the As:Fe ratio in a 1N HCl extract of surface sediment (Fig. 1A).

Mercury, on the other hand, has a high affinity for organic particles and concentrations in S. plana and M. balthica are related to the Hg: organic matter ratios in the < 100 μm fraction of surface sediments (Fig. 1B; Langston, 1982). This agrees with the work of Breteler et al. (1981) which showed that the highest concentrations of Hg in mussels Modiolus demissus and crabs Uca pugnax occurred in animals exposed to marsh sediments having low levels of organic matter.

Observations on other metals in S. plana and M. balthica have revealed important controls on sediment bioavailability. Thus the availability of sediment Ag to the clams appears to be reduced in high-Cu sediments, possibly through competition from Cu for uptake sites in the digestive system (Luoma & Bryan, 1982). Cu itself has a high affinity for Fe oxyhydroxide and an even higher affinity towards organic material. However, although it is possible to observe the influence of Fe oxyhydroxide in lowering the availability of Cu to the clams in some Fe contaminated sediments, it has so far proved impossible to obtain predictive relationships between Cu levels in the clams and surface sediment extracts (Luoma & Bryan, 1982). Unpredictably high levels of Cu have been found in clams from some sites having low-Cu sediments, the only unifying factor being the tendency of sediments at these sites to become especially anoxic. Although Cd is probably mainly associated with sediment organic matter

(Lion et al., 1982) it shows a tendency in estuaries to remain in
solution. In fact, a major source of dissolved Cd in some estuaries
appears to be the dissolution of particulate inputs including both
sewage and sewage sludge (Radford et al., 1981). Relationships between
concentrations of Cd in S. plana and M. balthica and those of the
sediments are comparatively weak: the most significant correlations
are found between levels in the clams and those of the seaweed Fucus
vesiculosus which, in turn, reflects the bioavailability of the
dissolved metal (Fig. 1C; Luoma & Bryan, 1982; Bryan, 1983b).

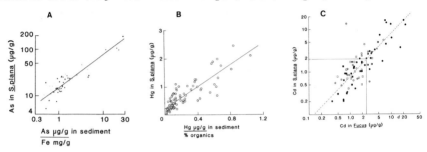

FIGURE 1. Correlations demonstrating the availability of A)
particulate arsenic, B) particulate mercury and C) dissolved cadmium
(as reflected by the seaweed F. vesiculosus) in the clam Scrobicularia
plana (from Langston, 1980, 1982).

It is clear from this work with deposit-feeding clams that the
availabilities of at least some sediment-bound trace metals can be
explained in terms of metal binding by major sediment components and
metal-metal competition, though there are undoubtedly other factors
which have not yet been identified.

5. AVAILABILITY OF METALS IN FOOD: FOOD CHAIN MAGNIFICATION

Accumulation of metals from food can be important for many aquatic
animals (see for example Fowler, 1982). Only for the smaller organisms
with large surface area/volume ratios, or those exposed to high water
concentrations is dissolved metal likely to be the primary route of
entry. For the majority of heterotrophs the proportion of metals
entering by different pathways will depend on the relative concentrations in various media together with the relative abundance of food.
The nature of the food may also influence metal assimilation. For

instance, experimental work with the polychaete Capitella capitata showed that the accumulation of Ag, Ni and Zn was increased when the quality of the detritus diet was increased by the addition of organic nitrogen (Windom et al., 1982).

Biomagnification of metals along food chains in natural communities has been established as a common feature only in the case of Hg, where magnification results from selective retention of methyl mercury at each trophic level. The major criteria responsible are the extremely high assimilation efficiency of methyl mercury and its long biological half life together with the greater longevity of most top predators. The importance of diet in influencing Hg levels is exemplified by the work of MacCrimmon et al. (1983) which showed that a change in diet from low Hg-containing invertebrates to high Hg-containing smelt resulted in a dramatic increase in Hg accumulation in lake trout, Salvelinus namaycush.

6. INTRACELLULAR MECHANISMS AND METAL COMPLEXATION

So far we have considered pathways of metal uptake into aquatic organisms and some of the external factors which influence this. It is important to realise, however, that the atomic properties of metals which influence their availability are the same ones which determine their retention and biochemical/toxicological roles within cells. Group IA, IIA and IIIA metals are exchanged readily from bound complexes and diffuse rapidly as simple ions. Na and K form such weak complexes that their retention within cells is established only by restriction within membranes and is maintained by selective energy pumps. Metals such as Ca, Mg, Sr and Mn show preference for ligands in the sequence O>N>S, as for example the charged oxygen atoms of pyrophosphate ions. Moving across the periodic table from Ca and Mg through the transition metals to B-sub-group metals there is an increasing tendency to be retained by the thermodynamic traps in polymers. These metals, which include Cu, Cd and Hg, show preference for the reverse sequence of ligands, S>N>O, notably the thiol groups of metallothionein type proteins (see Williams, 1981; Simkiss et al., 1982).

Movement of most heavy metals across membranes can thus be envisaged as transfer along a cascade of ligands of successively higher binding strengths as described earlier, until they are trapped within cells.

Some degree of control is available by regulation of ligand production, although in many cases this is not achieved, leading to accumulation of metals in excess of requirements. Confronted with such an excess to normal metabolic requirements an organism must metabolise, eliminate or otherwise detoxify the burden to prevent harmful complexation to, and inhibition of, its enzyme systems.

6.1. Elimination and metabolism

Where a large pool of easily exchangeable metal exists in the body, elimination may take place as rapidly as accumulation. This type of exchange often occurs when organisms are exposed to dissolved inorganic species (Fowler, 1982). Some organic species such as methyl mercury may, on the other hand, be eliminated at extremely slow rates as a result of strong affinity for -SH ligands and high lipid solubility.

Detoxication of arsenic in marine algae may be necessary due to indiscriminate uptake with PO_4^- described earlier. Methylation is usually involved together with conversion to organic compounds such as the arsenosugars produced in kelp, Ecklonia radiata (Edmonds & Francesconi, 1981). These compounds may act as important intermediates in the formation of arsenobetaine which appears to be fairly ubiquitous among marine organisms at higher trophic levels.

Metabolism and excretion can therefore be a successful means of removing metals from an organism or reducing their toxic threat. An alternative for many species is storage of the metal in isolated compartments by strong complexation to a suitable ligand.

6.2. Metal binding proteins

The presence of low molecular weight (6-10,000 daltons) proteins, characterised by high cysteine content (30%), and hence high group B metal-binding capacity, was first established in terrestrial mammals. There is now considerable evidence for the occurrence of similar proteins in a large number of aquatic organisms including many invertebrates (for review see Roesijadi, 1981), fish (Roch et al., 1982) birds (Brown et al., 1977) and mammals (Lee et al., 1977). Many invertebrate proteins have different absorbance characteristics to mammalian metallothionein as a result of lower cysteine content. Oyster Cd-binding protein for example has an amino acid composition dominated by aspartic and glutamic acids (Ridlington & Fowler, 1979).

The synthesis of metal-binding proteins can be induced by exposure to metals including Cd, Cu, Zn and Hg, and implies the ability of cells to recognise and respond to the presence of metal accumulation. They occur in organisms exposed to elevated metal levels in both the field and laboratory as well as in some individuals with no known history of metal exposure.

Metallothionein-like proteins are most frequently found in tissues which store metals (kidney, liver, hepatopancreas) and can only be detected in whole animal preparations following exposure to high metal concentrations. There may, therefore, be a threshold level for production of this type of protein enabling the organism to sequester excess metal and conferring a degree of tolerance at high concentrations. Should the complexing capacity of the protein be exceeded, spillover to other proteins may occur with resulting toxic effects.

6.3. Lysosomes, vesicles and granules

X-ray analytical studies on invertebrates exposed to dissolved metals have revealed the appearance of metals such as Hg, Fe, Pb and Cu in lysosomes or similar membrane-bound bodies (for review see Fowler et al., 1981). In addition to fulfilling a storage/detoxifying role, these vesicles may also represent an important pathway for metals into cells following pinocytosis of particulate, colloidal or dissolved metals at the cell surface (George et al., 1976, 1977).

Metal-rich granules are frequently found within membrane-limited vesicles suggesting lysosomal origins; however, their presence outside membranes may indicate alternative sources for initiation. Like metal binding proteins, granules appear to be a fairly ubiquitous means of metal storage and have been found in a variety of tissues from many invertebrate phyla (for review see Brown, 1982) fish, and aquatic mammals (Martoja & Viale, 1977).

Granules may serve a number of purposes in relation to the distribution of metals within cells and four principal types will be described. The first type are predominantly single-metal containing granules which may also contain sulphur, such as the Cu containing granules located within the epidermis of N. diversicolor. This type of granule is formed in response to high environmental levels and may help the organism to maintain a steady state relationship between stored and water soluble metal in the cytoplasm.

The second type of granule, based on calcium, can be subdivided in terms of function into two types. Carbonate-rich granules are important in buffering the Ca balance within the organism whilst those based on ortho- and pyrophosphate possess ideal characteristics to trap metals due to their insolubility, and may fulfil a role in intracellular control of metals (Mason and Nott, 1981).

The presence of a third type of granule containing Fe, in a large number of phyla indicates a role in regulating the organism's Fe requirements. A variety of protein ligands may be involved in Fe complexation although the ubiquitous occurrence of ferritin suggests that the basic mechanism in Fe regulation is common to most organisms (Simkiss et al., 1982).

The fourth type of granule, composed of pure mercuric selenide, has been found in the livers of cetaceans (Martoja & Viale, 1977). Their formation as non-biodegradable cell components is believed to be the final stage in a detoxification process resulting from the demethylation of organic mercury in the liver.

The mechanism of operation is still to be determined for most of the granule types, although they probably function as a kinetic trap for various metal forms. They may be transported in amoebocytes to be utilized at specific sites, or like those present in digestive tissues and kidney may be excreted, thereby fulfilling a detoxifying role.

Many of these granules are not purely inorganic and the presence of some organic material suggests a possible link with metal-binding proteins. It is important to emphasise, therefore, that none of these mechanisms of metal complexation are necessarily exclusive of each other.

7. CONCLUSIONS

There are clearly a formidable array of parameters, both biotic and abiotic, which influence metal uptake and retention. For many aquatic organisms, particularly the smaller types such as phytoplankton, uptake from water will be a dominant source of metal. Using the various speciation techniques now available, biological responses can often be predicted from a knowledge of the chemical properties of the metal and the complexation capacity of the surrounding media.

Many organisms can, however, derive metals from a range of sources and exhibit a variety of physiological responses to different metals, properties which make generalisations about metal species distributions and organisms difficult. Nevertheless, field and laboratory studies have established relationships between metal speciation, availability and toxicity. In many instances the major features influencing biological responses can be identified despite the difficulties in quantifying the heterogenous mixtures of ligands present in natural waters.

ACKNOWLEDGEMENTS

The support provided by the Department of the Environment (DGR 480/15) and the European Economic Community (ENV-686-UK(H)) is gratefully acknowledged.

REFERENCES

-Andrew RW, Biesinger KE, Glass GE. 1977. Effects of inorganic complexing on the toxicity of copper to Daphnia magna. Wat. Res. 11: 309-315.
-Breteler RJ, Valiela I, Teal JM. 1981. Bioavailability of mercury in several north eastern U.S. Spartina ecosystems. Estuar. cstl. shelf Sci. 12: 155-166.
-Brown BE. 1982. The form and function of metal-containing 'granules' in invertebrate tissues. Biol. Rev. 57: 621-667.
-Brown DA, Bawden CA, Chatel KW, Parsons TR. 1977. The Wildlife community of Iona Island Jetty, Vancouver B.C. and heavy metal pollution effects. Env. conserv. 4: 213-216.
-Bryan GW. 1983(a). Brown seaweed, Fucus vesiculosus, and the gastropod Littorina littoralis, as indicators of trace metal availability in estuaries. Sci. total Envir. 28: 91-104.
-Bryan GW. 1983(b). The biological availability and effects of heavy metals in marine deposits. Wastes in the sea Vol. 6: Nearshore Waste Disposal (In Press).
-Bryan GW, Hummerstone LG. 1971. Adaptation of the polychaete Nereis diversicolor to estuarine sediments containing high concentrations of heavy metals. I. General observations and adaptation to copper. J. mar. biol. Ass. U.K. 51: 845-863.
-Bryan GW, Hummerstone LG. 1973a. Adaptation of the polychaete Nereis diversicolor to manganese in estuarine sediments. J. mar. biol. Ass. U.K. 53: 859-872.
-Bryan GW, Hummerstone LG. 1973b. Adaptation of the polychaete Nereis diversicolor to estuarine sediments containing high concentrations of zinc and cadmium. J. mar. biol. Ass. UK 53: 839-857.
-Carpene E, George SG. 1981. Absorption of cadmium by gills of

390

Mytilus edulis (L.). Molecular Physiology, 1: 23-34.

-Chapnick SD, Moore WS, Nealson KH. 1982. Microbially mediated manganese oxidation in a freshwater lake. Limnol. Oceanogr. 27: 1004-1014.

-Chow TJ, Patterson CL, Settle D. 1974. Occurrence of lead in tuna. Nature (London) 251: 159-161.

-Crecelius EA, Hardy JT, Gibson CI, Schmidt RL, Apts CW, Gurtisen JM Joyce SP. 1982. Copper bioavailability to marine bivalves and shrimp: relationship to cupric ion activity. Mar. Envir. Res. 6: 13-26.

-Diks DM, Allen HE. 1983. Correlation of copper distribution in a freshwater-sediment system to bioavailability. Bull. envir. Contam. Toxicol. 30: 37-43.

-Edmonds JS, Francesconi KA. 1981. Arseno-sugars from brown kelp (Ecklonia radiata) as intermediates in cycling of arsenic in a marine ecosystem. Nature (London), 289: 602-604.

-Engel DW, Sunda WG, Fowler BA. 1981. Factors affecting trace metal uptake and toxicity to estuarine organisms. I. Environmental parameters. In: Vernberg FJ, Calabrese A, Thurberg FP, Vernberg WB (eds.) Biological monitoring of marine pollutants. p. 127-144 Academic Press.

-Fletcher CR. 1970. The regulation of calcium and magnesium in the brackish water polychaete Nereis diversicolor O.F.M. J. exp. Biol. 53: 425-443.

-Fowler BA, Carmichael NG, Squibb KS, Engel DW. 1981. Factors affecting trace metal uptake and toxicity to estuarine organisms II. Cellular mechanisms. In: Vernberg FJ, Calabrese A, Thurberg FP Vernberg WB (eds.) Biological monitoring of marine pollutants. p. 145-163. Academic Press.

-Fowler SW. 1982. Biological transfer and transport processes. In: Kullenburg G (ed.) Pollutant transfer and transport processes Vol. II, CRC Press, pp. 1-65.

-George SG, Pirie BJS, Coombs TL. 1976. The kinetics of accumulation and excretion of ferric hydroxide in Mytilus edulis (L.) and its distribution in the tissue. J. exp. mar. Biol. Ecol. 23: 71-84.

-George SG, Pirie BJS, Coombs TL. 1977. Metabolic characteristics of endocytosis of ferritin by gills of a marine bivalve mollusc. Biochem. Soc. Trans. 5: 136-137.

-Gjessing ET. 1981. The effect of aquatic humus on the biological availability of cadmium. Arch. Hydrobiol. 91: 144-149.

-Gutknecht J. 1981. Inorganic mercury (Hg^{2+}) transport through lipid bilayer membranes. The Journal of Membrane Biology, 6: 61-66.

-Hatcher PG, Breger IA, Mattingley MA. 1980. Structural characteristics of fulvic acids from Continental Shelf sediments. Nature (London), 285: 560-562.

-Huntsman SA, Sunda WG. 1981. The role of trace metals in regulating phytoplankton growth with emphasis on Fe, Mn and Cu. In: Morris I (ed.) The physiological ecology of phytoplankton, p. 285-328. Blackwell.

-Knezovich JP, Harrison FL, Tucker JS. 1981. The influence of organic chelators on the toxicity of copper to embryos of the Pacific oyster, Crassostrea gigas. Arch. Environ. Contam. Toxicol. 10: 241-249.

-Langston WJ. 1980. Arsenic in U.K. estuarine sediments and its availability to benthic organisms. J. mar. biol. Ass. U.K. 60: 869-881

-Langston WJ. 1982. The distribution of mercury in British estuarine sediments and its availability to deposit-feeding bivalves J mar. biol. Ass. UK. 62: 667-684.

-Lee SS, Mate BR, von der Trenk KT, Rimerman RA, Buhler DR. 1977 Metallothionein and the subcellular localization of mercury and cadmium in the California Sea Lion Comp. Biochem. Physiol. 57C 45-53.

-Lion LW, Altmann RS, Leckie JO. 1982 Trace metal adsorption characteristics of estuarine particulate matter: Evaluation of contributions of Fe/Mn oxide and organic surface coatings. Environ. Sci. Technol. 16: 660-666.

-Luoma SN, Bryan GW 1978. Factors controlling the availability of sediment-bound lead to the estuarine bivalve Scrobicularia plana. J. mar biol. Ass UK. 58 793-802

-Luoma SN, Bryan GW. 1982. A statistical study of environmental factors controlling concentrations of heavy metals in the burrowing bivalve Scrobicularia plana and the polychaete Nereis diversicolor. Estuar. cstl. Shelf Sci. 15 95-108.

-Luoma SN, Davis JA. 1983. Requirements for modeling trace metal partitioning in oxidized estuarine sediments. Mar. Chem 12 159-181.

-Luoma SN, Bryan GW Langston WJ. 1982 Scavenging of heavy metals from particulates by brown seaweed. Mar. Pollut. Bull. 13:   394-396

-MacCrimmon HR, Wren CD. Gots BL. 1983. Mercury uptake by lake trout Salvelinus namaycush, relative to age, growth and diet in Tadenac Lake with comparative data from other pre-Cambrian Shield lakes. Can. J. Fish. aquat. Sci. 40: 114-120.

-Mantoura RFC, Dickson A Riley JP. 1978. The complexation of metals with humic materials in Natural Waters. Estuar. cstl. Mar. Sci.   6. 387-408.

-Martoja R, Viale D. 1977. Accumulation de granules de séléniure mercurique dans le foie d'Odontocetes (Mammifères, Cétacés): un mécanisme possible de détoxification du méthyl-mercure par le sélénium. C.R Hebd Séances Acad. Sci. Paris   Sér. D 285 109-112.

-Mason AZ, Nott JA 1981. The role of intracellular biomineralised granules in the regulation and detoxification of metals in gastropods with special reference to the marine prosobranch Littorina littorea. Aquat. Toxicol. 1: 239-256.

-McKnight DM Morel FMM. 1980. Copper complexation by siderophores from filamentous blue-green algae. Limnol. Oceanogr. 25: 62-71.

-Noel-Lambot F. 1981. Presence in the intestinal lumen of marine fish of corpuscles with a high cadmium-, zinc- and copper-binding capacity: a possible mechanism of heavy metal tolerance. Mar Ecol. Prog. Ser., 4: 175-181.

-Ortner PB, Kreader C, Harvey GR. 1983. Interactive effects of metals and humus on marine phytoplankton carbon uptake. Nature, Lond., 301: 57-59

-Petersen R. 1982. Influence of copper and zinc on the growth of a freshwater alga, Scenedesmus quadricauda: the significance of chemical speciation. Envir. Sci. Technol. 16: 443-447.

-Radford PJ, Uncles RJ, Morris AW. 1981. Simulating the impact of technological change on dissolved cadmium distribution in the Severn estuary. Wat. Res. 15: 1045-1052.

-Ragan MA, Ragan CM, Jensen A. 1980. Natural chelations in sea water: detoxification of Zn^{2+} by brown algal polyphenols. J. expl. Mar. Biol. Ecol., 49: 261-267.

-Ragan MA, Smidstod O, Larsen B. 1979. Chelation of divalent metal ions by brown algal polyphenols. Mar. Chem. 7: 265-271.

-Rainbow PS, Scott AG, Wiggins EA, Jackson RW, 1980. Effect of chelating agents on the accumulation of cadmium by the barnacle Semibalanus balanoides and complexation of soluble Cd, Zn and Cu. Mar. Ecol. Prog. Ser. 2: 143-152.

-Ramamoorthy S, Cheng TC, Kushner CJ. 1983. Mercury speciation in water. Can. J. Fish. aquat. Sci. 40: 85-89.

-Ray S, McLeese DW, Pezzack D. 1979. Chelation and interelemental effects on the bioaccumulation of heavy metals by marine inverteb-rates. In: International conference: management and control of heavy metals in the environment, 1979. p. 35-38. Edinburgh, CEP Consultants.

-Renfro WC. 1973. Transfer of ^{65}Zn from sediments by marine polychaete worms. Mar. Biol. 21: 305-316.

-Ridlington JW, Fowler BA. 1979. Isolation and partial characterization of a cadmium binding protein from the American oyster (Crassostrea virginica). Chem. Biol. Interactions 25: 127-138.

-Roch M, McCarter JA, Matheson AT, Clark MJR, Olafson RW. 1982. Hepatic metallothionein in rainbow trout (Salmo gairdneri) as an indicator of metal pollution in the Campbell River system. Can. J. Fish aquat. Sci. 39: 1596-1601.

-Roesijadi G. 1981. The significance of low molecular weight, metallothionein-like proteins in marine invertebrates: current status. Mar. Envir. Res. 4: 167-179.

-Rudd JWM, Furutani A, Turner MA. 1980. Mercury methylation by fish intestinal contents. Appl. envir. Microbiol. 40: 777-782.

-Sanders JG, Windom HL. 1980. The uptake and reduction of arsenic species by marine algae. Estuar. Cstl. Mar. Sci. 10: 555-567.

-Simkiss K, Taylor M, Mason AZ. 1982. Metal detoxification and bioaccumulation in molluscs. Mar. Biol. Lett. 3: 187-201.

-Sunda W, Engel DW, Thuotte RM. 1978. Effect of chemical speciation on toxicity of cadmium to grass shrimp, Palaemonetes pugio: Importance of free cadmium ion. Environ. Sci. Technol. 12: 409-413.

-Wallace GT Jr, Seibert DL, Holzknecht SM, Thomas WH. 1982. The biogeochemical fate and toxicity of mercury in controlled experimental ecosystems. Estuar. cstl. Shelf Sci. 15: 151-182.

-Watling HR, Watling RJ. 1982. Comparative effects of metals on the filtering rate of the brown mussel (Perna perna). Bull. envir. Contam. Toxicol. 29: 651-657.

-Williams RJP. 1981. Natural selection of the chemical elements. Proc. R. Soc. (B), 213: 361-397.

-Windom HL, Tenore KT, Rice DL. 1982. Metal accumulation by the polychaete Capitella capitata: influences of metal content and nutritional quality of detritus. Can. J. Fish. aquat. Sci. 39: 191-196.

-Yeats PA, Campbell JA. 1983. Nickel, copper, cadmium and zinc in the Northwest Atlantic Ocean. Mar. Chem. 12: 43-58.

BIOASSAYS OF CUPRIC ION ACTIVITY AND COPPER COMPLEXATION

W. G. SUNDA, D. KLAVENESS and A. V. PALUMBO

1. INTRODUCTION

Because of its reactivity, copper is both toxic to organisms and tends to be complexed by organic ligands in natural waters (Sunda and Hanson 1979). The extent to which copper is complexed has been found to control its toxicity due to a dependence of copper toxicity on free cupric ion activity rather than on total or complexed copper concentrations (Sunda and Guillard 1976, Anderson and Morel 1978, Jackson and Morgan 1978).

Despite the importance of cupric ion activity in controlling the biological availability and geochemical behavior of copper, this variable has largely eluded measurement by chemists. Existing analytical techniques, by and large, have not been sensitive or selective enough to differentiate free from complexed copper ions at the low concentrations that exist in natural waters. For example, anodic stripping voltametry (ASV), a sensitive and widely used technique, usually can not differentiate between free metal ions and labile complexes (Chau et al. 1974). Because of this, attempts to use ASV to predict copper toxicity have met with only limited success (Gächter et al. 1973). Measurements of cupric ion activity with membrane electrodes, on the other hand, can be used to accurately predict copper toxicity in media of variable copper complexation to natural humic compounds (Sunda and Lewis 1978, Sunda and Gillespie 1979). These electrodes, however, have limited sensitivity (Sunda and Hanson 1979) and cannot be used in seawater due to interference by chloride ions (Westall et al. 1979).

The limitations in existing chemical and electrochemical techniques has prompted research into biological assays of free cupric ion activity in natural waters. In this paper we discuss the development, validity and limitations of these techniques, drawing both on previously published research and on studies whose results are presented here for the first time.

Kramer, C.J.M. and Duinker, J.C. (eds.), Complexation of Trace Metals in Natural Waters. ISBN 90-247-2973-4
© 1984 Martinus Nijhoff/Dr W. Junk Publishers, The Hague/Boston/Lancaster.
Printed in the Netherlands.

2. HISTORICAL REVIEW

Early bioassays of copper complexation were only semiquantitative. Davey e
al. (1973) noted that the addition of a synthetic chelator (ehtylenediaminete-
traacetic acid (EDTA)) or natural chelator (histidine) to synthetic seawater
markedly reduced the toxicity of copper to a marine diatom. By comparing thes
results with those for natural seawater, they were able to estimate the chelation capacity of the natural seawater samples. Gillespie and Vaccaro (1978)
developed a similar technique that was based on copper inhibition of glucose
uptake by a sensitive strain of bacteria. They calibrated the technique with
a series of EDTA additions to UV-treated seawater and found that chelation
capacity for seawater samples covaried with the dissolved carbon concentration
From comparisons of copper toxicity in filtered, ultrafiltered and UV-photooxidized coastal seawater, they concluded that the complexing material con-
sisted primarily of organic ligands of < 10,000 molecular weight.

A refinement of the bacterial bioassay of copper complexation was presented
by Sunda and Gillespie (1979). They calibrated the relationship between coppe
inhibition of ^{14}C-glucose incorporation and free cupric ion activity by using
copper-NTA buffers in UV-treated seawater. Cupric ion activities in these medi
were computed from chemical equilibria between copper and NTA (including side
reactions with calcium and magnesium) and between copper and inorganic ligands
By comparing the measured relationship between copper inhibition and total
copper in an estuarine seawater sample with the calibration curve for copper
inhibition as a function of cupric ion activity, they established the relation
ship between cupric ion activity and total copper concentration. Analysis of
this relationship provided an estimate of the chelation capacity (0.05 μmol·l^{-}
the conditional stability constant for copper chelation ($\geq 10^{10}$), and the ambie
cupric ion activity ($\leq 10^{-11}$ mol·l^{-1}). Sunda and Gillespie applied a similar
analysis to previously published data (Gillespie and Vaccaro 1978) for the tox
city of added copper to the same bacterial isolate in filtered seawater from
Nantucket Sound, Massachusetts and Saanich Inlet, British Columbia. They obta
conditional stability constants of $10^{9.0}$ and $10^{9.6}$ and chelation capacities
(organic ligand concentrations) of 0.16 and 0.20 μmol·l^{-1}. Using the same bac
terial bioassay technique, Anderson et al. (In press) reported similar results
for seawater from Vineyard Sound, Massachusetts. Apparent stability constants
for 10 samples ranged from $10^{8.8}$ to $10^{9.2}$ and chelation capacities ranged from
0.09 to 0.6 μmol·l^{-1}.

3. BACTERIAL CUPRIC ION BIOASSAYS UTILIZING INTERNAL CUPRIC ION CALIBRATION

The bacterial bioassay technique of Sunda and Gillespie (1979) was modified to utilize internal rather than external calibration of relationships between copper toxicity and cupric ion activity. As before, Cu-NTA buffers were used. In this technique, copper inhibition of tritium-labeled amino acid or glucose uptake by the Gillespie bacterial isolate was measured in two series of 25 ml subsamples of seawater. The first contained serial additions of copper and the second contained additions of copper plus a single addition of NTA (5-10 μmol$\cdot$l^{-1}). The seawater was collected in Teflon bottles from the lower Newport River Estuary, North Carolina and filtered through 0.2 μm pore size Nuclepore filters. In some cases, it also was exposed for four h to UV-light from a 1200 watt mercury lamp to photooxidize organic matter. After the addition of copper and copper-NTA buffers the subsamples were equilibrated for two h . They then received 0.1 ml aliquots of a washed suspension of the bacteria and were incubated for 2.5 to 5 h. Bac-

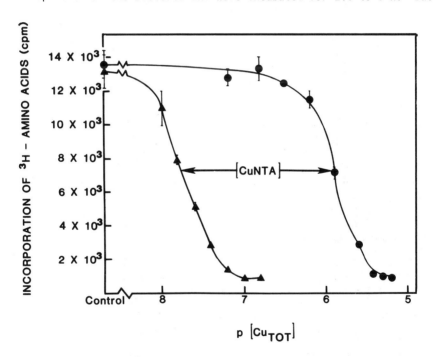

FIGURE 1. Relationship between uptake of ^{3}H-labeled amino acids by the bacterial isolate and the negative log of the total added copper concentration (p[Cu$_{TOT}$]) in UV-photooxidized seawater at 28.8 o/oo salinity, pH 8.1 and 25°C. ▲ Values with addition of CuSO$_4$ only and ● with addition of copper plus NTA (6.25 μmol$\cdot$l). Error bars represent ± SE for 3 replicate treatments.

terial incorporation of labeled substrate was measured during the final 0.5
h of these periods (Sunda and Gillespie 1979). The experiments were run at
in 30 ml acid washed Teflon vials. There were three replicates per treatmen
    Results of one such experiment conducted with UV-photooxidized seawater a
presented in Fig. 1. The addition of 6.25 µmol·l^{-1} NTA had no effect on am
acid incorporation, but displaced the toxicity curve to considerably higher
per concentrations due to the formation of CuNTA chelates, which are not dir
toxic (Sunda and Gillespie 1979). Equal levels of inhibition in seawater wi
and without NTA should occur at equal cupric ion activities, and at constant
activity, the concentration of copper species other than CuNTA, such as complexes with inorganic and natural organic ligands, should also be constant.
Therefore, for each combination of added copper plus NTA the concentration o
CuNTA complex should be equal to the total concentration of copper in that
solution minus the concentration that causes the same level of inhibition in
absence of NTA. These estimates of CuNTA concentration were then used to co
cupric ion activity from an equation derived from mass action equilibria bet

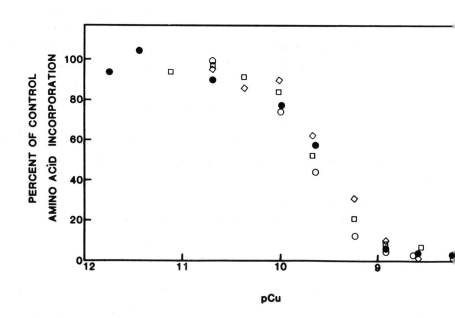

FIGURE 2. Relationship between ^{3}H-amino acid incorporation and pCu for the
bacterial clone. Sampling dates and copper exposure times, respectively, ar
□ January 23, 4 h; ● July 8, 3 h; o July 8, 5 h; ◇ July 18, 2.5 h.

copper and NTA, calcium and NTA, and magnesium and NTA:

$$[CuNTA^-] = \frac{[NTA_{TOT}]\, A_{Cu}\, K_{CuNTA}}{A_{Cu}K_{CuNTA} + A_{Mg}\, K_{MgNTA} + A_{Ca}K_{CaNTA}} \qquad (1)$$

In this equation square brackets denote concentration of the enclosed species and K_{CuNTA}, K_{CaNTA}, and K_{MgNTA} are stability constants taken from Martell and Smith (1974). Their respective values (I=0.1) are $10^{12.96}$, $10^{6.41}$ and $10^{5.41}$ at 20°C and $10^{12.94}$, $10^{6.39}$ and $10^{5.47}$ at 25°C. Correction of these constants in equation (1) to values valid at ionic strengths other than 0.1 are not necessary since the activity coefficient corrections should cancel out in the numerator and denominator of the equation. Values for the free ion activities of calcium and mangesium ions, A_{Ca} and A_{Mg}, were computed from published values of total concentrations (Wilson 1975) corrected for salinity and from total ion activity coefficients (Whitfield 1975).

From the above calculations, one can generate the relationship between amino acid incorporation and pCu (negative log of the cupric ion activity) (Fig. 2). This relationship was sigmoidal and showed fifty percent inhibition at a pCu of 9.6. Similar curves were obtained in other experiments in either natural seawater or UV-treated seawater. These curves show a slight increase in copper inhibition with increased time of exposure (2.5 to 5h) (Fig. 2).

By combining the relationships between inhibition of amino acid uptake and pCu and between inhibition of amino acid uptake and total copper (which is assumed here to equal the added copper concentration) we obtained relationships between pCu and $p[Cu_{TOT}]$ (-log total copper concentration). This latter relationship for the UV-treated 28.8 o/oo salinity seawater (Fig. 3A) was consistent with the equation:

$$pCu = p[Cu_{TOT}] + 1.9 \qquad (2)$$

In another experiment with UV-treated 35 o/oo seawater at pH 8.2, a similar bioassayed relationship was observed (Fig. 3B):

$$pCu = p[Cu_{TOT}] + 2.1 \qquad (3)$$

The constant difference between pCu and $p[Cu_{TOT}]$ is consistent with copper being complexed only by inorganic ions. By dividing the ratio $A_{Cu}/[Cu_{TOT}]$ (equation (3)) by an activity coefficient for copper of 0.25,* we compute that 3.2% of the copper

---

*The cupric ion activity coefficient is assumed to equal that for calcium (Whitfield 1975), an ion with the same charge and ionic radius (Stumm and Morgan 1970).

398

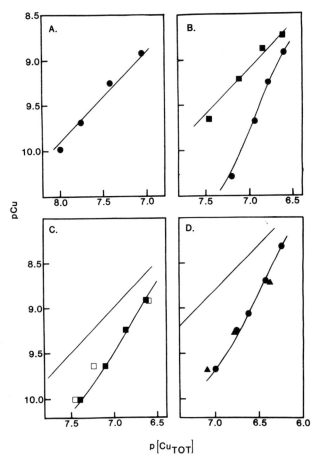

FIGURE 3.   Relationships between pCu and the negative log of total added copper
concentration (p[Cu_{TOT}]) as determined from the toxicity bioassays at 25°C.
Straight lines of unity slope are predicted relationships based on complexation
by inorganic ligands only.   Curves fit through data points (B, C and D) are
modeled relationships based on complexation of copper by both organic and inorganic ligands.   Stability constants and organic ligand concentrations used
calculate these curves are listed in Table 1.
A.   pCu vs p[Cu_{TOT}] for 28.8 o/oo salinity seawater at pH 8.1.   Seawater was
     collected January 23, 1980 and was exposed to ultraviolet light to photo-
     oxidize organic matter.
B.   Results from parallel experiments in filtered seawater ● and in filtered
     seawater that was UV-treated ■ .   Seawater was collected July 18, 1980 and
     had a salinity of 35.1 o/oo and a pH of 8.2.
C.   Results from parallel experimental runs with two different bacterial expo-
     times to copper: 3 h □ and 5 h ■.   Water was collected on July 8, 1980.
     pH and salinity were 8.2 and 35.0 o/oo.
D.   Results from parallel experiments conducted with two portions of the same
     seawater sample. ▲ Results obtained in an unfiltered aliquot using a nat
     community bioassay. ● Results from a bacterial isolate bioassay in a
     filtered (0.2 µm) portion of the same water.   Water was collected April
     1980 and had a pH and salinity of 8.0 and 31.7 o/oo.

in UV-treated seawater (35 o/oo salinity, pH 8.2, and 25°C) is present as free cupric ions; the rest is present as inorganic complexes. From an inorganic speciation model, Paulson (1978) computed an identical value (3.2% free) for seawater at the same pH, salinity and temperature. Paulson's calculations indicated that copper was primarily present as carbonate and hydroxide complexes, and therefore, inorganic complexation should increase with pH.

Bioassayed relationships between pCu and $p[Cu_{TOT}]$ in filtered estuarine seawater indicate a larger complexation of copper than in samples in which organic matter has been photooxidized by UV-light (Fig. 3A-D). This decrease in complexation with UV-photooxidation has been observed previously (Gillespie and Vaccaro 1978, Sunda and Hanson 1979, Anderson et al. in press) and is due to the destruction of organic ligands by UV-light.

By using Scatchard analysis (Sunda and Hanson 1979), we obtained estimates of conditional stability constants and ligand concentrations for the complexation of copper by organic ligands (Table 1). The data in all cases could be modeled on the basis of a single organic ligand whose stability constants ($10^{8.7}$ to $10^{9.6}$) and concentrations (0.14 to 0.3 $\mu mol \cdot l^{-1}$) are in excellent agreement with values obtained previously by bioassay in coastal seawater (Sunda and Gillespie 1979,

Table 1. Organic ligand concentrations (L_{TOT}) and conditional stability constants (K_c) as determined from bioassay data (Fig. 3).

Date collected	Salinity	pH	$L_{TOT}(\mu mol \cdot l^{-1})$	Log K_c*
April 17, 1980	31.7	8.0	0.30	8.7
July 8, 1980	35.0	8.2	0.14	8.9
July 18, 1980	35.1	8.2	0.11	9.6

$$*K_c = \frac{[CuL]}{[Cu^{2+}][L]}$$

Anderson et al. in press). Furthermore, conditional constants for humic materials extracted from coastal seawater with XAD-2 resins ($10^{8.9}$ to $10^{10.2}$, Mantoura et al. 1978) agree well with these results and our total ligand concentrations are similar to that (0.3 $\mu mol \cdot l^{-1}$) determined for North Sea seawater by ASV titration (Duinker and Kramer 1977).

4. BIOASSAYS OF COPPER COMPLEXATION UTILIZING NATURAL COMMUNITIES OF BACTERIA

The use of internal rather than external pCu calibration allowed for the development of bioassays with natural communities of bacteria, in which the

400

bacteria which naturally reside in the seawater replace the bacterial isolate
as the bioassay organisms. The procedure for these bioassays is the followin
(1) seawater is collected using clean technique, (2) it is subsampled into tw
or more series of aliquots, each of which, as before, receives additions of
copper or copper-NTA buffers, (3) the subsamples are incubated for ca. 4 h an
then receive additions of ^{3}H-labeled mixed amino acids or glucose, (4) the
uptake of ^{3}H-labeled substrate by the natural bacterial community is measured
by liquid scintillation counting.

Cupric ion bioassays with natural bacterial communities are simpler in pra
tice than those with bacterial isolates since one does not have to filter the
water nor culture, wash, and dispense bacteria. Bioassays with bacterial com
munities also yield information on the sensitivity of these communities to
copper additions and to cupric ion activity in addition to information on the
complexing characteristics of the water. Natural community bioassays, howeve
potentially can underestimate copper complexation if reaction kinetics betwee
copper and natural ligands are slow.

We conducted parallel bioassays with the isolate and with the natural bacterial community in aliquots of a single estuarine seawater sample. Essentia
identical results were obtained using the two techniques (Fig. 3D), and there
fore, both techniques appear to be equally valid. In the bacterial isolate v
natural community comparison experiment, ^{3}H-labeled glucose was used as a sub
strate rather than ^{3}H-labeled amino acids. The bacteria were exposed to coppe
for 2.5 h prior to the 0.5 h incubation with ^{3}H-glucose. Interestingly, both
bacterial isolate and the natural community showed similar responses to pCu v
50% inhibition at pCu of 9.2. This value is similar to that (pCu 9.1 to 9.2
observed previously with glucose as a substrate (Sunda and Gillespie 1979), i
is slightly lower than observed for copper inhibition of amino acid incorpor
(pCu 9.5 to 9.7) (Fig. 2).

Cupric ion bioassays with natural bacterial communities were conducted at
in samples from five stations in the Gulf of Mexico, ranging from low-pro-
ductivity oceanic to high-productivity coastal (Sunda and Ferguson 1983). T
natural community bioassay procedure was the same as that just described exce
that tritium labeled amino acids rather than glucose were used as substrates
Also each experiment employed cupric ion calibration with two different conc
trations of NTA (1 and 5 $\mu mol \cdot l^{-1}$). By doing this they were able to demonst
that the pCu calibration curves were independent of the concentration of NTA
before the bioassays were conducted with rigorous trace metal clean techniqu

Results of these bioassays indicated that microbial uptake of ^{3}H-labeled amino acids in low productivity seawater was inhibited by 2 μmol·1^{-1} additions of $CuSO_4$, a copper addition at or below the ambient concentration. Added copper was appreciably less toxic in high productivity coastal seawater. An analysis of the bioassay results indicated that the lower toxicity of copper in the coastal seawater could be attributed to both a higher degree of copper complexation (resulting in lower cupric ion activities) and to an apparent lower sensitivity of the coastal bacteria to cupric ion activity.

The bioassay estimates of copper complexation in the Gulf of Mexico samples were modeled by Scatchard Analysis. This analysis for high productivity coastal samples from the Mississippi Plume (pH 8.1 and salinity 28.9 o/oo) and from off Cape San Blas, Florida (pH 8.2 and salinity 26.8 o/oo) indicated the presence of at least two strong complexing ligands. (Two were the fewest number required to model the data, but the data also could have been modeled on the basis of more ligands. For simplicity, the data was fit to a two ligand model.) The first of these "ligands" was present at concentrations of 0.02 and 0.03 μmol·1^{-1} and had conditioned stability constants of $10^{11.1}$ and $10^{11.2}$ for the Mississippi and Cape San Blas samples. The second of these had higher concentrations (0.13 and 0.08 μmol·1^{-1}) and lower stability constants ($10^{8.9}$ and $10^{9.0}$), similar to those obtained from bacterial clone bioassays of coastal samples (Table 1, Sunda and Gillespie 1979, Anderson et al. in press.). Bacterial clone bioassays, however, lack the sensitivity to detect complexation of copper by high stability ligands present at low concentractions. Interestingly, the natural bacterial bioassay utilizing copper inhibition of glucose uptake also appears to suffer from a similar insensitivity.

From the bioassayed relationships between pCu and total copper concentration and the measured ambient copper concentrations, Sunda and Ferguson estimated ambient copper speciation and pCu for the five Gulf of Mexico samples. Cupric ion activity was estimated to be 10^{-12} mol·1^{-1} or less in all but one shallow (5 m) intermediate productivity sample which had an estimated activity of 10^{-11} mol·1-1. Soluble inorganic species of copper were estimated to represent only 0.9 to 18% of the ambient copper in both the coastal and oceanic samples. The remainder apparently was present as organic complexes.

5. COMPARISON OF NATURAL COMMUNITY BIOASSAY AND CUPRIC ION ELECTRODE MEASURE
    OF COPPER COMPLEXATION

The validity of bioassays of cupric ion activity was assessed by comparing
bioassay results with those obtained with a cupric ion electrode. The bioass
design was similar to that described previously except that a natural communi
of phytoplankton rather than bacteria was used and the assay was conducted in
lake water rather than seawater to avoid chloride ion poisoning of the cupric
ion electrode.

Water was collected from Lake Gjerojø near Oslo, Norway in August, 1979 an
was transferred to the laboratory in opaque polyethylene bottles. A portion
the water was filtered to remove phytoplankton and both filtered and unfilter
portions were bubbled with air to equilibrate the water with atmospheric CO_2.
(This was necessary to avoid pH changes during the experiment; the pH of the
water when sampled was 9.2 and after equilibration it was 7.9). The water wa

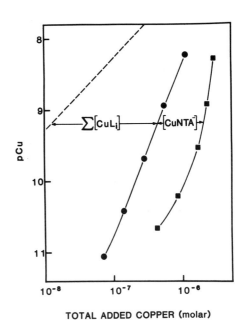

TOTAL ADDED COPPER (molar)

FIGURE 4. Relationship between pCu measured with a cupric ion electrode and
total concentration of added copper in lake water with ■ and without ● 1.8
$\mu mol \cdot l^{-1}$ NTA. The dottled line represents the relationship that would exist
copper were present only as inorganic species (e.g., Cu^{2+}, $CaCO_3$). $\Sigma [CuLi]$
the concentration of "natural organic" complexes of copper and $[CuNTA^-]$ is t
concentration of the copper complex with NTA.

subdivided into 300 ml portions in two series of polycarbonate flasks. One series received copper additions (0.07 to 1.1 $\mu mol \cdot l^{-1}$) and the second received copper (0.43 to 2.9 $\mu mol \cdot l^{-1}$) and NTA (1.8 $\mu mol \cdot l^{-1}$). The amended water was equilibrated for one h before addition of 50 ml of unfiltered water containing its natural assemblage of phytoplankton. The flasks were then incubated under fluorescent lighting for periods of 4 and 14 h before measurement of photosynthetic activity in 100 ml subsamples. Standard ^{14}C techniques were used and incubations were conducted for 2 h. The experiment was conducted at pH 7.9 and 20°C. Two replicates were run per treatment.

Cupric ion activity in the amended lake water samples was measured with an Orion cupric ion electrode (Sunda and Lewis 1978; Sunda and Hanson 1979) 6 to 10 h after additions of copper and NTA. These measurements indicated that copper was highly complexed and that complexation was in large excess of that predicted for the formation of inorganic complexes only (Fig. 4). The measured level of complexation in the lake water containing NTA was higher than that in water without NTA as expected.

Scatchard analysis (Sunda and Hanson 1979) was applied to the titration curves of lake water with and without NTA to determine stability constants for copper complexation by NTA and by natural "organic" ligands. In the analysis of Cu-NTA chelation we first had to determine the concentration of CuNTA complex. This concentration is equal to the concentration of total copper in lake water containing added NTA and copper minus the concentration of total copper in the lake water containing only added copper at points in the titrations where the measured cupric ion activity is constant (Fig. 4). The Scatchard plot of $[CuNTA^-]/A_{Cu}$ vs $[CuNTA^-]$ was linear as would be predicted for the formation of a single 1:1 complex (Fig. 5). The slope and the X-intercept gave values for the conditional stability constant and total NTA concentration of $10^{9.98}$ and 1.84 $\mu mol \cdot l^{-1}$ respectively. This conditional constant was in excellent agreement with the value ($10^{10.00}$) computed from published constants (Martell and Smith 1974) and the measured gross inorganic composition of the lake water (e.g., concentrations of Ca, Mg, Na, K, HCO_3, and SO_4 and pH). This agreement provides direct evidence for the validity of pCu measurements by electrode in natural fresh water samples. The agreement also indicates that mixed complexes are not formed between copper, NTA, and natural organic ligands. This is an important result in that cupric ion bioassays utilizing NTA for internal pCu calibration are based on the assumption that there is no formation of such mixed complexes.

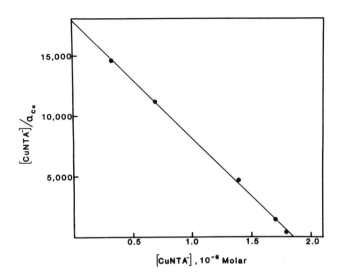

FIGURE 5. Scatchard plot showing the relationship between [CuNTA⁻]/A_Cu and [CuN for copper complexation by NTA in lake water. Linear regression line is shown.

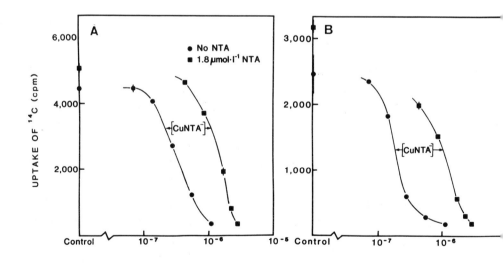

TOTAL ADDED COPPER (molar)

FIGURE 6. Relationship between photosynthetic carbon fixation and total concentration of added copper for the period 4-6 h (A) and 14-16 h (B) after exposure of phytoplankton to lake water containing added copper ● and copper-NTA mixtures ■. Each point represents the mean of two replicates and error bars represent the range.

Scatchard analysis was also applied to copper binding in lake water without added NTA (Sunda and Hanson 1979). The analysis showed that the titration data could be modeled on the basis of two "organic" ligands with concentrations of 0.17 and 1.2 $\mu mol \cdot l^{-1}$ and conditional stability constants at pH 7.9 of $10^{10.8}$ and $10^{8.6}$. These constants agree well with those for the complexation of copper at pH 8 by organic ligands in two rivers in North Carolina: the Newport and the Neuse (Sunda and Hanson 1979).

The relationships between ^{14}C-fixation by phytoplankton and added copper in the presence and absence of NTA are given in Fig. 6. As before, NTA decreased the toxicity of added copper. The rate of carbon fixation decreased with time in the controls and NTA added by itself was slightly stimulatory relative to controls. Such stimulation by NTA is problematic in that we do not know, a priori, its exact nature; that is, whether it results from alleviation of copper toxicity or whether there is some other effect, perhaps with another metal.

One of the primary criteria for a valid bioassay is that the added chelator used for pCu calibration must not itself affect the relationship between biological response and pCu. Plots of photosynthetic activity vs. measured pCu show that this criterion is in fact met (Fig. 7). The uptake of ^{14}C-bicarbonate

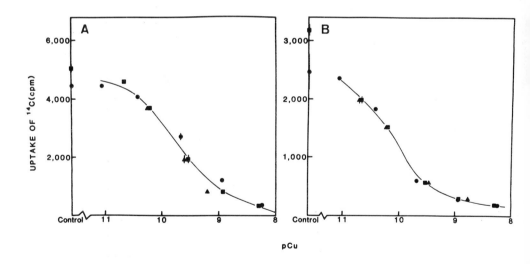

FIGURE 7. Relationship between ^{14}C-fixation and pCu measured with a cupric ion electrode in lake water with ■ and without ● NTA. Same relationship for pCu values determined by bioassay ▲. Each point represents the mean of two replicates and error bars represent the range. A. Results at 4-6 h. B. Results at 14-16 h.

is related to measured pCu independent of the presence of NTA. The fact that
the level of photosynthetic activity with addition of copper is related to pCu
strongly suggests that changes in pCu also explain the slight stimulatory effec
of added NTA in the absence of added copper; i.e., NTA addition alleviates the
toxicity of ambient copper in the water. It is possible, however, that this
toxicity is due to sample contamination.

Inhibition of photosynthetic activity by copper is greater at 14-16 h than
4-6 h. At 4-6 h inhibition occurs at cupric ion acitivities above $10^{-10.6}$ mol
with half inhibition at an activity of $\sim 10^{-9.6}$ mol·l^{-1} and total inhibition a
$\sim 10^{-8}$ mol·l^{-1}. At 14-16 h half inhibition occurs at a cupric ion activity of
$10^{-10.3}$ mol·l^{-1}, a factor of 5 below that observed at 4-6 h. Such increased
copper inhibition of photosynthesis with time also has been observed for unial
cultures of <u>Chlorella</u> (Steemann Nielsen et al. 1969).

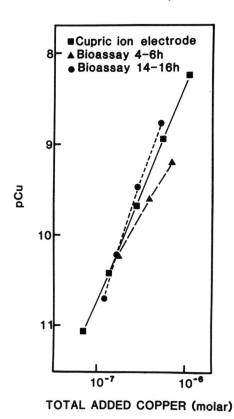

FIGURE 8. Relationship between pCu and total added copper measured by cupric i
electrode and by phytoplankton bioassay.

The observed dependence of toxic response on pCu agrees with previous findings (Sunda and Guillard 1976, Anderson and Morel 1978, Jackson and Morgan 1978, Sunda and Lewis 1978). In this case, however, we have shown directly that the relationship between phytoplankton response and pCu is not only independent of the total concentration of copper or the total level of copper complexation, but also that it is the same whether copper is complexed primarily by natural ligands or by a synthetic ligand (NTA).

The procedure for obtaining cupric ion activity from bioassay data is the same as that described for bacterial bioassays. As before we note that the concentration of CuNTA complex in the media containing added NTA should equal the total concentration of dissolved copper in that medium minus the concentration in lake water without NTA for which the photosynthetic response in both media is equal (Fig. 6). Cupric ion activity is then computed from the equilibrium relationship between cupric ion activity and the concentrations of NTA and CuNTA complex taking into account NTA side reactions with calcium, magnesium, and hydrogen ions.

A comparison of the relationship between ^{14}C-uptake and bioassayed pCu values with that between ^{14}C-uptake and pCu measured with a cupric ion electrode is given in Fig. 7. In general there is a fair agreement for the bioassay based on response at 4-6 h and an excellent agreement for that at 14-16 h. A comparison of the relationship between pCu and total copper concentration obtained from the bioassay with that obtained from electrode measurements also shows good agreement, particularly for bioassay values obtained from the 14-16 h toxicity data (Fig. 8).

These results confirm the validity of using natural populations of phytoplankton to bioassay cupric ion activity and also point to the importance of factors such as incubation time in bioassay results. Long incubation times favor equilibration, with respect to both the chemistry of copper in the medium and the interaction of organisms with the altered copper chemistry. On the other han w often are interested not only in the chemistry of copper in the water, b   o the natural population's response to increases in total copper concentra n d cupric ion activity. The longer a population is held in a container the greater the tendency for it to change from its initial state and composition. The choice of exposure time therefore, of necessity represents a compromise.

6. CONCLUSION

The results discussed here indicate that toxicity bioassays utilizing internal cupric ion calibration can be used to quantify relationships between and total copper concentration and cupric ion activity in natural water samples and, there-

408

fore, can be used to assess the complexing characteristics of samples. Unlike typical anodic stripping techniques, both labile (those with rapid dissociatio kinetics; e.g., $CuCO_3$) and nonlabile complexes are detected. The technique al has advantages over cupric ion electrodes in that it can be used in both sali and fresh water samples and with sensitive bioassay organisms, it is able, in many instances, to measure complexation at lower concentrations of total copp Bioassay techniques, however, are both cumbersome and time consuming and are limited to metals such as copper which are highly toxic to organisms. They a require good theoretical background and practial skills in specialized areas both chemistry and biology. Consequently, they do not readily lend themselve to routine analysis of the complexing characteristics of natural water sample

This research was funded by a contract from the Ocean Assessments Division, National Ocean Services, NOAA and by a grant from the Norwegian Research Coun for Science and Humanities.

REFERENCES

- Anderson, D.M., J.S. Lively, and R.F. Vaccaro, (in press) Copper complexation during spring phytoplankton blooms in coastal waters. J. Mar. Res.
- Anderson, D.M., and F.M.M. Morel, 1978. Copper sensitivity of Gonyaulax tamarensis. Limnol. Oceanogr. 23: 283-295.
- Chau, Y.K., R. Gächter, and K. Lum-Shue-Chan, 1974. Determination of the apparent complexing capacity of lake waters. J. Fish. Res. Board Can. 31: 1515-1519.
- Davey, E.W., J.J. Morgan, and S.J. Erickson, 1973. A biological measurement of the copper complexation capacity of seawater. Limnol. Oceanogr. 18: 993-997.
- Duinker, J.C., and C.J.M. Kramer, 1977. An experimental study on the speciation of dissolved zinc, cadmium, lead, and copper in river Rhine and north seawater by differential pulsed anodic stripping voltammetry. Mar. Chem. 5: 207-228.
- Gächter, R., Lum-Shue, K. Chan, and U.K. Chau, 1973. Complexing capacity of the nutrient medium and its relation to inhibition of algal photosynthesis by copper, Schweiz. Z. Hydrol. 35: 252-261.
- Gillespie, P.A., R.F. Vaccaro, 1978. A bacterial bioassay for measuring the copper-chelation capacity of seawater. Limnol. Oceanogr. 23: 543-548.
- Jackson, G.A., and J.J. Morgan, 1978. Trace metal-chelator interactions and phytoplankton growth in seawater media: Theoretical analysis and comparison with reported observations. Limnol. Oceanogr. 23: 268-282.
- Mantoura, R.F.C., A. Dickson, and J.P. Riley, 1978. The complexation of metals with humic materials in natural waters. Estuarine Coastal Mar. Sci. 6: 387-408.
- Martell, A.E., and R.M. Smith, 1974. Critical stability constants, Vol. 1, Amino acids. Plenum Press, 469 pp.

- Paulson, A.J., 1978. Potentiometric studies of cupric hydroxide complexation. M.S. thesis, Univ. Rhode Island, Kingston, 102 pp.
- Steemann Nielsen E., Kamp-Nielsen L. & Wium-Anderson S. 1969. The effect of deleterious concentrations of copper on the phytosynthesis of Chlorella pyrenoidosa. Physiologia Pl. 22, 1121-33.
- Stumm W. and Morgan, J. J. 1970. Aquatic Chemistry: an Introduction emphasizing Chemicial Equilibria in Natural Waters. Interscience, New York.
- Sunda, W.G., and R.L. Ferguson, 1983. Sensitivity of natural bacterial communities to additions of copper and to cupric ion activity: a bioassay of copper complexation in seawater. In C.S. Wong, E. Boyle, K.W. Bruland, J.D. Burton and E.D. Goldberg (Eds), Trace metals in seawater. Plenum Press, 920 pp.
- Sunda, W.G., and P.A. Gillespie, 1979. The response of a marine bacterium to cupric ion and its use to estimate cupric ion activity. J. Mar. Res. 37: 761-777.
- Sunda, W.G. and R.R.L. Guillard, 1976. The relationship between cupric ion activity and the toxicity of copper to phytoplankton. J. Mar. Res. 34: 511-529.
- Sunda, W.G., and P.J. Hanson, 1979. Chemical speciation of copper in river water: Effect of total copper, pH, carbonate and dissolved organic matter. In E.A. Jenne (Ed), Chemical modeling in aqueous systems. American Chemical Society, Washington, 914 pp.
- Sunda, W.G., and J.M. Lewis, 1978. Effect of complexation by natural organic ligands on the toxicity of copper to a unicellular alga, Monochrysis lutheri. Limmnol. Oceanogr. 23: 870-876.
- Westall, J.C., F.M.M. Morel, and D.N. Hume, 1979. Chloride interference in cupric ion-selective electrode. Anal. Chem. 51: 1792-1798.
- Whitfield, M., 1975. Seawater as an electrolyte solution. In: J.P. Riley and G. Skirrow (Ed.), Chemical oceanography, Vol. 1, Academic Press, New York, 606 pp.
- Wilson, T.R.S. 1975, Salinity and the major elements of seawater. In: J.P. Riley and G. Skirrow (Eds), Chemical oceanography, Vol. 1, Academic Press, New York, 606 pp.

ALGAE AS INDICATORS OF COPPER SPECIATION

T.M. FLORENCE, B.G. LUMSDEN and J.J. FARDY

1. INTRODUCTION

It is usually assumed that the free, hydrated cupric ion is the
physico-chemical form of copper most toxic to aquatic organisms, and that
complexed copper is much less toxic (Magnuson et al., 1979). A recent
study (Florence, 1982) of the effect of model ligands on the speciation
of copper, lead, cadmium and zinc, led to the conclusion that in most
natural waters a large fraction of each of these metals is bound in
highly stable complexes by unidentified organic ligands. In addition,
many of the water samples were found to contain lipid-soluble metal.
Florence (1982; 1982a) pointed out that lipid-soluble metal complexes are
often neglected in biotoxicity studies, even though some such compounds
are known to be strongly bioaccumulated and highly toxic.

This investigation with copper complexes in seawater was undertaken
to establish the relationship between the fraction of total copper which
was toxic to Nitzschia closterium, and the concentration of labile copper
as measured by anodic stripping voltammetry (a.s.v.), chelating resins,
or solvent extraction. The mechanism of copper toxicity was also
studied, in particular the role played by oxygen free radicals.

2. EXPERIMENTAL

2.1 Apparatus and reagents

An E.G. and G. Princeton Applied Research Model 384 Polarographic
Analyzer, with a Model 303 static mercury drop electrode assembly, was
used for all a.s.v. measurements. The Chelex-100 and thiol resin columns
were prepared and maintained as described previously (Florence, 1982).
Pacific Ocean water was filtered on the day of collection and stored at

Kramer, C.J.M. and Duinker, J.C. (eds.), Complexation of Trace Metals in Natural Waters. ISBN 90-247-2973-4
© 1984 Martinus Nijhoff/Dr W. Junk Publishers, The Hague/Boston/Lancaster.
Printed in the Netherlands.

4°C. The filtered seawater contained 0.2 to 0.6 μgL^{-1} total copper, and 0.03 to 0.3 μgL^{-1} a.s.v.-labile copper.

2.2 Algal assays

The marine diatom Nitzschia closterium was cultured in seawater containing the following nutrients: $NaNO_3$, 0.15 gL^{-1}; $NaH_2PO_4.2H_2O$, 0.01 gL^{-1}; SiO_2, 6.0 mgL^{-1}; iron (III) citrate, 9 μgL^{-1}; citric acid, 9 μgL^{-1}; thiamine hydrochloride, 200 μgL^{-1}; biotin, 1 μgL^{-1}; vitamin B_{12}, 1 μgL^{-1}. No trace metals were added. The axenic cultures were illuminated with daylight fluorescent light at 6,000 lux, and were maintained on a 12 h/12 h light/dark cycle. To avoid complexing of copper by components of the culture medium (phosphate, silicate, colloidal iron), the algal assays were carried out in pure seawater by washing the algae three times with seawater using centrifugation, then adding aliquots of the washed algae suspension to 50 mL of seawater in 250-mL conical flasks to give an initial cell density of 2-4 x 10^4 cells mL^{-1} (Lumsden and Florence, 1983). The flasks were illuminated on a light box at 16,000 lux ($23\pm1^{\circ}$C) on a 12 h/12 h light/dark cycle, and counted daily for four days using a haemocytometer. Logarithmic growth was maintained for this period. Depression of growth rate was used as a measure of toxicity.

Copper and other toxicants were added to the assay flasks just before the algae, from either aqueous or ethanolic stock solutions. Ethanol, up to 0.5 mL per 50 mL of seawater, had no effect on algal growth rate.

3. RESULTS AND DISCUSSION

3.1 Effect of complexing agents on toxicity

The mediation of the toxicity of copper to Nitzschia closterium by some naturally-occurring complexing agents is shown in Table 1. The toxicity index for copper complexes is defined as the ratio of the apparent copper concentration (as determined from a growth rate-copper concentration curve), to the added copper concentration. The growth rates were corrected for the effect of excess ligand, although free ligand was always much less toxic then its copper complex (Florence et al., 1983). A high toxicity index implies that the complex is more toxic than inorganic copper(II), and vice versa.

The results of a similar study, but using synthetic ligands, are shown in Table 2. Oxine (8-hydroxyquinoline) and its derivatives, and

Table 1. THE TOXICITY OF COPPER COMPLEXES WITH SOME NATURALLY-OCCURRING LIGANDS IN SEAWATER.

Ligand [a]	Ligand Concn.	a.s.v.-labile copper, [b] %	Chelex-100 labile copper, [c] %	Copper on algae [d] %	Toxicity index of copper complex [e]
Fulvic acid	1.0×10^{-5} mol/L	1.5	66	<5	0.08
Humic acid	6.4 mgL^{-1}	81	100	63	0.70
Fe-humic colloid	f	70	100	53	0.60
Fe(III) colloid	2.0×10^{-5} mol/L	92	100	51	0.85
Tannic acid	5.9×10^{-7} mol/L	5.5	100	77	0.13
Alginic acid	10 mgL^{-1}	100	95	58	1.6
Desferal	2.0×10^{-6} mol/L	100	95	60	0.55

a   Each solution contained 20 µgCuL^{-1}.
b   pH8.2, deposition potential of -0.6V vs. Ag/AgCl.
c   Per cent of copper removed by chelating resin column.
d   With copper alone, 42% of Cu was adsorbed on the algae.
e   For definition, see text. Copper (II) alone has a toxicity index of 1.0.
f   1.0 mgL^{-1} Fe(III) and 5.3 mgL^{-1} humic acid.

Table 2. THE TOXICITY OF COPPER COMPLEXES WITH SOME SYNTHETIC LIGANDS IN SEAWATER.

Ligand [a]	a.s.v.-labile copper, %	Chelex-100 labile copper, %	Solvent extractable copper, [b] %	Copper on algae %	Toxicity index of copper complex
Oxine [c]	64	98	92	>90	20
PAN [c]	<0.5	72	84	>90	>25
TAN [c]	<0.5	65	92	>90	>25
DDTC [d]	<0.5	70	89	80	1.9
EXA [d]	10.5	63	93	80	3.5
NTA [d]	100	100	<10	40	0.20
LAS [e]	65	95	13	65	0.25

a   Abbreviations: PAN, 1-(2-pyridylazo)-2-naphthol;
    TAN, 1-(2-thiazolylazo)-2-naphthol;
    DDTC, diethyldithiocarbamate;
    EXA, ethyl xanthogenate;
    NTA, nitrilotriacetic acid;
    LAS, linear alkylbenzene sulphonate.
b   20% n-butanol in n-hexane.
c 5×10^{-8} mol/L ligand + 2 µg CuL^{-1}.
d 2×10^{-6} mol/L ligand + 20 µg CuL^{-1}.
e   0.5 mg L^{-1}.

the diethyldithiocarbamates are widely applied as fungicides, while
alkylxanthogenates are used in large quantities as mineral flotation
agents.

It is evident from Table 1 that many naturally-occurring complexing
agents decrease the toxicity of copper, and that for fulvic, tannic and
humic acids and the iron colloids, a.s.v.-labile copper (at a low
deposition potential) is a reasonable estimate of the toxic fraction of
copper (Florence et al., 1983). In natural waters, therefore, a.s.v. may
provide a useful determination of bioavailable, or toxic, copper,
although even in these samples much of the copper may be strongly bound
as porphyrin or metallothionein complexes (Florence, 1982a; Florence,
1983; Florence et al., 1983). Chelex-100 (Table 1) and thiol (Florence
et al., 1983) resins, on the other hand, grossly overestimate the toxic
fraction of copper, and their use for this purpose should be discouraged.

The situation with synthetic ligands is completely different (Table
2). Here a.s.v. provides no indication whatsoever of the toxicity of the
copper complex; solvent extractable (lipid soluble) complexes being much
more toxic than inorganic copper(II), and NTA and LAS decreasing the
toxicity to a greater extent than would be expected from the
a.s.v.-labile measurement. Even though NTA forms a strong complex with
copper(II), the complex is highly labile, and dissociates completely at a
mercury electrode (Figura and McDuffie, 1980). Lipid solubility is
obviously important to high toxicity, since the addition of a sulphonate
group to a ligand substantially reduced or eliminated its toxic effect
with copper (Table 3). Strong chelation with copper was also essential
for high toxicity. The copper complexes of 8-hydroxyquinoline, PAN, TAN
and 8-aminoquinoline were all very toxic, whereas the non-chelating
isomers, 4-hydroxyquinoline, 1-(3-pyridylazo)-2-naphthol,
1-(2-pyridylazo)-4-naphthol, 1-(3-thiazolylazo)-2-naphthol, and
3-aminoquinoline caused no increase in the toxicity of 20 μgL^{-1} of
copper.

The toxicity of the copper complexes of the phenanthrolines and some
related compounds provided further support for the theory that lipid
solubility is an essential requirement for toxicity. Table 4 lists the
solvent extraction and toxicity data, while Table 5 shows a.s.v.,
stability constant (Martell and Smith, 1974), and redox potential (James
and Williams, 1961) data for some of these complexes.

Table 3. THE EFFECT OF LIPID SOLUBILITY ON TOXICITY OF COPPER COMPLEXES IN SEAWATER.

Ligand [a]	Solvent extractable copper, % [b]	Toxicity index of copper complex
Oxine	92	20
Oxine-5-sulphonate	<10	0.35
Bathocuproine [c]	91	2.5
Bathocuproine disulphonate	<10	<0.1
1-(2-pyridylazo)-2-naphthol (PAN)	84	>25
4-(2-pyridylazo)-resorcinol (PAR)	<10	<0.1
1-(2-thiazolylazo)-2-naphthol (TAN)	92	>25
4-(2-thiazolylazo)-resorcinol (TAR)	<10	<0.1

a   Concentrations used were 5×10^{-8} mol/L ligand and 2 μgCuL^{-1} for the lipid-soluble complexes, and 2×10^{-6} mol/L ligand and 20 μgCuL^{-1} for the non-lipid soluble.
b   20% n-butanol in n-hexane.
c   2,9-dimethyl-4,7-diphenyl-1,10-phenanthroline.

Table 4.   THE TOXICITY OF THE COPPER COMPLEXES OF 1,10-PHENANTHROLINE AND RELATED LIGANDS IN SEAWATER.

Ligand	Solvent extractable copper, % [a]	Toxicity index of copper complex
1,10-phenanthroline [b]	57	1.2
5,6-dimethyl-1,10-phenanthroline [c]	59	3.0
2,9-dimethyl-1,10-phenanthroline [c]	91	>25
4,7-diphenyl-1,10-phenanthroline [c]	84	5.0
Bathocuproine [c,d]	91	2.1
2,2'-bipyridine [b]	49	0.35
2,2-biquinoline [c]	73	3.9

a   20% n-butanol in n-hexane plus 7.6×10^{-4} mol/L H$_2$O$_2$ in seawater.
b 2×10^{-6} mol/L ligand + 20 μgCuL^{-1}.
c 5×10^{-8} mol/L ligand + 2 μgCuL^{-1}.
d   2,9-dimethyl-4,7-diphenyl-1,10-phenanthroline.

Hydrogen peroxide, present in algal cultures, was found to be an efficient reductant for copper(II) to (I) when certain phenanthroline derivatives were present (Table 4). Neocuproine (2,9-dimethyl-1,10 -phenanthroline) is widely used as a spectrophotometric reagent for copper(I). Steric hindrance by the methyl groups in this compound favours the formation of a copper(I) complex, which is reflected in the high β_2, ΔE_p and E_f values (Table 5), and its high extractability (Table 4).

Table 5. SOME PROPERTIES OF COPPER COMPLEXES WITH PHENANTHROLINE-TYPE LIGANDS.

Ligand	Logβ_1 or β_2			ΔE_p Cu(0)/Cu(I), mV[a]	E_f Cu(II)/Cu(I v.vs. N.H.E
	Cu(II)L$_1$	Cu(II)L$_2$	Cu(I)L$_2$		
1,10-phenanthroline	7.4	15.7	15.8	−52	0.174
5,6-dimethyl-1,10 -phenanthroline	8.7	15.7	−	−42	0.13 [c]
2,9-dimethyl-1,10 -phenanthroline	5.2	11.0	19.1	−150	0.594
4,7-diphenyl-1,10 -phenanthroline	−	−	−	−149	0.11 [c]
2,2'-bipyridine	6.3	13.6	13.2	−76	0.120
2,2'biquinoline	−	7.7	16.5	−149	0.65 [c]
Bathocuproine	−	−	−	−132	−

a   Shift in a.s.v. peak potential of copper(II) on addition of ligand.
b   Standard reduction potential for the copper(II)/copper(I) complex in water.
c   Estimated from values in 50% dioxan (James and Williams, 1961).

Neocuproine is by far the most toxic of the copper-complexing phenanthrolines (Table 4), a property which is probably a result of the high lipid solubility of the copper(I) complex.   Lipid solubility, howeve may not be the only factor affecting toxicity, since bathocuproine (2,9-dimethyl-4,7-diphenyl-1,10-phenanthroline) also forms a stable, lipid-soluble copper(I) complex, yet it is far less toxic than neocuproin (Table 4).

3.2 Mechanisms of the toxic effect

It is apparent from the results presented that strong chelation with copper plus high lipid solubility are essential for toxicity.   In the case of oxine, PAN, TAN, and similar ligands, the complex almost certainly enters the algal cell as copper(II), whereas with neocuproine and, possibly, sulphur donor ligands such as diethyldithiocarbamate and the xanthogenates, lipid-soluble copper is in the I valency state. Toxicity is therefore not dependent on the valency of copper.   What, then, is the injurious action of these copper complexes inside the cell? Inhibition of algal growth can be observed with some ligands (e.g. PAN,

TAN, and neocuproine) at concentrations of the copper complex as low as 2×10^{-9} mol/L in seawater (Florence et al., 1983). The liberation (from the complex) of toxic levels of free cupric ion in the cell is therefore unlikely to be the toxic event, especially as these complexes are very stable. Some candidate mechanisms for the toxic effect are: (a) catalysis by the copper complexes of the formation of hydrogen peroxide and oxygen free radicals from molecular oxygen, (b) intercalation of the complexes into DNA, and (c) inhibition of DNA or RNA polymerase.

The production of oxygen free radicals in the cell could lead to the oxidation of glutathione, lipids, and other sensitive materials. Copper complexes are known (Walling, 1975) to catalyze the Fenton reaction.

$$O_2 + e^- \rightarrow O_2^- \text{ (superoxide radical)}$$
$$O_2^- + 2H^+ + e^- \rightarrow H_2O_2$$
$$Cu^{II}L_2 + O_2^- \rightarrow Cu^{I}L_2 + O_2$$
$$Cu^{I}L_2 + H_2O_2 \rightarrow Cu^{II}L_2 + OH^- + OH \cdot \text{ (hydroxyl radical)}$$

The hydroxyl radical, OH·, reacts at almost diffusion-controlled rates with most biological substances, and would cause devastating damage if present in significant concentrations in the cell. Hydrogen peroxide, probably formed from superoxide radical precursor, was measured at concentrations above 1×10^{-4} mol/L in illuminated Nitzschia culture solutions (Florence and Stauber, unpublished results). Catalase (20 µg mL^{-1}) or superoxide dismutase (40 µg mL^{-1}) caused a 50% increase in growth rate of a Nitzschia culture illuminated at 16,000 lux, and it is possible that the production of hydrogen peroxide and oxygen free radicals in surface waters (Cooper and Zika, 1983) is a natural control mechanism on algal growth.

However, some of the most toxic copper complexes, e.g. those of oxine and PAN, are unlikely catalysts for redox reactions, since an effective catalyst must itself undergo oxidation-reduction, and should have an intermediate redox potential. Studies are at present underway to measure the catalytic effect of several of the toxic copper complexes on biological oxidation reactions.

Copper complexes are known (Mikelens et al., 1976; White, 1977) to form intercalative complexes with DNA. It is possible that the large, and unexpected, difference in toxicity between the copper complexes of 2,9-dimethyl-1,10-phenanthroline and 2,9-dimethyl-4,7-diphenyl-1,10-phenanthroline (Table 4) is due to the inability of the bulkier

418

4,7-diphenyl derivative to form an intercalative complex. The copper(I)-
1,10-phenanthroline complex is a potent inhibitor of DNA and RNA
polymerase (D'Aurora et al., 1978). However, the 2,9-dimethyl-1,
10-phenanthroline complex, although much more toxic than the
unsubstituted ligand (Table 4), was found to be completely
non-inhibitory.

REFERENCES

- Cooper, W.J. and Zika, R.G., 1983. Photochemical formation of
  hydrogen peroxide in surface and ground waters exposed to sunlight.
  Science 220: 711-712.
- D'Aurora, V., Stern, A.M. and Sigman, D.S., 1978. 1,10-phenanthroline
  -cuprous ion complex, a potent inhibitor of DNA and RNA polymerases.
  Biochim.Biophys.Res.Comm. 80: 1025-1032.
- Figura, P. and McDuffie, B., 1980. Determination of the labilities of
  soluble trace metal species in aqueous environmental samples by
  anodic stripping voltammetry and Chelex column and batch methods.
  Anal.Chem. 52: 1433-1440.
- Florence, T.M., 1982. Development of physico-chemical speciation
  procedures to investigate the toxicity of copper, lead, cadmium and
  zinc towards aquatic biota. Anal.Chim.Acta 141: 73-94.
- Florence, T.M., 1982a. The speciation of trace elements in waters.
  Talanta 29: 345-364.
- Florence, T.M., 1983. Trace element speciation and aquatic
  toxicology. Trends Anal.Chem. 2: 162-166.
- Florence, T.M., Lumsden, B.G. and Fardy, J.J., 1983. Evaluation of
  some physico-chemical techniques for the determination of the
  fraction of dissolved copper toxic to the marine diatom Nitzschia
  closterium. Anal.Chim.Acta 151: 281-295.
- James B.R. and Williams, R.J., 1961. The oxidation-reduction
  potentials of some copper complexes. J.Chem.Soc. 2007-2019.
- Lumsden, B.G. and Florence, T.M., 1983. A new algal assay procedure
  for the determination of the toxicity of copper species in seawater.
  Environ.Tech.Lett. 4: 271-276.
- Magnuson, V.R., Harris, D.K., Sun, M.S., Taylor, D.K. and Glass,
  G.E., 1979. Relationships of activities of metal-ligand species to
  aquatic toxicity. In: E.A. Jenne (Ed.), Chemical modelling in aqueous
  systems. American Chemical Society, Washington D.C., p.635.
- Martell, A.E. and Smith, R.M., 1974. Critical stability constants.
  Plenum Press, New York.
- Mikelens, P.E., Woodson, B.A. and Levinson, W.E., 1976. Association
  of nucleic acids with complexes of N-methyl isatin-β-thiosemi-
  carbazone and copper. Biochem.Pharmacol. 25: 821-827.
- Walling, C., 1975. Fenton's reagent revisited. Accounts Chem.Res. 8:
  125-131.
- White, J.R., 1977. Streptonigrin-transition metal complexes: binding
  to DNA and biological activity. Biochem.Biophys.Res.Comm. 77:
  387-391.

METAL SPECIATION - BIOLOGICAL RESPONSE.

An evaluation of the assumed close connection between metal
speciation and biological response.

M. LAEGREID, J. ALSTAD, D. KLAVENESS, H.M. SEIP

1. INTRODUCTION

The prevailing hypothesis concerning metal toxicity towards
phytoplankton is that a given free metal ion activity causes a
certain toxic effect. This has been shown by phytoplankton
toxicity experiments in synthetic media (Davey et al. 1973, Sunda
& Guillard 1976, Anderson & Morel 1978) and in humic localities
(Gächter et al. 1978, Sunda & Lewis 1978, Gjessing 1980, Toledo
et al. 1980, Sedlacek et al. 1983, Laegreid et al. 1983). Therefore
it is inferred that biotests can be used as a method for measuring
the metal complexing capacity in natural waters in general.

In less humic lakes, however, the situation may be somewhat
more complicated. As shown by the works of Gächter (1976),
Steemann-Nielsen & Bruun-Laursen (1976) and Laegreid et al. (1983),
in natural lake waters there may be an increased toxicity during
summer months when the water chemistry undergoes large changes
due to increased solar radiation, biological activity or other
seasonal processes. In our work from lake Gjersjøen (Laegreid
et al. 1983) we found that the toxicity of a given amount of
cadmium increases more than ten-fold during summer months. The
same effect was also observed for nickel (Laegreid 1983).

A plausible explanation would be the depletion of essential
trace elements (like Fe and Mn) during growth season, and thereby
an increased uptake of the toxic elements.

The addition of 10^{-7} M Fe or Mn did in fact reduce the toxic
response, but the same result was obtained just by storing the
filtered lake water in a cold chamber for 1-2 weeks before
performing the toxicity tests. Thus, the observation of the
increased toxic effect can not be explained by a lack of
essential trace elements only.

Kramer, C.J.M. and Duinker, J.C. (eds.), Complexation of Trace Metals in Natural Waters. ISBN 90-247-2973-4
© *1984 Martinus Nijhoff/Dr W. Junk Publishers, The Hague/Boston/Lancaster.*
Printed in the Netherlands.

The experiments presented in this paper were performed to explore the possible causes of the results referred to above, that apparently were inconsistent with the prevailing hypothesis. As will be shown, the question of what is toxic among metal spec is probably more complicated to answer than previously thought.

2. MATERIALS AND METHODS

Biotest experiments with cadmium and nickel were performed i the artificial medium "Fraquil" (Morel et al. 1979) omitting EDTA and copper, and with the addition of the ligand citric acid (10^{-7} - 10^{-5} M) or nitrilotriacetic acid (NTA) (10^{-7} - 10^{-6} M). The green alga <u>Selenastrum capricornutum</u> Printz are used as a test organism in the experiments. To ensure that the algae are adapted to the given medium, the algae were transferred, by sterile techniques, to subsamples of the medium where they would adapt before the final transfer to the test solutions. The experimental conditions were otherwise as described in Laegreid et al. (1983).

The computer program MINEQL (Westall et al. 1976) was used to estimate metal speciation. The metal-inorganic complexation nev exceeded 5% of the total metal concentration except for nickel at a concentration above 3.4×10^{-5} M, where $Ni(OH)_2$ makes an increasing contribution. The metal-organic complexation is estimated to be less than 30%.

3. RESULTS AND DISCUSSION

Figures 1 and 2 show the reduction of short-term ^{14}C-uptake after 24 h as a function of calculated free metal ion concentrat in the presence of different types and concentrations of organic ligands. As shown in fig. 1, in the concentration range 10^{-7} - 10^{-6} M, there is an increasing tolerance toward <u>free</u> metal ions with increasing concentration of NTA. This is especially pronounced in the case of nickel, but it is also clear in the case of cadmium. The results are indipendent of the stability constants used for the calculations. Fig. 2 shows the dose-response-curves at 10^{-7} - 10^{-5} M citric acid. The dose-response-curves at 10^{-7} - 10^{-6} M citric acid, where a maximum of 2% of Cd are organically complexed, the curves are coinciding,

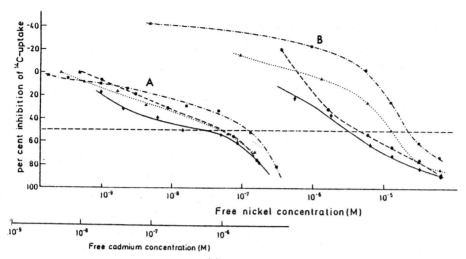

FIGURE 1. Percent reduction of ^{14}C-uptake as a function of free metal ion activity in Fraquil solution at varying concentrations of NTA.——◆—— 10^{-7} M NTA, ---●--- 3×10^{-7} M NTA, ...▲.... 6×10^{-7} M NTA, --■--·--. 10^{-6} NTA. A. cadium B. nickel.

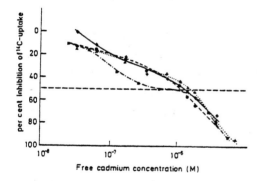

FIGURE 2. Per cent reduction of ^{14}C-uptake as a function of free cadmium activity in Fraquil solution at varying concentrations of citric acid.——◆——10^{-7}M, --●-- 6×10^{-7} M, ...▲...10^{-6}M, --·■-·10^{-5} M.

whereas at 10^{-5} M citric acid, where about 25% of Cd are organically complexed, there seems to be a somewhat increased toxicity at the lower Cd concentrations. This is in agreement with Guy &Ross Kean (1980) who found that (Cu-citrate-OH)$^{2-}$ was inhibitory toward Selenastrum.

Variations in toxic response not explainable in terms of changes in free metal concentration have long been known for mercury; several organomercurials are 100 times as toxic as inorganic mercuric chloride towards phytoplankton (Harriss 1971). A similar behaviour is found for cadmium and the ligand diethyl-dithiocarbamate towards zooplankton (Poldoski 1979). Wong et al.

422

(1982) found organic tin compounds to be much more toxic than inorganic tin - they also found a relationship between toxicity and partition coefficients upon extraction into octanol.

In our laboratories, Fadum (1983) found that Selenastrum was able to incorporate citric acid into the cell as an active proces her results are entirely comparable to those of Bollmann & Robins (1977) employing citric acid and other low molecular weight organic substances, and a number of different green algae (closely related to Selenastrum). Sverdrup (1982) also found indications on a different reaction towards free copper ions when 3,4-dihydroxybenzoic acid served as an organic ligand rathe than NTA. Several authors (Giesy & Paine 1977, Giesy et al. 1977a,b, Sedlacek et al. 1982) have shown that the lowest molecu weight fraction of naturally occuring organics may increase the metal uptake into organisms. These results, in connection with the published results refered to above, indicate that the assumed close connections between free metal ion activity and toxicity are not generally valid (toxicity is not necessarily a sole function of free metal ion activity), but that the toxic respons in certain cases also may be a function of type and concentratio of organic ligand present. It appears therefore that interpreti anything about metal speciation from biotest experiments should be done with great care. And biotests may not be suitable as a method for estimating the total metal complexing capacity in natural waters.

REFERENCES
- Anderson, D.M., Morel, F.M.M., 1978. Copper sensitivity of Gonyaulax tamarensis. Limnol. Oceanogr. 23(2): 283-295.
- Bollman, R.C., Robinson, G.G.C. 1977. The kinetics of organic acid uptake by three Chlorophyta in axenic culture. J. Phycol. 13: 1-5.
- Davey, E.W., Morgan, M.J., Erickson, S.J., 1973. A biological measure-ment of copper complexation capacity of seawater. Limnol. Oceanogr. 18: 993-997.
- Fadum, E., 1983. Undersøkelse av opptak av citrat hos Selenastrum capricornutum Printz. Thesis, University of Oslo.
- Gächter, R., 1976. Untersuchungen über die Beeinflussung der plank-tischen Photosynthese durch anorganische Metallsalze im eutrophen Alpnachersee und der mesotrophen Horver Burcht. Schweitz. Z. Hydrol. 38: 97-120.
- Gächter, R., Davis, J.S., Mares, A., 1978. Regulation of copper availability to phytoplankton by macromolecules in lake water. Environ. Sci. Technol. 12: 1416-1422.

- Giesy, J.P. Jr., Leversee, G.T., Williams, D.R., 1977a. Effects of naturally occuring aquatic organic fractions on cadmium toxicity to Simocephalus serrulatus (Daphnidae) and Gambusia affinis (Poecilliidae). Water Res. 11: 1013-1020.
- Giesy, J.P. Jr., Paine, D., 1977. Effects of naturally occuring aquatic organic fractions on ^{241}Am uptake by Scenedesmus obliguus (Chlorophyceae) and Aeromonas hydrophila (Pseudomonadaceae). Appl. & Environ. Microbiol. 33(1): 89-96.
- Giesy, J.P.Jr., Paine, D., Hersloff, L.W., 1977b. Effect of naturally occuring organics on plutonium-237 uptake by algae and bacteria. In: Transuranics in Natural Environment. M.G. White & P.B. Dunaway (eds.): 531-543.
- Gjessing, E.T., 1980. The effect of aquatic humus on the biological availability of cadmium. Arch. Hydrobiol. 91(2): 144-149.
- Guy, R.D., Ross Kean, C.L., 1980. Algae as a chemical speciation monitor - I. A comparison of algal growth and computed calculated speciation. Water Res. 14: 891-899.
- Harriss, R.C., 1971. Ecological implications of mercury pollution in aquatic systems. Biol. Conserv. 3(4): 279-283.
- Laegreid, M., 1983. En vurdering av kadmiums og nikkels kjemiske tilstandsform og gifteffekt overfor algen Selenastrum capricornutum Printz i to forskjellige innsjøsystemer. Thesis. University of Oslo.
- Laegreid, M., Alstad, J., Klaveness, D., Seip, H.M., 1983. Seasonal variation of cadmium toxicity towards the alga Selenastrum capricornutum Printz in two lakes with different humus content. Environ. Sci. Technol. 17(6): 357-361.
- Morel, F.M.M., Reuter, J.G., Anderson, D.M., Guillard, R.R.L., 1979. AQUIL: A chemical defined phytoplankton culture medium for trace metal studies. J. Phycol. 15: 135-141.
- Poldoski, J.E., 1979. Cadmium bioaccumulation assays. Their relationships to various ionic equilibria in lake Superiour water. Environ. Sci. Technol. 13: 701-706.
- Sedlacek, J., Källqvist, T., Gjessing, E.T., 1982. Drikkevannsrapport 5/82, NTNFs utvalg for drikkevannsforskning. Norwegian institute for water research, 1. nat. rapport.
- Sedlacek, J., Källqvist, T., Gjessing, E.T., 1983. Effect of aquatic humus on the uptake and toxicity of cadmium to Selenastrum capricornutum Printz. In: Aquatic and Terrestrial Humic Materials; Christman, R.F., Gjessing, E.T. (eds.). Ann Arbor Sci.: 495-516.
- Steemann-Nielsen, E., Bruun-Laursen, H., 1976. Effect of $CuSO_4$ on the photosynthetic rate of phytoplankton in four Danish lakes. Oikos 27: 239-242.
- Sunda, W., Guillard, R.R.L., 1976. The relationship between cupric ion activity and the toxicity of copper to phytoplankton. J. Mar. Res. 34 (4): 511-529.
- Sunda, W., Lewis, J.A.M., 1978. Effect of complexation by natural organic ligands on the toxicity of copper to a unicellular alga, Monochrysis lutheri. Limnol. Oceanogr. 23(5): 870-876.
- Sverdrup, A.C., 1982. Effekter av kobber på Selenastrum capricornutum Printz med vekt på forskjeller i veksthastighet og ^{14}C-opptak. Thesis, University of Oslo.
- Toledo, A.P.P., Tundisi, J.G., D'Aquini, V.A., 1980. Humic acid influence on the growth and copper tolerance of Chlorella sp. Hydrobiol. 71: 261-263.

- Westall, J.C., Zachary, J.L., Morel, F.M.M., 1976. MINEQL - A computer program for the calculation of chemical equilibrium composition of aqueous systems. Technical note no. 18. R.M. Parsons laboratory for water resources and environmental engineering, Massachusetts Institute of Technology, Cambridge, Massachusetts. 91 pp.
- Wong, P.T.S., Chau, Y.K., Kramar, O., Bengert, G.A., 1982. Structure - toxicity relationship of tin compounds on algae. Can. J. Fish. Aquat. Sci. 39: 483-488.

COPPER AND CADMIUM SPECIATION IN DIFFERENT PHYTOPLANKTON CULTURE MEDIA

M. GNASSIA-BARELLI, M. HARDSTEDT-ROMEO AND E. NICOLAS

1. INTRODUCTION

Phytoplankton is generally considered as the main source of dissolved organic matter in the sea. Organic matter may change the chemical state of metals and, therefore, their bioavailability and toxicity to marine producers. In order to study this phenomenon in vitro, we investigated the chemical state of dissolved copper and cadmium in phytoplankton media which are rich in natural organic exudates. Copper was shown to be less toxic to Hymenomonas elongata in presence of exudates (Gnassia-Barelli et al., 1978).

2. MATERIAL AND METHODS

The batch cultures (18°C, continuous light) were grown aseptically in sea water,enriched with 20 mg KNO_3, 10 mg K_2HPO_4 and 1 mg Na_2SiO_4 per liter and filtered over 0.2 µm Sartorius filters. The enrichment is sufficient to allow a large growth rate without introducing potential complexing agents. The Haptophyceae Hymenomonas elongata (Droop) Braarud, the Prasinophyceae Prasinocladus marinus (Ceink.) Waern, the Dinophyceae Amphidinium carterae Hulburt and the 2 Bacillariophyceae Chaetoceros curvisetum Cleve and Chaetoceros protuberans Gran and Yendo were studied. At the end of the log-phase of the growth, the media were rich in organic exudates. Then, phytoplankton cells were discarded by filtration on 0.2 µm filters. Media were incubated with Cu and Cd during 24 h and transfered to an Amicon 2000 Ultrafiltration cell (dark room at 4°C). The different molecular weight (MW) fractions from ultrafiltration (UF) were collected, acidified with HCl and UV irradiated with two low pressure mercury Philips lamps (15 W) during 12 h. Dissolved Cu and Cd in the fractions were analyzed by atomic absorption spectrometry (Härdstedt-Roméo and Gnassia-Barelli, 1980).

Kramer, C.J.M. and Duinker, J.C. (eds.), Complexation of Trace Metals in Natural Waters. ISBN 90-247-2973-4
© *1984 Martinus Nijhoff/Dr W. Junk Publishers, The Hague/Boston/Lancaster.*
Printed in the Netherlands.

3. RESULTS

No metal retention was observed on the ultrafilters. Moreover, the higher MW material was concentrated not more than 10 times to avoid under- and over-estimation of the low and high MW fractions, respectively.

Table 1 shows the results concerning Cu fractionation in H. elongata mediu

TABLE 1. Cu concentrations in initial H. elongata medium and in MW fractions after UF. Cu is introduced into the medium at ca 25 and 20 µg 1^{-1}.

Fraction studied		Cu(μg 1^{-1})		Cu(μg 1^{-1})	
H. elongata	medium	26.2		20.7	
> 10,000	MW	2.6		1.0	
1000 - 10,000	MW	7.2 }	19.2	12.5 {	18.7
500 - 1000	MW	9.4		5.2	
< 500	MW	7.5		2.3	

The major part of Cu seems to be associated with the MW fraction ranging from 500 to 10,000 with a significant percentage in the 500-1000 fraction. Th comparison between experiments indicates that Cu in the lowest MW fraction varies as a function of the total concentration whereas Cu concentration in t fraction > 500 (ca 19 µg Cu 1^{-1} = 0.3 µM) does not depend on the total concentration. This value may be considered as the "complexing capacity" of H. elongata medium. Around 10 mg 1^{-1} carbohydrates and 6 mg 1^{-1} proteins were observed in H. elongata medium just before UF. Gelfiltration of the 500-10,000 MW fraction on Sephadex G 25 shows that Cu is eluted with the products absorbing at 280 nm. Only a small fraction of Cu is associated with the major part of carbohydrates which are eluted before proteins. Therefore, the complexation of Cu may be due to proteins present in H. elongata medium. Moreover, before UF, the physiological state of the cells was checked. The decay products are in low quantity | since the phaeophytin : chlorophyll a ratio never exceeds 10%.

As regards to the four other culture media taken into consideration, the incubated media (20 µg Cu 1^{-1}) of the same species were separated into 2 part UF was performed on the first part with a membrane of 1000 MW cut-off whereas a 500 MW membrane was used for the second part. The results (Table 2) show that, as for H. elongata medium, Cu is mainly associated with organic substances of MW > 500. The concentration of Cu linked to the 500-1000 fracti seems to depend on the phytoplankton species studied.

TABLE 2. Copper concentrations ($\mu g\ l^{-1}$) in the MW fractions collected from UF of the different media. Cu is introduced into the medium at ca 20 $\mu g\ l^{-1}$

Fractions		P. marinus	A. carterae	C. curvisetum	C. protuberans
> 1000	MW	17.4	8.9	10.9	12.3
500 - 1000	MW	0.5	7.9	4.6	2.0
< 500	MW	2.6	3.7	3.0	5.7

For a study of Cd, the different media are ultrafiltered on a 500 MW membrane after addition of 10 and 20 μg Cd l^{-1} (Table 3).

TABLE 3. Cd concentration ($\mu g\ l^{-1}$) in the MW fractions in the media.

Media	MW fractions	10 $\mu g\ l^{-1}$	20 $\mu g\ l^{-1}$
H. elongata	> 500	2.0	1.5
	< 500	8.0	19.0
P. marinus	> 500	2.9	5.7
	< 500	6.5	15.0
A. carterae	> 500	2.4	2.2
	< 500	9.4	17.8
C. curvisetum	> 500	2.1	2.8
	< 500	7.6	17.2
C. protuberans	> 500	2.0	3.4
	< 500	8.0	16.6

The link of Cd to organic substances of MW > 500 seems rather limited since it is mainly found in the lowest MW fraction. Even if Cd in the medium increases (x2), Cd associated to the MW > 500 remains constant ca 3 $\mu g\ l^{-1}$ i.e. 0.03 μM. This may be the Cd "complexing capacity" of the media, it is much lower than Cu "complexing capacity" evaluated for H. elongata medium. P. marinus medium seems to be able to complex more Cd than the other media.

4. DISCUSSION

Ultrafiltration (with 500 MW filters) was also carried out on non-inocculated enriched sea water, contaminated with 20 $\mu g\ l^{-1}$ Cu or Cd. Cu is found equally distributed between the 2 fractions in enriched sea water. It is, therefore, less complexed than in phytoplankton media. The complexation of Cd is low in enriched sea water where 95% of the metal are recovered in the lowest MW fraction. Thus, exudates from the phytoplankton species studied appear to have a complexing capacity towards Cu and to a lesser extent towards Cd. The observations are in agreement with results of Hasle and Abdullah (1981), who fractionated dissolved Cu and Cd in coastal water by UF and DPASV. Cd was present in low MW < 1000 labile species whereas Cu distribution was irregular with extensive organic and colloidal association. Fisher and Fabris (1982) studied the metal complexing capacity of exudates from 3 marine diatoms by DPASV. Cu was the most complexed metal, being ca 10 times more reactive than Cd. The Cu capacity ranged from 0.2 to 0.4 μM. Van den Berg et al. (1979), using an ion exchange method, noted that 3 algal species produced Cu complexing ligands in varying quantities from 0.66 to 6.73 μM. Nevertheless, Swallow et al. (1978) found by potentiometric titration that, among the 8 marin algae studied, only one could produce Cu complexing materials at a concentration higher than 1 μM.

Cu and Cd introduced as ions into phytoplankton media behave differently. Cu forms complexes with exudates whereas Cd remains essentially ionic. Cu and Cd exhibited also different patterns as regards their accumulation by H. elongata as shown in previous experiments (Gnassia-Barelli and Härdstedt-Roméo, 1982). Metal uptake by the other species and the relationship with the chemical state of the metal are both being studied.

REFERENCES

-Fisher, N.S. and J.G. Fabris, 1982. Complexation of Cu, Zn and Cd by metabolites excreted from marine diatoms. Mar. Chem. 11: 245-255.
-Gnassia-Barelli, M., M. Roméo, F. Laumond and D. Pesando, 1978. Experimenta studies on the relationship between natural copper complexes and their toxicity to phytoplankton. Mar. Biol. 47: 15-19.
-Gnassia-Barelli, M. and M. Härdstedt-Roméo, 1982. Short-term time series study of copper and cadmium uptake by Cricosphaera elongata (Droop) Braarud. J. Exp. Mar. Biol. Ecol. 61: 287-298.
-Härdstedt-Roméo, M. and M. Gnassia-Barelli, 1980. Effect of complexation by natural phytoplankton exudates on the accumulation of cadmium and copper by the Haptophyceae Cricosphaera elongata. Mar. Biol. 59: 79-84.
-Hasle, J.R. and M.I. Abdullah, 1981. Analytical fractionation of dissolved copper, lead and cadmium in coastal seawater. Mar. Chem. 10: 487-503.
-Swallow, K.C., J.C. Westall, D.M. McKnight, N.M.L. Morel and F.M.M. Morel, 1978. Potentiometric determination of copper complexation by phytoplankton exudates. Limnol. Oceanogr. 23: 538-542.
-Van den Berg, C.M.G., P.T.S. Wong and Y.K. Chau, 1979. Measurement of complexing materials excreted from algae and their ability to ameliorate copper toxicity. J. Fish. Res. Board Can. 36/ 901-905.

COMPEXATION BY DIATOM EXUDATES IN CULTURE AND IN THE FIELD

B. IMBER, M.G. ROBINSÖN and F. POLLEHNE

1. INTRODUCTION

The toxic effects of trace metals to marine phytoplankton have been
extensively studied for almost forty years but it is only recently that the
possible limiting nature of some essential trace metals in the environment
has been realized.

The response of algae to trace metals seems to be related to the free
ion activity of the element (Sunda and Guillard 1976, Jackson and Morgan
1978, Anderson et al., 1978). Recent measurements in open ocean
surface water have shown dissolved zinc concentrations of approximately
1 x 10^{-11}M (Bruland 1980, Bruland et al.,1978) and calculations using the
significant ionic interactions in sea water (Whitfield 1975) suggest a
total activity coefficient for this essential micro-nutrient of 0.19.
This would indicate that the zinc ion activities in ocean surface waters
are in the order of 10^{-11}M. Anderson et al.,(1978) have shown that the
optimal growth of the diatom Thalassiosira weissflogii (Grun.) occurs with
zinc ion activities in the range 10^{-9} to 10^{-11}M. Under these conditions
any additional complexation by organic ligands may have important
ecological effects.

In coastal waters however, the ambient zinc ion activity is somewhat
higher ($\approx 10^{-10}$M) than for open ocean waters and so some degree of organic
complexation may not effect the growth of coastal organisms. It is
possible that in such environments, organic chelation may act as a kind of
'buffer' to the biologically available metal fraction (Mantoura 1982). If
a high 'free' zinc concentration occurs, more organic chelation results,
while uptake of metal during low ambient conditions would lead to an
equilibrium release from organic complexes. With these factors in mind,
the concentration of metal that is organically complexed, the nature of
such complexes and the mechanisms for their formation are all critical for

Kramer, C.J.M. and Duinker, J.C. (eds.), Complexation of Trace Metals in Natural Waters. ISBN 90-247-2973-4
© *1984 Martinus Nijhoff/Dr W. Junk Publishers, The Hague/Boston/Lancaster.*
Printed in the Netherlands.

a true understanding of metal budgets and their biological impact in coastal environments.

Quantification of organic complexation (or potential complexation) can be made by complexiometric titrations with a chosen metal (Chau et al., 1974). However, several interpretations of such data have been made (Matson 1964, Shuman and Woodward 1977, Ruzic 1982).

The purpose of our investigations was to ascertain the existence and extent of complexation of zinc by diatom exudates and to evaluate the usefulness of complexing capacity (C_L) measurements in relation to such biological systems in the environment. Initially, experiments were carried out with diatom cultures and subsequently an enclosed "natural spring bloom" was investigated.

2. CULTURE EXPERIMENTS

2.1.Methods

The first experiments were performed on Thalassiosira fluviatilis. Cultures were grown in medium based on F/2 but initially free of zinc complexing ligands (Imber and Robinson 1983). Aliquots of culture were removed at regular intervals over a 37 day period and were analyzed for DOC, chlorophyll a and zinc.

The zinc complexing capacities were determined by complexiometric titration with added Zn^{2+} in the range 10^{-8} to 10^{-5}M. Additionally aliquots of these samples were photo-oxidized prior to determination and the slope of the latter determination was taken as being due to "free zinc ions" (Fig. 1). Initially, titrations showed one inflection point and C_L and conditional stability constant (K') were calculated using the procedure of Shuman and Woodward.(1977). At the beginning of the stationary phase a second site was observed making the interpretation of C_L more complex. The Shuman and Woodward procedure was again used to calculate the two analytically distinguishable ligand sites C_{L_1} and C_{L_2} (Fig.1) (Imber and Robinson 1983).

However, the interpretation of the magnitude of these sites is dependent on the relative magnitude of C_{L_1} x K'_1 and C_{L_2} x K'_2. If C_{L_2} x K'_2 $\gg$ C_{L_1} x K'_1 then from Fig.1, A = C_{L_2} and B = C_{L_2}. If, however, this criterion is not satisfied, and the two products are almost equal in magnitude, then initially sites from both ligand groups will be "filled" until the ligand of lower concentration is completely filled. Experiments with Thalassiosira fluviatilis produced values such that

$C_{L2} \times K'_2 \simeq 3 \times C_{L_1} \times K'_1$. Accordingly the determined values of C_L represent, in the case of C_{L_2}, maximum values and for C_{L_1} minimum values.

2.2. Results and Discussion

The complexing capacities (C_L) calculated by this procedure, together with Chlorophyll a concentration are depicted in Fig. 2. A rapid increase of both C_{L_1} and C_{L_2} was observed after the stationary phase. Subsequent experiments with Skeletonema costatum (Imber et al., 1983a) have shown similar results and it is postulated that C_{L_2} may be due to intracellular material derived from lysed cells. For all our culture experiments ultra-filtration showed no complexing below 10^4 nominal molecular weight (NMW). Stability constant data showed a logarithmic decrease throughout our culture experiments. This decrease was partly due to complex dissociation during the determination but when normalized for C_L the exudates from Thalassiosira fluviatilis still showed a logarithmic decrease (Fig. 3) indicating some change in the nature of the ligands that are produced. However, each ligand assemblage is still defined analytically as a separate site.

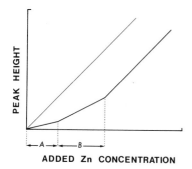

FIGURE 1. Titration curve showing analytically defined ligand sites

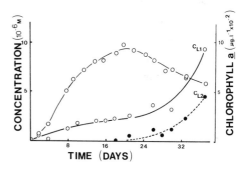

FIGURE 2. Variation of C_L and chlorophyll a with time from a Thalassiosira fluviatilis culture.

3. ENVIRONMENTAL EXPERIMENTS

3.1 Methods

In order to investigate the zinc complexation associated with a spring phytoplankton bloom and its possible environmental importance, a small CEPEX type bag was deployed in Patricia Bay, off Saanich Inlet in British Columbia. This bag was 2.5m in diameter and 16m deep (Grice & Reeve 1982). Samples were collected on a daily basis by peristaltic pump (see Pollehne and Imber 1983a) over a 15 day period.

Analyses for salinity, DOC, POC, dissolved nutrients, PON, PP, Chlorophyll a, cell counts productivity (C^{14}), labile zinc and zinc complexing capacity were performed. In general, integrated samples were collected from 0-3, 3-6, and 6-12 metres. Samples for zinc analysis were, however, integrated from 0-6 metres. This reduced the number of these samples facilitating the analysis of an aliquot of each sample on the day of collection.

Zinc samples were filtered through 0.45μ Millipore membrane filters in clean room conditions and under two atm. of nitrogen. The filtrate was then divided into three aliquots each of approximately 100 ml. One aliquot was analyzed immediately for labile zinc and complexing capacity (C_L), one was subjected to ultra-violet photo-oxidation using a Technicon U.V. digestor (188-B097-02), and the third passed through a PLGG Millipore membrane ultra-filter with a nominal molecular weight (NMW) cut off of 10^4. The filtrate from this process was frozen to await further analysis.

3.2 Results and Discussion

From salinity profiles turbulent mixing and convective mixing was generally of minor importance with the exception of day 10 when vertical mixing of more saline nutrient rich water caused enhanced production throughout the bag (Pollehne and Imber 1983a).

Production in the top 6 metres was essentially uniform and increased

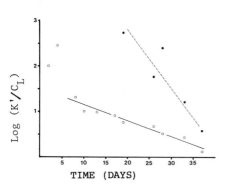

FIGURE 3. Variation of Log (K'/C_L) with time, from a Thalassiosira fluviatilis culture.

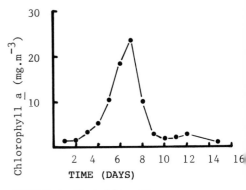

FIGURE 4. Variation of chlorophyll a with time, from an enclosed spring bloom.

steadily over the first seven days from 80 mg C m^{-3} d^{-1} to 500 mg C m^{-3} d^{-1}. After day 7 production was generally low 30-50 mg C m^{-3} d^{-1} and centric diatoms accounted for 85-95% of the total primary production. At maximum production the species composition was <u>Thalassiosira</u> <u>spp.</u> 42% <u>Skeletonema</u> <u>costatum</u> 35% and <u>Chaetoceros</u> sp. 10%. The cells became nitrate limited after 7 days and the biomass quickly declined (Fig. 4).

The complexing capacities were determined by the methods of Matson (1964), Duinker and Kramer (1977), (see also Chau et al., 1974) and Ruzic (1982). Similar temporal variations were observed for each method (Fig. 5). Only one complexing site was found and ultra-filtration showed that the moieties responsible for complexation had nominal molecular weights > 10^{4}.

From the C_L data shown in Fig. 5 the maximum concentration was observed on day 6, prior to the biomass peak on day 7. By day 6 nitrogen limitation began (Pollehne and Imber 1983a) and the subsequent physiological changes of the cells coincided with a decrease in C_L. A nutrient perturbation from the bottom layer of the bag was observed on day 10 and this stimulated growth throughout the bag. This too reflected in the C_L concentration with an increase on days 10 and 11.

No build up of C_L was observed after the peak of the bloom, indicating that some removal process must be operating. However, no significant heterotrophic activity was observed (Pollehne and Imber, 1983a)

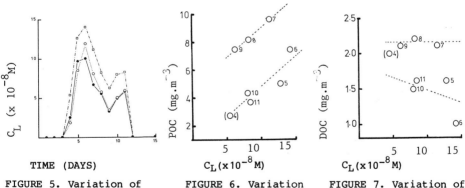

FIGURE 5. Variation of C_L with time

FIGURE 6. Variation of POC with time Numbers refer to sample day

FIGURE 7. Variation of DOC with time Numbers refer to sample day

and the only significant removal process from the upper water layer was sedimentation of the phytoplankton cells (Pollehne et al., 1983).

Throughout the entire experiment the complexation of zinc closely followed the concentration of the phytoplankton cells as portrayed by the chlorophyll a concentrations. Accordingly the observed complexation would appear to be phytoplankton induced and it seems that changes in C_L can be explained by particulate phase measurements. The observed chelation is probably not a truly dissolved phase phenomenon but is in some way connected to the particulate phase. This theory is reinforced by the extremely rapid response of the C_L/C ratio compared to C/N ratio as will be shown later. The most reasonable explanation is that exuded matter is in some way loosely bound to the exterior surfaces of the cells and during the filtration process is removed into solution. The extent of such a removal is not known but the data indicates that the percentage removal is approximately constant.

From the POC vs C_L data two types of relationship are observed (Fig. 6). On days 4, 5, 6, 10 and 11, more C_L/POC was produced than on days 7, 8 and 9. With the exception of day 4, similar patterns were obtained with PON vs C_L, PP vs C_L and with DOC. In the case of DOC vs C_L however, there is more scatter in the data (Fig. 7). The concentration of C_L determined on day 4 was close to the detection limit and also appears at the time of maximum variation in C_L. A

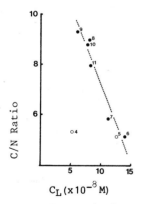
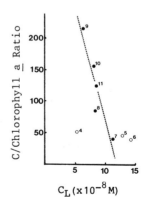

FIGURE 8. Variation with C_L of a) C/N ratio and b) C/chlorophyll a ratio. Numbers refer to sample day. Open symbols not used for regression (see Imber et al 1983).

combination of these facts makes the value of C_L on day 4 the least dependable. Thus for the following discussion the data for day 4 has generally not been included.

The POC content of the seston was assumed to be predominantly composed of viable phytoplankton cells. Microscopic investigation showed that the size distribution and species composition remained quite uniform throughout the bloom with greater than 85% of the biomass being composed of centric diatoms, chiefly Thalassiosira sp. Skeletonema sp., and Chaetoceros sp., which have similar complexing properties (Imber and Robinson 1983), Imber et al., 1983). If then, relative POC is taken as an indication of the relative numbers of cells, it appears that days 7, 8 and 9 can be identified as producing a smaller ligand to cell ratio than on the other days. The cells on these days are, however, nitrogen deficient and it appears that it is the poor condition of these cells that reduces their ligand production.

In this study the relative physiological condition of the cells can be approximated by the C/chlorophyll a and the C/N ratios of the seston. (Thomas and Dodson 1972, Caperon and Mayer 1978) as the phytoplankton represent almost all the seston matter. Accordingly, plots of

TABLE 1. Variation of Log K' during an enclosed spring bloom.

Days	Log K'		
	1*	2*	3*
4	7.5	7.3	7.0
5	7.4	7.4	7.4
6	7.7	7.9	7.7
7	7.4	7.1	7.6
8	7.5	7.5	7.6
9	7.4	7.3	8.0
10	7.4	7.1	7.5
11	7.5	7.4	7.8

*Method of determination
1= Shuman and Woodward (1977)
2= Ruzic (1982) and 3=
Scatchard (1949))

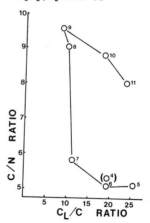

FIGURE 9. Variation of C/N ratio with C_L/POC ratio.

C/chlorophyll a and C/N vs C_L are presented (Fig. 8) and are interpreted as relative physiological condition of the phytoplankton cells vs relative ligand production.

Since the cells were observed to be nitrogen deficient after 6 to 7 days, the C/N ratio would be expected to give a better indication of the physiological condition of the cells, particularly as the C/chlorophyll a ratio is modified by self shading (Pollehne and Imber 1983b). It is, however, conceivable that the C/N ratio could give a misleading indication of the cells' ability to produce complexing ligands as cells could be only nitrogen deficient and nitrogen may not be a major component of the zinc chelators. However, this does not seem to be the case and the best correlation was obtained for C/N ratio vs C_L (r = 0.99, n = 7).

An alternate way of looking at the physiological condition of the cells that produce zinc complexing ligands is to plot C/N ratio against the C_L/POC ratio (C_L/C). This plot approximates to a measure of the physiological condition of the cells vs the relative concentration of ligand produced "per cell" and is presented in Fig. 9. From day 4 to day 7 the cells are in good condition as expressed by their C/N ratio, but the concentration of complexing ligand produced per cell decreases with time. From day 7 to day 9 the ligand production per cell remains low but the condition of the cells abruptly worsens. Towards the end of the experiment (day 9 to 11) the condition of the cells improved slightly, producing a marked improvement in the ligand production per cell. The reason for the reduction in ligand production per cell between days 4 and 6 is not immediately clear. Nitrogen limitation was not observed until day 6 and while this mechanism is most likely responsible for the reduction in the C_L/C ratio from day 6 to day 7 there appeared to be no appreciable physiological change until after day 6.

From day 6 to day 7 the physiological change of the cells was small but the accompanying C_L/C variation was large, indicating that the exuded ligands were more quickly effected by the decreasing ambient nutrient concentrations. The faster response of this process may account for the drop in C_L/C ratio with no apparent change in the C/N ratio between days 5 and 6. From day 7 to day 9 the cells were nitrogen starved and their condition deteriorated appreciably. However, by day 7 the C_L/C ratio was already low and only a small reduction occurred between days 8 and 9, when the cells were in their worst physiological state. On days 10 and 11 the addition of nutrients from the lower water stimulated some growth. Again the C_L/C ratio increased before the cells' condition improved appreciably By day 11, the C/N ratio was considerably higher than on days 5 and 6, but the C_L/C ratio was close to that of day 5.

Values for K' were calculated by the methods of Shuman and Woodward 1977), Ruzic (1982) and Scatchard (1949). In addition, K' values were also computed using the Shuman and Woodward (1977) procedure but with C_L values determined by the Matson (1964) approach. These values are presented in Table 1. Each method produces similar values for K' but after normalization for ligand concentration (C_L) no temporal variation was observed.

4. CONCLUSIONS

Recently, Fisher and Fabris (1982) investigated the complexation of several metals at two points from the growth cycle of several organisms. They observed higher zinc complexation in the stationary phase than in the log growth phase. We have observed similar phenomenon from our own culture work but on a "per cell" basis both our own and the work of Fisher and Fabris (1982) indicate that more zinc complexing ligand production occurred during the log growth phase. These findings are also supported by the data from our bag experiment where ligand production seems to be highest when cells are in their best physiological condition.

The characterization of ligands from phytoplankton in terms of their conditional stability constants has been attempted by several authors (see Hart 1982). Previous estimations of K' for copper complexing by diatoms are in the range log K' = 7.6 to 11.8 (Gachter et al., 1978, Van den Berg et al., 1979, Mantoura 1982, Hirose et al., 1982) while K' with zinc are somewhat lower log K' = 6.5 to 9.2 (Imber and Robinson 1983, Mantoura 1981, Hirose et al., 1982). The values from our bag study, (log K' = 7.4 to 7.6), fall within this range. In our culture experiments we observed a decrease in K' during the growth of Thalassiosira fluviatilis in culture and Van den Berg et al., (1979) observed higher K' values with the lowest concentration of diatom exudates. In the bag experiment, however, no significant temporal variation in K' could be discerned when normalized to C_L. The determined magnitudes of C_L and K' suggest that significant concentrations of metal can be potentially bound by diatom exudates. Such a phenomenon may be of local importance in areas of low ambient metal concentration and may contribute to the controlling factors effecting species succession.

Our work has confirmed that the exudates of diatoms complex trace metals. The K' and C_L values obtained indicate that such complexation could be ecologically significant.

The determination of relative C_L is of use in understanding algal biological systems. However, it is possible that, instead of C_L determinations, the ratio of organically bound to total metal in the dissolved fraction may be as informative and certainly easier to measure.

Production of zinc complexing exudates by marine diatoms is related to their physiological condition.

REFERENCES

- Anderson, M.A., F.M.M. Morel and R.R.L. Guillard., 1978. Growth limitation of a coastal diatom by low zinc ion activity. Nature, 276. 71-76.
- Bruland, K.W., 1980. Oceanographic distributions of cadmium, zinc, nickel and copper in the North Pacific. Earth Planet. Sci. Lett. 47, 176-198.
- Bruland, K.W., G.A. Krauer and J.H. Martin., 1978. Zinc in north-east Pacific Water. Nature 271, 741-743.
- Caperon, J. and J. Mayer., 1972. Nitrogen-limited growth of marine phytoplankton. Deep Sea Res., 19. 619-632.
- Chau, Y.K., R. Gachter and K. Lum-Shue-Chan., 1974. Determination of the apparent complexing capacity of lake waters. J. Fish. Res. Board Can. 31, 1515-1519.
- Duinker, J.C. and C.J.M. Kramer., 1977. An experimental study on the speciation of dissolved zinc, cadmium, lead and copper in river Rhine and North Sea water, by differential pulsed anodic stripping voltametry. Mar. Chem., 5, 207-228.
- Fisher, N.S., and J.G. Fabris., 1982. Complexation of Cu, Zn and Cd by metabolites excreted from marine diatoms. Mar. Chem. 11, 245-255.
- Gachter, R., J.S. Davis and A. Mares., 1978 Regulation of copper availability to phytoplankton by macromolecules in lake water. Envir. Sci. Technol., 12, 1416-1421.

- Grice, G.D. and M.R. Reeve., 1982: Introduction and description of experimental ecosystems. In G.D. Grice and M.R. Reeve (ed) Marine Mesocosms. Springer, New York p. 1-9.
- Hart, B.T., 1982. Trace metal complexing capacity of natural waters: A Review. Environ. Tech. Let. 2: 95-110.
- Hirose, K., Y. Dokiya and Y. Sugimura., 1982. Determination of conditional stability constant of organic copper and zinc complexes dissolved in sea water using ligand exchange method with EDTA. Mar. Chem. 11. 343-354.
- Imber, B.E. and M.G. Robinson., 1983. Complexation of zinc by exudates of Thalassiosira fluviatilis grown in culture. Mar. Chem. 13. (in press) .
- Imber, B.E., M.G. Robinson and F. Pollehne., 1983. An enclosed spring bloom in Saanich Inlet. IV Phytoplankton induced zinc complexation. Submitted to Mar. Biol.
- Imber, B.E., M.G. Robinson, A.M. Ortega and J.D. Burton., 1983a. Complexation of zinc by exudates from Skeletonema costatum grown in culture. Submitted to Mar. Chem.
- Jackson, G.A. and J.J. Morgan., 1978. ª Trace metal-chelator interactions and phytoplankton growth in sea water media and theoretical analysis and comparison with reported observations. Limnol. Oceanogr., 23, 268-282.
- Mantoura, R.F.C., 1982. Organo-metallic interactions in natural waters. In Marine organic chemistry, E.K. Duursma and R. Dawson, Eds. Elsevier Scientific Pub. Company., Amsterdam. p. 179-223.
- Matson, W.R., 1964. Trace metal complexes in natural media. Thesis, Mass. Inst. of Technol. 258pp.
- Pollehne, F. and B.E. Imber., 1983a. An enclosed spring bloom in Saanich Inlet. I. Nutrient uptake, growth rate and elemental composition of the phytoplankton. Submitted to Progress in Marine Ecology.
- Pollehne, F. and B.E. Imber., 1983b. An enclosed spring bloom in Saanich Inlet III. Changes in pigmentation and photoadaptation of the phytoplankton. Submitted to Progress in Marine Ecology.
- Pollehne, F., F.A. Whitney and C.S. Wong., 1983. An enclosed spring bloom in Saanich Inlet. II. Sedimentation and sinking characteristics of the phytoplankton. Submitted to Progress in Marine Ecology.
- Ruzic, I., 1982. Theoretical aspects of the direct titration of natural waters and its information yield for trace metal speciation. Chem. Acta 140: 99-113.
- Scatchard, G., 1949. The attractions of proteins for small molecules and ions. N.Y. Acad. Sci. Ann. 51, 660-672.
- Shuman, M.S. and G.D. Woodward Jr., 1977. Stability constants of copper-organic chelates in aquatic samples. Envir. Sci. Technol. 11, 809-813.
- Sunda, W. and R.R.L. Guillard., 1976. The relationship between cupric ion activity and the toxicity of copper to phytoplankton. J. Mar. Res. 34: 511-529.

440

- Thomas, W.H. and A.N. Dodson., 1972. On nitrogen deficiency in tropical Pacific Phytoplankton, II. Photosynthetic and cellular characteristics of a chemostat-grown diatom. Limnol. Oceanogr. 17 515-523.
- Van den Berg, C.M.G., P.T.S. Wong and Y.K. Chau., 1979. Measurement of complexing materials excreted from algae and their ability to ameliorate copper toxicity. J. Fish. Res. Board Can. 36: 901-905.
- Whitfield, M., 1975. The extension of chemical models for sea water to include trace components at 25°C and 1 atm pressure. Geochim. Cosmochim. Acta 39: 1545-1557.

THE EFFECT OF NATURAL COMPLEXING AGENTS ON HEAVY METAL TOXICITY IN
AQUATIC PLANTS

MADELIJN VAN DER WERFF

1. INTRODUCTION

Large amounts of organic substances, such as fulvic and phenolic acids may occur
in surface waters, either dissolved or in collodial form (De Haan 1972a,b;
Reuter and Perdue 1977; Lawrence 1980). These substances will influence the
speciation of heavy metals in waters aside from pH-dependent processes (Stumm
and Morgan 1970; Wagemann and Barica 1979). Due to the interaction of metal
speciation with bioavailability the effect of several natural complexing agents
have been tested on growth of and uptake of Zn and Cu by Elodea nuttallii.

2. MATERIALS AND METHODS

Shoots of Elodea nuttallii (Planch) St. John were collected from uncontaminated
ditches. Experiments were run in filtered ditch water, pH 8. As complexing
compounds were used, FA, HA, phenolic acids, hydroquinone, citric acid and the
artificial EDTA. The metals Zn and Cu were mixed with the ligand solution in
advance. After a growth period of 7–10 days, carbonate precipitation was removed
from the adaxial sides of the leaves. HA and FA precipitation was removed by
washing in cold NaOH 0.01 mol/l for not more than 15 seconds. Plant material was
dried, wet ashed ($HClO_4/HNO_3$), and analysed by AAS for Zn and Cu, as described
elsewhere (Van der Werff 1981).

3. RESULTS AND DISCUSSION

Despite nearly similar growth reduction by the presence of the metals Cu and Zn
as ion, citrate, FA or HA complex, there is a significant difference in metal
concentration in the shoots (Fig. 1). FA and HA reduced the concentration of Cu
by 20 and 40% respectively and of Zn by about 50%; EDTA reduces the Cu and Zn
concentration by 80% of the level in the ionic situation all in agreement with
results on terrestrial plants (Ernst 1968 ; Marquenie-van der Werff and Out 1981).

In Fig. 2a the effect of FA on growth and Zn concentration in the plant is to
be seen at different Zn levels. Where the Zn concentration is low, there is a
steady increase in biomass as the FA increases. With 5 μmol Zn/l the beneficial

Kramer, C.J.M. and Duinker, J.C. (eds.), Complexation of Trace Metals in Natural Waters. ISBN 90-247-2973-4
© 1984 Martinus Nijhoff/Dr W. Junk Publishers, The Hague/Boston/Lancaster.
Printed in the Netherlands.

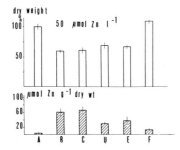
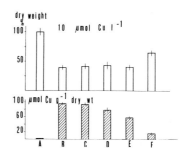

FIGURE 1. The effect of chemical binding of Zn and Cu on metal uptake by and growth of <u>Elodea nuttallii</u>. Citric acid and EDTA were given in equimolar amounts With Zn or Cu the concentration of FA and HA was 100 mg/l and 20 mg/l respective Biomass is given in percentages to the control, to which no Zn or Cu compounds were added. Mean values with standard errors from three replicates. A: contro B: ionic; C: citric acid complex; D: FA complex; E: HA complex; F: EDTA comple

effect of FA is quite clear. Increasing concentrations of FA however; produce no further growth stimulation. With 25 μmol Zn/l the beneficial effect of FA up to 25 mg/l. Fig. 2b shows the Zn concentration of the plant as a function the Zn and FA concentration in the medium. There is a decrease in Zn concen- tration in the tissue, with increasing FA concentration.

FIGURE 2. Effect of Zn and FA on growth and Zn uptake by Elodea nuttallii. Zn is added in 0 (—), 5 (...) and 25 (---) μmol/l, mixed with each 0, 10, 25 and 50 mg FA/l.
Figure 2a. Final biomass in mean values (from three replicates) per vessel. Each value with a specific letter differs significantly (P<0.05) from the value with different letters.
Figure 2b. Final tissue concentration of Zn in mean values (from three replicates). Within each FA concentration each point differs significant- ly (P<0.05) from the other points.

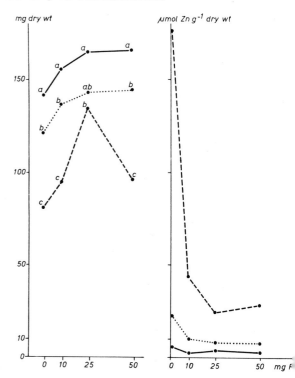

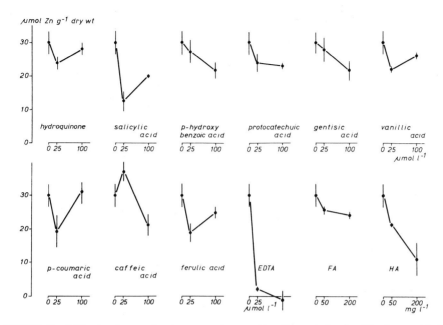

FIGURE 3. Effect of complexing agents on the Zn uptake by Elodea nuttallii. Zn is added in 25 µmol/l. The Zn concentration in the tissue is given as the increase in comparison to the control to which no Zn compounds were added. Mean values with standard deviations from three replicates.

The effect of interaction between organic compounds and Zn on Zn uptake by Elodea shoots is given in fig. 3. Despite no significant difference in biomass between the experimental groups, there was once more a significant effect of the complexing agents on Zn uptake, EDTA, FA, and HA decrease the Zn concentration, whereas the effect of the phenolic acids on Zn uptake is not homogeneous. All phenolic acids except caffeic acid, tend to decrease the Zn concentration, if given in an equimolar amount. More than equimolar quantities of complexing agents produced quite different effects. Gentisic and p-hydroxybenzoic acid decreased the Zn concentration further; the caffeic acid concentration is probably then high enough to show a clear decrease by complexing the Zn ions. p-Coumaric acid, ferulic acid, salicylic acid, vanillic acid and hydroquinone may be toxic in the higher concentrations and are so tending to produce a higher Zn concentration in the leaves by causing tissue damages.

Phenolic acids may be beneficial where there is an ion excess by complexing them, and so inhibiting heavy metal uptake. However when the phenolic acid concentration increases above the ideal combination with the heavy metal, the phenolic acids may produce a toxic effect. Phenolic acids may then act toxific-

444

ally by causing physiological disorders e.g. in oxidative phosphorylation, enzyme activities and protein synthesis (Horsley 1977; Rice 1979).

4. CONCLUSIONS

1. Natural complexing agents as HA, FA and phenolic acids are beneficial when there is an ion excess by complexing them and so inhibiting heavy metal uptake.
2. When the phenolic acid concentration increases above the ideal combination with the heavy metal, the phenolic acids may produce a toxic effect.
3. Toxicity of heavy metals is determined not only by their being complexed or not, but also by each specific complexing compound.
4. The concentration of heavy metals in the medium does not limit plant growth. The internal concentration of the heavy metal within the plant tissue is the most essential factor.

REFERENCES

-Ernst, W., 1968. Der Einfluss der Phosphatversorgung sowie die Wirking von ionogenen und chelatisierten Zink auf die Zink und Phosphataufname einiger Schwermetallpflanzen. Physiol. Plant. 21: 323-333.
-Haan, H. de, 1972a. Some structural and ecological studies on soluble humic compounds from Tjeukemeer. Verh. Internat. Verein. Limnol. 18: 685-699.
-Haan, H. de, 1972b. Molecule-size distribution of soluble humic compounds from different natural waters. Freshwat. Biol. 2: 235-241.
-Horsley, S.B., 1977. Allelopathic interference among plants. II. Physiological modes of action. Proc. 4th N. Amer. For. Biol. Workshop H.E. Wilcox and A. Hamer (eds) pp. 93-136. Syracuse Univ. Press Syracuse N.
-Lawrence, J., 1980. Semi-quantitative determination of fulvic acid, tannin and lignin in natural waters. Water Research 14: 373-377.
-Marquenie-van der Werff, M. and Out, T., 1981. The effect of humic acid as Zn complexing agent on water cultures of Holcus lanatus. Biochem. Physiol. Pflanzen 176: 274-282.
-Reuter, J.H. and Perdue, E.M., 1977. Importance of heavy metal-organic matter interactions in natural waters. Geochimica et Cosmochimica acta 41:325-334.
-Rice, E.L., 1979. Allelopathy - an update. The Botanical Review 45: 15-109.
-Stumm, W. and Morgan, J.J., 1970. Aquatic chemistry - An introduction emphasizing chemical equilibria in natural waters. Wiley-Interscience. New York, London, Sydney, Toronto.
-Wagemann, R. and Barica, J. 1979. Speciation and rate of loss of copper from lakewater with implication to toxicity. Water Research 13: 515-523.
-Werff, M. van der 1981. Ecotoxicity of heavy metals in aquatic and terrestrial higher plants. Academic Thesis, Vrije Universiteit, Amsterdem 1981.

SPECIATION OF HEAVY METALS AND THE IN-SITU ACCUMULATION BY DREISSENA
POLYMORPHA: A NEW METHOD

P. del Castilho, R.G. Gerritse, J.M. Marquenie and W. Salomons

1. INTRODUCTION

In the Dutch struggle against the sea a large dam (Enclosure dike)
was built in 1932 closing off a coastal lagoon (fig. 1). This operation
created a large freshwater body containing 5.5 km^3 of water, mainly fed
by the river Rhine. The estuarine ecosystem had disappeared within 3
years. Large amounts of contaminants are discharged in the lake and
accumulate in sediments.

The route and fate of heavy metals during the summer of 1982 in the
river Rhine was investigated by a team of biologists and chemists. In
summer massive algal blooms occur in the lake and to a smaller extent in
the river.

This paper will discuss the biological monitoring of cadmium (a non-
essential element) and copper (an essential element) in relation to
aquatic metal speciation in the region studied.

For comparison with the
river Rhine system one location in the
river Meuse was included. A previous field
study (Marquenie, 1981) showed a weak correlation between total metal in filtrate water
and soft tissues for cadmium and a poor
correlation for copper in river sections.
In the study reported here apart from total
metal in filtrate water, "free" metal in
in-situ dialysates was determined.

Fig. 1. Map of The Netherlands
showing the sampling locations.

Kramer, C.J.M. and Duinker, J.C. (eds.), Complexation of Trace Metals in Natural Waters. ISBN 90-247-2973-4
©*1984 Martinus Nijhoff/Dr W. Junk Publishers, The Hague/Boston/Lancaster.*
Printed in the Netherlands.

2. MATERIALS AND METHODS

The freshwater bivalve Dreissena polymorpha responds in a similar
way to cadmium pollution as other organisms and is suitable for biological monitoring studies (Marquenie, 1981).
Dreissena polymorpha (15-20 mm) was sampled at a relatively clean area,
the center of the Enclosure dike, randomised to samples of about 100 individuals and exposed during 40 or 60 days at selected locations. The animals were put in hard-polythene baskets, 1-2 metres below the water
surface.

After collection samples were stored frozen (- 20 ^{o}C). The soft tissues were homogenised with an Ultra Turrax R (deldrin cutters) and stored
in acid-cleaned glassware until analyses. After wet ashing cadmium and
copper were determined with ETAAS (Perkin Elmer 430, HGA 500). The percentage of ash-free dry material was determined by drying (105 ^{o}C) followed by ashing in a muffle oven (600 ^{o}C).

A new method was developed to monitor the metal concentrations during
the exposure period. Use was made of dialysis tubes. Due to the slow response of dialysis to rapid changes in external concentrations, an integrated metal fraction is obtained.

The dialysis cells (fig. 2) consisting of PTFE tubes, with ultra
filtration membranes (Amicon YM5, MW cut-off 5.000 D) at both ends, filled
with 0.01 M KCl, were put in baskets next to Dreissena polymorpha. Every
7 days during a six-week period, 25 1 of water was collected in jerrycans
and dialysis cells were collected and replaced by new ones. The raw water
samples were collected for e.g. filtration purposes. The pH was measured
on the spot. The last date of sampling coincided with day 60 of the
exposure of Dreissena polymorpha.

The content of the cells was analysed for "free" cadmium and copper
by means of DPASV in 0.09 M acetate-buffer, pH 4.65. Deposition time 2.5
minutes, rest 0.5 minute, scan-
ning from -1.3 V to + 0.2 V
with a rate of 5 mV/sec and a
modulation amplitude of 10 mV.
The samples were deoxygenated
with nitrogen prior to analysis
(10 min.). Apparatus PAR 303,
174 A combination.

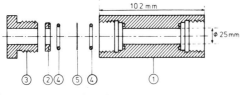

Fig. 2. Schematic diagram of the dialysis cell. 1. Body, 2. Holder, 3. Screwcap, 4. "Viton" ring, 5. "Amicon" filtre.
Body, holder and screwcap from PTFE.

3. RESULTS AND DISCUSSION

In the present study findings of the previous study (Marquenie, 1981) were confirmed: metal levels in filtrates (0.45 μm) yield a poor estimate of biological uptake. Because of the uptake kinetics of these organisms for these metals in the present study uptake of cadmium by mussels was related to the last 3 weeks of exposure, and copper to the last week. Dialysis cells yield information about metal species with M < 5000 D.

It was noted that downstream the river Rhine both "free" metal and H$^+$ concentrations gradually decrease (pH rises from 7.5 to 8.7). Dialysable "free" cadmium shows a strong linear relation with cadmium accumulation (r = 0.84, n = 19). This relation is shown in figure 3.

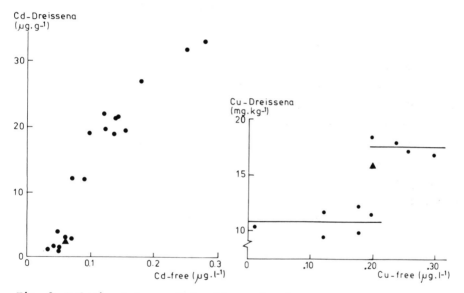

Fig. 3. Relation between dialysable "free" cadmium in water and biological uptake.

Fig. 4. Relation between dialysable "free" copper in water and biological uptake

This suggests that the measured "free" cadmium must be strongly related to the actual available cadmium species in the water.

Copper showed another type of exposure-accumulation relationship at very low "free" copper levels occurring in summer. The low levels are a result of high pH (precipitation) caused by algal productivity (Salomons and Mook, 1980).

The relation seems to be S-shaped (fig. 4). At the lowest exposure levels the biological levels remained in a narrow range (9-12 mg/kg) whereas from a certain exposure level, the biological levels rose quite abruptly to a narrow range at a higher level (16-18 mg/kg).
This suggests a homeostatic mechanism.

Dose-response relationships for both metals at a location of the river Meuse (with different characteristics e.g. lower pH and chlorinity) fit in well with the overall picture of the river Rhine system (Symbol: delta, in figures 3 and 4). This indicates that "free" metal levels of dialysates may determine biological uptake irrespective of the nature of the system.

4. CONCLUSIONS

Our new method for prediction of biological uptake of cadmium seems promising. The prediction of copper levels is more complicated as biological regulating mechanisms and/or competition with other metals at the same binding sites in the organism may influence the dose-response relationship.

ACKNOWLEDGEMENTS

This study was partly financed by the Department of Public Works.

REFERENCES

-Marquenie, J.M., 1981. The fresh water molusc Dreissena polymorpha as a potential tool for assessing bioavailability of heavy metals in aqueous systems. Proc. of "Heavy metals in the environment", CEP Consultants Ltd, Edinburg, U.K. p 409.
-Salomons, W. and Mook, W.G., 1980. Biogeochemical processes affecting metal concentrations in lake sediments (IJsselmeer, The Netherlands), Sci. of the Tot. Environ. 16, p 217.

seen, in *Remembrance* all signs are inhabited by an inherent doubleness and operate in disguised and inverted ways. The ubiquity of the lie stems from its reference to an invisible underside of signification that, for Proust, is consistently associated with homosexuality. Finally, Albertine's lies are paradigmatic because they seem to indicate secret lesbian liaisons.

A passage from *The Captive* makes clear these various points and, hence, bears full quotation.

If we were not obliged, in the interests of narrative tidiness, to confine ourselves to frivolous reasons, how many more serious reasons would enable us to demonstrate the mendacious flimsiness of the opening pages of this volume in which, from my bed, I hear the world awake, now to one sort of weather, now to another! Yes, I have been forced to whittle down the facts, and to be a liar, but it is not one universe, but millions, almost as many as the number of human eyes and brains in existence, that awake every morning.

To return to Albertine, I have never known any woman more amply endowed than herself with the happy aptitude for a lie that is animated, coloured with the very hues of life. . . . [Despite] those painful moments, those furious innuendoes, which were frequent with my mistress, [she] . . . was however charming . . . when she invented a story which left no room for doubt . . . Verisimilitude alone inspired Albertine, never the desire to make me jealous . . . [She] used to invent, in order to exculpate herself, confessions as natural as these stories which I never doubted and one of which was her meeting with Bergotte when he was already dead. (*R* 3:189–91)

The jocular violation of convention in the beginning of the passage – the hero stepping out of the story in order to comment, as narrator, upon his technique – appears to be an assertion of his authority over the text, enabling him to condescend to Albertine's narrative abilities. The problem with her stories is their "furious innuendoes" – are they not rather "infuriating" innuendoes? The hero's projection occurs when he is reduced to a character (the jealous lover) in a tale told by Albertine. The prototypical reader in Proust is, of course, the inquisitive lover who tries to know the beloved. However, the hero claims that he prefers a mollifying tale or a happy lie. Albertine's pretended confessions which he pretends to swallow make him seem like the straight man of the text, like Swann, who "loved sincerity, but only as he might love a pimp who could keep him in touch with the daily life of his mistress" (*R* 1:391). But, also like Swann, who "lied as much as did Odette," especially in his resolute naïveté, the hero is exposed – or exposes himself – at the moment when he claims his innocence. A story he never doubted, he says with tongue in cheek, was her meeting with the already-dead Bergotte. As

everyone knows, one of the slips in *Remembrance* is the fact that Bergotte
dies twice. Proust pokes fun at the hero's presumed authority over the
story he tells and at his belief in his moral superiority over the deceitful
Albertine.

The mendacity to which the hero confesses for the wildly implausible
reason of narrative economy ("to whittle down the facts") pales next to
Albertine's lies, which open up millions of possible worlds to the jealous
reader. The fecundity of her lies springs from their ambiguous reference
to a homosexual world that lies hidden from sight, a world never absent
from the hero's awareness yet of which he continually disavows knowledge. Immediately before commenting on the "mendacious flimsiness"
of his story, he offers the general principle that "the evidence of the
senses is also an operation of the mind in which conviction creates the
facts." As an example, he cites his butler's story about seeing Charlus
stand for a "good hour" in the *pissotiere*, from which the butler naively
concludes that he "must have caught a disease" from "running after the
girls" (*R* 3:188–89). The butler's ignorance of Charlus's homosexual
assignations, related with a knowing wink to the reader, throws suspicion
upon the hero's supposed ignorance of Albertine's sexuality. Indeed, in
the final volumes of *Remembrance*, he is cast in an increasingly ironic light,
as a poor liar.

The spectacle in the *pissotiere* is a footnote to the scenes of homosexual
voyeurism that play a central role in the novel. Many critics have commented on the decisive importance of these scenes and on the hero's
ambivalent fascination and disgust with them. What has largely gone
unnoticed, though, is that these scenes comprise two distinct series:
lesbian and gay male. As Bal points out, the first two scenes are between
women: Mlle Vinteuil and her lover at Montjouvain, and Albertine and
Andrée in the casino at Incarville. Bal focuses on the second of these
scenes, commonly referred to as the "dance against breasts," as exemplary of the lesbian libidinal and epistemological models of the novel: a
nonpenetrative eroticism of "flat, rubbing flesh" and a knowledge of
flat, opaque surfaces without depth or interiority.[59] By contrast, in the
famous encounter between Charlus and Jupien, the hero trains his inter-
pretive lens upon decoding the scene, investigating the natural history of
inverts and peering into the meaning of a sight that he does not actually
see. The overheard spectacle, he says, "opened my eyes by a transformation in M. de Charlus as complete . . . as if he had been touched by a
magician's wand" (*R* 2:635). The one who has been transformed,
however, is the hero for whom the universe is rewritten: "everything

that hitherto had seemed to my mind incoherent became intelligible," he claims. Although lesbian desire remains indecipherable, male same-sex desire is like a secret code which the scene between Charlus and Jupien suddenly explains; "just as a sentence which presents no meaning so long as it remains broken up in letters arranged at random expresses, if these letters can be rearranged in the proper order, a thought which one can never afterwards forget" (*R* 2:637).

The "proper order" of Charlus's homosexuality is rendered visible late in the novel in the spectacle of sadomasochism in Jupien's brothel. These elaborately staged scenes of Sodom, accompanied by the hero's commentaries on the nature and etiology of inversion, offer an in-depth view of male homosexuality that is conspicuously lacking for lesbianism. Indeed, visualizations of lesbianism typically engender more questions than answers. Thus, when Albertine remarks that she plans to meet the best friend of Vinteuil's daughter, the hero is stricken with the horrified recollection of the scene he had witnessed at Montjouvain. The "noxious power" of that sight had "fatal consequences" for him, who had "done no more . . . than look on at a curious and entertaining spectacle" (*R* 2:1152).

Protesting his innocence, the hero has been contaminated by the secret of homosexuality. Indeed, in *Remembrance* Proust depicts the homophobic construction of modern male homosexuality through the eyes of a hero who asserts his heterosexuality while constantly revealing the hidden perversions of others. The novel is laced with his disavowals of knowledge of homosexuality yet he cannot keep from gazing at the sight. "Grieved" that the reader be "offended" by his portrayal of Jupien and Charlus, he allows that "considerable interest . . . may be found in actions inspired by a cast of mind so remote from anything we feel . . . that they remain incomprehensible to us, displaying themselves before our eyes like a spectacle without rhyme or reason" (*R* 3:40). Of course, he also claims that homosexual desire makes "rhyme and reason" out of Charlus's otherwise senseless behavior. In short, his insistence upon homosexuality as a *visible secret* enables him, he believes, to remain in the closet. Yet Proust "outs" the hero by the very means through which the latter hides himself. Posing as a brave doctor who has descended "through the circles of an inferno of sulphur and brimstone" in order to investigate the "inhabitants of Sodom," thereby risking contagion, he asks for the reader's "pity." "The subject of a psychiatrist's study often rebounds on him. But . . . what obscure inclination, what dreadful fascination had made him choose that subject?" (*R* 3:205–6). Far from concealing the narrator's queerness, Proust exposes his homophobic mendacity.

A prominent theme of *The Captive* and the subtext of the hero's social observations therein is lying. The passage quoted above regarding Albertine's and the hero's lies is followed by a celebration of "the perfect lie."

The lie, the perfect lie, about people we know, about the relations we have had with them, about our motive for some action, formulated in totally different terms, the lie as to what we are, whom we love, what we feel with regard to people who love us and believe that they have fashioned us in their own image because they keep on kissing us morning, noon and night – that lie is one of the few things in the world that can open windows for us on to what is new and unknown, that can awaken in us sleeping senses for the contemplation of universes that otherwise we should never have known. (*R* 3:213)

The hermeneutic richness of the lie is attested to by the entirety of *Remembrance*. What is commonly considered to be the hypertrophy of the novel following the revelation of Albertine's possible homosexual liaison is not simply due to the hero's suspicion that she lies but to his collusion with her lies. Despite his jealous desire for disclosure, he typically refuses to follow up his leads when he thinks that Albertine is lying (*R* 2:827–30, 3:175–76). Although he appears to search for the truth, he prefers to believe in her duplicity.[60]

Albertine's putative lies pose a narrative challenge, for they are not so much falsifications as openings within the reality the narrator has constructed. "What she said . . . had to such a degree the same characteristics as the formal evidence of the case – what we see with our own eyes . . . – that she sowed thus in the gaps of her life episodes of another life" (*R* 3:187). The imaginary, other life of Albertine, lived in the interstices of her visible, legitimate existence, is inaccessible to him; he can neither fill in the holes of her stories nor form a story out of her lies.

Her lies, her admissions, left me to complete the task of elucidating the truth: her innumerable lies, because she was . . . by nature . . . a liar (and so inconsistent, moreover that, even if she told me the truth every time about, for instance, what she thought of other people, she would say something different every time); her admissions . . . left between them . . . huge blanks over the whole expanse of which I was obliged to retrace . . . her life. (*R* 3:93)

Albertine's truthfulness poses a greater problem than her supposed lies, for the fragmentary nature of her remarks frustrates the narrator's apparent desire to compose her life into a narrative. Just as her visual images fail to cohere into a complete picture because they are projections of the narrator's desires, so her statements are invariably lies, regardless of their content, because they all allude, one way or another, to what he imagines to be her other life as a lesbian.

For this reason, the lie that organizes *Remembrance* and that generalizes deception is homosexuality. In a narrow, Sodomite sense, homosexuality is the concealed identity of certain characters who are revealed as such in the course of the novel. The passage on "the perfect lie" occurs in the context of the concert at Mme Verdurin's when Charlus's homosexuality is exposed. The Baron's appearance has altered, for "the vice" that he had once "confined in the most secret recesses of his being" now "overflowed into his speech," his campy gestures and "little squeaks" symptoms that he has "reached a certain stage in [his] malady" (*R* 3:206, 209). Inadvertently, Charlus has slipped into a style that publicly marks him as queer, a condition that "certain scandalmongers or certain over-dogmatic theorists," such as the hero, "aver . . . is complemented by an innate taste, a knowledge and feeling for female dress" (*R* 3:220). His effeminacy indicates the "lie" he has been living, a deception previously disclosed by indirect signs known only to the cognoscenti of Sodom.

The secret language of the Sodomites is one of the hero's favorite topics. The Jews of sexuality, the "race of inverts" are supposed to possess a secret code through which they communicate with each other and thus form an "oriental colony" invisible to the eyes of the world (*R* 2:655). Indeed, "even members who do not wish to know one another recognize one another immediately by . . . involuntary or deliberate signs" (*R* 2:640). To decipher their code requires inverting its signs; similarly, the "invert reader" misreads but gains the "full import" of a heterosexual novel by reversing the gender of its characters (*R* 3:948–49). Just as the hero explains the nature of homosexual attraction as a gender inversion ("that race of beings . . . whose ideal is manly precisely because their temperament is feminine" [2:637]), so the private language of homosexuals is marked by a transposition of gender. (In this sense, the hero anticipates the practice of transposing the gender of the characters in *Remembrance*, notably of reading Albertine as Albert in disguise.) To the straight reader, played by the hero, this code is an "ideographic script" interpreted by taking signs as symptoms, or as displaced or anagrammatic signifiers. According to the hero, "the words themselves did not enlighten me unless they were interpreted in the same way as a rush of blood to the cheeks of a person who is embarrassed, or as a sudden silence" (*R* 3:83). Like Freud's slips of the tongue, these signs are involuntary betrayals of the truth, such as the Duc's comment on his brother's "peculiar" "tastes" (*R* 2:744), or an ellipsis in a statement or a "word read aloud in mistake for another" (*R* 3:177). Often they operate by negation; for example, Charlus's feigned gestures of indifference outside the Grand Hotel at Balbec signify the opposite. The hero tells us that sometimes "the script from which I deciphered Albertine's lies, without being ideographic, needed simply to be read backwards" (*R* 3:85). Despite his presumed expertise in decoding Albertine's lies, his clever translations may well be mistranslations.

In a larger, Gomorrahan sense, the lie that animates *Remembrance* is not homosexuality as a definite sexual orientation but an opaque alterity that produces a paranoid proliferation of seeming lies. The uncertain, aberrant signs of lesbianism are, in the semiotics of the novel, the prototype of all signs, ranging from the "hieroglyphic characters" of the narrator's impressions of Combray to the "inner book of unknown symbols" (*R* 3:912–13). In short, they are "perfect lies" insofar as they refer to another universe hidden within the straight world yet strangely enclosing it. In contrast to Sodom's cursed men, revealed by the telltale marks of gender inversion, lesbians in the novel remain for the most part unmarked by a reversal of gender; like Molly's same-sex desires in *Ulysses*, theirs is a diffuse, polymorphous perversion.

Charlus and Albertine thus dominate two, contradictory trajectories in the text. The elaborate portrayal of the former's homosexuality incites the hero's horror, yet he shares a specular relationship with him; Charlus is the Other by whom he is simultaneously repelled and attracted, who represents his forbidden desires. According to Sedgwick, the "lavish production of M. de Charlus – as spectacle; as, to be specific, the spectacle of the closet – enables the world of the novel to . . . turn [upon] . . . his distance from the differently structured closet of the narrative and its narrator."[61] Compagnon is harsher in his assessment of the way in which the novel is "protective and indulgent of the hero." The latter watches homosexual sex but "never questions the meaning of his recurrent voyeurism . . . The narrator lies, he keeps the hero under cover instead of exposing his 'vices.' Since the hero is riddled with vice and the narrator hides it from the reader, the novel is based upon a lie."[62] The hero's denial of his complicity with the homosexuality for which he has contempt points to the collusion between hero and narrator.

Charlus is transformed into increasingly grotesque shapes as his vice is progressively exposed. His most bizarre presentation occurs in the flagellation scene at Jupien's brothel, where the hero watches with disgusted fascination through a concealed window (*R* 3:837–84). The scene parallels that at Montjouvain, where the hero comments that "sadists of Mlle Vinteuil's sort are creatures so . . . naturally virtuous" that they must "impersonate . . . the wicked" in order to escape into "the inhuman world of pleasure" (*R* 1:179). Like the spectacle at Montjouvain, the theater of pleasure at Jupien's brothel is pathetic not only because Charlus's kindhearted tormentors are unable to give him sufficiently brutal beatings, but because the Baron, like Mlle Vinteuil, needs to perform a sadomasochistic role in order to have sex. Similarly, watching them enables the hero to imagine an escape into another world, a Venice or Nighttown of perverse pleasure. Throughout the second half of the novel, Charlus has clearly becomes the hero's alter ego. His subtle awareness of the duplicity of appearances and the play of signs gives him the ability to write. Nicknamed "the dressmaker" (*R* 3:220) and transformed into a virtual drag queen at the end, Charlus is, like the hero, a gossip who "presents to our gaze . . . the reverse side of the fabric" of society (*R* 2:1082).

On the opposing trajectory of the text, Albertine dies but she never disappears because, as an emblem of his own perverse possibilities, she has lodged herself within the hero's body. His initial association of her with Balbec was an "error in localisation" (*R* 3:507). "In leaving Balbec, I had imagined that I was leaving Gomorrah; . . . in reality, alas, Gomorrah was disseminated all over the world" (*R* 3:15) – indeed, within himself. She defeats his belief that homosexuality and heterosexuality are as distinct and as separate as the Méséglise and Guermantes Ways; that he is the detached social observer whose desire for exposure, projected onto Albertine, does not implicate himself; and, finally, that her lesbianism is the truth of her "lie." For her lesbianism, pursued by him as a question that he could, if he only had enough information, answer, is the symmetrical opposite of his all-too-knowing abjection of Charlus. He never locates Albertine, though, who at her death is left in the "doubly crepuscular" zone of a liminal, errant (bi)sexuality.

In her sexual indeterminancy, Albertine is linked to Morel, the musician whose bisexuality causes Charlus, his lover, considerable confusion and which occasions the hero's discourse on "the perfect lie." By mistake, Charlus opens a letter sent to Morel from the actress Lea, the contents of which plunge him into "grief and stupefaction." It is a love letter, "written in the most passionate terms" by a woman "notorious for her exclusive taste for women," to Morel, whom Charlus has believed to be exclusively homosexual (*R* 3:211–12). He is confounded by the notion

of a sexual liaison between a lesbian and a gay man. "What most disturbed the Baron was the phrase, 'one of us,'" which Lea used in
addressing Morel – "'Of course you're one of us, you pretty sweet-
heart.'"

Ignorant at first of its application, he had eventually . . . learned that he himself
was "one of them." And now this notion that he had acquired was thrown back
into question. When he had discovered that he was "one of them," he had supposed this to mean that his tastes . . . did not lie in the direction of women. And
here was this expression taking on, for Morel, an extension of meaning of which
M. de Charlus was unaware, so much so that Morel gave proof . . . of being
"one of them" by having the same taste as certain women for other women . . .
So, to be "one of them" meant not simply what he had hitherto assumed, but
to belong to a whole vast section of the inhabitants of the planet, consisting of
women as well as of men, of men loving not merely men but women also. (*R*
3:212)

In their sexual relations, Morel (a gay man) plays the role of a lesbian
while Lea (a lesbian) plays that of a gay man. This dizzying alteration of
gender and sexual roles had been previously described by the hero as the
nadir of vice and jealousy. Men like Morel "seek out those women who
love other women," with whom they "enjoy . . . the same pleasure as with
a man"; "in their relations with women, they play, for the woman who
loves her own sex, the part of another woman and she offers them at the
same time more or less what they find in other men" (*R* 2:645–46). For
the Baron, though, such an arrangement provokes shock and vertigo.
Having with great difficulty learned to distinguish two sexual identities
– homosexuality and heterosexuality – he is now confronted by at least
four and possibly more, and is thus thrown into an anxiety born of "the
sudden inadequacy of a definition."

His aporia bears directly upon "the perfect lie," which immediately
follows, as well as upon the hero's pursuit of the knowledge of lesbianism which animates the second half of the novel. The Baron's perplex-
ity throws the latter's investigative efforts into comic relief. The perfect
lie, which "can open windows for us on to what is new and unknown,"
is poised ironically between the revelation of Morel's bisexuality and
Albertine's dubious professions of love for the hero. Moreover, it suggests
a continuity between the despised Morel and the attractive Albertine,
thereby introducing complications into the novel's opposition between
Sodom and Gomorrah.

Although Albertine is represented as sexually indefinite, the suspicion
of her lesbianism arises from her friendship with the women associated

with Mlle Vinteuil who, far from exuding a diffuse sensuality like the
little band at Balbec, are hard-core inverts who, like Charlus, practice
sadomasochism. In a corresponding reversal, the denizens of Sodom
turn out to include the handsome, heroic, and otherwise normal Saint-
Loup. As Bal observes, he is characterized by the same fluidity of movement and shimmering surfaces as Albertine.[63] These complications
indicate the inadequacy of the hero's definitions of the cities of the plain
and cast doubt upon the notion that, as Michael Sprinker claims,
"lesbianism is an unassimilable value" in the novel. Instead of disrupting
"the smooth functioning of the sex/gender system,"[64] lesbianism, as an
unfathomable mystery, ensures its continued operation. Positing female
same-sex desire as "absolutely Other" to heterosexuality, in Gray's
words, not only reiterates the hero's depiction of Gomorrah but sustains
the normativity of the heterosexuality from which it is excluded.[65]

Far from existing outside discursive codes, Proust's lesbians are thor-
oughly embedded within a specific historical context that had a quite
conventional, even clichéd image repertoire of female same-sex eroticism. The degraded Mlle Vinteuil and the alluring Albertine roughly
approximate the twin versions of lesbianism in turn-of-the-century
French and English literature and culture, which may be characterized
as the perverse and the angelic. The locus classicus of lesbian representation for Proust was Baudelaire's *Les Fleurs du mal*. In his 1921 essay
on Baudelaire, Proust praises the poet's depiction of love between
women as among his greatest works and announces that, in the not yet
published final volumes of his novel, the link between Sodom and
Gomorrah would be "entrusted to a brute, Charles Morel."[66]
Baudelaire's "femmes damnées," Morel-like in their seductive wicked-
ness, were highly influential in the production of the fin-de-siècle image
of lascivious sapphists, often feminine yet with a streak of virile cruelty.
For instance, Pierre Louÿs's *Songs of Bilitis* (1894), purportedly a translation of Greek erotic poems composed by a female contemporary of
Sappho, and Alfred de Vigny's poetry, which Proust cites in his essay on
Baudelaire and from which he took the epigraph to *Cities of the Plain*,
derive from Baudelaire and contribute to the typical modernist representation of lesbianism as decadent. This figure, which informs Eliot's
poetry, reaches its modernist apogee in Djuna Barnes's *Nightwood*, which
Eliot almost single-handedly published and for which he wrote the introduction. Mlle Vinteuil plays this part in *Remembrance*, a role for which
sadism is de rigueur. Her profanation of her father's photograph, as
most critics agree, is a displacement of the aggression generally aimed

at mothers by sons in Proust's work. Most of all, though, the spectacle of her sadism at Montjouvain is the pretext for the hero's voyeurism – another commonplace of lesbian representation is the presence of a third party, a man who witnesses the exotic scene.

The other face of sapphic love, sometimes combined in the same portrait, is that of a refined, aestheticized sensuality, typically feminine or androgynous. Observing the "fascination that the lesbian exerted on the male imagination" at the time, Alain Corbin describes her as a "paragon of gentleness [and] delicacy."[67] This pure version is typically associated with the lesbian salons of early modern Paris, such as the one that Natalie Clifford Barney conducted in her "Temple of Friendship." Although Albertine may not appear as "poeticized" as Corbin suggests, Proust apparently thought her to be so. In a conversation with Barney shortly before his death, he assured her, in Barney's words, "that in fact his Sodomites were horrible, but that his Gomorrhans would all be charming." Barney found them above all unrealistic! Not everyone can penetrate the Eleusinian mysteries ". . . Proust could not have had any contact with these mysteries."[68] In her own writings, Barney depicts an idealized sapphism as inaccessible to men as Albertine is to Proust's hero.

To return to the question with which we began, why is Albertine the cause of the novel's epistemological and textual errancy? Her unknowability is, in effect, a ruse of the narrative, an incitement to the endless pleasures of probing, prying, and even, like Swann, "spying . . . outside a window" or "listening at doors." Such methods, the hero observes, are "legitimately employable in the search for truth" (R 1:299). The enigma of Albertine is the pretext that engages and sanctions the hero's voyeurism in the guise of a quest for knowledge; indeed, the two are indistinguishable. The mystery of her sexuality has no solution, not because of her profound alterity but because as a person, even as a woman, she figures only as a question mark. The hero's tireless investigations produce no answer because he does not want one; rather, he wants the desire provoked by the question.

To try to figure out if Albertine really is a lesbian or not is to miss the point of the hero's relentless inquiry, which is not to get to the bottom of things but to prolong the painful pleasure of the inquiry itself. In this sense, the hero's voyeurism is the engine that drives the narrative. By contrast, to read the ambiguity of Albertine's sexuality as the fundamental undecidability of desire itself, while somewhat closer to the mark, accepts the ruse that the hero is in pursuit of a truth that eludes him. Moreover, it obscures the ways in which Albertine's sexuality is constructed in terms of the conventional modern image of the lesbian as mysteriously exotic, by turns an angel and a pervert.

Finally, and perhaps most importantly, Albertine is an incitement to the reader's voyeurism. We read through the thousands of pages of *Remembrance* because, seduced by its disclosive desires, we want to see as well. The hero's continual peering into the secret lives of others invites the reader to do the same. His pleasure at watching scenes of perversion, his "enraptured tourism" at Balbec and Venice, and even Aunt Leonie's "life of constant drama," lived "by looking out the window," lure the reader while blocking an in-depth view. Above all, the novel incites the desire to expose the hero's – and Proust's – homosexuality. Intrigued that Proust may have transcribed his homosexual affairs into the novel, critics have been encouraged to do so by his comment to Gide that "in order to provide material for the heterosexual part of his novel . . . he had transposed to the budding grove of girls the feelings of tenderness and charm supplied by his homosexual memories."[69] Outraged, Gide accused him of duplicity, of camouflaging his true desires, yet dissimula-tion is precisely Proust's aim in *Remembrance*. The wish to force a confes-sion is inscribed within the text, but as a pretense; the desire for revelation is the "perfect lie" that sustains the narrative in its endless deferrals.

As everyone knows, *Remembrance* has a happy ending in "time regained." The magical anamnesia of *mémoire involuntaire* seems to give the long-awaited answers, dissolving the problems of textual and sexual error, but such a resolution is scarcely plausible in view of the preceding pages of the novel. The miracle that Proust dangles before readers who are desperate for a cure for "this perpetual error" that is "life" and who long to return to the madeleine and Combray is an evasion. Nonetheless, it remains a powerful temptation and, indeed, is crucial to the canonical interpretation of *Remembrance* as offering a redemption of life through art; thus, Leo Bersani asserts that, "ideally," for Proust, "art would be truth liberated from phenomena." Although he allows that "its redemp-tive project – the petrifying translation of life and truth" – is ultimately defeated by the noncoincidence of *Remembrance* to the promised fiction, Bersani claims that Proust wanted to compose an intransitive novel.[70] Far from being Proust's intention, this is a deceit, entertained by his hero, that is unmasked in the text. As Walter Benjamin realized, Proust deglamorizes not only romantic love, high society, and the sovereignty of the ego, but also the veils of art that permit and sustain these illusions.[71] Similarly, Sprinker cogently argues that "Proust's text holds no illusions

about the one ideology most often ascribed to it: the ideology of art";
Remembrance "refuses the temptation of aestheticizing history while
demonstrating society's ineluctable entanglement in the material processes of historical change."[72] Proust's strategy of disillusionment consists of ironically miming the aesthetic ideology voiced by the hero, a
strategy similar to that by which he undermines the latter's sexual ideology. His hero's faith in the consolations and transcendence of works of
art are as dubious as his phobic views of male homosexuality and his
belief in the fundamental unknowabililty of Albertine. Rather than
attaining the goal of a quest for truth, Proust's text asks that we examine
the errancy, even the perversity, of the desires that motivate such a
search. Instead of giving us answers, Proust gives us desires.

Conclusion

What a paradox was to me in the sphere of thought,
perversity became to me in the sphere of passion.

Oscar Wilde, *De Profundis*

It has become a truism that modernist writers espoused an ideology that promoted the belief that the work of art is an autonomous, organically unified whole, transcending its historical moment to achieve a putatively universal, uniquely aesthetic value. Eliot's remark concerning Joyce's use of Homeric myth as a way "of controlling, of ordering, of giving a shape . . . to the . . . anarchy which is contemporary history" (*SP* 177) has itself been ripped out of context in order to serve as a touchstone for this interpretation of modernism in which Eliot exemplifies the error of its ways. However, a closer analysis of Eliot's writings, as of Joyce's and Proust's, fails to sustain this coarse-grained and often polemical critique. On the contrary, their works demonstrate a fundamental skepticism toward such an aesthetic ideology.

Throughout his career, Eliot was deeply suspicious of claims for the transcendent value of art and its "higher" truth because he believed that such matters were, at bottom, theological, not aesthetic. To make art into a religion, to substitute it for faith, or to take comfort in the former as a consolation for the loss of the latter was, in his view, to commit Arnold's mistake. Even in *Four Quartets*, the most explicit poetic statement of his religious beliefs, Eliot does not look to poetry for deliverance from social alienation but presents his faith as a desperate, austere necessity. In his prose criticism, Eliot's strictures against error stress the limits and the limitations of poetry. Literature is an expression, and further evidence, of our fallen estate.

Proust was equally hard on those who fetishize works of art, although he was more alive to the temptations of what he called "idolatry" than was Eliot. The academic fetishization of *Remembrance of Things Past* has

217

blinded readers to the ways in which it fails to live up to the status of a literary icon. To be sure, Proust shared with Joyce and Eliot considerable anxiety concerning the coherence of the book. Antoine Compagnon points out that Proust presents himself at times as "the standard-bearer of [the] premeditated unity" of the work of art and "often reaffirms the architectural unity of his book."[1] At its end, however, the narrator, speculating on the book that he plans to write, confesses that his method of composition would be like sewing, not architecture: "pinning here and there an extra page, I should construct my book, I dare not say ambitiously like a cathedral, but quite simply like a dress" (*R* 3:1090). The retrospective unity imposed by the conclusion of *Time Regained*, especially the recurrence of involuntary memory – the slip on the paving stone recalling the taste of the madeleine, comprising the matching bookends of the text – resembles Joyce's not altogether convincing attempts to provide an architectural unity for *Ulysses* in the schema he gave Linati, not to mention the host of reading instructions he offered for it and for the minutely organized chaos of *Finnegans Wake* in his letters to Harriet Weaver. Eliot's notes to *The Waste Land*, presented with a possum-like straight face, were at best an afterthought, written either in response to puzzled readers (like Joyce's schema) or to fill up some blank pages or both. In all three cases, critics have typically overlooked the disjunctions between these, in some respects, occasional or aleatory assertions of a governing structure and the disarray of the actual texts. To read *Time Regained* as a sort of nostos, the return home being the conclusion of a circular, teleological itinerary, requires one to pass over – indeed, to cover over – the endless digressions that pull *Ulysses* and *Remembrance* in errant directions or the threads in *The Waste Land* that are never caught up into a comprehensive web. These errors of construction or failures of organic unity, such as the loose thread of the "hyacinth girl" that is left hanging in *The Waste Land* or the jumbled citations in "Ode," give access to what Margot Norris, referring to Joyce, terms the textual "unraveling" of the ideology of modernist formalism.[2] These so-called literary monuments may be usefully compared to Pound's epic ruin, the *Cantos*, to Richardson's baggy monster, *Pilgrimage*, and to Stein's failure to produce the epic that she wanted to write.

Stephen Dedalus's pronouncement – "A man of genius makes no mistakes. His errors are volitional and are the portals of discovery" (*U* 9:238–39) – has often been read as though it came out of Joyce's mouth. This supposedly authorial endorsement of the author's infallibility has underwritten inflated notions of Joyce's aesthetic claims for the ideal work of art as a "total book" and of *Ulysses* as just such a book. Interpretations of *Ulysses*, including Joyce's, tend to fall into opposing camps: on the one hand, he referred to it as "a sort of encyclopedia" (*SL* 271), manifesting what Hermann Broch terms a "*Totalitätsanspruch*," or an urge for totality; on the other hand, its multiplicity leads Stephen Heath to call it an unfinalized "assemblage," citing Joyce's description of himself as a "scissors and paste man" (*LJ* 1:297).[3] This polarization is even more pronounced in readings of *Finnegans Wake*, which is commonly viewed either as a comprehensive book that consumes its readers, anticipating and foreclosing anything anyone could ever say about it, or as a complete mess, a disintegration of language into a babel of polyglot nonsense. In both cases, the text is accounted illegible for virtually the same reasons, but with an opposite valence attached to intelligibility and (w)holeness. Was Joyce a genius, or are his readers paranoid? Instead of adopting either of these positions – which are inverted images of each other – I argue that the errancy of Joyce's texts has particular motives and effects, traceable in no small part to his historically specific articulation of sexual differences and desires.

In a somewhat different fashion than *Ulysses*, Proust's novel is marked by pervasive metacommentaries, culminating in the narrator's announcement of the book he intends to write. As the realization of its hero's literary aspirations and the fulfillment of its own artistic doctrines, *Remembrance* is thus typically understood as the achievement of aesthetic autonomy and autotelic coherence. Maurice Blanchot advanced this view of *Remembrance* as a total book, claiming that it "attempts to achieve . . . the essence of art," which is to be purely self-referential, purged of nonliterary elements, with its end in its beginning, and vice versa.[4] The fact that one reads *Remembrance* and not its hero's imagined novel should be sufficient evidence that the two are not identical and that the hero's conception of art as a means of mastering time and history is neither Proust's aim nor his accomplishment. Michael Sprinker argues that "the theory of art embraced by Marcel and projected for the work he intends cannot be the benchmark for judging the *Remembrance*," which is, instead, "resistant to aesthetic idealization." Indeed, it demonstrates "the failure of this aestheticizing strategy."[5] Proust undermines the narrator's aesthetic beliefs just as he subverts his sexual ideology, including both his homophobia and his faith in Albertine's fundamental alterity.

The famous search for a truth that is supposedly discovered in *Time Regained* is led astray by all that comes between the question and its answer, especially by the novel's more productive elaboration of errors of perception and judgment. Proust's engagement with error resembles that of the early Eliot, whose study in the philosophy of Henri Bergson and F. H. Bradley led him to conclude in his thesis, *Knowledge and Experience*, that "the life of a soul does not consist in the contemplation of one consistent world but in the painful task of unifying . . . jarring and incompatible ones."[6] Like Eliot, who argued for the need for a "higher" perspective that would "somehow include" and connect these discordant viewpoints, Proust's hero continually searches for general principles that would link discrete temporal moments and their disjunctive presentations. The epistemological skepticism of Bradley's concept of consciousness as a "finite centre," endorsed by Eliot, is consistent with what Gilles Deleuze describes as Proust's mondadology, in which individuals are more or less locked into solipsistic worlds.[7] By contrast to Eliot, Proust seems to offer a way out through art, specifically, through metaphor. "Truth will begin," the narrator claims, "when the writer takes two different objects, poses their relation, . . . and encloses them within the necessary rings of a beautiful style, . . . uniting them in order to extract them from the contingencies of time, in a metaphor" (*R* 3:889). In practice, though, this "truth" remains an abstraction or, at best, a promise, just as the conjunction of two moments effected by involuntary memory offers an evanescent déjà vu and the promise of a book; meanwhile, we read a book whose digressions and postponements reflect that equally errant, "inner book" of the mind which, like Morel's, resembles "an old book of the Middle Ages, full of mistakes, of absurd traditions, of obscenities; . . . extraordinarily composite" (*R* 2:1066).[8]

Virginia Woolf and Dorothy Richardson share Proust's focus on interiority, Woolf applauding the Frenchman's "perfectly receptive" sensibility in terms similar to what she calls Richardson's "feminine" style, that is, "a sentence which we might call the psychological sentence of the feminine gender."[9] Although she praised Proust more than any other contemporary writer, according to Lyndall Gordon, her compliment that he "was wholly androgynous, if not perhaps a little too much of a woman,"[10] introduces doubts raised by what one might call Gertrude Stein's *écriture masculine*. Does Proust go too far in writing like a woman, his sentences, like his desires, straying off the track? Woolf's "feminine" sentence, like Hélène Cixous's *écriture féminine*, is bound to unquestioned sexual verities.

"For male modernist writers seeking an avant-garde dislocation of forms, a recasting of given identity into multiplicity," according to Stephen Heath, following Cixous's logic, "writing differently has seemed to be naturally definable as writing feminine, as moving across into a woman's place."[11] The emphasis in Heath's sentence should fall on "naturally," for gender transgressions in modern Western societies often have the paradoxical effect of affirming while denying the irreducible naturalness of sexual difference. Explorations of feminine alterity by modernist male writers, like the current interest in cross-dressing and transsexuality, attest to the decidedly equivocal operations of such "drag" gender styles. The condition of possibility for modernist male writers of discovering an internal femininity – for Pound, of attaining the "chaotic fluidities" of women, or for Bloom, of becoming a "womanly man" – was the absolute otherness of femininity. The desire to move into a woman's place, or, as Beckett's Malloy put it, to occupy mother's room, to sleep in her bed, or to speak in her voice, offered the promise not simply of a retreat to the womb but of escape from phallic fixity, from masculine social norms as well as from traditional literary forms. Yet the price for this incorporation of female alterity, as we have seen in Joyce's work, is typically the mystification of femininity. By contrast, Eliot, who fulfilled his mother's ambitions by becoming a writer, taking her place by becoming the poet she longed to be, effected a near-total erasure of feminine influence from his work. However, the unwitting eruptions of sexual anxiety in his poetry and his strayings from the patriarchal citation in what I have called their maternal intertexts are muted but eloquent witnesses to the contemporary crisis in heterosexual masculinity. From a different angle, culturally and stylistically, Proust posits but rejects the essentializing premises of contemporary sexual ideology in a manner that combines discretion, deceit, and disillusionment concerning homosexuality. The promise and threat of feminine errancy is apparent in many other texts by modernist male writers, especially those of D. H. Lawrence and Ernest Hemingway.

Lawrence's novels are populated by powerful mothers with aspirations to artistic culture who dominate and defeat their brutish, working-class husbands. These failed fathers, dead or wounded, come to embody for their sons a nostalgic masculinity – crude, a bit dumb, but strong and handsome. For instance, Gertrude Morel of *Sons and Lovers*, like Charlotte Eliot, imposes upon her son Paul her frustrated aesthetic ambitions, with the result that Paul not only grows up to be the artist his mother wants him to be but that he nourishes an abiding ambivalence toward her. This love/hate relationship of sons to mothers is a feature of many modernist texts, from *The Family Reunion* to *Remembrance*, with its theme of the "Profanation of the Mother" (*R* 2:939). More explicitly,

Proust's story "A Young Girl's Confession" concerns the death-by-desecration of the mother, and his essay "Filial Sentiments of a Parricide" sympathetically recounts Henri van Blarenburgh's murder of his mother.[12] The resentful Paul Morel rebels against Gertrude by celebrating the animal-like masculinity that she had degraded; Lawrence's work, in general, is marked by outbursts of gynophobia. However, insofar as they adopt a "feminine" position in relation to attractive, virile men, Lawrence's heroes identify equivocally with their mothers. Some of Lawrence's poems overtly represent the male body as an erotic object; for instance, "Eloi, Eloi, Lama Sabachthani?" describes the pleasure of penetrating an enemy soldier:

> . . . I knew he wanted it.
> Like a bride he took my bayonet, wanting it,
> Like a virgin the blade of my bayonet, wanting it
> And it sank to rest from me in him,
> And I, the lover, am consummate.[13]

Lawrence's poem elaborates his description in his essay, "The Crown," of "the enemy" as "the bride, whose body we will reduce with rapture of agony and wounds. We are the bridegroom, engaged with him in the long, voluptuous embrace."[14]

That embrace appears again in the "Man to Man" chapter of *Women in Love* between the refined, intellectual Rupert Birkin and the "coarse, insensitive," and blonde, Gerald Crich. The *Blutbrüderschaft* that Birkin fantasizes with the latter is the answer to "the problem of love and eternal conjunction between two men" which suddenly confronts Birkin when he realizes that he is "deeply bondaged in fascinated attraction" to Crich, "resenting the bondage, hating the attraction."[15] He and Gerald consummate their union in a famous wrestling match that fades out at just the right moment: "He slid forward quite unconscious over Gerald, and Gerald did not notice. . . . [E]verything was sliding off into the darkness" (263). As Christopher Craft observes, the condition of Birkin's desire for Gerald is the latter's pronounced masculinity, which resists contamination by the "feminine" cast of homosexual inversion or even by the feminizing potential of heterosexual marriage, thus ennobling the blood brotherhood that Birkin imagines with him. Moreover, it is made possible by Lawrence's phobic projection of homosexuality onto the detested faggot Loerke, who is doubly despised as a Jew and is described as a sewer rat in accordance with the popular imagery of anti-Semitism in the early twentieth century, a rhetoric that Eliot also employed.[16] Of course, Lawrence's heroes are well known for their coprophilia or, rather, for the anal locus of their erotic desires. Sex between Birkin and Ursula is a sodomite's dream come true, celebrating the "deepest, strangest life-source of the human body, at the back and base of the loins," from whence issue "floods of ineffable darkness and ineffable riches" (306) – a more lavish fantasy than Joyce's Bloom ever entertained. At its most thrilling, heterosexuality in Lawrence is sodomical.

Where does this leave women? Principally as a vehicle for desire between men and as a displacement for homoerotic bonds, according to Craft, and as a figure for castration anxiety, but also – again connecting Lawrence with Joyce – as a means for his male heroes to take the "female" role in intercourse. Notwithstanding his frequent fulminations against women, Lawrence speaks in the persona of a woman in love with and having sex with a man in poems such as "Love on the Farm," "A Youth Mowing," and "Two Wives."[17] Unlike Joyce, Lawrence shows little interest in writing from a woman's perspective. However, the torrents of his rhapsodic prose, with its melting dissolves and its unrestrained, turgid gushes, have a perversely feminizing linguistic effect; his style is exemplary of what Gregory Woods calls, in a different context, *écriture efféminée*.[18]

Worlds apart stylistically from Lawrence, Hemingway was nonetheless also intrigued by heterosexual sodomy and attracted to virility in both men and women. David Bourne, the hero of his posthumously published novel, *The Garden of Eden*, has a surprise on his honeymoon when his bride, Catherine, turns out to be bisexual, with an erotic trick up her sleeve: "He lay there and felt something and then her hand holding him and searching lower and he . . . lay back . . . and only felt the weight and the strangeness inside and she said, 'Now you can't tell who is who can you?'"[19] When Catherine offers to be his "boy" and try it again, David demurs, "Not if you're a boy and I'm a boy" (67). The permutations of sex and gender in *The Garden of Eden* cast light on the many depictions of impotent men and potent women in Hemingway's previous novels. Perhaps because women have the phallus in these fictions, he is all the more eager to show that a man without the phallus is still a real man. Although Jake Barnes in *The Sun Also Rises* may be maimed and in love with a woman, Lady Brett Ashley, who looks like a lesbian and hangs out with gay men, Jake at least is not an effeminate homosexual.

Hemingway's oft-remarked infatuation with bullfighters – tough guys in tight pants and flashy costumes who spend an inordinate amount of time dressing – would imply the homoeroticization of masculinity in his novels were they not written in a resolutely hard-nosed style. His laconic prose and studied refusal to plumb the psyches of his characters, whose most eloquent expression is swearing, parade what Woods calls "stunted interiority" as the mark of "scarred masculinity." Real men, he points out, do not "gush."[20] Yet such a suppression of sentiment calls attention, in the form of paralipsis, to the emotions of which it ostentatiously omits mention; such "virile display," Lacan observes, "itself seem[s] feminine."[21] Moreover, Hemingway learned these lessons from Gertrude Stein, herself a better man, stylistically, than many of the writers who championed a masculinist aesthetic – the Hulme/Lewis/Pound current of modernism. Hemingway's heroes are not only the tragic victims of their masculinist credo; they are also the victims of rapacious females, like Margot Macomber, who shoots her husband with a 6.5 Mannlicher rifle in "The Short Happy Life of Francis Macomber." His disavowal of feminine identification reverses the modernist deployment of the feminine as a means of errancy, yet his texts frequently represent gender exchanges and are troubled by their homosexual implications. Indeed, they demonstrate the contradictory construction of modern male homosexuality, which can result from liking men too much (so that you desire them) as much as from liking women too much (so that you're one of them).

As the examples of Lawrence and Hemingway show, the significance of reading male modernist texts from the perspective of contemporary queer theory is the light it sheds upon the construction of heterosexual masculinity during a period in which the latter became subject to serious doubt. "Nobody was born a man," Norman Mailer observes; "you earned manhood provided you were good enough, bold enough."[22] Echoing Simone de Beauvoir's aphorism, "One is not born, but becomes a woman,"[23] Mailer's remark points toward what Judith Butler calls the performative constitution of gender identity, especially the inevitability of failed performances – that is, of not living up to or, as Mailer puts it, "earning manhood."[24] Lesbians and gay men are notably "bad copies" of heterosexual gender norms, and none more so, in modernist literature, than Oscar Wilde. The figure of Wilde focuses attention on problems of homosexual denomination and visibility, both literary and popular inscriptions at the time and in current queer theory,

Wilde's 1895 trial was spectacularly successful in assigning a name to the love that had hitherto "dared not speak" it. Subsequently, "men of the Oscar Wilde type," in E. M. Forster's phrase, became identifiable as such, both to themselves and to their fellows. When in *Ulysses* Stephen thinks of Mr. Best, "Tame essence of Wilde" (*U* 9.532), we are left in no doubt as to the kind of man he is. However, the coerced labeling of Wilde as homosexual had the effect of suppressing discussion of the love to which his name was attached. "The Wilde trial had done its work," Edward Carpenter complained afterward, "and silence must henceforth reign on sex-subjects."[25] The volubility of that silence, though, testifies to the paradoxical relation of homosexuality to discursive denomination. Since the late nineteenth century, male same-sex desire has been the site of a nomenclative fury, called all kinds of names, scientific and demotic, from "Uranism" to "inversion" and worse. This appellative urge is coextensive with an equally powerful prohibition against any reference to the offense cited in English common law as "a crime not fit to be named among Christians" – "*peccatum illud horribile, inter christianos non nominandum*" – in Blackstone's memorable phrase.[26]

Modernist literature is notable for its reluctance to utter this unspeakable desire despite pervasive and occluded allusions to it. Even that modernist text devoted to identifying homosexuals – *Remembrance* – is remarkably reticent about Wilde. Referring to "the poet one day fêted in every drawing-room and applauded in every theatre in London, and the next driven from every lodging" (*R* 2:638), Proust's sympathy is guarded by the danger of too close an association with Wilde. His Charlus is closely linked to Wilde's Dorian Gray, both modeled after Robert de Montesquiou (in the latter instance, through the textual mediation of Des Esseintes in Huysman's *A Rebours*), and the parallels between Wilde's and Charlus's public humiliations are unmistakable.[27] The ambivalent distance that Proust maintained from Wilde is also evident in his response to the 1906 Eulenbourgh Affair in Berlin, a homosexual scandal about which he proposed writing an essay friendly to Eulenbourgh, and which perhaps influenced his characterization of Charlus as a Germanophile. Proust never wrote the essay, though, just as he never made a public admission of homosexuality. Proust's rejection of that self-designation is equivocally poised between (in Gide's view) protective camouflage and a Wildean, postmodern sense of the paradoxical perversity of desire.[28]

The thrust of contemporary lesbian and gay studies has been to tear away such evasive masks and to insist upon naming homosexuality, confident that such a denomination expresses a heretofore suppressed truth. The fundamental performative act of lesbian and gay studies is thus "coming out," both for oneself and for others. In an assertion that peculiarly echoes the intentions of Wilde's inquisitors, the editors of *Professions of Desire: Lesbian and Gay Studies in Literature* claim that "this is a field that one does not enter so much as come out in."[29] The confessional imperative is based upon the humanist belief that the endorsement of a sexual identity is an ethical act; being true to oneself, of course, presumes a self to be true to, a consistent identity founded in a natural or essential self. The institutionalization of lesbian and gay literary study enforces the practice of "naming names," that is, of fingering lesbian and gay authors or texts in order to construct a canon.

By contrast, my aim in examining a group of authors with varying inclinations and identifications has been to reject the binary division between heterosexuality and homosexuality together with its questionable ontological assumptions and its universalization of historically relative typologies of desire. Foucault's attack on the "repressive hypothesis," his arguments concerning the invention of the homosexual as a "personage" in the late nineteenth century, and his skepticism toward sexuality as a privileged means of self-knowledge are especially useful in reading modernist texts in which such ideas had powerful effects. Indeed, in the twentieth century, sexuality has been central to inquiry into the nature of subjectivity; it has become an epistemological problem in a variety of disciplines and popular discourses. In Foucault's words, "we demand that it tell us our truth, or rather, the deeply buried truth of that truth about ourselves which we think we possess in our immediate consciousness" – in short, an unconscious truth that underlies and undermines the conscious mind.[30] Sexuality seems to be a part of ourselves that escapes us, about which we are not only ignorant but which produces that ignorance, and which divides us from ourselves even as it determines our destiny. Such a belief, according to Foucault, draws its allure from the putative unknowability of sexuality. Hence, the will to knowledge concerning sex is bound to err, yet, like Marcel with his Albertine, we take great pleasure in the indefatigable production of discourses that still promise to tear away the ultimate veil.

The perpetual failure of inquiry into sexuality – its seemingly inherent errancy – is historically bound up with Christian beliefs in the essential errancy of sexual desire itself. The disclosure of sexual secrets, like the injunction to sexual honesty, is energized by the thrill of confession or, in scientific discourses, by the discovery of aberrant types. The intertwining of theological with epistemological norms in the production of homosexuality in the twentieth century and its imbrication with questions concerning the nature of masculinity and femininity, have meant that a host of apparently unrelated issues, including aesthetic ones, has gravitated around homosexuality. Moreover, because modern sexual discourses are so heterogeneous and unstable, putting homosexuality explicitly into play has multivalent, often unpredictable, and wayward effects. For instance, emancipatory political ideologies, especially feminism and gay liberation, have sometimes resulted in the confirmation of homophobic and misogynistic sexual codes.[31] Hence, instead of formulating an overarching, "correct" reading of the gender and sexual forces at work in modernist literature, I have focused upon the conjunction of those forces and textual errors, mistaken interpretations, and slips of the tongue or pen. Rather than disclosing the truth of (homo)sexuality, modernist texts offer misspelled telegrams and Venetian labyrinths, errors that manifest not an irreducible textual indeterminacy or an essential sexual errance (as in Leo Bersani's interpretation of the death drive) but a culturally contingent confusion with specific historical determinants. Finally, by analyzing the relation between textual and sexual errancy in the works of Eliot, Joyce, and Proust, I am participating in the modern phenomenon that Foucault describes, but with the aim not of eliciting inadvertent admissions from these works but of interrogating the conceptual structure of sexuality upon which they rest and of disrupting the coherence of the logic of homophobia that, far from being the personal problem of these writers, is a common sense that we continue to share with them.

Wilde was the highly visible incarnation of homosexuality in his day. The public scene of his humiliation has foregrounded the problem in queer theory of the contradictory (in)visibility of homosexuality as an open secret in which occlusion and exposure go hand in hand. The equivocally homophobic and homophilic fascination with revelation and the ways in which the knowledge of homosexuality is bound up with its concealment and disguise have been the subject of groundbreaking work by Eve Sedgwick and D. A. Miller and of subsequent scholarship. Moreover, the spectacle of Wilde's degradation made him the center of a collective voyeuristic gaze at the time and continues to call attention to the specularization of homosexuality. In effect, Wilde taught us to "see" homosexuality in a certain way.

Wilde himself argued that vision is the product of discursive constructions. "People see fogs," he argued, "not because there are fogs, but because poets and painters have taught them the mysterious loveliness of such effects."[32] Proust, for whom seeing is knowing, was intrigued by Wilde's observation. In a letter explaining why he decided not to write an essay in defense of the Prince of Eulenbourgh, he referred to Wilde's trivialization of sorrow in his pretrial works, a criticism that he repeated in *Contre Sainte-Beuve*:

> Oscar Wilde, whom life, alas, would teach that there are sorrows more piercing than those we get from books, said in his first period (the period of his remark, "Before the Lake poets there were no fogs on the Thames"), "The greatest grief of my life? The death of Lucien de Rubempré."[33]

Proust's misquotation from Wilde's "Decay of Lying" praises the impressionistic aesthetic to which Proust gives ample expression in *Remembrance* but also implies Wilde's superficial decadence. If there is no difference between real fogs and those of poets, what is the difference between real tears and those evoked by the death of Lucien de Rubempré? One answer, of course, is prison. But the tears Wilde shed in Pentonville were those of a man, while those he spilled over Balzac's novel were women's tears, copiously sentimental and artificial.

Wilde marks the decisive conjunction of male homosexuality and femininity; according to Alan Sinfield, "the image of the queer cohered at the moment when the leisured, effeminate, aesthetic dandy was discovered in same-sex practices . . . with lower-class boys."[34] Proust's remarks on Wilde's tears reveal his uneasiness with the aestheticism, the effeteness, and fin-de-siècle decadence which, after Wilde's trial, immediately defined homosexuality and against which many male modernist writers reacted with homophobic assertions of virility. The mainstream of lesbian and gay studies continues to fight the notion that gay men are not real men, while current debates concerning "butch" lesbian sexual styles recapitulate the modernist controversy over the virile, inverted heroine of Hall's *Well of Loneliness*.[35] However, the refusal of lesbians and gay men to follow appropriate gender scripts has complex social significations within gay subcultures and in heterosexual contexts, and, for queer theorists, it implies the contingency of sexual difference in general. Thus, Monique Wittig brazenly asserts that "lesbians are not women" because "'woman' has meaning only in heterosexual systems of thought and heterosexual economic systems."[36] Butler's claims for the refiguration of gender and sexual identities through distorting, performative iterations draw upon Wittig's sense, taken from de Beauvoir, that women (and men) are constructed, not born, and thus potentially subject to change.[37]

Examining male modernist texts from the perspective of postmodern queer theory does not mean assimilating the former to the latter but means examining the ways in which modern constructions of hetero-sexual masculinity were (and are) called into question. The emergence of homosexuality as a sexual category aligned with femininity in the early twentieth century cast into doubt the normality of heterosexuality; or, to be precise, homosexuality became the name attached to the unset-tling of masculinity, a way of fixing in the person of the pervert certain abjected desires and identifications. The usefulness of inquiry into what was quickly placed under the heading of "homosexuality" in the modern period is thus not confined to – indeed, it resists – categoriza-tion as "homosexual," but instead opens theoretical and empirical examination of the production of sexed bodies and sexed desires, including those identified as heterosexual. For feminist as well as lesbian and gay theorists, this means giving up what has been the moral high ground of both their enterprises: the claim to the "truth" of experience and the victim status of women and homosexuals.[38] Queer theory is not only about homosexuals, just as this study is designed not to discover same-sex desire within the texts of canonical male modernism but to challenge the division between heterosexuality and homosexuality at the moment of its inscription.

The textual and sexual errancies in modern literature do not, finally, amount to men writing as or wanting to be like women, or to the dis-ruptive emergence of a repressed, primal maternal force within patriar-chal discourse, as Kristeva argues. Rather, these errancies issue from the particular pressures and constraints brought to bear upon men and women in the early twentieth century which rendered sexual difference – and potential indifference – a powerfully motivated blind spot, a sig-nificant site of ideological formations and deformations, or, quite simply, a place where one is apt to go wrong. Of course, going astray, dis-cursively and sexually, has its rewards. The waywardness of Eliot, Joyce, and Proust bore strange and wonderful fruit in the work of Djuna Barnes, whose *Nightwood* was published by Eliot, who was a friend of Joyce, and whose *Ladies Almanack* was, she said, "neaptide to the Proustian chronicle." Her erratic texts, manifestly unamenable to aes-thetic or sexual norms, are the lucky heirs of their errors.

Notes

INTRODUCTION

1 Among the many attempts to list the defining terms of modernism, see Malcolm Bradbury and James McFarlane, "The Name and Nature of Modernism," in *Modernism 1890–1930*, ed. Bradbury and McFarlane (Harmondsworth: Penguin, 1976); and Maurice Beebe, "What Modernism Was," *Journal of Modern Literature* 3 (July 1974): 1065–84.

2 The most influential argument that the modernist movement was a reactionary, masculine backlash against women is Sandra M. Gilbert and Susan Gubar's three-volume *No Man's Land: The Place of the Woman Writer in the Twentieth Century* (New Haven: Yale University Press, 1988–94).

3 See Suzanne Clark, *Sentimental Modernism: Women Writers and the Revolution of the Word* (Bloomington: Indiana University Press, 1991), 35; and Marianne DeKoven, *Rich and Strange: Gender, History, Modernism* (Princeton: Princeton University Press, 1991), 11.

4 See Theodor Adorno and Max Horkheimer, *Dialectic of Enlightenment* (London: Verso, 1979).

5 See Gillian Hanscombe and Virginia L. Smyers, *Writing for Their Lives: The Modernist Women, 1910–1940* (London: Women's Press, 1987); Kathleen Wheeler, *"Modernist" Women Writers and Narrative Art* (New York: New York University Press, 1994); *Breaking the Sequence: Women's Experimental Fiction*, ed. Ellen G. Friedman and Miriam Fuchs (Princeton: Princeton University Press, 1989); Rachel Blau DuPlessis, *Writing beyond the Ending: Narrative Strategies of Twentieth-Century Women Writers* (Bloomington: Indiana University Press, 1985); Shari Benstock, *Women of the Left Bank: Paris, 1900–1940* (Austin: University of Texas Press, 1985); and Bonnie Kime Scott, *Refiguring Modernism*, vols. 1 and 2 (Bloomington: Indiana University Press, 1995).

6 See Hélène Cixous and Catherine Clément, *The Newly Born Woman*, tr. Betsy Wing (Minneapolis: University of Minnesota Press, 1986); Alice Jardine, *Gynesis: Configurations of Women and Modernity* (Ithaca: Cornell University Press, 1985); Shari Benstock, *Textualizing the Feminine: On the Limits of Genre* (Norman: University of Oklahoma Press, 1991); Suzette Henke, *James Joyce and the Politics of Desire* (London: Routledge, 1990); Susan

Rubin Suleiman, *Subversive Intent: Gender Politics and the Avant-Garde* (Cambridge, MA: Harvard University Press, 1990).

7 See Terry Eagleton, *Marxism and Literary Criticism* (London: Methuen, 1976).

8 See Paul Morrison, *The Poetics of Fascism: Ezra Pound, T. S. Eliot, Paul de Man* (New York: Oxford University Press, 1996).

9 Georg Lukács, *The Meaning of Contemporary Realism*, tr. John and Necke Mander (London: Merlin Press, 1962). For an unusual perspective, see Michael North, "Eliot, Lukács, and the Politics of Modernism," in *T. S. Eliot: The Modernist in History*, ed. Ronald Bush (Cambridge: Cambridge University Press, 1991), 169–89.

10 See David Trotter, "Modernism and Empire: Reading *The Waste Land*," *Critical Quarterly* 28: 1 and 2 (spring, summer 1986): 143–53; Andreas Huyssen, *After the Great Divide: Modernism, Mass Culture, Postmodernism* (Bloomington: Indiana University Press, 1986).

11 See Pierre Bourdieu, *Distinction: A Social Critique of the Judgement of Taste*, tr. Richard Nice (Cambridge, MA: Harvard University Press, 1984); and Barbara Herrnstein Smith, *Contingencies of Value: Alternative Perspectives for Critical Theory* (Cambridge, MA: Harvard University Press, 1988).

12 Astradur Eysteinsonn, *The Concept of Modernism* (Ithaca: Cornell University Press, 1990), 100.

13 Michael H. Levenson, *A Genealogy of Modernism: A Study of English Literary Doctrine 1908–1922* (Cambridge: Cambridge University Press, 1984), 86, 104–5.

14 F. T. Marinetti, *Observer* (7 June 1914): 7.

15 Levenson, *A Genealogy of Modernism*, 148.

16 Wyndham Lewis, *Tarr* (New York: Knopf, 1926), 334.

17 Bonnie Kime Scott, *Refiguring Modernism*, vol. 1: *The Women of 1928* (Bloomington: Indiana University Press, 1995), 104.

18 Wyndham Lewis, *Blast* 1 (1914), 11.

19 Rebecca West, "Imagisme," *New Freewoman* 1.5 (15 August 1913): 86–87.

20 Ezra Pound, *The Cantos* (New York: New Directions, 1970), 144.

21 Ezra Pound, *New Age* (1 August 1918); quoted in Scott, *Refiguring Modernism*, 1:93.

22 Ezra Pound, "Postscript," in *The Natural Philosophy of Love*, by Remy de Gourmont, tr. Ezra Pound (New York: Collier, 1961), 170.

23 Ezra Pound, quoted in *The Gender of Modernism: A Critical Anthology*, ed. Bonnie Kime Scott (Bloomington: Indiana University Press, 1990), 362–63.

24 See Christine Battersby, *Gender and Genius* (London: Women's Press, 1989), chapters 11 and 12.

25 John Fletcher, "Forster's Self-Erasure: *Maurice* and the Scene of Masculine Love," in *Sexual Sameness: Textual Differences in Lesbian and Gay Writing*, ed. Joseph Bristow (London: Routledge, 1992), 73, 83.

26 Christopher Lane, *The Ruling Passion: British Colonial Allegory and the Paradox of Homosexual Desire* (Durham, NC: Duke University Press, 1995), 2.

27 Friedrich Nietzsche, *Beyond Good and Evil: Prelude to a Philosophy of the Future*, tr. Walter Kaufmann (New York: Random House, 1966), 1; and *The Gay Science*, tr. Walter Kaufmann (New York: Random House, 1974), 125.

28 Eve Kosofsky Sedgwick, *Epistemology of the Closet* (Berkeley: University of California Press, 1989), 159–60.

29 See Madelon Sprengnether, *The Spectral Mother: Freud, Feminism, and Psychoanalysis* (Ithaca: Cornell University Press, 1990), chapter 4; and Rita Felski, *The Gender of Modernity*, (Cambridge, MA: Harvard University Press, 1995), chapter 2.

30 Samuel Beckett, *Molloy* in *Three Novels* (New York: Grove Press, 1965), 7, 64–65.

31 Julia Kristeva, "The Father, Love and Banishment" in *Desire in Language*, ed. and tr. Leon Roudiez (New York: Columbia University Press, 1980), 157; and "Stabat Mater," in *Tales of Love*, tr. Leon Roudiez (New York: Columbia University Press, 1987). For a critique, see Domna C. Stanton, "Difference on Trial: A Critique of the Maternal Metaphor in Cixous, Irigaray, and Kristeva," in *The Poetics of Gender*, ed. Nancy K. Miller (New York: Columbia University Press, 1986), 156–82.

32 See Karen Lawrence, "Joyce and Feminism," in *The Cambridge Companion to James Joyce*, ed. Derek Attridge (Cambridge: Cambridge University Press, 1990), 237–58.

33 Serge Doubrovsky, *Writing and Fantasy in Proust: La Place de la madeleine*, tr. Carol Mastrangelo Bové (Paris, 1974; Lincoln: University of Nebraska Press, 1986), 48.

34 See Proust's essay, "Filial Sentiments of a Parricide," concerning the case of Henri van Blarenberghe, who murdered his mother, for a sympathetic account of the love of a guilt-ridden, vengeful son for a devoted, reproachful mother *"Pleasures and Days," and Other Writings*, ed. F. W. Dupee, tr. Louise Varese, et al. [Garden City, NY: Doubleday, 1957], 293–304).

35 Gilles Deleuze, *Proust and Signs*, tr. Richard Howard (New York: George Braziller, 1972).

36 Although at some points in his work Freud detached same-sex desire from gender inversion, notably in *Three Essays on the Theory of Sexuality*, at others he links homosexuality to a reversal of gender identity.

37 Jeffrey Weeks, *Sexuality and Its Discontents: Means, Myths and Modern Sexuality* (London: Routledge, 1985).

1 STRAIGHTENING OUT LITERARY CRITICISM: T. S. ELIOT AND ERROR

1 Jacques Derrida, "Signature Event Context," in *Limited Inc.*, tr. Samuel Weber and Jeffrey Mehlman (Evanston: Northwestern University Press, 1988), 15, 17.

2 I am indebted to Barbara Herrnstein Smith for discussions of these matters.

3 Judith Butler, *Bodies That Matter: On the Discursive Limits of "Sex"* (New York and London: Routledge, 1993), 14.

4 "The Progress of Error," lines 4–8, in *The Poetical Works of William Cowper*, ed. H. S. Milford, 3d edn. (London: Oxford University Press, 1926).

5 *The Poetical Works of Edmund Spenser* (New York: Thomas Y. Crowell, n.d.), 31. Spenser's monster is the source for Milton's representation of Sin and her many offspring in *Paradise Lost*, book 2, lines 746–814.

6 Michel Foucault, *History of Sexuality*, vol. 1: *An Introduction*, tr. Robert Hurley (New York: Viking, 1980), 101.

7 Margaret Soltan, "Night Errantcy: The Epistemology of the Wandering Woman," *New Formations* 5 (1988): 110.

8 *Webster's Third International Dictionary* lists the Old French *error* or *errour*, from the Latin *error* or *errare*, meaning "to err," as the source of the English *error*. The origin of *errantcy*, according to *Webster*, resides in the Old French *errer*, "to travel or wander," itself derived from the Middle Latin *iterare*, "to journey". But *Webster* also cites the Latin *errare*, which signifies both "to wander" and "to err."

9 Jonathan Dollimore, *Sexual Dissidence: Augustine to Wilde, Freud to Foucault* (Oxford: Oxford University Press, 1991), 130.

10 Milton, *Prose Works*, 2:514–15; quoted by Dollimore, *Sexual Dissidence*, 126.

11 Graham Hough rightly points out the "cross-winds and inconsistencies" in Eliot's critical writing, yet these variations do not vitiate a consideration of the major issues that preoccupied Eliot throughout the course of his career ("The Poet as Critic," in *The Literary Criticism of T. S. Eliot*, ed. David Newton-De Molina [London: Athlone Press, 1977], 45).

12 Sigmund Freud, *Three Essays on the Theory of Sexuality*, in *The Standard Edition of the Complete Psychological Works*, vol. 7, tr. James Strachey (London: Hogarth Press, 1953), 150.

13 Maud Ellmann, *The Poetics of Impersonality: T. S. Eliot and Ezra Pound* (Cambridge, MA: Harvard University Press, 1987), 40.

14 "Testing the Razor: T. S. Eliot's *Poems 1920*," in *Engendering the Word: Feminist Essays in Psychosexual Poetics*, ed. Temma F. Berg et al. (Urbana: University of Illinois Press, 1989), 168.

15 I am indebted to Barbara Herrnstein Smith for discussions of these ideas.

16 Eliot elsewhere discovers a kinship between Donne and Joris-Karl Huysmans, who served as a metonymy for nineties, and especially French, licentiousness. In the 1926 Clark Lectures he writes that "an attitude like that of Donne leads . . . to the collapse of the hero of Huysmans' *En Route*: 'Mon dieu, que c'est donc bête!'"(*V* 114–15). However, Ronald Schuchard points out that Eliot mistakenly quotes from Huysmans's *Là-bas* [Down There], whose title Eliot echoes at the conclusion of the paragraph: "It leads in fact to most of modern literature; for whether you seek the Absolute in marriage, adultery or debauchery, it is all one – you are seeking in the wrong place. *Donna è laggiù* [Lady is down there]" (*V* 115). Eliot's reversal of the Italian Renaissance elevation of women, epitomized by the

phrase *Donna è lassù* [Lady is up there], implies that Donne, Huysmans, and most modern writers wrongly seek female carnality. Eliot's harsh judgment of Donne corresponds to his sharp but highly conventional split between good and bad women, virgins and whores, Beatrice-figures (e.g., in "Ash Wednesday") and the host of degraded Eves that populates his poems.

17 Eliot's 1923 review of Marianne Moore's *Poems* and *Marriage* calls attention to her femininity: "And there is one final, and 'magnificent' compliment: Miss Moore's poetry is as 'feminine' as Christina Rossetti's, one never forgets that it is written by a woman; but with both one never thinks of this particularity as anything but a positive virtue" (*Dial* 75 [July–December 1923]: 597).

18 In a letter to Pound (23 September 1917), Eliot frets that the presence of "too many women . . . lowers the tone" of a favorite literary gathering, and he suggests that "there should be a special evening for males only" (*L* 198). He commiserates with Scofield Thayer, then editor of the *Dial*, on the tribulations of having a female "superior officer" (30 June 1918), counseling Thayer, "Be PATIENT, I say PATIENT. Be Sly, INSIDIOUS, even UNSCRUPULOUS, . . . concealing the Paw of the Lion . . . beneath the Pelt of the ASS. . . . I WILL REPAY, saith the LORD. I speak from experience, as asst. (I say ASST.) Editor of the *Egoist* [owned by Miss Weaver] . . . I am the only male, and three (3) women, incumbents, incunabula, incubae" (*L* 236).

19 "The Pater of Joyce and Eliot," in *Addressing Frank Kermode: Essays in Criticism and Interpretation*, ed. Margaret Tudeau-Clayton and Martin Warner (London: Macmillan, 1991), 169, 179.

20 *Selected Letters of Conrad Aiken*, ed. Joseph Killorin (New Haven: Yale University Press, 1978), 109 (emphasis Eliot's).

21 Poirier, "The Pater of Joyce and Eliot," 180; Perry Meisel, *The Myth of the Modern: A Study of British Literature and Criticism after 1850* (New Haven: Yale University Press, 1987), 71–86; and Louis Menand, *Discovering Modernism: T. S. Eliot and His Context* (New York: Oxford University Press, 1987), 75–94.

22 Richard Jenkyns, *The Victorians and Ancient Greece* (Oxford: Blackwell, 1980), 150. By contrast, Linda Dowling examines the erotic aspects of Pater's Hellenism and claims that he "was romantically involved with a nineteen-year-old Balliol undergraduate named William Money Hardinge" in the 1870s, thus confirming a long-standing rumor (*Hellenism and Homosexuality in Victorian Oxford* [Ithaca: Cornell University Press, 1994], 100). She also argues that Pater had become a "Uranian" hero within homosexual circles at the turn of the century (136–37). Both Pater and Symonds were forced to withdraw their candidacies for the chair of Professor of Poetry at Oxford in 1877 because of a homophobic attack by Richard St. John Tyrwhitt. For an account of the affair, of Pater's sexual-aesthetic ideal, and of the homosexual implications of Pater's aestheticism in late Victorian and modern literary culture, see Richard Dellamora, *Masculine Desire: The Sexual Politics of Victorian Aestheticism* (Chapel Hill: University of North Carolina Press, 1990), 158–64 and *passim*.

23 Alan Sinfield, *The Wilde Century: Effeminacy, Oscar Wilde and the Queer Moment* (New York: Columbia University Press, 1994), 84–108.

24 Francis Bacon, "Advertisement Touching an Holy Warre," in *The Works*, ed. J. Spedding and R. L. Ellis, vol. 7 (Stuttgart: Frommann, 1961–63), 33–34.

25 "Sexual inversion" typically implied a reversal of gender as well as of sexual object choice, so that the male invert, in Karl Heinrich Ulrich's formulation, possessed a woman's soul trapped in a man's body: *anima muliebris virili corpore inclusa*. The term "inversion" first appeared in English in John Addington Symonds's privately printed *A Problem in Greek Ethics* (1883), and it entered wide circulation with Havelock Ellis's *Sexual Inversion* (1898), where he defined it as "sexual instinct turned by inborn constitutional abnormality toward persons of the same sex" (*Studies in the Psychology of Sex*, vol. 1, part 4 [New York: Random House, 1936], 1).

26 Critics such as Graham Hough have noticed that Eliot's juvenilia, nonetheless, "are quite obviously based on Swinburne" ("The Critic as Poet," 56); Eliot's subsequent abjection of Swinburne, like his repression of his debt to Pater, may have had to do with his association of both with perverse eroticism.

27 For an examination of the perverse eroticism in Swinburne's work and his ties to Baudelaire, who shared similar interests, see Dellamora, *Masculine Desire* (70–85). Swinburne occupies an ambiguous but crucial position in the articulation of sexual deviance; although he acknowledged that he had "some points in common with Walt Whitman" (*The Swinburne Letters*, ed. Cecil Y. Lang, vol. 1 [New Haven: Yale University Press, 1959–62], 208), he later shrank from any homosexual identification while his name became a byword for decadence. Eliot does not openly charge Swinburne with sexual vice but hints at it slyly. The "characteristic insinuating tone" of Eliot's essays, which succeed not so much by logical persuasiveness as by "the careful dropping of certain names in the margin," is ably described by Roger Sharrock ("Eliot's Tone," in *The Literary Criticism of T. S. Eliot*, 175).

28 Grover Smith, "Eliot and the Ghost of Poe," in *T. S. Eliot: A Voice Descanting: Centenary Essays*, ed. Shyamal Bagchee (New York: St. Martin's Press, 1990), 149.

29 Harold Bloom, *The Breaking of the Vessels* (Chicago: University of Chicago Press, 1982), 21; James Loganbach repeats this charge as an accepted fact in "'Mature Poets Steal': Eliot's Allusive Practice," in *The Cambridge Companion to T. S. Eliot*, ed. A. David Moody (Cambridge: Cambridge University Press, 1996), 185.

30 Quoted by Hough, "The Poet as Critic," 48.

31 Christopher Craft discusses the discomfort occasioned by Tennyson's poem among contemporary reviewers and subsequent critics in *Another Kind of Love: Male Homosexual Desire in English Discourse 1850–1920* (Berkeley: University of California Press, 1994), 44–70. See also Jeff Nunokawa, "*In Memoriam* and the Extinction of the Homosexual," *ELH* 58:2 (1991): 427–38.

32 Ronald Schuchard makes this connection in his notes to *The Varieties of Metaphysical Poetry* by T. S. Eliot, ed. Ronald Schuchard (London: Faber & Faber, 1993), 114, n.45.

33 John Peter calls *The Waste Land* "something like the order of *In Memoriam*" ("A New Interpretation of *The Waste Land*," *Essays in Criticism* 19 [1969]: 168).

34 Eve Kosofsky Sedgwick, in *Between Men: English Literature and Male Homosocial Desire* (New York: Columbia University Press, 1985), uses the term in order "to draw the 'homosocial' back into the orbit . . . of the potentially erotic," so as "to hypothesize the potential unbrokenness of a continuum between homosocial and homosexual – a continuum whose visibility, for men, in our society, is radically disrupted" (1–2).

35 Eve Kosofsky Sedgwick, *Epistemology of the Closet* (Berkeley: University of California Press, 1990), 185. Sedgwick calls "The Love Song of J. Alfred Prufrock" a "manifesto of male homosexual panic" (240).

36 Gregory S. Jay, *T. S. Eliot and the Poetics of Literary History* (Baton Rouge: Louisiana State University Press, 1983), 170. In the introduction to *Ezra Pound: Selected Poems* (1928; reprint, London: Faber and Faber, 1943), Eliot admits that he "had to conquer an aversion to [Whitman's] form as well as to much of his matter" before he could read him at all. Eliot's claims that he "did not read Whitman until much later in life" and that "Pound owes nothing to Whitman" (8) are disingenuous. In "Ezra Pound: His Metric and Poetry" (1917), Eliot quotes approvingly de Bosschere's remark that Pound's "poems incite man to . . . a becoming egotism . . . The virile complaint . . . – that is poetry'" (*CC* 178). Eliot's antipathy to Whitman's subject matter and his wish to separate himself and Pound from Whitman's influence may stem from his phobic reaction to the latter's outspoken celebration of "manly love." In "Whitman and Tennyson," Eliot again dismisses Whitman's influence on contemporary poetry (*The Nation and the Athenaeum* 40: 11 [December 1926]: 426), yet he sees "a fundamental resemblance" between Whitman and Tennyson, including what he criticizes as their idealization of sexual relations. "There is, fundamentally, no difference between the Whitman frankness [about sex] and the Tennyson delicacy" (426); Eliot disapproves of both. The basis of this unusual conjunction may be traced to Eliot's peculiar remark that "Whitman had the ordinary desires of the flesh" (426). Despite Eliot's oft-expressed hostility to Whitman and his denials of Whitman's presence in his own or other modernist poetry, his critics have discovered otherwise. See Betsy Erkkila, *Walt Whitman among the French: Poet and Myth* (Princeton: Princeton University Press, 1980), 232–33; Sydney Musgrove, *T. S. Eliot and Walt Whitman* (Wellington, New Zealand, 1952); and Jay, *T. S. Eliot and the Poetics of Literary History*, 168–86.

37 Sedgwick, *Epistemology*, 186; emphasis Sedgwick's.

38 T. S. Eliot, "The Education of Taste," *The Athenaeum*, no. 4652 (27 June 1919): 521.

39 T. S. Eliot, "Reflections on Contemporary Poetry," *The Egoist* 6:3 (July 1919): 39.

40 T. S. Eliot, "The Hawthorne Aspect," *The Little Review* (August 1918): 50.

41 See Freud, *Three Essays on the Theory of Sexuality*, 7:145.

42 See, for instance, Eliot's doggerel, to be sung to the tune of "C. Columbo lived in Spain," concerning a German "cabin boy [who] was sav'd alive / And bugger'd, in the sphincter," contained in a 30 September 1914 letter to Conrad Aiken (*L* 59). The cabin boy, Columbus, and "his friend King Bolo" were a standing joke in Eliot's letters to Aiken and Ezra Pound (*L* 42, 86, 125, 206, 568); he even tried to amuse or impress James Joyce with his priapic verse (*L* 455). The cabin boy makes a reappearance in a January 1926 letter to Aiken, in which Eliot included the following lines: "Blue eyed Claude the cabin boy, / the clever little nipper / who filled his ass with broken glass / and circumcised the skipper" (*Selected Letters of Conrad Aiken*, 110). See also Wayne Koestenbaum's analysis of the homoerotic banter between Eliot and Pound, including the latter's "SAGE HOMME," a squib on himself as the male midwife-muse of *The Waste Land* (*Double Talk: The Erotics of Male Literary Collaboration* [New York: Routledge, 1989], 120–23).

43 T. S. Eliot, "The Idea of a Literary Review," *The New Criterion* 4:1 (January 1926): 4.

44 Mary Douglas, *Purity and Danger* (1966; reprint, London: Routledge & Kegan Paul, 1985), 35, 4.

45 Anthony Julius, *T. S. Eliot, Anti-Semitism, and Literary Form* (Cambridge: Cambridge University Press, 1995), 173.

46 T. S. Eliot, "A Note on Poetry and Belief," *The Enemy: A Review of Art and Literature* 1 (January 1927): 16.

47 Alfred Kazin, *New York Jew* (New York: Vintage, 1979), 212.

48 Gregory Jay, "Postmodern in *The Waste Land*: Women, Mass Culture, and Others," in *Rereading the New: A Backward Glance at Modernism*, ed. Kevin Dettmar (Ann Arbor: University of Michigan Press, 1992), 223; see also Paul Morrison, *The Poetics of Fascism: Ezra Pound, T. S. Eliot, Paul de Man* (New York: Oxford University Press, 1996).

49 "The Borderline of Prose," *New Statesman* 9 (19 May 1917): 158–59.

50 Eliot's notion of the "dissociation of sensibility" introduces some ambiguity into his injunction against the promiscuous mingling of thought and feeling. The "direct sensuous apprehension of thought, or a recreation of thought into feeling" (*SE* 286) by the metaphysical poets represented for Eliot a prelapsarian ideal. However, when later writers, such as Wordsworth and Lawrence, try to mix thoughts and sentiments, or philosophy and poetry, they are "great heretics" (*UPUC* 91).

51 T. S. Eliot, "The Idea of a Literary Review," *The New Criterion* 4:1 (January 1926): 3–4 (emphasis mine).

52 John Peter, "A New Interpretation of *The Waste Land*" (*Essays in Criticism* 2 [July 1952]: 242–66). In a postscript to the 1969 reprint of the essay, Peter explains that "at Eliot's insistence all copies of [the 1952] issue on hand after publication were destroyed, and when the run of volumes I–XVI was subsequently re-issued . . . he refused to sanction the reprinting of my essay" (*Essays in Criticism* 19 [1969]: 165).

53 Christopher Ricks credits Eliot "for at least some rescinding of the book,"

and dwells on his later work, especially *Four Quartets*, as evidence of Eliot's "humility" and humane "sympathy" (*T. S. Eliot and Prejudice* [Berkeley: University of California Press, 1988], 40, 240).

54 T. S. Eliot, Review of John Middleton Murry, *The Son of Woman: The Story of D. H. Lawrence*, in *The Criterion* 10:41 (July 1931): 771; Julius, *Eliot, Anti-Semitism, and Literary Form*, 157.

55 T. S. Eliot, "The Post-Georgians," *Athenaeum* (11 April 1919): 171.

56 Michel Foucault, "What Is an Author?" in *Language, Counter-Memory, Practice: Selected Essays and Interviews*, ed. Donald F. Bouchard, tr. Bouchard and Sherry Simon (Ithaca: Cornell University Press, 1977), 113–38. Eliot's view is close to that of Steven Knapp and Walter Benn Michaels, who defend the idea of an intending, governing author as a necessary pre-supposition of reading ("Against Theory: 1," *Critical Inquiry* 8 [1982]: 723–42).

57 Friedrich Nietzsche, *Twilight of the Idols*, in *The Portable Nietzsche*, ed. and tr. Walter Kaufmann (New York: Viking, 1976), 485.

58 Immanuel Kant, *The Critique of Judgement*, Part One: "Critique of Aesthetic Judgement," sections 21 and 22, tr. James Creed Meredith (Oxford: Oxford University Press, 1952), 83–85.

59 Friedrich Nietzsche, "On Truth and Lies in a Nonmoral Sense," in *Philosophy and Truth: Selections from Nietzsche's Notebooks of the Early 1870's*, ed. and tr. Daniel Breazeale (New Jersey and London: Humanities Press International, 1979), 81.

60 Nietzsche, "On Truth and Lies," 92.

61 T. S. Eliot, "The Three Provincialities," *The Tyro* 2 (1922): 11.

62 T. S. Eliot, "London Letter," *Dial* 73 (July–December 1922): 329–31. A few years earlier, Eliot described *Ulysses* as "terrifying . . . But this attractive terror repels the majority of men" ("Tarr," the *Egoist* 5:8 [September 1918]: 105–6).

63 The relationship between Eliot's essay, Joyce's novel, and *The Waste Land* has been an ongoing topic of critical debate in which Grover Smith's argument – that Eliot was "privately laying claim" to the "mythical method" as his own – remains canonical (*The Waste Land* [London: George Allen & Unwin, 1983], 59, 61). The argument for Eliot's appropriation of Joyce's method was first made by Joyce himself.

64 See Barbara Herrnstein Smith, *Contingencies of Value: Alternative Perspectives for Critical Theory* (Cambridge, MA: Harvard University Press, 1988), 40–41 and *passim* for an analysis of the ways in which contingencies are standard-ized or normativized – and thus rendered apparently noncontingent – through the self-justificatory exercise of evaluative authority.

65 Bernard Sharratt, "Eliot: Modernism, Postmodernism, and After," in Moody, *The Cambridge Companion*, 233.

66 C. S. Lewis, letter to Paul Elmer More (23 May 1935), quoted in Ricks, *Eliot*, 197–98; Eric Sigg, "Eliot as the Product of America," in Moody, *The Cambridge Companion*, 20; Ronald Bush, "T. S. Eliot and Modernism at the

Present Time: A Provocation," in *T. S. Eliot: The Modernist in History*, ed. Ronald Bush (Cambridge: Cambridge University Press, 1991), 199.

67 *Knowledge and Experience in the Philosophy of F. H. Bradley* (New York: Farrar, Straus, 1964), 161; Richard Schusterman, "Eliot as Philosopher," in Moody, *The Cambridge Companion*, 33, 41.

68 Frank Kermode, *The Classic* (London: Faber & Faber, 1975), *passim*.

2 END OF POETRY FOR LADIES: T. S. ELIOT'S EARLY POETRY

1 *James Joyce's Scribbledehobble: The Ur-Workbook for "Finnegans Wake,"* ed. Thomas Connolly (Evanston: Northwestern University Press, 1961), 11. From his dating of the notebook (1922–23), Connolly concludes that Joyce's remark refers to *The Waste Land*, a copy of which Eliot had sent to Joyce.

2 Michael Levenson, *A Genealogy of Modernism: A Study of English Literary Doctrine 1908–1922* (Cambridge: Cambridge University Press, 1984), 154–55.

3 Ezra Pound, "The Hard and the Soft in French Poetry," *Poetry* 11 (February 1918): 264.

4 T. E. Hulme, "Romanticism and Classicism," in *Speculations: Essays on Humanism and the Philosophy of Art*, ed. Herbert Read (New York: Harcourt, Brace and Co., 1924), 126; "A Lecture on Modern Poetry," in *Further Speculations*, ed. Samuel Hynes (Minneapolis: University of Minnesota Press, 1955), 69. For Hulme's influence on Eliot's critical stance in the 1910s and 1920s, see Erik Svarny, *"The Men of 1914": T. S. Eliot and Early Modernism* (Milton Keynes: Open University Press, 1988), 13–43.

5 T. S. Eliot, "Reflections on Contemporary Poetry," *Egoist* 4:8 (September 1917): 118; and "Reflections on Contemporary Poetry," *Egoist* 4:9 (October 1917): 133.

6 Sandra M. Gilbert and Susan Gubar, *No Man's Land: The Place of the Woman Writer in the Twentieth Century*, vol. 1: *The War of the Words* (New Haven: Yale University Press, 1988), 147, 154.

7 Carol Christ, "The Feminine Subject in Victorian Poetry," *ELH* 54:2 (summer 1987): 385.

8 Carol Christ, "Gender, Voice, and Figuration in Eliot's Early Poetry," in *T. S. Eliot: The Modernist in History*, ed. Ronald Bush (Cambridge: Cambridge University Press, 1991), 25, 27.

9 The major exceptions are Gilbert and Gubar's *No Man's Land*, Carol Christ's essays, Bonnie Kime Scott's *Refiguring Modernism*, vol. 1 (Bloomington: Indiana University Press, 1995), and Maud Ellmann's *The Poetics of Impersonality: T. S. Eliot and Ezra Pound* (Cambridge, MA: Harvard University Press, 1987).

10 Levenson, *A Genealogy of Modernism*, 200–2.

11 Gilbert and Gubar, *No Man's Land*, 1:36.

12 In a letter written that same day to his mother, Eliot does not mention the

publication of *Ara Vos Prec*, nor can I find any reference to the volume in his subsequent letters to her, or hers to him.

13 The edition of "Ode" that Ricks reprints is the manuscript version among Eliot's papers in the Berg Collection of the New York Public Library. Entitled "Ode on Independence Day, July 4th 1918," it lacks the epigraph that accompanied it in the *Ara Voc Prec* edition, on which my reading of the poem is based (*I* 383).

14 James Loganbach, "*Ara Vos Prec*: Eliot's Negotiation of Satire and Suffering," in Bush, *T. S. Eliot: The Modernist in History*, 43.

15 See Sigmund Freud, *The Psychopathology of Everyday Life*, ed. James Strachey (New York: W. W. Norton, 1960), 106–33.

16 For the concept of textual divergence, see Jacques Derrida, *Of Grammatology*, tr. Gayatri Chakravorty Spivak (Baltimore: Johns Hopkins University Press, 1976), 102.

17 J. L. Austin, *How to Do Things with Words* (New York: Oxford University Press, 1962), 21–22.

18 Jacques Derrida, "Signature, Event, Context," in *Limited Inc*, tr. Samuel Weber and Jeffrey Mehlman (Evanston: Northwestern University Press, 1988), 17, 12.

19 A. Walton Litz, "The Allusive Poet: Eliot and His Sources," in Bush, *T. S. Eliot: The Modernist in History*, 140; and Bloom, introduction, *T. S. Eliot: Modern Critical Views*, ed. Harold Bloom (New York: Chelsea House, 1985), 6. Among others who argue that Eliot's allusions function as a validation of the literariness of his poems and of himself as a poet, see Louis Menand, *Discovering Modernism: T. S. Eliot and His Context* (New York: Oxford University Press, 1987), 16.

20 See Derrida, *Limited Inc*, 7, 12.

21 Don Gifford with Robert Seidman, *Notes for Joyce: An Annotation of James Joyce's "Ulysses"* (New York: Dutton, 1974), xi–xii.

22 Fritz Senn, "Protean Inglossabilities: 'To No End Gathered,'" in *European Joyce Studies*, ed. Christine van Boheemen (Amsterdam: Rodopi, 1989), 40–65.

23 Peter Brooker, *A Student's Guide to "The Selected Poems of Ezra Pound"* (London: Faber & Faber, 1979), 11–12.

24 Susan Stewart, "The Pickpocket: A Study in Tradition and Allusion," *Modern Language Notes* 95 (1980): 1128.

25 Walter Benjamin, quoted by Hannah Arendt, introduction to Walter Benjamin, *Illuminations*, tr. Harry Zohn (New York: Schocken Books, 1969), 39.

26 John Hollander, *The Figure of Echo: A Mode of Allusion in Milton and After* (Berkeley: University of California Press, 1981), 103.

27 Gregory Jay, *T. S. Eliot and the Poetics of Literary History* (Baton Rouge: Louisiana State University Press, 1983), 147. See also Perry Meisel, *The Myth of the Modern: A Study in British Literature and Criticism after 1850* (New Haven: Yale University Press, 1987), 88.

28 Ellmann, *The Poetics of Impersonality*, 95.

29 See Hermann Meyer, *The Poetics of Quotation in the European Novel*, tr. Theodore and Yetta Ziolkowski (Princeton: Princeton University Press, 1968), 76.

30 Stewart, "The Pickpocket," 1128 (emphasis mine). Jay places *The Waste Land* in the tradition of the elegy (*Eliot and the Poetics of Literary History*, 156).

31 Jean-François Lyotard, "Answering the Question: What Is Postmodernism?" in *The Postmodern Condition: A Report on Knowledge*, tr. Geoff Bennington and Brian Massumi (Minneapolis: University of Minnesota Press, 1984), 81.

32 Arendt, introduction, *Illuminations*, 38.

33 Lyotard, *The Postmodern Condition*, 78–79.

34 Jacques Derrida, *Dissemination*, tr. Barbara Johnson (Chicago: University of Chicago Press, 1981), 109 (emphasis Derrida's).

35 Eric Partridge, *Origins: A Short Etymological Dictionary of Modern English* (New York: Macmillan, 1959), 733.

36 Anonymous review of *The Waste Land Dial* 73 (December 1922): 685–87.

37 T. S. Eliot, "A Note on Ezra Pound," *To-Day* 4, no. 19 (September 1918) 4–6.

38 Harold Bloom, *The Anxiety of Influence: A Theory of Poetry* (New York: Oxford University Press, 1973), 5; Hollander, *The Figure of Echo*, 102.

39 Hollander, *The Figure of Echo*, 133–49; Bloom, *Anxiety*, 139; John Guillory, *Poetic Authority: Spenser, Milton, and Literary History* (New York: Columbia University Press, 1983), 131.

40 For Eliot's continuity with Romantic poets, see works by Frank Kermode, Edmund Wilson, Monroe Spears, C. K. Stead, Robert Langbaum, George Bornstein, Ronald Bush, Perry Meisel, and, especially, John Paul Riquelme, *Harmony of Dissonances: T. S. Eliot, Romanticism, and Imagination* (Baltimore: Johns Hopkins University Press, 1991). Bloom claims that Eliot's "actual forerunners are Whitman and Tennyson" (introduction to *T. S. Eliot: Modern Critical Views*, 2), and Jay argues for Eliot's "American genealogy" (*Eliot and the Poetics of Literary History*, chapter 1 and *passim*). Ronald Bush sees Emerson as Eliot's "private encumbrance" ("T. S. Eliot: Singing the Emersonian Blues," in *Emerson: Prospect and Retrospect*, ed. Joel Porte [Cambridge, MA: Harvard University Press, 1982], 180).

41 Graham Hough, "The Poet as Critic," in *The Literary Criticism of T. S. Eliot*, ed. David Newton-de Molina (London: Athlone Press, 1977), 52–56. See also Bernard Sharratt, "Eliot: Modernism, Postmodern, and After," in *The Cambridge Companion to T. S. Eliot*, ed. A. David Moody (Cambridge: Cambridge University Press, 1994), who points to Eliot's "social awkwardness" and his miscellaneous reviewing as reasons for the profuse and diverse quotations in his early poetry (226–27).

42 Jorge Luis Borges, *Labyrinths: Selected Stories and Other Writings*, ed. Donald A. Yates and James E. Irby (New York: New Directions, 1964), 201 (emphasis Borges's).

43 Bloom, *Anxiety*, 19. Roger Sharrock ("Eliot's Tone," in *The Literary Criticism of T. S. Eliot*, 171) and Leonard Unger (*Eliot's Compound Ghost: Influence and*

Confluence [University Park: Pennsylvania State University Press, 1981], 99) have argued that Eliot is thus Bloom's precursor.

44 For a cogent analysis of the relation between the concepts of influence and intertextuality that includes some discussion of Eliot, see Jay Clayton and Eric Rothstein, "Figures in the Corpus: Theories of Influence and Intertextuality," in *Influence and Intertextuality in Literary History*, ed. Jay Clayton and Eric Rothstein (Madison: University of Wisconsin Press, 1991), 3–36.

45 Roland Barthes, "The Death of the Author," in *Image–Music–Text*, tr. Stephen Heath (New York: Hill and Wang, 1977), 146.

46 Ibid., 159–61.

47 Ibid., 160 (emphasis Barthes's).

48 This notion of intertextuality is defined and developed by Michael Riffaterre, for whom "intertextuality is . . . the deciphering of the text by the reader in such a way that he identifies the structures to which the text owes its quality of a work of art," specifically, the requisite "intertexts" ("Syllepsis," *Critical Inquiry* 6 [summer 1980]: 625).

49 See Jonathan Culler, "Presupposition and Intertextuality," in *The Pursuit of Signs: Semiotics, Literature, Deconstruction* (Ithaca: Cornell University Press, 1981), 103, 118.

50 Jonathan Culler, "Textual Self-Consciousness and the Textual Unconscious," *Style* 18:3 (summer 1984): 371.

51 Julia Kristeva, "Word, Dialogue, and Novel," in *Desire in Language: A Semiotic Approach to Literature and Art*, ed. Leon S. Roudiez, tr. Thomas Gora et al. (New York: Columbia University Press, 1980), 65–66.

52 Kristeva, "The Bounded Text," in *Desire in Language*, 36.

53 See Shoshana Felman, "Turning the Screw of Interpretation," *Yale French Studies* 55/56 (1977): 94–207; and Barbara Johnson, *The Critical Difference: Essays in the Contemporary Rhetoric of Reading* (Baltimore: Johns Hopkins University Press, 1981).

54 Barthes, "The Death of the Author," 146.

55 Ellmann, *The Poetics of Impersonality*, 101.

56 For the link between sexual desire and money, see Sigmund Freud, "On Transformations of Instinct as Exemplified in Anal Eroticism," in *The Standard Edition of the Complete Psychological Works*, vol. 17, ed. James Strachey (London: Hogarth Press, 1953), 127–33.

57 F. R. Leavis, *New Bearings in English Poetry* (London: Chatto and Windus, 1932), 90–91.

58 George Williamson, *A Reader's Guide to T. S. Eliot* (New York: Noonday, 1953), 135.

59 Peter Middleton, "The Academic Development of *The Waste Land*," *Demarcating the Disciplines: Philosophy, Literature, Art*, Glyph Textual Studies 1 (Minneapolis: University of Minnesota Press, 1986), 159.

60 Bloom's "family romance" of literary influence is underwritten by homophobia (*Anxiety*, 26–27); he tosses off an aside about father–son incest but

fails to pursue the topic (95). Susan Stanford Friedman astutely notes the phallicism of the figure of influence but overlooks its homoeroticism ("Weavings: Intertextuality and the (Re)Birth of the Author," in Clayton and Rothstein, *Influence and Intertextuality*, 151–52).

61 See Wayne Koestenbaum, *Double Talk: The Erotics of Male Literary Collaboration* (New York: Routledge, 1989), 112–40. Harry Trosman insightfully compares the Eliot/Pound relationship to that between Freud and Fleiss; in particular, Eliot's "narcissistic transference" onto the "authoritative" Pound crucially allayed anxieties arising from the death of his father ("T. S. Eliot and *The Waste Land*: Psychopathological Antecedents and Transformations," *Archives of General Psychiatry* 30 [May 1974]: 715).

62 Steve Ellis, "*The Waste Land* and the Reader's Response," in *The Waste Land*, ed. Tony Davies and Nigel Wood (Buckingham: Open University Press, 1994), 104.

63 Christine Froula, "Eliot's Grail Quest, or the Lover, the Police, and *The Waste Land*," *Yale Review* 78:2 (1989): 253 (emphasis Froula's).

64 Jay, *Eliot and the Poetics of Literary History*, 74–75. Jay accepts the phobic cliché that homosexuality is based upon narcissism, a view with which I disagree but which Eliot apparently espoused in "The Death of Saint Narcissus."

65 Ronald Bush, *T. S. Eliot: A Study in Character and Style* (New York: Oxford University Press, 1984), 7.

66 Sigmund Freud, "The Ego and the Id," in *The Standard Edition of the Complete Psychological Works*, vol. 19, ed. James Strachey (London: Hogarth Press, 1955), 34.

67 Ibid., 37.

68 Freud, "The Economic Problem in Masochism," in *The Standard Edition*, vol. 12, 169.

69 Lyndall Gordon, *Eliot's Early Years* (New York: Oxford University Press, 1977), 11.

70 T. S. Eliot, *For Lancelot Andrewes: Essays on Style and Order* (London: Faber & Gwyer, 1928), ix. This volume in which he avows his allegiance to paternal authority is dedicated to his mother.

71 Gordon, *Eliot's Early Years*, 131. In a letter to William Force Stead (10 April 1928), Eliot expresses his desire for a spiritual advisor who would give him "something more ascetic, more violent, more 'Ignatian'" (131).

72 Undated letter, quoted in Gordon, *Eliot's Early Years*, 62; "A Note on Richard Crashaw," in Eliot, *For Lancelot Andrewes*, 124–25.

73 Peter Ackroyd, *T. S. Eliot: A Life* (New York: Simon and Schuster, 1984), 44

74 Freud, "A Child Is Being Beaten," in Strachey, *The Standard Edition*, vol. 17. The convergence of masochism and sadism in Eliot's poetry places him squarely within Freud's theoretical paradigm, as opposed to Gilles Deleuze's revisionist theory of masochism, which I will discuss in the following chapter.

75 Herbert Howarth, *Notes on Some Figures behind T. S. Eliot* (Boston: Houghton Mifflin, 1964), 31; Gordon notes the similarities between Mrs. Eliot's poetry and her son's "saint" poems (*Eliot's Early Years*, 5, 60, 70).

76 Howarth, *Notes on Some Figures*, 1.

77 According to Trosman, he was "completely deaf" when Eliot was born ("Eliot and *The Waste Land*," 710).

78 Mrs. Eliot's correspondence with Russell was probably instigated by Eliot in order to justify his presence in England; for much the same reason, Eliot persuaded Pound to write a lengthy letter to his father (28 June 1915).

79 Gordon, *Eliot's Early Years*, 4; Ackroyd, *T. S. Eliot*, 20.

80 T. S. Eliot, review, *Son of Woman: The Story of D. H. Lawrence*, by John Middleton Murry, *Criterion* 10:41 (July 1931): 770.

81 Ackroyd, *T. S. Eliot*, 20.

82 Sigmund Freud, "The Most Prevalent Form of Degradation in Erotic Life," in *Sexuality and the Psychology of Love* (New York: Collier, 1963), 62.

83 Donald Hall, *Remembering Poets: Reminiscences and Opinions: Dylan Thomas, Robert Frost, T. S. Eliot, Ezra Pound* (New York: Harper & Row, 1977), 100.

84 John T. Mayer, *T. S. Eliot's Silent Voices* (New York: Oxford University Press, 1989), 184 (emphasis mine).

85 Grover Smith, *T. S. Eliot's Poetry and Plays* (Chicago: University of Chicago Press, 1956), 33; see also Mayer, *T. S. Eliot's Silent Voices*, 182.

86 Tony Pinkney, *Women in the Poetry of T. S. Eliot: A Psychoanalytic Approach* (London: Macmillan, 1984), 20–21.

87 Jacques Lacan, *Seminars*, vol. 2 (New York: W. W. Norton, 1975), 164.

88 Koestenbaum, *Double Talk*, 113, 155.

89 Barbara Ehrenreich and Deirdre English, *For Her Own Good: 150 Years of the Experts' Advice to Women* (New York: Doubleday, 1979), 137.

90 Freud, quoted in ibid., 138.

91 Sigmund Freud, *Dora: An Analysis of a Case of Hysteria* (New York: Macmillan, 1963), 59.

92 Stephen Heath, *The Sexual Fix* (New York: Schocken Books, 1984), 27–31.

93 Alan Krohn, *Hysteria: The Elusive Neurosis* (New York: International Universities Press, 1978), 179.

94 Bertrand Russell, letter to Ottoline Morrell (10 November 1915), in *The Autobiography of Bertrand Russell*, vol. 2, *1914–1944* (Boston: Little, Brown, 1968), 64.

95 Ackroyd, *T. S. Eliot*, 61–62.

96 Ibid., 62.

97 Anthony E. Fathman, "Viv and Tom: The Eliots as Ether Addict and Co-Dependent," *Yeats Eliot Review* 11:2 (fall 1991): 35.

98 Aldous Huxley, *Letters of Aldous Huxley* (London: Chatto and Windus, 1969), 232; John Pearson, *The Sitwells: A Family's Biography* (New York: Harcourt Brace Jovanovich, 1978), 277; Virginia Woolf, *The Letters of Virginia Woolf*, ed. Nigel Nicolson, vol. 5 (New York: Harcourt Brace Jovanovich, 1975–80), 71; Cyril Connolly, "The Poet's Workshop," *Sunday Times*, 7 November 1971.

99 Fathman, "Viv and Tom," 35.

100 Gordon, *Eliot's Early Years*, 74.

101 Bush, *T. S. Eliot*, 54.

102 Ackroyd, *T. S. Eliot*, 66; Gordon, *Eliot's Early Years*, 75–76.

103 Bush, *T. S. Eliot*, 53; Richard Ellmann, *Golden Codgers: Biographical Speculations* (New York: Oxford University Press, 1973), 160. Ellmann borrows Eliot's own words; describing his wife in a 1925 letter to Russell, he writes, "I can never escape from the spell of her persuasive (even coercive) gift of argument" (quoted in Bush, 161).

104 Gordon, *Eliot's Early Years*, 79.

105 Ibid., 26, 76–77.

106 For fetishism, see Robert J. Stoller, *Observing the Erotic Imagination* (New Haven: Yale University Press, 1985), 155–56; and Emily Apter, *Feminizing the Fetish: Psychoanalysis and Narrative Obsession in Turn-of-the-Century France* (Ithaca: Cornell University Press, 1991), 1–14.

107 F. O. Matthiessen, *The Achievement of T. S. Eliot* (Boston: Houghton Mifflin, 1935), 25. Gertrude Patterson argues that the poem "presents Eliot's 'apology' for his own metaphysical . . . manner, and the need for contemporary poets to avoid 'dissociation of sensibility'" (*T. S. Eliot: Poems in the Making* [Manchester: Manchester University Press, 1971], 118–19); A. D. Moody, *Thomas Stearns Eliot, Poet* (Cambridge: Cambridge University Press, 1979), 63, 72; Bush, *T. S. Eliot*, 171; and Mayer, who claims that "the poem's first half is an act of . . . [unified] sensibility" (*T. S. Eliot's Silent Voices*, 193).

108 Grover Smith, *T. S. Eliot's Poetry and Plays*, 41–42.

109 William Empson points out these mutually exclusive interpretations in *Seven Types of Ambiguity* (New York: New Directions, 1966), 79.

110 Pinkney, *Women in the Poetry of T. S. Eliot*, 84.

111 For Eliot's cats, see Kerry Weinberg, *T. S. Eliot and Charles Baudelaire* (The Hague: Mouton, 1969), 62–64; and Marianne Thormahlen, *Eliot's Animals* (Lund: CWK Glerrup, 1984), 39–55.

112 Leonard Unger mentions the importance of smells in Eliot's poems but fails to discuss their implications for their representation of women (*T. S. Eliot: Moments and Patterns* [Minneapolis: University of Minnesota Press, 1967], 179–80).

113 Freud, "Fetishism," in *Sexuality and the Psychology of Love*, ed. Philip Rieff (New York: Collier, 1963), 215.

114 Manuscripts of the poem, annotated by Ezra Pound, are preserved in the Berg Collection of the New York Public Library; Ricks mentions some variants (*I* 365–67).

115 Maud Ellmann, *Poetics of Impersonality*, 98

116 Smith, *T. S. Eliot's Poetry and Plays*, 38.

117 Bloom, *T. S. Eliot's "The Waste Land,"* 1.

118 Stephen Spender, *T. S. Eliot* (Harmondsworth: Penguin, 1975), 53.

119 Mayer, *T. S. Eliot's Silent Voices*, 196.

120 Quoted by T. S. Matthews, *Great Tom: Notes toward the Definition of T. S. Eliot* (New York: Harper & Row, 1974), 44.

121 Ibid., 44.

122 Ibid., 45.

123 Alan Sinfield, *The Wilde Century: Effeminacy, Oscar Wilde and the Queer Moment* (New York: Columbia University Press, 1994), 29.

124 Gareth Reeves, *T. S. Eliot: A Virgilian Poet* (London: Macmillan, 1989), 166.

125 Neil Hertz, *The End of the Line: Essays on Psychoanalysis and the Sublime* (New York: Columbia University Press, 1985), 222.

126 Paul de Man, "Autobiography as De-facement," in *The Rhetoric of Romanticism* (New York: Columbia University Press, 1984), 70–71 (emphasis de Man's).

127 Quoted in Gordon, *Eliot's Early Years*, 76.

128 *Plato: The Collected Dialogues*, ed. Edith Hamilton and Huntington Cairns (Princeton: Princeton University Press, 1961), 276c, 275e.

129 Cleo McNeally Kearns, "Eliot, Russell, and Whitman: Realism, Politics, and Literary Persona in *The Waste Land*," in *T. S. Eliot's "The Waste Land": Modern Critical Interpretations*, ed. Harold Bloom (New York: Chelsea House, 1986), 144.

130 Mayer, *T. S. Eliot's Silent Voices*, 196–203; James E. Miller, Jr., *T. S. Eliot's Personal Waste Land: Exorcism of the Demons* (University Park: Pennsylvania State University Press, 1977), 46–58.

131 *The Poems of Catullus*, tr. Peter Whigham (Los Angeles: University of California Press, 1969), 121–30.

132 Christine Froula draws a connection between the quotation from Verlaine's sonnet "Parsifal" in *The Waste Land* and this line in "Ode," claiming that "the bridegroom of 'Ode' is a failed Parsifal" ("Eliot's Grail Quest," 243). Although Froula does not point it out, "Ode" also links Parsifal and Perseus. Unlike Verlaine's hero, who conquers the woman by piercing her womb with his sword, Eliot and Laforgue's Perseus fails either to kill the dragon or to get the woman.

133 Stewart, "The Pickpocket," 1143.

134 W. K. Wimsatt claimed that "it would not much matter if Eliot had invented his sources (as Sir Walter Scott invented chapter epigraphs from 'old plays' and 'anonymous' authors, or as Coleridge wrote marginal glosses for *The Ancient Mariner*)" (*The Verbal Icon: Studies in the Meaning of Poetry* [New York: Noonday, 1954], 15).

135 The agrammaticality of "Succuba eviscerate" gives grounds for speculation. As a transitive verb, "eviscerate," meaning "to gut," "to devitalize," or "to remove an organ from a body" (*Webster's Third International Dictionary*), should be "eviscerates" to agree with its singular subject. As an intransitive verb (although also grammatically incorrect in the poem), "eviscerate" refers to the protrusion of a body part through a surgical incision, or, of a person, it means "to suffer such a protrusion of a part through an incision." The transitive and intransitive forms of the verb thus reverse directions: the succuba either cuts out an organ from her victim, or some part of her protrudes or sticks out through a cut, thus further complicating the castrater/castrated relation. "Eviscerate" may also be read as an errant

adjectival form of "eviscerated," in which case the succuba is herself dis-
emboweled. Vicki Mahaffey points out a third sense of the phrase, as the
imperative "Succuba, eviscerate!" She also notes that the adjectival form
of the word implies that the eviscerated succuba is "drawn from the bowels
of the earth" ("'The Death of Saint Narcissus' and 'Ode': Two
Suppressed Poems by T. S. Eliot," *American Literature* 50:4 [January 1979]:
610).

136 Hertz, *The End of the Line*, 222–23.
137 Pinkney, *Women in the Poetry of T. S. Eliot*, 18–19; Maud Ellmann, *Poetics of Impersonality*, 94.
138 Pinkney, *Women in the Poetry of T. S. Eliot*, 24–25.
139 Julia Kristeva, *Powers of Horror: An Essay on Abjection*, tr. Leon S. Roudiez (New York: Columbia University Press, 1982), 13, 53.
140 The most virulent statement of Eliot's revulsion at women's bodies is his response to Conrad Aiken's congratulations on the publication of Eliot's *Poems 1920–1925*. According to Aiken, Eliot replied by sending him "a page torn out of the *Midwives Gazette*: instructions to those about to take exams for nursing certificates. At the top, T. S. E. had underlined the words *Model Answers*. Under this was a column descriptive of various forms of vaginal discharge, normal and abnormal. Here the words *blood*, *mucus*, and *shreds of mucus* had been underlined with a pen, and lower down also the phrase *purulent offensive discharge*. Otherwise, no comment" (letter to Robert N. Linscott, 4 January 1926), *Selected Letters of Conrad Aiken*, ed. Joseph Killorin (New Haven: Yale University Press, 1978), 109. That Eliot compared his own poetry to this feminine discharge suggests his sense of emasculation or evisceration.
141 C. D. Daly, "The Menstruation Complex in Literature," *Psychoanalytic Quarterly* 4 (1935): 307–40.
142 Adrienne Auslander Munich, *Andromeda's Chains: Gender and Interpretation in Victorian Literature* (New York: Columbia University Press, 1989), 18, 25.
143 Ibid., 33.
144 Jules Laforgue, *Six Moral Tales*, ed. and tr. Frances Newman (New York: Liveright, 1928), 247.
145 Ibid., 292.
146 "Charles' Wagon" is another name for the more commonly known "Charles' Wain" or the "Great Bear."
147 Ackroyd, *T. S. Eliot*, 84.
148 Miller, *T. S. Eliot's Personal Waste Land*, 55.
149 Harriet Davidson, "Improper Desire: Reading *The Waste Land*," in Moody, *The Cambridge Companion to T. S. Eliot*, 122.
150 John Crowe Ransom, "Waste Lands," *New York Evening Literary Review* (14 July 1923); reprinted in *T. S. Eliot: The Critical Heritage*, ed. Michael Grant, vol. 1 (London: Routledge & Kegan Paul, 1982), 178.
151 Tony Pinkney, "*The Waste Land*, Dialogism and Poetic Discourse," in Davies and Wood, *The Waste Land*, 131, 133.

152 Terry Eagleton, *Criticism and Ideology: A Study in Marxist Literary Theory* (London: Verso, 1978), 146, 148.

153 Paul Morrison, *The Poetics of Fascism: Ezra Pound, T. S. Eliot, Paul de Man* (New York: Oxford University Press, 1996), 94–95.

154 John Bowen, "The Politics of Redemption: Eliot and Benjamin," in *The Waste Land*, ed. Davies and Wood, 46–47.

155 Harriet Davidson, "The Logic of Desire: The Lacanian Subject of *The Waste Land*," in *The Waste Land*, ed. Davies and Wood, 68.

156 Ellis, "*The Waste Land* and the Reader's Response," 96, 98; Pinkney, "*The Waste Land*, Dialogism and Poetic Discourse," 133.

157 See Robert Langbaum, "New Modes of Characterization in *The Waste Land*," in *Eliot in His Time: Essays on the Occasion of the Fiftieth Anniversary of "The Waste Land"*, ed. A. Walton Litz (Princeton: Princeton University Press, 1973), 107.

158 Calvin Bedient, *He Do the Police in Different Voice: "The Waste Land" and Its Protagonist* (Chicago: University of Chicago Press, 1986), 130–31, ix.

159 Morrison also points out that "Philomel's story is Lil's life," raising the possibility of a more expansive female identification on Eliot's part (*Poetics of Fascism*, 91, 93).

160 Anthony Julius, *T. S. Eliot, Anti-Semitism, and Literary Form* (Cambridge: Cambridge University Press, 1995), 139.

161 T. S. Eliot, Review of John Middleton Murry, *The Son of Woman: The Story of D. H. Lawrence*, in *Criterion* 10:41 (July 1931), 773.

162 Apter, *Feminizing the Fetish*, 170; Richard Dellamora, *Masculine Desire: The Sexual Politics of Victorian Aestheticism* (Chapel Hill: University of North Carolina Press, 1990), 187.

163 John Peter, "Postscript," *Essays in Criticism* 19 (April 1969); Miller, *Eliot's Personal Waste Land*, 71, 76; Mayer, *Eliot's Silent Voices*, 199–201.

164 Richard Ellmann, *Golden Codgers*, 164; Jay, *Eliot and the Poetics of Literary History*, 104.

165 G. Wilson Knight, "Thoughts on *The Waste Land*," *Denver Quarterly* 7:2 (summer 1972): 3; Bedient, *He Do the Police*, 33.

166 See E. W. F. Tomlin, "T. S. Eliot: A Memoir," *Agenda* 23:1–2 (spring–summer 1985): 141.

167 Hugh Kenner, *The Invisible Poet: T. S. Eliot* (New York: Citadel, 1964), 333–34.

168 Ackroyd, *T. S. Eliot*, 143; Julius, *Anti-Semitism*, 23.

169 Quoted in Jean MacVean, "*The Family Reunion*," *Agenda* 23:1–2 (spring–summer 1985): 119.

170 Quoted in ibid., 123 (emphasis Eliot's).

3 TEXT OF ERROR, TEXT IN ERROR: JAMES JOYCE'S *ULYSSES*

1 Fritz Senn, "'All the errears and erroribosse': Joyce's Misconducting Universe," in *International Perspectives on James Joyce*, ed. Gottlieb Gaiser (Troy, NY: Whitson Pub. Co., 1986), 161.

2 Fritz Senn, *Joyce's Dislocations: Essays on Reading as Translation*, ed. John Paul Riquelme (Baltimore: Johns Hopkins University Press, 1984), 207.

3 Karen Lawrence, *The Odyssey of Style in "Ulysses"* (Princeton: Princeton University Press, 1981), 10.
4 David Hayman, *"Ulysses": The Mechanics of Meaning* (Englewood Cliffs, NJ: Prentice Hall, 1970), 70.
5 Hugh Kenner, *Ulysses* (Baltimore: Johns Hopkins University Press, 1980), 64.
6 See, for instance, Clive Hart, "Wandering Rocks," in *James Joyce's "Ulysses": Critical Essays*, ed. Clive Hart and David Hayman (Berkeley: University of California Press, 1974), 190.
7 Frederic Jameson, *"Ulysses* in History," in *James Joyce and Modern Literature*, ed. W. J. McCormack and Alistair Stead (London: Routledge & Kegan Paul, 1982), 138.
8 Senn, "'All the errears,'" 161, 164.
9 Ibid., 169.
10 Vicki Mahaffey, *Reauthorizing Joyce* (Cambridge: Cambridge University Press, 1988), chapter 1, 23–50.
11 Senn, *Joyce's Dislocutions*, 208.
12 This view has been most famously stated by Jacques Derrida, according to whom "[e]verything we can say about *Ulysses* . . . has already been anticipated" by Joyce, whose work is like "a hypermnesis machine . . . Yes, everything has already . . . been signed in advance by Joyce" ("Ulysses Gramophone: *Hear say yes in Joyce*," in *James Joyce: The Augmented Ninth*, ed. Bernard Benstock [Syracuse: Syracuse University Press, 1988], 48). Derrida similarly comments regarding *Finnegans Wake* that "everything we can say after it looks in advance like a minute self-commentary with which this work accompanies itself. It is already comprehended by it" ("Two Words for Joyce," in *Post-structuralist Joyce: Essays from the French*, ed. Derek Attridge and Daniel Ferrer [Cambridge: Cambridge University Press, 1984], 149). Such remarks call for a study of the particularities of transference onto Joyce.
13 Phillip F. Herring, *Joyce's Uncertainty Principle* (Princeton: Princeton University Press, 1987), 101.
14 According to Hugh Kenner, "Penelope" has "no style" (*Ulysses*, 148). A. Walter Litz asserts its "formlessness" ("Ithaca," in Hart and Hayman *James Joyce's "Ulysses"*, 404).
15 See Christine van Boheemen, *The Novel as Family Romance: Language, Gender, and Authority from Fielding to Joyce* (Ithaca: Cornell University Press, 1987), 174. I will take up this issue in the following section.
16 Eve Kosofsky Sedgwick, *Epistemology of the Closet* (Berkeley: University of California Press, 1990), 1 and *passim*.
17 See Michel Foucault's argument for the *"incorporation* of perversions" and for the articulation of "the homosexual" at the end of the nineteenth century in *The History of Sexuality*, vol. 1: *An Introduction*, tr. Robert Hurley (New York: Vintage, 1980), 42. For an elaboration of Foucault's argument, see Jeffrey Weeks, *Sexuality and Its Discontents: Meanings, Myths and Modern Sexuality* (London: Routledge, 1985); and for a discussion of the ensuing

dispute over its implications, see *Forms of Desire: Sexual Orientation and the Social Constructionist Controversy*, ed. Edward Stein (New York: Routledge, 1992).

18 Gabler in fact "reports over 5,000 departures in the first [1922] edition from the author's own text in the documents of composition" (Foreword, *Ulysses: A Critical and Synoptic Edition*, ed. Hans Walter Gabler [New York and London: Garland Publishing, 1984], vii). For a discussion of the disputed number of errors in the 1922 and subsequent editions see Charles Rossman, "The Critical Reception of the 'Gabler *Ulysses*': Or, Gabler's *Ulysses* Kidd-napped," *Studies in the Novel* 21:2 (summer 1989), 176, n. 6.

19 According to Jerome McGann, "Gabler's edition alters the text of 1922 in more than five thousand instances, but the 1922 text does not contain more than five thousand mistakes, nor anywhere near that number. The number of errors in the first edition is several hundred at the most." The difference between these two figures is due to the fact that the Gabler and the 1922 editions are based upon "two different conceptions of 'the text of *Ulysses*'" ("*Ulysses* as a Postmodern Text: The Gabler Edition," *Criticism* 27:3 [summer 1985], 290).

20 McGann, "*Ulysses* as a Postmodern Text," 290, 287.

21 John Kidd, "An Inquiry into *Ulysses: The Corrected Text*," *The Papers of the Bibliographical Society of America* 82 (December 1988): 411–584. Kidd raises other objections, such as alleged errors in the transcription of the Rosenbach manuscript on the part of Gabler's assistants.

22 See Patrick McGee, "The Error of Theory," *Studies in the Novel* 22:2 (summer 1990): 157.

23 Hans Walter Gabler, "The Text as Process and the Problem of Intentionality," *Text* 3 (1987): 111–12. Gabler's retrospective account of his editing practice may have been influenced by McGann's defense of it.

24 A good example of intersection of the errancy of Joyce's text, the irretrievability of his intentions, and the equivocality of his representation of homosexual desire is a line that Gabler inserts in which Stephen thinks, "His arm: Cranly's arm" (*U* 3.451–52). The debate over the significance of this line and whether it should be included at all in *Ulysses* testifies to the homophobia which has constrained the interpretation of the text and, perhaps, Joyce's own production of it. According to Rosa Maria Bollettieri Bosinelli, "Gabler's restoration is very important, because despite the fact that the existence of the omitted sequence had been pointed out by Harry Levin, . . . very few had noticed it or fully understood its implications" ("Joyce the Scribe and the Right Hand Reader," in *Assessing the 1984 'Ulysses*,'" ed. C. George Sandulescu and Clive Hart [Totowa, N.J.: Barnes and Noble Books, 1986], 9). However, Richard Brown argues that, because "the phrase, in its context, clearly implies a suspicion of homosexual attachment between Stephen and Cranly," Joyce may have been "conscious of leaving this small but *contentious hint* out of his book" ("To Administer Correction," *James Joyce Broadsheet* 15 [October 1984]: 1

[emphasis mine]). Clearly, the imputation of authorial intention to Joyce has much to do with the interests of his readers.

25 Vicki Mahaffey, "Intentional Error: The Paradox of Editing Joyce's *Ulysses*," in *Representing Modernist Texts: Editing as Interpretation*, ed. George Bornstein (Ann Arbor: University of Michigan Press, 1991), 183.

26 Ibid., 184.

27 Sedgwick, *Epistemology*, 185 (emphasis Sedgwick's).

28 Patrick McGee, *Paperspace: Style as Ideology in Joyce's "Ulysses"* (Lincoln: University of Nebraska Press, 1988), 171.

29 For a critique of such national myths in terms of their "phrasing" and structures of address, see Jean-François Lyotard, *The Differend: Phrases in Dispute*, tr. Georges van den Abbeele (Minneapolis: University of Minnesota Press, 1988), 147.

30 Bonnie Kime Scott, *James Joyce* (Brighton: Harvester, 1977), 129.

31 McGee, *Paperspace*, 103–4.

32 Scott, *James Joyce*, 109, 117.

33 McGee, *Paperspace*, 101, 68.

34 Ibid., 104.

35 Ibid., 95. Similar readings abound. However, for a defense of Gerty's "one dimensionality" in terms of the commodification of turn-of-the-century Irish popular culture, see Thomas Karr Richards, "Gerty MacDowell and the Irish Common Reader," *English Literary History* 52 (fall 1985): 755–76. Jules David Law argues that Gerty is a "specular projection" of Bloom's, which enables him to see himself "as object, rather than as subject, of an erotic fantasy," but women in general and, especially, Gerty still "'can't see themselves'" ("'Pity They Can't See Themselves': Assessing the 'Subject' of Pornography in 'Nausicaa,'" *James Joyce Quarterly* 27 [winter 1990]: 220, 234).

36 Colin MacCabe, *James Joyce and the Revolution of the Word* (London: Macmillan, 1978), 128.

37 Ibid., 123.

38 McGee, *Paperspace*, 147 (emphasis McGee's).

39 Philippe Sollers, quoted by McGee, *Paperspace*, 146. The theme of woman as "the untruth of truth" is developed most fully in Nietzsche's texts – e.g., *Beyond Good and Evil* and *The Gay Science* – and is analyzed at length by Jacques Derrida in *Spurs: Nietzsche's Styles*, tr. Barbara Harlow (Chicago: University of Chicago Press, 1979), 54–55.

40 van Boheemen, *The Novel as Family Romance*, 184.

41 "Intermediate sex" is borrowed by Richard Brown from turn-of-the-century sexologists (*James Joyce and Sexuality* [Cambridge: Cambridge University Press, 1985], 103, 106–7). McGee argues that "the only desire speaking in 'Penelope' is Joyce's masquerading as female" (*Paperspace*, 175), although he does not pursue the implications of that performance. Cheryl Herr notes the "pantomime" transvestism of "Circe" and claims that, "instead of yearning for a true androgyny, Bloom seems always to have

gravitated toward only a stereotyped femaleness composed of clothing, makeup, hairstyle and mannerism." However, she fails to analyze the difference between true and simulated "femaleness," or Bloom's motives for such simulation (*Joyce's Anatomy of Culture* [Chicago: University of Illinois Press, 1986], 152). For a sophisticated, Lacanian reading of the masquerade of gender in *Ulysses*, see Kimberley J. Devlin, "Pretending in 'Penelope': Masquerade, Mimicry, and Molly Bloom," *Novel* 25:1 (fall 1991): 71–89, and "Castration and Its Discontents: A Lacanian Approach to *Ulysses*," *James Joyce Quarterly* 29:1 (fall 1991): 117–44.

42 Gilles Deleuze, *Coldness and Cruelty*, in *Masochism*, tr. Jean McNeil (New York: Zone Books, 1989), 63.

43 Ibid., 66, 68 (emphasis Deleuze's).

44 Frances L. Restuccia, *Joyce and the Law of the Father* (New Haven: Yale University Press, 1989), 138, 174. According to Deleuze, the paternal law, far from being abolished, is "invested upon the mother," whose subsequent exercise of phallic authority in the masochist's fantasy is grounds for humor, irony, and incestuous pleasure (*Coldness and Cruelty*, 88–90). Kaja Silverman's caution against a "'utopian' rereading of masochism" following Deleuze is useful here (*Male Subjectivity at the Margins* [New York: Routledge, 1992], 211).

45 Brown, *James Joyce and Sexuality*, 101. He describes Molly as a "phallic woman" in these terms but does not link Molly with Bella Cohen.

46 See Samuel Weber, "Afterword," in *Just Gaming*, by Jean François Lyotard and Jean-Loup Thébaud, tr. Wald Godzich (Minneapolis: University of Minnesota Press, 1985), 105.

47 Lyotard argues that "[w]hen a 'feminist' is reproached for confusing the phallus, symbolic operator of meaning, and the penis, empirical sign of sexual difference, it is admitted without discussion that the metalinguistic order (the symbolic) is distinct from its domain of reference (realities). But if the women's movement has an immense impact . . . it is that this movement solicits and destroys the (masculine) belief in meta-statements independent of ordinary statements" ("On One of the Things at Stake in Women's Struggles," tr. Deborah J. Clarke et al., *Sub-stance* 20 [1978]: 15).

48 Jane Gallop, *The Daughter's Seduction: Feminism and Psychoanalysis* (Ithaca: Cornell University Press, 1982), 99.

49 Sigmund Freud, "'A Child Is Being Beaten': A Contribution to the Study of the Origin of Sexual Perversions," in *The Standard Edition of the Complete Psychological Works*, ed. James Strachey, vol. 17 (London: Hogarth Press, 1955), 198 (emphasis Freud's).

50 Sigmund Freud, "The Ego and the Id," in *The Standard Edition of the Complete Psychological Works*, ed. James Strachey, vol. 19 (London: Hogarth Press, 1955), 37.

51 Freud, "A Child Is Being Beaten," in *The Standard Edition* vol. 17: 199–200.

52 Daniel Ferrer, "Circe, Regret, and Regression," in *Post-Structuralist Joyce*, 137.

53 Marilyn French, *The Book as World: James Joyce's "Ulysses"* (Cambridge, MA: Harvard University Press, 1976), 249.

54 Harry Levin, *Memories of the Moderns* (New York: New Directions, 1980), 50. Among those who have commented in passing on homosexuality in *Ulysses* are Richard Ellmann, Hugh Kenner, David Hayman, Stanley Sultan, and Darcy O'Brien.

55 Frank Budgen, *James Joyce and the Making of "Ulysses"* (1934; reprint, Bloomington: University of Indiana Press, 1960), 146, 315.

56 Brown, *James Joyce and Sexuality*, 79, 83, 84.

57 David Fuller, *James Joyce's "Ulysses"* (New York: St. Martin's Press, 1992), 93.

58 See Eve Kosofsky Sedgwick, "Privilege of Unknowing: Diderot's *The Nun*," in *Tendencies* (Durham, NC: Duke University Press, 1993), 23–51.

59 See Judy Grahn's *Another Mother Tongue* for descriptions of these and other phrases in gay and lesbian sociolects. See Randy Shilts for an account of the way U.S. military investigations into homosexuality are conducted, often by implicating service members who somehow "know" about homosexuality (*Conduct Unbecoming: Gays and Lesbians in the U.S. Military* [New York: Random House, 1993]).

60 Marcel Proust, *Remembrance of Things Past*, trans. C. K. Scott Moncrieff and Terence Kilmartin, vol. 3 (New York: Vintage, 1982), 206.

61 D. A. Miller, *The Novel and the Police* (Berkeley: University of California Press, 1988), 206.

62 Sedgwick, *Epistemology of the Closet*, 85–87.

63 Baron Richard von Krafft-Ebing, *Psychopathia Sexualis*, tr. Franklin S. Klaf (New York: Stein and Day, 1965), 53–54 (emphasis Krafft-Ebing's). For his distinction between "congenital" and "acquired" homosexuality, see also 188.

64 John Addington Symonds, *Male Love: A Problem in Greek Ethics and Other Writings* (1901; reprint, New York: Pagan Press, 1983).

65 Sigmund Freud, "Some Neurotic Mechanisms in Jealousy, Paranoia and Homosexuality," in *The Standard Edition of the Complete Psychological Works*, ed. James Strachey, vol. 18 (London: Hogarth Press, 1955), 226–27.

66 Ibid., 226, 228.

67 Harold Beaver, "Homosexual Signs," *Critical Inquiry* 8:1 (autumn 1981), 105.

68 The source of much of Joyce's information and opinions regarding Wilde's sexuality was a contemporary biography by Robert Harborough Sherard. Joyce summarizes Sherard's claim that Wilde descended from "those 'wild O'Flaherty's'" and reiterates his argument that Wilde was feminized by his mother (*The Life of Oscar Wilde* [1908; reprint, New York: Brentano's, 1911], 5, 77). Sherard also quotes Wilde's remark about her in *De Profundis*, that "[h]er death was terrible to me; but I, once a lord of language, have no words in which to express my anguish" (6). Stephen Dedalus also quotes Wilde's phrase "a lord of language" in "Scylla and Charybdis," during a scene in which he repeatedly tries to suppress discussion of Wilde's theory of Shakespeare's homosexual attachment to Willie Hughes (*U* 9.454).

69 See Joseph Valente, "Thrilled by His Touch: Homosexual Panic and the Will to Artistry in *A Portrait of the Artist as a Young Man*," *James Joyce Quarterly* 31:3 (spring 1994), 167–88.

70 *The Literary Criticism of Oscar Wilde*, ed. Stanley Weintraub (Lincoln: University of Nebraska Press, 1968), 243.

71 Sedgwick's "Privilege of Unknowing" is an exemplary instance of such attentiveness (*Tendencies*).

72 The structural implications of the father/son relation for the narrative of *Ulysses* have been amply explored in Joyce criticism, while Jean-Michel Rabaté has very ably examined the homosexual dynamics of that relation in *James Joyce, Authorized Reader* (Baltimore: Johns Hopkins University Press, 1991). Because my focus in this chapter is principally upon the errant knowledge of homosexuality in *Ulysses*, I do not discuss the father/son relation in detail.

73 Oscar Wilde, "The Portrait of Mr W. H.," *The Complete Works of Oscar Wilde*.

74 George Brandes and Frank Harris were, according to William Schutte, two of the principal sources for Stephen's Shakespeare theory (*Joyce and Shakespeare* [New Haven: Yale University Press, 1957], 153). Brandes allows that Shakespeare's "passionate attachment" to Pembroke was sexual but denies that it was homosexual; their friendship had "an erotic fervour such as we never find in our century manifested between man and man," a "friendship . . . which is now unknown" (*William Shakespeare* [New York: Macmillan, 1899], 296, 291). Harris denounces the homosexual reading of the sonnets at length and quotes Hallam's remark that "it would have been better for Shakespeare's reputation if the sonnets had never been written" than that they be read in such a fashion (*The Man Shakespeare and His Tragic Life-Story* [New York: Mitchell Kennerley, 1909], 227). Schutte's own stake in the matter is evident in his attack on Best, whom he describes as a "mincing," "mannered young aesthete" whose "enthusiasms clearly are confined to those deemed appropriate in a disciple of Pater and Wilde" (37–38). As in T. S. Eliot's essays, Paterian aestheticism leads to Wildean degeneracy.

75 According to Joyce, "Even the best Englishmen seem to love a lord" (quoted by Budgen, *James Joyce and the Making of "Ulysses,"* 75); Harris, *The Man Shakespeare*, 230.

76 Harris, *The Man Shakespeare*, 229, 243–44.

77 See Ian Gibson, *The English Vice: Beating, Sex, and Shame in Victorian England and After* (London: Duckworth, 1978); and Mario Praz, *The Romantic Agony*, tr. Angus Davidson, appendix 1, "Swinburne and 'Le Vice Anglais'" (Cleveland and New York: World Publishing Co., 1956), 413–34. See also François Caradec, *Dictionnaire du français argotique et populaire* (Paris: Librairie Larousse, 1977).

78 Mulligan quotes from William Dowden's introduction to his edition of *The Sonnets of William Shakespeare* (New York: D. Appleton, 1881), which offers the arguments for the poet's love for a youth that Wilde took up in his 1889 "Portrait of Mr W. H."

79 Stephen's envoi to Cranly, "God speed. Good hunting" (*U* 9.39–40), refers to the faithless wife of John Synge's play *In the Shadow of the Glen*, who goes "hunting" after her husband throws her out of the house. See Don Gifford with Robert Seidman, *"Ulysses" Annotated: Notes for James Joyce's "Ulysses,"* rev. edn. (Berkeley: University of California Press, 1988), 195. In a letter to Nora about Byrne, the prototype of Cranly, Joyce wrote, "When I was younger I had a friend to whom I gave myself freely – in a way more than I give myself to you and in a way less. He was Irish, that is to say, he was false to me" (*LJ* 2:50).

80 Freud, *"Some* Neurotic Mechanisms in Jealousy, Paranoia, and Homosexuality," in *The Standard Edition*, vol. 18, 227.

81 Mark Shechner, *Joyce in Nighttown: A Psychoanalytic Inquiry into "Ulysses"* (Berkeley: University of California Press, 1974), 101, 55.

82 French, *The Book as World*, 44.

83 Devlin, "Castration and Its Discontents," 137.

84 Foucault, *The History of Sexuality*, vol. 1, 21, 61–2.

85 Ibid., vol. 1, 44.

86 Richard Burton, "Terminal Essay," in *The Thousand Nights and a Night* (New York: Limited Editions Club, 1934), 3653–3817; Krafft-Ebing, *Psychopathia Sexualis*, 187.

87 Bloom is referred to in the feminine third person upon his metamorphosis (*U* 15.2852–2912), after which he is again taken as a man except for the moment of his penetration by Bella/o.

88 Krafft-Ebing, *Psychopathia Sexualis*, 186–90; Otto Weininger, *Sex and Character*, (New York: G. P. Putnam's Sons, 1906), 45, 306; Sigmund Freud, *Three Essays on the Theory of Sexuality*, *Standard Edition*, vol. 7, 141–48.

89 Gifford, *"Ulysses" Annotated*, 497.

90 An early sexologist, Carl Westphal coined the phrase in 1869. See Vern L. Bullough, *Homosexuality: A History* (New York: New American Library, 1979), 7–8.

91 Richard Ellmann, *Ulysses on the Liffey* (New York: Oxford University Press, 1972), 154.

92 Hugh Kenner, *Ulysses*, 130; Gerald L. Bruns, "Eumaeus," in Hart and Hayman, *James Joyce's "Ulysses": Critical Essays*, 368–69; McGee, *Paperspace*, 151.

93 Gabler's edition removed approximately six hundred of the commas that had appeared in the 1922 edition but which are not present in the Rosenbach manuscript. These commas, along with other alterations, such as the addition of quotation marks and other devices to elucidate the syntax, were likely introduced by a new typist, who typed only "Eumaeus" (*Ulysses: The Critical and Synoptic Edition*, 1749).

94 Fritz Senn, "Inherent Delicacy: Eumaean Questions," *Studies in the Novel* 22:2 (summer 1990): 180.

95 Derek Attridge, *Peculiar Language: Literature as Difference from the Renaissance to James Joyce* (Ithaca: Cornell University Press, 1988), 182.

96 Ibid., 183.

97 "Queer" has for centuries implied a "strange, odd, [or] peculiar" person,
    particularly one "of questionable character, suspicious, dubious," and,
    according to the *OED*, has been especially frequent in Ireland and in nau-
    tical contexts. The first recorded use of "queer" to refer to a homosexual
    person dates from 1922, in the United States (*OED*).
        The *OED* notes a later date (1935) for the homosexual designation of
    "gay," although since the seventeenth century it has commonly connoted
    a "loose or immoral life," as in "a gay lothario." This connotation
    expanded in the nineteenth century to include women who led an
    immoral, or a harlot's, life. Wayne R. Dynes points out that "it is safe to
    assume that the [homosexual] usage [of "gay"] must have been circulat-
    ing orally in the United States for a decade or more" before being recorded
    in Noel Ersine's 1933 *Dictionary of Underworld Slang* as referring to "a homo-
    sexual boy" (*Encyclopedia of Homosexuality*, ed. Wayne R. Dynes [New York:
    Garland Publishing Co., 1990], 456).
        "Faggot" entered the argot of Anglo-American homosexual subcultures
    at about the same time, probably as an extension of "fag," a term that in
    the nineteenth century commonly denoted a British schoolboy who did
    menial work for one in a higher form (Partridge). The word thus carries
    the connotation of homosexual behavior, which Joyce believed was preva-
    lent in British public schools.
        Joyce's use of "gay," "fag," and, especially, "queer" thus stands on the
    cusp between their older, broader reference to persons of dissolute morals
    or questionable integrity and the specific homosexual connotation that so
    quickly and phobically attached to them in the 1920s and 1930s, concomi-
    tant with the diffusion of sexological knowledge concerning homosexual-
    ity. The unusual concentration of these terms in conjunction with "the
    genus *homo*" in this episode suggests that Joyce was overcoding his text,
    implying not only Murphy's suspicious character but also his possession of
    equally dubious, sexual secrets, along with Bloom's uncertain complicity
    in those secrets.

98 Sedgwick defines the term as the endemic, fearful response of hetero-
    sexually identified men to the potential "blackmailability" of all male
    bonds "through the leverage of homophobia" (*Between Men: English
    Literature and Male Homosocial Desire* [New York: Columbia University Press,
    1985], 89).

99 See Sedgwick, *Between Men*, on the homosocial triangle.

100 Gifford, *"Ulysses" Annotated*, 544. Gifford offers no source for this informa-
    tion. Stuart Gilbert associates Murphy's tattoo with the duplicity of signa-
    tures of identity, for "tattoo marks have played an important part in the
    solution of such problems of identity as the Tichbourne case," also alluded
    to in "Eumaeus" (*James Joyce's "Ulysses"* [New York: Vintage, 1952], 364).

101 For a discussion of the Criminal Law Amendment Act of 1885 and the link
    it forged between prostitution and homosexuality, see Jeffrey Weeks,
    "Inverts, Perverts, and Mary-Annes: Male Prostitution and the Regulation
    of Homosexuality in England in the Nineteenth and Early Twentieth

Centuries," in *Hidden from History: Reclaiming the Gay and Lesbian Past*, ed. Martin Duberman, Martha Vicinus, and George Chauncey, Jr. (New York: Penguin, 1989), 195–211. Weeks's article also explains the split in the male homosexual subculture of England around the turn of the century between, on the one hand, upper- and middle-class men who purchased sex and, on the other hand, the lower-class men who sold it, including virile guardsmen and sailors as well as transvestite "mary-annes." The dichotomy in the representation of homosexuality in *Ulysses* between Wilde the aesthete and Best, his effeminate avatar, and the hypermasculine Mulligan and Murphy plays out the split that Weeks documents.

102 Weldon Thornton explains Bloom's and probably Joyce's confusion between sections 2 and 11 of the Criminal Law Amendment Act (*Allusions in "Ulysses"* [Chapel Hill: University of North Carolina Press, 1968], 447–48.

103 Joyce seems to have linked his coprophilic interest in women's (soiled) underwear to homosexuality. Richard Ellmann describes Joyce's behavior during a meeting with an openly homosexual man, in which Joyce taunted the unnamed man by taking out a pair of women's drawers, placing his fingers in them, and "walking" them across the table (*James Joyce*, rev. edn. [New York: Oxford University Press, 1982], 438).

104 Gifford, *"Ulysses" Annotated*, 559.

105 Budgen, *James Joyce and the Making of "Ulysses,"* 149.

106 Joyce, *Dubliners*, 125; Fuller, *James Joyce's "Ulysses,"* 90.

107 For discussions of the difficulty as well as attempts at reading female same-sex desire in this period see Carol Smith-Rosenberg, "Discourses of Sexuality and Subjectivity: The New Woman, 1870–1936," in Duberman et al., *Hidden from History*, 264–80, and Martha Vicinus, "Distance and Desire: English Boarding School Friendships, 1870–1920," in *Hidden from History*, 212–32.

108 Kitty O'Shea was Parnell's mistress, another woman whose sexual agency defied the norm of female erotic passivity.

109 Krafft-Ebing, *Psychopathia Sexualis*, 264.

4 SEXUAL/TEXTUAL INVERSION: MARCEL PROUST

1 Roger Shattuck, *Proust's Binoculars: A Study of Memory, Time and Recognition in "A la recherche du temps perdu"* (Princeton: Princeton University Press, 1962), 96, 99, 113.

2 Proust, quoted in Vincent Descombes, *Proust: Philosophy of the Novel*, tr. Catherine Chance Macksey (Stanford: Stanford University Press, 1992), 4–5.

3 Ibid., 4 (emphasis Descombes's).

4 Antoine Compagnon, *Proust between Two Centuries*, tr. Richard E. Goodkin (New York: Columbia University Press, 1992), 271.

5 F. R. Leavis, *New Bearings in English Poetry* (Harmondsworth: Penguin, 1932), 81.

6 Gerard Genette, "Discours du recit," *Figures* III (Paris: Seuil, 1972), 260, 244.

7 Compagnon, *Proust between Two Centuries*, 9, 11.

8 Richard E. Goodkin, *Around Proust* (Princeton: Princeton University Press, 1991), 28.

9 Ibid., 6, 86–87.

10 Michel Foucault, *History of Sexuality*, vol. 1: *An Introduction*, tr. Robert Hurley (New York: Vintage, 1980), 43.

11 See Otto Weininger, *Sex and Character* (New York: G. P. Putnam's Sons, 1906).

12 Compagnon, *Proust between Two Centuries*, 158, 240–48.

13 Malcolm Bowie, *Freud, Proust, and Lacan: Theory as Fiction* (Cambridge: Cambridge University Press, 1987), 95.

14 Descombes, *Proust*, 241; he calls this "the Dostoyevski side of Mme de Sévigné," which he proceeds to dispute. See J. E. Rivers, *Proust and the Art of Love: The Aesthetics of Sexuality in the Life, Times, and Art of Marcel Proust* (New York: Columbia University Press, 1980), 215f., for a discussion of the "principle of inversion as style."

15 Leo Bersani, "'The Culture of Redemption': Marcel Proust and Melanie Klein," *Critical Inquiry* 12 (winter 1986), 416.

16 Eve Kosofsky Sedgwick, *Epistemology of the Closet* (Berkeley: University of California Press, 1990), 217.

17 Margaret E. Gray, *Postmodern Proust* (Philadelphia: University of Pennsylvania Press, 1992), 99.

18 The locus classicus of this practice is Justin O'Brien's "Albertine the Ambiguous: Notes on Proust's Transposition of Sexes," *PMLA* 64 (1949): 933–52.

19 Leo Bersani, *Marcel Proust: The Fictions of Life and of Art* (New York: Oxford University Press, 1965), 193. Bersani remarks that "in much of the novel Proust's authorial privilege of omniscience is given to Marcel," whereas in *The Captive* and *The Fugitive* his knowledge is limited (196).

20 Bowie, *Freud, Proust and Lacan*, 58.

21 Gilles Deleuze, *Proust and Signs*, tr. Richard Howard (New York: George Braziller, 1972), 120, 122 (emphasis Deleuze's).

22 Compagnon, *Proust between Two Centuries*, 4.

23 David R. Ellison, *The Reading of Proust* (Baltimore: Johns Hopkins University Press, 1984), 77.

24 John Ruskin, *The Stones of Venice*, 3 vols. (London: J. M. Dent, 1907), 2:356 (emphasis Ruskin's).

25 John Ruskin, *Sesame and Lilies* (New York: Charles E. Merrill & Co., 1891), 21–22, 53, 47.

26 Ellison, *The Reading of Proust*, 83.

27 Roland Barthes, "Proust and Names," in *New Critical Essays*, tr. Richard Howard (Berkeley: University of California Press, 1990), 61, 66.

28 Tony Tanner, *Venice Desired* (Cambridge, MA: Harvard University Press, 1992), 242.

29 Bowie, *Freud, Proust and Lacan*, 85, 93.

30 For an account of Gide's pederastic eroticism in the context of contemporary French culture, see Christopher Robinson, *Scandal in the Ink: Male and Female Homosexuality in Twentieth-Century French Literature* (London: Cassell, 1995). Jonathan Dollimore's *Sexual Dissidence: Augustine to Wilde, Freud to Foucault* (Oxford: Oxford University Press, 1991) critiques the colonialist investments of early twentieth-century male homosexuality.

31 Ruskin, *The Stones of Venice*, 3:150–51.

32 Ruskin, *Sesames and Lilies*, 42. George Painter terms the essay an aberration that is "far from representing Proust's normal opinion of the nature of reading" (*Marcel Proust: A Biography*, 2 vols. [New York: Random House, 1978], 2:36).

33 For modern American (water) closets, see Lee Edelman, "Tearooms and Sympathy; or The Epistemology of the Water Closet," *Homographesis: Essays in Gay Literary and Cultural Theory* (New York and London: Routledge, 1994), 148–70.

34 Gilles Deleuze, *Proust and Signs*, tr. Richard Howard (New York: George Braziller, 1972), 103–4 (emphasis Deleuze's).

35 See David M. Halperin, *One Hundred Years of Homosexuality* (New York: Routledge, 1990), and Arnold I. Davidson, "Sex and the Emergence of Sexuality," *Critical Inquiry* 14:1 (autumn 1987): 16–48.

36 Bowie, *Freud, Proust and Lacan*, 75 (emphasis Bowie's).

37 Ruskin, *Sesame and Lilies*, 10–11.

38 Deleuze, *Proust and Signs*, 107.

39 Paul de Man, *Allegories of Reading* (New Haven: Yale University Press, 1979), 67.

40 Quoted in Ellison, *The Reading of Proust*, 177–78.

41 Serge Doubrovsky, *Writing and Fantasy in Proust: La Place de la madeleine*, tr. Carol Mastrangelo Bové (Lincoln and London: University of Nebraska Press, 1986), 43.

42 Deleuze, *Proust and Signs*, 104.

43 Walter Benjamin, "The Image of Proust," in *Illuminations*, tr. Hannah Arendt (New York: Schocken Books, 1969), 202.

44 Descombes, *Proust*, 30 (emphasis Descombes's).

45 Ibid., 267–70, 274, 291.

46 Shattuck, *Proust's Binoculars*, 18 (emphasis Shattuck's).

47 Bersani, "'The Culture of Redemption,'" 404.

48 Gerard Genette, *Figures of Literary Discourse*, tr. Alan Sheridan (New York: Columbia University Press, 1982), 66.

49 Gray, *Postmodern Proust*, 42, 64.

50 Genette, *Figures*, 70.

51 Maurice Blanchot, *The Siren's Song: Selected Essays*, ed. Gabriel Josipovici, tr. Sacha Rabinovitch (Bloomington: Indiana University Press, 1982), 66, 70.

52 Genette, *Figures*, 221, 223.

53 Ibid., 213.

54 Samuel Beckett, *Proust* (New York: Grove Press, 1978), 32 (emphasis Beckett's).
55 Ibid., 37.
56 Deleuze, *Proust and Signs*, 9.
57 Ibid., 78, 10.
58 Mieke Bal, "Bird Watching: Visuality and Lesbian Desire in Marcel Proust's *A la recherche du temps perdu*," *Thamyris* 2:1 (1995), 45, 49.
59 Ibid., 64–65.
60 Gray examines in detail the hero's self-defeating investigation of Albertine's sexuality (97–99).
61 Sedgwick, *Epistemology of the Closet*, 223.
62 Compagnon, *Proust between Two Centuries*, 271.
63 Bal, "Bird-Watching," 59–64.
64 Michael Sprinker, *History and Ideology in Proust: "A la recherche du temps perdu" and the Third French Republic* (Cambridge: Cambridge University Press, 1994), 148.
65 Gray, *Postmodern Proust*, 99.
66 Proust, *Essais et articles* (Paris: Gallimard, 1971), 327–29. See Painter, *Proust*, 2:323. The translation is Painter's.
67 Alain Corbin, quoted in Sprinker, *History and Ideology in Proust*, 144.
68 Natalie Clifford Barney, *Adventures of the Mind*, tr. John Spalding Gatton (New York: New York University Press, 1992), 67.
69 André Gide, *Journal*, quoted in Painter, *Proust* 2:313.
70 Leo Bersani, "Death and Literary Authority," in *A New History of French Literature*, ed. Denis Hollier et al. (Cambridge, MA: Harvard University Press, 1989), 863–65.
71 Walter Benjamin, "The Image of Proust," *Illuminations*, tr. Harry Zohn (New York: Schocken Books, 1969), 210.
72 Sprinker, *History and Ideology*, 184.

CONCLUSION

1 Antoine Compagnon, *Proust between Two Centuries*, tr. Richard Goodkin (New York: Columbia University Press, 1992), 29, 31.
2 Margot Norris, *Joyce's Web: The Social Unraveling of Modernism* (Austin: University of Texas Press, 1992), 9.
3 Stephen Heath, "Ambiviolences," in *Post-Structuralist Joyce: Essays from the French*, ed. Derek Attridge and Daniel Ferrer (Cambridge: Cambridge University Press, 1984), 39; Broch is quoted by Heath, 38.
4 Maurice Blanchot, *The Space of Literature*, tr. Ann Smock (Lincoln: University of Nebraska Press, 1982), 219.
5 Michael Sprinker, *History and Ideology in Proust: "A la recherche du temps perdu" and the Third French Republic* (Cambridge: Cambridge University Press, 1994), 178.
6 T. S. Eliot, *Knowledge and Experience in the Philosophy of F. H. Bradley* (London: Faber & Faber, 1964), 147–48.

7 Gilles Deleuze, *Proust and Signs*, tr. Richard Howard (New York: George Braziller, 1972), 41.

8 See Richard Goodkin, *Around Proust* (Princeton: Princeton University Press, 1991), 65–88.

9 Virginia Woolf, "Phases of Fiction," in *Collected Essays*, ed. Leonard Woolf, vol. 2 (London: Hogarth Press, 1967), 83; "Romance and the Heart," in *Essays*, ed. Andrew McNeillie, vol. 3 (London: Hogarth Press, 1986), 367.

10 Lyndall Gordon, *Virginia Woolf: A Writer's Life* (New York: Oxford University Press, 1984), 914; Virginia Woolf, *A Room of One's Own* (1929; reprint New York: Harcourt Brace Jovanovich, 1989), 103.

11 Stephen Heath, "Male Feminism," in *Men in Feminism*, ed. Alice Jardine and Paul Smith (New York and London: Methuen, 1987), 27.

12 Marcel Proust, *"Pleasures and Days" and Other Writings*, tr. Louise Varese et al. (Garden City: Doubleday, 1957), 79–90, 293–304.

13 D. H. Lawrence, *The Complete Poems of D. H. Lawrence*, ed. Vivian de Sola Pinto and Warren Roberts (New York: Viking, 1964), 741–43.

14 D. H. Lawrence, quoted in Gregory Woods, *Articulate Flesh: Male Homo-Eroticism and Modern Poetry* (New Haven: Yale University Press, 1987), 131.

15 D. H. Lawrence, *Women in Love* (Harmondsworth: Penguin, 1976), 198–99. Further references are included in the text.

16 Christopher Craft, *Another Kind of Love: Male Homosexual Desire in English Discourse, 1850–1920* (Berkeley: University of California Press), 179; see Anthony Julius, *T. S. Eliot, Anti-Semitism, and Literary Form* (Cambridge: Cambridge University Press, 1995).

17 Woods, *Articulate Flesh*, 133.

18 Gregory Woods, "The Injured Sex: Hemingway's Voice of Masculine Anxiety," in *Textuality and Sexuality: Reading Theories and Practices*, ed. Judith Still and Michael Worton (Manchester: Manchester University Press, 1993), 161.

19 Ernest Hemingway, *The Garden of Eden* (New York: Scribner, 1995), 17. Further references are included in the text.

20 Woods, "The Injured Sex," 163–65.

21 Jacques Lacan, "The Signification of the Phallus," *Ecrits: A Selection*, tr. Alan Sheridan (New York: W. W. Norton, 1977), 291.

22 Norman Mailer, *Armies of the Night* (New York, 1968), 25.

23 Simone de Beauvoir, *The Second Sex* (New York: Bantam, 1952), 249.

24 Judith Butler, *Gender Trouble: Feminism and the Subversion of Identity* (New York: Routledge, 1990), 146.

25 Edward Carpenter, quoted in Alan Sinfield, *The Wilde Century: Effeminacy, Oscar Wilde and the Queer Moment* (New York: Columbia University Press, 1994), 124.

26 Sir William Blackstone, *Commentaries on the Laws of England*, ed. James DeWitt Andrewes, vol. 2 (Chicago, 1899), 316.

27 Robert Fraser, *Proust and the Victorians: The Lamp of Memory* (New York: St. Martin's Press, 1994), 209.

28 On the Wilde–Gide relation as exemplary of the current split between

normalizing versus radically transgressive versions of homosexuality, see Jonathan Dollimore, *Sexual Dissidence: Augustine to Wilde, Freud to Foucault* (Oxford: Oxford University Press, 1991), chapter 1.

29 George E. Haggerty and Bonnie Zimmerman, Introduction to *Professions of Desire: Lesbian and Gay Studies in Literature*, ed. George E. Haggerty and Bonnie Zimmerman (New York: Modern Language Association, 1995), 2.

30 Michel Foucault, *The History of Sexuality*, vol. 1: *An Introduction*, tr. Robert Hurley (New York: Random House, 1978), 69.

31 For a discussion of the legibility of homosexual inscriptions and their varying effects, see Lee Edelman, *Homographesis: Essays in Gay Literary and Cultural Theory* (New York: Routledge, 1994).

32 Oscar Wilde, "The Decay of Lying," in *Literary Criticism of Oscar Wilde*, ed. Stanley Weintraub (Lincoln: University of Nebraska Press, 1968), 187.

33 Marcel Proust, *Marcel Proust on Art and Literature: 1896–1919*, tr. Sylvia Townsend Warner (New York: Meridian Books, 1958), 180.

34 Sinfield, *The Wilde Century*, 121.

35 On the Hall debate, see Jean Radford, "An Inverted Romance: The Politics of *The Well of Loneliness* and Sexual Ideology," in *The Progress of Romance*, ed. Jean Radford (New York: Routledge, 1986), 97–112. For the butch/femme debate, see Esther Newton, "The Mythic Mannish Lesbian: Radclyffe Hall and the New Woman," in *Hidden from History: Reclaiming the Gay and Lesbian Past*, ed. Martin Duberman, Martha Vicinus, and George Chauncey, Jr. (New York: Penguin, 1989), 281–93; and essays in *Lesbian Erotics*, ed. Karla Jay (New York: New York University Press, 1995).

36 Monique Wittig, *The Straight Mind and Other Essays* (Boston: Beacon Press, 1992), 32.

37 Judith Butler, "Contingent Foundations: Feminism and the Question of 'Postmodernism,'" in *Feminists Theorize the Political*, ed. Judith Butler and Joan Scott (New York: Routledge, 1992), 3–21.

38 See Joan W. Scott, "The Evidence of Experience," in *The Lesbian and Gay Studies Reader*, ed. Henry Abelove et al. (New York: Routledge, 1993), 397–415; and Wendy Brown, "Feminist Hesitations, Postmodern Exposures," *differences* 3:1 (1991): 63–84.

Index

263